Application of the Bayesian Method in Statistical Modeling

Application of the Bayesian Method in Statistical Modeling

Guest Editor

Diana Mindrila

MDPI

Basel • Beijing • Wuhan • Barcelona • Belgrade • Novi Sad • Cluj • Manchester

Guest Editor
Diana Mindrila
Department of Educational
Leadership, Research, and
School Improvement
University of West Georgia
Carrollton
United States

Editorial Office
MDPI AG
Grosspeteranlage 5
4052 Basel, Switzerland

This is a reprint of the Special Issue, published open access by the journal *Mathematics* (ISSN 2227-7390), freely accessible at: https://www.mdpi.com/journal/mathematics/special_issues/4D34PCSB1H.

For citation purposes, cite each article independently as indicated on the article page online and as indicated below:

Lastname, A.A.; Lastname, B.B. Article Title. *Journal Name* **Year**, *Volume Number*, Page Range.

ISBN 978-3-7258-3499-0 (Hbk)
ISBN 978-3-7258-3500-3 (PDF)
https://doi.org/10.3390/books978-3-7258-3500-3

Contents

About the Editor

Diana Mindrila

Dr. Mindrila is a Professor of Educational Research at the University of West Georgia. Her teaching experience includes courses in quantitative research methodology, research design, and classroom assessment. Her research interests include the development of behavior typologies using multivariate classification procedures. Dr. Mindrila also published methodological studies on topics such as cluster analysis, latent class analysis, structural equation modeling, and factor analysis.

Preface

The landscape of modern research is increasingly defined by complexity and uncertainty. As scholars and practitioners across diverse fields grapple with intricate data and ambiguous outcomes, Bayesian analysis has emerged as a powerful tool for making sense of the world around us. This reprint brings together a collection of studies that showcase the versatility and depth of Bayesian methods, offering readers both theoretical insights and practical applications.

In compiling these studies, the goal was to highlight the unique strengths of Bayesian analysis across a range of disciplines. Each study presents a detailed exploration of how Bayesian methods can be applied to real-world problems, demonstrating their capacity to incorporate prior knowledge, handle uncertainty, and update beliefs in light of new data.

The studies included in this reprint are diverse in their subject matter but united by a common thread: the application of Bayesian analysis to answer pressing research questions. By walking through these case studies, readers will see firsthand how Bayesian methods can model complex phenomena, make predictions, and inform decision-making in contexts where traditional methods may fall short.

This reprint is intended for researchers, students, and professionals who are interested in deepening their understanding of Bayesian analysis through applied examples. While some familiarity with Bayesian concepts will be helpful, the studies are presented with sufficient detail and explanation to be accessible to those new to the field. Each study not only delves into the technical aspects of Bayesian analysis but also provides context, discussing the implications of the findings and the advantages and limitations of the Bayesian approach.

Bayesian analysis is more than just a set of statistical tools; it is a way of thinking about data and uncertainty. This reprint will illustrate the power of that perspective through concrete examples, showing how Bayesian methods can lead to more nuanced and informed conclusions. Whether you are looking to apply Bayesian analysis in your own work or simply want to learn more about its potential, I hope that these studies will inspire you to explore the rich possibilities that this approach offers.

In an era where data-driven insights are crucial, the ability to reason effectively under uncertainty is invaluable. The studies presented here serve as a testament to the growing importance of Bayesian analysis in modern research, and I am excited to share them with you.

Diana Mindrila
Guest Editor

mathematics

Article

Tampered Random Variable Analysis in Step-Stress Testing: Modeling, Inference, and Applications

Hanan Haj Ahmad [1,2,*], Dina A. Ramadan [3] and Ehab M. Almetwally [4,5,*]

1 Department of Basic Science, The General Administration of Preparatory Year, King Faisal University, Hofuf 31982, Al Ahsa, Saudi Arabia
2 Department of Mathematics and Statistics, College of Science, King Faisal University, Hofuf 31982, Al-Ahsa, Saudi Arabia
3 Department of Mathematics, Faculty of Science, Mansoura University, Mansoura 33516, Egypt; dinaramadan21@mans.edu.eg
4 Department of Mathematics and Statistics, Faculty of Science, Imam Mohammad Ibn Saud Islamic University (IMSIU), Riyadh 11432, Riyadh, Saudi Arabia
5 Faculty of Business Administration, Delta University of Science and Technology, Gamasa 11152, Egypt
* Correspondence: hhajahmed@kfu.edu.sa (H.H.A.); emalmetwally@imamu.edu.sa (E.M.A.)

Abstract: This study explores a new dimension of accelerated life testing by analyzing competing risk data through Tampered Random Variable (TRV) modeling, a method that has not been extensively studied. This method is applied to simple step-stress life testing (SSLT), and it considers multiple causes of failure. The lifetime of test units under changeable stress levels is modeled using Power Rayleigh distribution with distinct scale parameters and a constant shape parameter. The research introduces unique tampering coefficients for different failure causes in step-stress data modeling through TRV. Using SSLT data, we calculate maximum likelihood estimates for the parameters of our model along with the tampering coefficients and establish three types of confidence intervals under the Type-II censoring scheme. Additionally, we delve into Bayesian inference for these parameters, supported by suitable prior distributions. Our method's validity is demonstrated through extensive simulations and real data application in the medical and electrical engineering fields. We also propose an optimal stress change time criterion and conduct a thorough sensitivity analysis.

Keywords: tampered random variable; competing risk; step stress; censoring scheme; maximum likelihood estimation; Bayes estimation; bootstrap method; simulation analysis

MSC: 62E10; 62F15; 62N05; 60E05; 62P30

Citation: Ahmad, H.H.; Ramadan, D.A.; Almetwally, E.M. Tampered Random Variable Analysis in Step-Stress Testing: Modeling, Inference, and Applications. *Mathematics* **2024**, *12*, 1248. https://doi.org/10.3390/math12081248

Academic Editors: Diana Mindrila and Jose Luis Vicente Villardon

Received: 13 March 2024
Revised: 11 April 2024
Accepted: 17 April 2024
Published: 20 April 2024

1. Introduction

With ongoing improvements in the manufacturing sector, numerous industrial products, known for their high reliability and complex designs, are becoming increasingly usable in everyday life. Accelerated life testing (ALT) addresses the challenge of evaluating such products by exposing them to stress levels higher than their usual operating conditions, producing rapid failures in turn. These growing stress factors—such as temperature, voltage, and humidity—significantly influence the lifespan of electronic equipment, including electric bulbs, fans, computers, toasters, and more. By employing these high-stress factors in ALT experiments, valuable insights concerning product reliability can be rapidly acquired within a condensed experimental time frame. Analyzing reliability and making inferences from it have gained significant interest in the literature, as illustrated by references [1–3].

ALT experiments can be conducted in two ways: with a starting constant high-stress level or with a changeable stress factor that can be varied during different time intervals. In the realm of ALT, there exists a specific class known as step-stress life testing (SSLT). This method permits experimenters to incrementally increase the stress levels at predetermined time points during the experiment. A basic form of SSLT is exemplified in experiments involving only two stress levels, denoted as s_1 and s_2, along with a single, pre-determined point in time, τ, at which the stress level shifts.

To understand how lifetime distributions vary under different stress levels, some basic modeling assumptions are typically discussed:

- Cumulative Exposure models (CE). In this method, specific restrictions are applied to ensure that the lifetime distributions at each progressive stress level align at their designated transition points, maintaining continuity. This approach is detailed in works by Sedyakin [4] and Nelson [5].
- Tampered Failure Rate (TFR) modeling. This technique involves adjusting the failure rates, increasing them at each subsequent stress level. Key references for TFR modeling include Bhattacharyya and Soejoeti [6], and Madi [7].
- The Tampered Random Variable (TRV) model. Here, the focus is on reducing the remaining lifetime for each new stress level. For more information on this approach, you can refer to Goel [8] and DeGroot and Goel [9].
- Step-stress partially accelerated life testing with a large amount of censored data. This approach addresses the gap in estimating non-homogeneous distribution and acceleration factor parameters under multiple censored data conditions. For more details, one can refer to Khan and Aslam [10].

Additionally, Sultana and Dewanji [11] explored the relationships between the TRV model and the two other models, TFR and CE, within a multi-step stress environment. They noted that TRV modeling aligns with CE and TFR when the fundamental lifetime distribution is exponential and the distributions at each stress level adhere to a scale-based parametric family. Thus, it is observed that the above three models converge when the fundamental distribution is exponential. TRV modeling stands out for its ability to be generalized to multiple-step-stress situations more effectively than the other two models. It also offers advantages in terms of modeling discrete and multivariate lifetimes, which are more complex tasks for the CE and TFR models.

Comparing factors that lead to risk model failures is essential for comprehending the contributing factors, detecting common changes, assessing model performance, and influencing decision making and risk management. It assists in identifying important issues that must be resolved to increase the precision and dependability of the model. The development of more reliable models can be made possible by recognizing similar patterns among various occurrences or outcomes that can be identified by understanding these components. Additionally, it offers useful information for model creators and validators, enabling them to improve model development processes, assumptions, and validation processes for more accurate and dependable models. The competing risks concept refers to the possibility of individual failure in a specific field owing to distinct factors. The cause-of-failure indication and the individual failure time are examples of observable data in this approach. When examining data on competing risks, the failure variables are typically unrelated to one another which means that the two risk factors are statistically independent. In the industrial and mechanical domains, fatigue and aging deterioration can lead to an assembly device failing due to electrical/optical signal (voltage, current, or light intensity) falling to an intolerable level. Numerous studies in the existing literature utilize CEM and TFR modeling within competing risk scenarios. However, to the best of our knowledge, research incorporating TRV modeling into the context of competing risk data is notably scarce. See for example Sultana et al. [12], Ramadan et al. [13] and Tolba et al. [14].

In this work, TRV is used with the SSLT model under two independent competing risk factors where the failure times follow the Power Rayleigh distribution. The sample is observed under the Type-II censored scheme. The censoring schemes have been introduced to solve the lack of information in lifetime experiments, saving time and costs. Type-I censoring has a predetermined time, while Type-II censoring has predetermined failure units.

The main goals of this study are summarized below:

- Performing an inferential analysis to obtain point and interval estimation of the unknown parameters of the distribution and the acceleration factor using both the maximum likelihood estimator and the Bayesian method.
- Applying numerical methods like Monte Carlo simulation to assess the performance of estimators obtained from Maximum Likelihood Estimation (MLE) and Bayesian methods, focusing on their bias, mean squared error, and the coverage probability (CP) for the confidence intervals.
- Evaluating real-world data sets from the medical field concerning AIDS infection, alongside another study from electrical engineering involving the causes of the failure of electronic components, serves to empirically assess the effectiveness of the newly proposed model.

The structure of the remainder of this document is as follows: Section 2 outlines the SSLT model under the TRV framework with the Power Rayleigh distribution. Section 3 details the methodologies used for point estimation, specifically using maximum likelihood and Bayesian methods. Section 4 is dedicated to interval estimation, exploring three distinct methods. Section 5 focuses on simulation analysis and presents the results in tabular form. The determination of the optimal time for stress change and an analysis of sensitivity are discussed in Section 6. An application using real-world data is examined in Section 7. The paper concludes with a summary of the findings in Section 8.

2. Model Description

In this study, we consider the SSLT model with random failure time variables denoted by U_1 and U_2 along with the stress levels s_1 and s_2 that are assumed to follow a Power Rayleigh distribution with a common shape parameter γ and distinct scale parameters λ_1 and λ_2. The two risk factors are called cause I and cause II and both are performed using Type-II censored samples. At a prefixed time τ, the stress level moves from s_1 to s_2. During the first stress level, the s_1 units will operate until a specific time τ, following which any remaining survivals that have not failed by time τ are moved to be tested under accelerated conditions with an acceleration factor β. Consequently, the system will operate under the second stress level s_2 until we obtain the required failure times. The effect of stress transition from the first stress to the accelerated condition may be explained by multiplying the remaining lifetime by the acceleration factor β . Hence the TRV for U_1 and U_2 is expressed as

$$\tilde{U_1} = \begin{cases} U_1 & ,0 < t \leq \tau \\ \tau + \beta(U_1 - \tau) & ,t > \tau \end{cases} \tag{1}$$

and

$$\tilde{U_2} = \begin{cases} U_2 & ,0 < t \leq \tau \\ \tau + \beta(U_2 - \tau) & ,t > \tau, \end{cases} \tag{2}$$

where τ is the time at which the stress changes and the acceleration parameter is $0 < \beta < 1$.

We consider Power Rayleigh distribution as a lifetime model. The Rayleigh distribution, a continuous distribution of significant practical relevance, has been the subject of extensive study by various authors who have explored its statistical properties, inference methods, and reliability analysis. Additionally, a variety of extended versions of the Rayleigh distribution have been introduced. For example, Rosaiah and Kantam [15] applied the inverse Rayleigh to failure times data. Merovci [16] introduced the transmuted Rayleigh and modeled the amount of nicotine in blood. Cordeiro et al. [17] studied the beta-generalized Rayleigh distribution and its application. More generalizations of Rayleigh distribution can be found in the literature and one may refer to [18–24].

The Power Rayleigh (PR) distribution was first introduced by Neveen et al. [25]. It is a versatile and flexible statistical model known for its ability to handle a wide range of data types. This distribution is particularly useful due to its capability to model data that exhibit a skewed pattern, which is common in many practical situations. The Power Rayleigh distribution is characterized by two parameters that allow it to adapt to various data shapes and sizes, making it more flexible than the standard Rayleigh distribution. Its applications are diverse, ranging from reliability engineering and survival analysis to modeling wind speed and signal processing. The flexibility in shape and scale provided by the Power Rayleigh distribution makes it a valuable tool for analyzing and interpreting real-world data in various scientific and engineering fields. We assume that the Power Rayleigh distribution has a shape parameter γ and scale parameter λ, where both have positive support, and then the cumulative distribution function (CDF) becomes

$$F(t) = 1 - e^{-\frac{t^{2\gamma}}{2\lambda^2}},$$

and the probability density function (PDF) is

$$f(t) = \frac{\gamma}{\lambda^2} t^{2\gamma-1} e^{-\frac{t^{2\gamma}}{2\lambda^2}}.$$

Consider a set of n units subjected to a life test starting at stress level s_1. Failures and their corresponding risks are documented over time. At a designated moment τ the stress level shifts from s_1 to s_2, and the test runs until r (with $r < n$) failures are noted. If r equals n, a complete dataset is collected as in a simple SSLT without data truncation. We assume that each unit's failure is attributable to one of two competing risks, each described by a Power Rayleigh distribution with a consistent shape parameter γ but distinct scale parameters λ_j for $j = 1, 2$, aligned with the TRV model.

The CDF for the lifetime U_j associated with risk j for $j = 1, 2$ is then expressed as follows:

$$F_j(t) = F_j(t; \gamma, \lambda_j) = \begin{cases} 1 - e^{-\frac{t^{2\gamma}}{2\lambda_j^2}} & \text{if } 0 < t \leq \tau \\ 1 - e^{-\frac{(\tau+\beta^{-1}(t-\tau))^{2\gamma}}{2\lambda_j^2}} & \text{if } t > \tau \end{cases} \tag{3}$$

and the corresponding PDF of U_j is given by

$$f_j(t) = f_j(t; \gamma, \lambda_j) = \begin{cases} \frac{\gamma}{\lambda_j^2} t^{2\gamma-1} e^{-\frac{t^{2\gamma}}{2\lambda_j^2}} & \text{if } 0 < t \leq \tau \\ \frac{\gamma}{\lambda_j^2 \beta} (\tau + \beta^{-1}(t-\tau))^{2\gamma-1} e^{-\frac{(\tau+\beta^{-1}(t-\tau))^{2\gamma}}{2\lambda_j^2}} & \text{if } t > \tau \end{cases} \tag{4}$$

Let us denote the overall failure time of a unit under test as U, which is obtained by $U = min\{U_1, U_2\}$. Then, the CDF and PDF are easily obtained as

$$F(t) = F(t; \gamma, \lambda) = 1 - (1 - F_1(t))(1 - F_2(t)) = \begin{cases} 1 - e^{-\Lambda t^{2\gamma}} & \text{if } 0 < t \leq \tau \\ 1 - e^{-\Lambda(\tau+\beta^{-1}(t-\tau))^{2\gamma}} & \text{if } t > \tau \end{cases} \tag{5}$$

and

$$f(t) = f(t; \gamma, \lambda) = \begin{cases} \gamma \Lambda t^{2\gamma-1} e^{-\Lambda t^{2\gamma}} & \text{if } 0 < t \leq \tau \\ \frac{\gamma}{\beta} \Lambda (\tau + \beta^{-1}(t-\tau))^{2\gamma-1} e^{-\Lambda(\tau+\beta^{-1}(t-\tau))^{2\gamma}} & \text{if } t > \tau, \end{cases} \tag{6}$$

respectively, where $\lambda = (\lambda_1, \lambda_2)$ and $\Lambda = \frac{1}{2\lambda_2^2} + \frac{1}{2\lambda_1^2}$. Furthermore, let C denote the indicator for the cause of failure. Then, under our assumptions, the joint PDF of (U, C) is given by

$$f_{U,C}(t; \gamma, \lambda) = f_j(t)[1 - F_k(t)] = \begin{cases} \frac{\gamma}{\lambda_j^2} t^{2\gamma-1} e^{-\Lambda t^{2\gamma}} & \text{if } 0 < t \leq \tau \\ \frac{\gamma}{\lambda_j^2 \beta} (\tau + \beta^{-1}(t-\tau))^{2\gamma-1} e^{-\Lambda(\tau+\beta^{-1}(t-\tau))^{2\gamma}} & \text{if }, t > \tau, \end{cases} \tag{7}$$

for $j, k = 1, 2,\ j \neq k$.

In competing risk models, the assumption of independence is often considered to be impractical. Identifiable issues may emerge if dependencies exist within the model or due to a lack of covariates in the data. To mitigate these issues, we postulate a latent failure time model and treat the risks U_1 and U_2 as independent. Let N_{j1} represent the number of units failing from risk j before time τ and N_{j2} after τ, with $N_j = N_{j1} + N_{j2}$, ensuring $N_1 + N_2 \leq r$. The sequence of observed failure times is $0 < t_{1:n_1} < t_{2:n_2} < \cdots < t_{r:n}$. Let $\hat{n}_1$ denote the observed value for N_1, $\hat{n}_2$ denote the observed value for N_2, and let $N = (N_1, N_2)$ be the vector of these counts.

In the next section, classical and Bayesian estimation methods are constructed to estimate the unknown parameters for the Power Rayleigh and the accelerated constant β under the two competing risk factors with the Type-II censoring scheme.

3. Point Estimation

In this study, two approaches to estimation are examined: the frequentist maximum likelihood estimation (MLE) and the Bayesian estimation method. Section 4 is dedicated to conducting a simulation analysis and applying numerical techniques to evaluate the efficacy of these estimation strategies.

3.1. Maximum Likelihood Estimation

In this section, the maximum likelihood estimation MLE method is employed to determine the unknown parameters of the Power Rayleigh distribution within the TRV framework. Numerical methods, including the renowned Newton–Raphson technique, are utilized to compute the necessary estimators. Subsequently, assuming the TRV model, we construct the likelihood function of $\psi = (\gamma, \lambda_1, \lambda_2, \beta)$ based on Type-II censored data as

$$L(\psi|(t,c)) = \frac{n!}{(n-r)!} \prod_{i=1}^{\hat{n}_1} f_{U,C}(t_{i:n}, c_i) \prod_{i=\hat{n}_1+1}^{r} f_{U,C}(t_{i:n}, c_i)[1 - F(t_{r:n})]^{n-r}.$$

Here $r = \hat{n}_1 + \hat{n}_2 = n_{11} + n_{12} + n_{21} + n_{22}$. By substituting Equations (5) and (7) into the above likelihood equation we obtain

$$L(\psi|(t,c)) = \frac{n!}{(n-r)!} \left(\frac{\gamma^r \beta^{-\hat{n}_2}}{\lambda_1^{2\hat{n}_1} \lambda_2^{2\hat{n}_2}} \right) \prod_{i=1}^{\hat{n}_1} t_i^{2\gamma-1} e^{-\Lambda t_i^{2\gamma}} \prod_{i=\hat{n}_1+1}^{r} (\tau + \beta^{-1}(t_i - \tau))^{2\gamma-1} e^{-\Lambda(\tau+\beta^{-1}(t_i-\tau))^{2\gamma}} e^{-\Lambda(n-r)t_{r:n}^{2\gamma}}. \tag{8}$$

The log-likelihood function can be written as

$$\begin{aligned} \ell(\psi) =& r \log \gamma - \hat{n}_2 \log \beta - 2\hat{n}_1 \log \lambda_1 - 2\hat{n}_2 \log \lambda_2 + (2\gamma - 1) \left[\sum_{i=1}^{\hat{n}_1} \log(t_i) + \sum_{i=n_1+1}^{r} \log(\tau + \beta^{-1}(t_i - \tau)) \right] - \\ & \Lambda \left[\sum_{i=1}^{\hat{n}_1} t_i^{2\gamma} + \sum_{i=\hat{n}_1+1}^{r} (\tau + \beta^{-1}(t_i - \tau))^{2\gamma} \right] + (n-r) t_{r:n}^{2\gamma}. \end{aligned} \tag{9}$$

The maximum likelihood estimations of the parameters $(\gamma, \lambda_1, \lambda_2, \beta)$ are obtained by differentiating the log-likelihood function $\ell(\psi)$ with respect to the parameters $(\gamma, \lambda_1, \lambda_2, \beta)$ and setting the result to zero, so we have the following normal equations.

$$\begin{aligned} \frac{\partial \ell(\psi)}{\partial \gamma} =& \frac{r}{\gamma} + 2 \left[\sum_{i=1}^{\hat{n}_1} \log(t_i) + \sum_{i=\hat{n}_1+1}^{r} \log(\tau + \beta^{-1}(t_i - \tau)) \right] - \\ & \Lambda \left[\sum_{i=1}^{\hat{n}_1} t_i^{2\gamma} \log(t_i) + \sum_{i=\hat{n}_1+1}^{r} (\tau + \beta^{-1}(t_i - \tau))^{2\gamma} \log(\tau + \beta^{-1}(t_i - \tau)) \right] + (n-r) t_{r:n}^{2\gamma} \log(t_{r:n}), \end{aligned} \tag{10}$$

$$\frac{\partial \ell(\psi)}{\partial \beta} = -\frac{\hat{n}_2}{\beta} - (2\gamma-1)\sum_{i=\hat{n}_1+1}^{r} \frac{\beta^{-2}(t_i-\tau))}{(\tau+\beta^{-1}(t_i-\tau))} + \Lambda \sum_{i=\hat{n}_1+1}^{r} \gamma\beta^{-2}(t_i-\tau)(\tau+\beta^{-1}(t_i-\tau))^{2\gamma-1}, \tag{11}$$

$$\frac{\partial \ell(\psi)}{\partial \lambda_1} = \frac{-2\hat{n}_1}{\lambda_1} + \lambda_1^{-3}\left[\sum_{i=1}^{\hat{n}_1} t_i^{2\gamma} + \sum_{i=\hat{n}_1+1}^{r} (\tau+\beta^{-1}(t_i-\tau))^{2\gamma} + (n-r)t_{r:n}^{2\gamma}\right] \tag{12}$$

and

$$\frac{\partial \ell(\psi)}{\partial \lambda_2} = \frac{-2\hat{n}_2}{\lambda_2} + \lambda_2^{-3}\left[\sum_{i=1}^{\hat{n}_1} t_i^{2\gamma} + \sum_{i=\hat{n}_1+1}^{r} (\tau+\beta^{-1}(t_i-\tau))^{2\gamma} + (n-r)t_{r:n}^{2\gamma}\right]. \tag{13}$$

For known γ and β, the MLEs of λ_1 and λ_2 are given by

$$\hat{\lambda}_1 = \frac{1}{\sqrt{2}(\hat{n}_1)^{\frac{1}{2}}}\left[\sum_{i=1}^{\hat{n}_1} t_i^{2\gamma} + \sum_{i=\hat{n}_1+1}^{r} (\tau+\beta^{-1}(t_i-\tau))^{2\gamma} + (n-r)t_{r:n}^{2\gamma}\right]^{\frac{1}{2}}$$

and

$$\hat{\lambda}_2 = \frac{1}{\sqrt{2}(\hat{n}_2)^{\frac{1}{2}}}\left[\sum_{i=1}^{\hat{n}_1} t_i^{2\gamma} + \sum_{i=\hat{n}_1+1}^{r} (\tau+\beta^{-1}(t_i-\tau))^{2\gamma} + (n-r)t_{r:n}^{2\gamma}\right]^{\frac{1}{2}}.$$

To address the system of nonlinear equations presented in Equations (10)–(13), numerical approaches are essential. Various numerical methods have been applied in existing research; in this instance, we employ the Newton–Raphson method. The outcomes of this application are detailed in Section 5.

3.2. Bayesian Inference

In this section, we apply the Bayesian estimation technique to determine the unknown parameters of the Power Rayleigh distribution. The fundamental principle of the Bayesian approach posits that the model's parameters are random variables with a predefined distribution, referred to as the prior distribution. Given the availability of prior knowledge, selecting an appropriate prior is crucial. We opt for the gamma conjugate prior distribution for the parameters for many reasons, such as the flexibility in its nature with a non-informative domain and the calculations' simplicity making analytical or computational updates to the posterior easier. Also, the positive of the domain makes it suitable for modeling parameters. We perform the Bayesian inference method for estimating the unknown parameters $\psi = (\gamma, \lambda_1, \lambda_2, \beta)$. We assume independent gamma priors for γ, λ_1, and λ_2 and a uniform prior for β. That is, γ, λ_1 and λ_2 have $Gamma(c_i, d_i)$, where $c_i, d_i > 0$, $i = 1, 2, 3$, are non-negative hyperparameters, and β follows uniform prior as follows:

$$\pi(\beta) = 1, \ 0 < \beta < 1.$$

The estimates have been developed under the square error loss function (SELF) and the linear exponential loss function (LLF). Hence, the joint prior density of the independent parameters is given by

$$\pi(\psi) = \pi(\gamma)\pi(\lambda_1)\pi(\lambda_2)\pi(\beta),$$

$$\pi(\psi) = \gamma^{c_1-1}e^{-d_1\gamma}\lambda_1^{c_2-1}e^{-d_2\lambda_1}\lambda_2^{c_3-1}e^{-d_3\lambda_2}, \quad \gamma > 0, \lambda_1 > 0, \ \lambda_2 > 0, 0 < \beta < 1. \tag{14}$$

The joint posterior density function for the parameters can be derived by incorporating the observed censored samples, and the prior distributions of these parameters are as follows:

$$\pi^*(\psi,t,c) = \pi(\psi)\ L(\psi|t,c)$$
$$= \gamma^{c_1-1}e^{-d_1\gamma}\lambda_1^{c_2-1}e^{-d_2\lambda_1}\lambda_2^{c_3-1}e^{-d_3\lambda_2}\frac{\gamma^r\beta^{-\hat{n}_2}}{\lambda_1^{2\hat{n}_1}}\lambda_2^{2\hat{n}_2}\prod_{i=1}^{\hat{n}_1}t_i^{2\gamma-1}e^{-\Lambda t_i^{2\gamma}}\times \tag{15}$$
$$\prod_{i=\hat{n}_1+1}^{r}(\tau+\beta^{-1}(t_i-\tau))^{2\gamma-1}e^{-\Lambda(\tau+\beta^{-1}(t_i-\tau))^{2\gamma}}e^{-\Lambda(n-r)t_{r:n}^{2\gamma}}.$$

Thus, the conditional posterior densities of the parameters $\gamma, \lambda_1, \lambda_2$, and β can be obtained by simplifying Equation (15) as follows

$$\pi_1^*(\lambda_1|\lambda_2,\gamma,\beta,t,c) = \lambda_1^{-2\hat{n}_1+c_2-1}e^{-d_2\lambda_1}\prod_{i=1}^{\hat{n}_1}e^{-\Lambda t_i^{2\gamma}}\prod_{i=\hat{n}_1+1}^{r}e^{-\Lambda(\tau+\beta^{-1}(t_i-\tau))^{2\gamma}}e^{-\Lambda(n-r)t_{r:n}^{2\gamma}}, \tag{16}$$

$$\pi_2^*(\lambda_2|\lambda_1,\gamma,\beta,t,c) = \lambda_2^{2\hat{n}_2+c_3-1}e^{-d_3\lambda_2}\prod_{i=1}^{\hat{n}_1}e^{-\Lambda t_i^{2\gamma}}\prod_{i=\hat{n}_1+1}^{r}e^{-(\tau+\beta^{-1}(t_i-\tau))^{2\gamma}\Lambda}e^{-\Lambda(n-r)t_{r:n}^{2\gamma}}, \tag{17}$$

$$\pi_3^*(\gamma|\lambda_1,\lambda_2,\beta,t,c) = \gamma^{r+c_1-1}e^{-d_1\gamma}\prod_{i=1}^{\hat{n}_1}t_i^{2\gamma-1}e^{-\Lambda t_i^{2\gamma}}\prod_{i=\hat{n}_1+1}^{r}(\tau+\beta^{-1}(t_i-\tau))^{2\gamma-1}e^{-\Lambda(\tau+\beta^{-1}(t_i-\tau))^{2\gamma}}e^{-\Lambda(n-r)t_{r:n}^{2\gamma}} \tag{18}$$

and

$$\pi_4^*(\beta|\lambda_1,\lambda_2,\gamma,t,c) = \beta^{-\hat{n}_2}\prod_{i=\hat{n}_1+1}^{r}(\tau+\beta^{-1}(t_i-\tau))^{2\gamma-1}e^{-\Lambda(\tau+\beta^{-1}(t_i-\tau))^{2\gamma}}. \tag{19}$$

Since the Equations (16)–(19) cannot be computed explicitly, numerical techniques are employed. One of the most powerful numerical techniques in Bayesian estimation is the Monte Carlo Markov Chain method (MCMC). In this scenario, we suggest employing the Metropolis–Hastings (M-H) sampling method within the Gibbs algorithm, utilizing a normal proposal distribution as recommended by Tierney [26]. The procedure for Gibbs sampling incorporating the *(M-H)* approach is outlined as follows:

(1) Set initial values $\left(\lambda_1^{(0)},\lambda_2^{(0)},\gamma^{(0)},\beta^{(0)}\right)$.

(2) Set $j = 1$.

(3) Using the following M-H algorithm, from $\pi_1^*(\lambda_1^{(j-1)}|\lambda_2^{(j-1)},\gamma^{(j-1)},\beta^{(j-1)},t,c)$ $\pi_2^*(\lambda_2^{(j-1)}|\lambda_1^{(j)},\gamma^{(j-1)},\beta^{(j-1)},t,c)$, $\pi_3^*(\gamma^{(j-1)}|\lambda_1^{(j)},\lambda_2^{(j)},\beta^{(j-1)},t,c)$, and

$\pi_4^*(\beta^{(j-1)}|\lambda_1^{(j)},\lambda_2^{(j)},\gamma^{(j)},t,c)$ generate $\lambda_1^{(j)},\lambda_2^{(j)},\gamma^{(j)}$, and $\beta^{(j)}$ with the normal proposal distributions

$$N\left(\lambda_1^{(j-1)},var(\lambda_1)\right),\ N\left(\lambda_2^{(j-1)},var(\lambda_2)\right),N\left(\gamma^{(j-1)},var(\gamma)\right),\text{ and }N\left(\beta^{(j-1)},var(\beta)\right),$$

and from the main diagonal in the inverse Fisher information matrix we obtained $var(\lambda_1), var(\lambda_2)$, $var(\gamma)$, and $var(\beta)$.

(4) Generate a proposal for λ_1^* from $N\left(\lambda_1^{(j-1)},var(\lambda_1)\right)$, λ_2^* from $N\left(\lambda_2^{(j-1)},var(\lambda_2)\right)$, γ^* from $N\left(\gamma^{(j-1)},var(\gamma)\right)$, and β^* from $N\left(\beta^{(j-1)},var(\beta)\right)$.

(i) The acceptance probabilities are

$$\left.\begin{aligned}
\mu_{\lambda_1} &= \min\left[1, \frac{\pi_1^*(\lambda_1^*|\lambda_2^{(j-1)},\gamma^{(j-1)},\beta^{(j-1)},t,c)}{\pi_1^*(\lambda_1^{(j-1)}|\lambda_2^{(j-1)},\gamma^{(j-1)},\beta^{(j-1)},t,c)}\right], \\
\mu_{\lambda_2} &= \min\left[1, \frac{\pi_2^*(\lambda_2^*|\lambda_1^{(j)},\gamma^{(j-1)},\beta^{(j-1)},t,c)}{\pi_2^*(\lambda_2^{(j-1)}|\lambda_1^{(j)},\gamma^{(j-1)},\beta^{(j-1)},t,c)}\right], \\
\mu_{\gamma} &= \min\left[1, \frac{\pi_3^*(\gamma^*|\lambda_1^{(j)},\lambda_2^{(j)},\beta^{(j-1)},t,c)}{\pi_3^*(\gamma^{(j-1)}|\lambda_1^{(j)},\lambda_2^{(j)},\beta^{(j-1)},t,c)}\right] \\
&\textit{and} \\
\mu_{\beta} &= \min\left[1, \frac{\pi_4^*(\beta^*|\lambda_1^{(j)},\lambda_2^{(j)},\gamma^{(j)},t,c)}{\pi_4^*(\beta^{(j-1)}|\lambda_1^{(j)},\lambda_2^{(j)},\gamma^{(j)},t,c)}\right].
\end{aligned}\right\}$$

(ii) From a Uniform $(0,1)$ distribution u_1, u_2, u_3, and u_4 are generated.
(iii) If $u_1 < \mu_{\lambda_1}$, accept the proposal and set $\lambda_1^{(j)} = \lambda_1^*$, otherwise set $\lambda_1^{(j)} = \lambda_1^{(j-1)}$.
(iv) If $u_2 < \mu_{\lambda_2}$, accept the proposal and set $\lambda_2^{(j)} = \lambda_2^*$, otherwise set $\lambda_2^{(j)} = \lambda_2^{(j-1)}$.
(v) If $u_3 < \mu_{\gamma}$, accept the proposal and set $\gamma^{(j)} = \gamma^*$, otherwise set $\gamma^{(j)} = \gamma^{(j-1)}$.
(vi) If $u_4 < \mu_{\beta}$, accept the proposal and set $\beta^{(j)} = \beta^*$, otherwise set $\beta^{(j)} = \beta^{(j-1)}$.

(5) Set $j = j + 1$.
(6) Steps (3)–(5), are repeated N times to obtain $\lambda_1^{(j)}, \lambda_2^{(j)}, \gamma^{(j)}$, and $\beta^{(j)}$, j = 1, 2, . . . N.

To guarantee convergence and eliminate the impact of initial value selection, the first M simulated variants are eliminated. For a sufficiently high N, the chosen samples are then $\psi_k^{(j)}, j = M+1, \ldots N$. The SEL function-based approximate BEs of ψ_k are generated using

$$\hat{\psi}_k^{(j)} = \frac{1}{N-M} \sum_{j=M+1}^{N} \psi^{(j)}, k = 1, 2, 3, 4. \tag{20}$$

The approximate Bayes estimates for ψ_k, under the Entropy loss function are given as

$$\hat{\psi}_k^{(j)} = \left[\frac{1}{N-M} \sum_{j=M+1}^{N} (\psi^{(j)})^{-q}\right]^{\frac{-1}{q}}, k = 1, 2, 3, 4. \tag{21}$$

4. Interval Estimation

Confidence interval estimation is a fundamental statistical method used to indicate the reliability of an estimate. It provides a range of values, derived from sample data, that is likely to contain the true value of an unknown population parameter. The concept is central to inferential statistics and has numerous applications across various fields such as engineering, economics, medicine, and the social sciences. Among its key properties, the asymptotic interval is notable for its reliance on large sample sizes, where the distribution of the estimate approaches a normal distribution, making it increasingly accurate as the sample size grows. This property is particularly useful for electrical engineering projects where large data sets are analyzed for reliability and performance assessments.

Credible intervals, on the other hand, are used in Bayesian statistics and represent the range within which a parameter lies with a certain probability, given the observed data and a prior belief about the parameter's distribution. This approach is valuable in research and development projects within electrical engineering, where prior knowledge or expert opinions can be quantitatively incorporated into the analysis, offering a more nuanced understanding of uncertainty.

Bootstrap intervals utilize resampling techniques to generate an empirical distribution of the estimator by drawing samples with replacements from the original dataset.

This method does not assume a specific distribution, making it versatile and robust, especially in complex engineering studies where theoretical distributions are hard to justify. The bootstrap approach is particularly important for evaluating the uncertainty of estimates derived from small or non-standard datasets, providing a powerful tool for uncertainty quantification in both academic research and practical applications.

The application and importance of these intervals lie in their ability to quantify the uncertainty in estimates, guiding decision making and hypothesis testing. In electrical engineering, for example, they can be used to assess the reliability of system parameters, evaluate the performance of new designs, or validate models against empirical data. By understanding and applying these concepts, researchers, and practitioners can enhance the rigor and credibility of their findings, contributing to more reliable and effective solutions in their respective fields. The following subsections work out the previously mentioned interval estimations.

4.1. *Asymptotic Confidence Interval*

This subsection presents the observed Fisher information matrix, commonly employed for the construction of asymptotic confidence intervals (ACIs).

The MLEs $(\hat{\lambda}_1, \hat{\lambda}_2, \hat{\gamma}, \hat{\beta})$ are approximately normal with a mean of $(\hat{\lambda}_1, \hat{\lambda}_2, \hat{\gamma}, \hat{\beta})$ and a variance matrix $I^{-1}(\hat{\lambda}_1, \hat{\lambda}_2, \hat{\gamma}, \hat{\beta})$. Here, $\hat{I}(\lambda_1, \lambda_2, \gamma, \beta)$ is the observed Fisher information matrix, and it is defined as

$$\hat{I}(\lambda_1, \lambda_2, \gamma, \beta) = \begin{pmatrix} -\frac{\partial^2 \ell}{\partial \lambda_1^2} & -\frac{\partial^2 \ell}{\partial \lambda_1 \partial \lambda_2} & -\frac{\partial^2 \ell}{\partial \lambda_1 \partial \gamma} & -\frac{\partial^2 \ell}{\partial \lambda_1 \partial \beta} \\ -\frac{\partial^2 \ell}{\partial \lambda_2 \partial \lambda_1} & -\frac{\partial^2 \ell}{\partial \lambda_2^2} & -\frac{\partial^2 \ell}{\partial \lambda_2 \partial \gamma} & -\frac{\partial^2 \ell}{\partial \lambda_2 \partial \beta} \\ -\frac{\partial^2 \ell}{\partial \gamma \partial \lambda_1} & -\frac{\partial^2 \ell}{\partial \gamma \partial \lambda_2} & -\frac{\partial^2 \ell}{\partial \gamma^2} & -\frac{\partial^2 \ell}{\partial \gamma \partial \beta} \\ -\frac{\partial^2 \ell}{\partial \beta \partial \lambda_1} & -\frac{\partial^2 \ell}{\partial \beta \partial \lambda_2} & -\frac{\partial^2 \ell}{\partial \beta \partial \gamma} & -\frac{\partial^2 \ell}{\partial \beta^2} \end{pmatrix}_{(\lambda_1, \lambda_2, \gamma, \beta) = (\hat{\lambda}_1, \hat{\lambda}_2, \hat{\gamma}, \hat{\beta})}, \quad (22)$$

where the second partial derivatives are as follows:

$$\frac{\partial^2 \ell}{\partial \gamma^2} = \frac{-r}{\gamma^2} - 2\Lambda \left[\sum_{i=1}^{\hat{n}_1} t_i^{2\gamma} (\log(t_i))^2 + \sum_{i=\hat{n}_1+1}^{r} (\tau + \beta^{-1}(t_i - \tau))^{2\gamma} (\log(\tau + \beta^{-1}(t_i - \tau)))^2 \right] + (n-r) t_{r:n}^{2\gamma} (\log(t_{r:n}))^2,$$

$$\begin{aligned} \frac{\partial^2 \ell}{\partial \beta^2} =& (2\gamma - 1) \sum_{i=\hat{n}_1+1}^{r} \frac{\beta^{-4}(t_i - \tau)^2 - 2\beta^{-3}(t_i - \tau)(\tau + \beta^{-1}(t_i - \tau))}{(\tau + \beta^{-1}(t_i - \tau))^2} - \\ & \Lambda \sum_{i=\hat{n}_1+1}^{r} \gamma \beta^{-3} (t_i - \tau)(\tau + \beta^{-1}(t_i - \tau))^{2\gamma - 1} \left[2 + (2\gamma - 1)(t_i - \tau)\beta^{-1}(\tau + \beta^{-1}(t_i - \tau))^{-1} \right], \end{aligned}$$

$$\frac{\partial^2 \ell}{\partial \lambda_1^2} = \frac{2\hat{n}_1}{\lambda_1^2} - 3\lambda_1^{-4} \left[\sum_{i=1}^{n_1} t_i^{2\gamma} + \sum_{i=n_1+1}^{r} (\tau + \beta^{-1}(t_i - \tau))^{2\gamma} + (n-r) t_{r:n}^{2\gamma} \right],$$

$$\frac{\partial^2 \ell}{\partial \lambda_2^2} = \frac{2\hat{n}_2}{\lambda_2^2} - 3\lambda_2^{-4} \left[\sum_{i=1}^{n_1} t_i^{2\gamma} + \sum_{i=n_1+1}^{r} (\tau + \beta^{-1}(t_i - \tau))^{2\gamma} + (n-r) t_{r:n}^{2\gamma} \right],$$

$$\frac{\partial^2 \ell}{\partial\gamma\partial\beta} = -2\sum_{i=n_1+1}^{r}\frac{\beta^{-2}(t_i-\tau)}{(\tau+\beta^{-1}(t_i-\tau))^2}+$$
$$\Lambda\left[\sum_{i=n_1+1}^{r}\beta^{-2}(t_i-\tau)\left[2\gamma(\tau+\beta^{-1}(t_i-\tau))^{2\gamma-1}\log(\tau+\beta^{-1}(t_i-\tau))-(\tau+\beta^{-1}(t_i-\tau))^{2\gamma-2}\right]\right],$$

$$\frac{\partial^2 \ell}{\partial\gamma\partial\lambda_1} = \frac{-2}{\lambda_1^3}\left[\sum_{i=1}^{n_1} t_i^{2\gamma}\log(t_i)+\sum_{i=n_1+1}^{r}(\tau+\beta^{-1}(t_i-\tau))^{2\gamma}\log(\tau+\beta^{-1}(t_i-\tau))\right]+(n-r)t_{r:n}^{2\gamma}\log(t_{r:n})$$

$$\frac{\partial^2 \ell}{\partial\gamma\partial\lambda_2} = \frac{-2}{\lambda_2^3}\left[\sum_{i=1}^{n_1} t_i^{2\gamma}\log(t_i)+\sum_{i=n_1+1}^{r}(\tau+\beta^{-1}(t_i-\tau))^{2\gamma}\log(\tau+\beta^{-1}(t_i-\tau))\right]+(n-r)t_{r:n}^{2\gamma}\log(t_{r:n})$$

$$\frac{\partial^2 \ell}{\partial\beta\partial\lambda_1} = -2\gamma\lambda_1^{-3}\sum_{i=n_1+1}^{r}\beta^{-2}(t_i-\tau)(\tau+\beta^{-1}(t_i-\tau))^{2\gamma-1},$$

$$\frac{\partial^2 \ell}{\partial\beta\partial\lambda_2} = -2\gamma\lambda_2^{-3}\sum_{i=n_1+1}^{r}\beta^{-2}(t_i-\tau)(\tau+\beta^{-1}(t_i-\tau))^{2\gamma-1}$$

$$\frac{\partial^2 \ell}{\partial\lambda_1\partial\lambda_2} = 0.$$

Consequently, the estimated asymptotic variance–covariance matrix $\hat{V}$ for the MLEs can be obtained by taking the inverse of the observed information matrix $\hat{I}(\lambda_1,\lambda_2,\gamma,\beta)$ which is given by

$$[\hat{V}] = \hat{I}^{-1} = \begin{pmatrix} \widehat{Var}(\hat{\lambda}_1) & cov(\hat{\lambda}_1,\hat{\lambda}_2) & cov(\hat{\lambda}_1,\hat{\gamma}) & cov(\hat{\lambda}_1,\hat{\beta}) \\ cov(\hat{\lambda}_1,\hat{\lambda}_2) & \widehat{Var}(\hat{\lambda}_2) & cov(\hat{\lambda}_2,\hat{\gamma}) & cov(\hat{\lambda}_2,\hat{\beta}) \\ cov(\hat{\lambda}_1,\hat{\gamma}) & cov(\hat{\lambda}_2,\hat{\gamma}) & \widehat{Var}(\hat{\gamma}) & cov(\hat{\gamma},\hat{\beta}) \\ cov(\hat{\lambda}_1,\hat{\beta}) & cov(\hat{\lambda}_2,\hat{\beta}) & cov(\hat{\gamma},\hat{\beta}) & \widehat{Var}(\hat{\beta}) \end{pmatrix}. \quad (23)$$

The $100(1-\zeta)\%$ two-sided confidence interval can be written as

$$\widehat{\lambda_1} \pm Z_{\frac{\zeta}{2}}\sqrt{\widehat{Var}(\hat{\lambda}_1)},\ \widehat{\lambda_2} \pm Z_{\frac{\zeta}{2}}\sqrt{\widehat{Var}(\hat{\lambda}_2)},\ \hat{\gamma} \pm Z_{\frac{\zeta}{2}}\sqrt{\widehat{Var}(\hat{\gamma})},\ \text{and}\ \hat{\beta} \pm Z_{\frac{\zeta}{2}}\sqrt{\widehat{Var}(\hat{\beta})}, \quad (24)$$

where $Z_{\frac{\zeta}{2}}$ is the percentile of the standard normal distribution with right-tail probability $\frac{\zeta}{2}$.

4.2. Credible Interval

Using the Metropolis–Hastings algorithm within the Gibbs sampling framework, we determined the credible confidence interval (CCI). For clarity, we refer to subsection 3.2, and the algorithm steps mentioned there. Proceeding after step (6), the $100(1-\zeta)\%$ CCIs of ψ_k where $(\psi_1,\psi_2,\psi_3,\psi_4)=(\lambda_1,\lambda_2,\gamma,\beta)$ with $\psi_k^{(1)} < \psi_k^{(2)} \ldots < \psi_k^{(N)}$, is given by

$$\left(\psi_{k(N\ (\zeta/2))},\psi_{k(N\ (1-\zeta/2))}\right).$$

4.3. Bootstrap Interval

Bootstrap confidence intervals offer a versatile approach to estimating the uncertainty of an estimator when the underlying distribution is unknown or complex. There are two main types: the bootstrap-t and the bootstrap percentile (bootstrap-p) methods.

4.3.1. Parametric Boot-p

The bootstrap percentile (p) method involves generating a large number of bootstrap samples from the original data. For each sample, the statistic of interest is calculated, creating a distribution of these statistics. The confidence interval is then directly obtained by taking percentiles from this empirical distribution. The following steps describe the algorithm of this method:

(1) Based on $x = x_{1:n}, x_{2:n}, \ldots, x_{m:n}$, obtain $\hat{\lambda}_1, \hat{\lambda}_2, \hat{\gamma}$, and $\hat{\beta}$ by maximizing Equations (10)–(13).
(2) Generate $x^* = x^*_{1:n}, x^*_{2:n}, \ldots, x^*_{m:n}$ from the PR distribution with parameters $\hat{\lambda}_1, \hat{\lambda}_2, \hat{\gamma}$, and $\hat{\beta}$ based on Type-II censoring under TRV, by considering the algorithm presented in [27].
(3) Obtain the bootstrap parameter estimation $\hat{\psi}^*_i = \left(\hat{\lambda}^*_{1i}, \hat{\lambda}^*_{2i}, \hat{\gamma}^*_i, \hat{\beta}^*_i, \right)$, with $i = 1, 2, 3, \ldots, N$ boots using the MLEs under the bootstrap sampling.
(4) Repeat steps (2) and (3) N boot times, and obtain $\hat{\psi}^*_1, \hat{\psi}^*_2, \ldots, \hat{\psi}^*_{N\,boot}$
(5) Obtain $\hat{\psi}^*_{(1)}, \hat{\psi}^*_{(2)}, \ldots, \hat{\psi}^*_{(N\,boot)}$ by arrange $\hat{\psi}^*_i, i = 1, 2, 3, \ldots, N\,boot$ in ascending orders.

Define $\hat{\psi}_{boot-p} = G_1^{-1}(z)$ for a given z, where $G_1(z) = P(\hat{\psi}^* \le z)$ denotes the cumulative distribution function of $\hat{\psi}^*$. The $100(1-\zeta)\%$ approximate bootstrap-p CI of $\hat{\psi}$ is given by

$$\left[\hat{\psi}_{boot-p}\left(\frac{\zeta}{2}\right), \hat{\psi}_{boot-p}\left(1-\frac{\zeta}{2}\right)\right]. \tag{25}$$

4.3.2. Parametric Boot-t

The bootstrap-t method is an adaptation of the traditional t-interval, designed to handle situations where the sample size is small or the data do not meet the assumptions of normality. It involves resampling the original data with replacements to generate a large number of bootstrap samples. These are used to calculate a t-statistic, analogous to the one used in traditional t-tests but derived from the bootstrap distribution. This collection of t-statistics forms a distribution from which confidence intervals can be derived, the bootstrap-t algorithm is itemized as follows:

(1) Repeat the initial three steps of the parametric Boot-p procedure.
(2) Calculate the variance–covariance matrix I^{-1*} utilizing the delta method.
(3) Define the statistic $T^{*\psi}$ as

$$T^{*\psi} = \frac{(\hat{\psi}^* - \hat{\psi})}{\sqrt{\widehat{var}(\hat{\psi}^*)}}$$

(4) Generate $T_1^{*\psi}, T_2^{*\psi}, \ldots, T_{N\,boot}^{*\psi}$ from repeating steps $2-5$ N Boot times
(5) Sort the sequence $T_{(1)}^{*\psi}, T_{(2)}^{*\psi}, \ldots, T_{(N\,boot)}^{*\psi}$ by arranging $\hat{\psi}^*_i, i = 1, 2, 3, \ldots, N\,boot$ in $T_1^{*\psi}, T_2^{*\psi}, \ldots, T_{N\,boot}^{*\psi}$ in ascending order.

Define $\hat{\psi}_{boot-t} = \hat{\psi} + G_2^{-1}(z)\sqrt{\widehat{var}(\hat{\psi}^*)}$, where $G_2(z) = P(T^* \le z)$ is the cumulative distribution function of T^* for a given z.

Then, the approximate bootstrap-t $100(1-\zeta)\%$ CI of $\hat{\psi}$ is obtained by

$$\left[\hat{\psi}_{boot-t}\left(\frac{\zeta}{2}\right), \hat{\psi}_{boot-t}\left(1-\frac{\zeta}{2}\right)\right]. \tag{26}$$

5. Simulation Analysis

In this section, we present various simulation methods to demonstrate the theoretical results. Initially, we create accelerated PR datasets using the inverse transformation technique. To achieve this, we employ a quantile function derived from the equation where V represents a random sample from the uniform distribution. Consequently, we generate random samples of sizes 40, and 100 using Equation (27).

$$F_j^{-1}(v,\lambda_1,\lambda_2,\gamma,\beta)=\begin{cases}\left[-2\lambda_j^2\ln(1-v)\right]^{2\gamma}\\ \tau+\left(\beta\left[-2\lambda_j^2\ln(1-v)\right]^{2\gamma}-\tau\right)\end{cases} \tag{27}$$

where $j = 1,2$. Moreover, within the Type-II censoring framework, we employed two distinct predetermined numbers of failures for each sample size. Thus, we selected m = 25 and m = 35 for n = 40, and r = 75 and r = 90 for n = 100, respectively. We examined two different sets of actual parameter values in this context. In the initial approach, we set $(\lambda_1,\lambda_2,\gamma,\beta)$ = (1.5, 1.8, 1.2, 0.8), (1.5, 1.8, 1.2, 0.3), (0.6, 0.7, 2, 0.3), and (0.6, 0.7, 2, 0.8) with two distinct stress transition points: τ = 0.60 and τ = 0.90. In all scenarios, we determined the stress transition points based on the ranges of the generated samples, which varied depending on the chosen actual parameter values.

We employed the software developed by R Team et al. [28] for computational tasks. For MLE computations, we utilized the "L-BFGS-B" method within the "optim" function to optimize the profile log-likelihood function described in Equation (9) within the restricted area of $0 < \beta < 1$. We set the significance level to 0.05 for approximate confidence intervals. Subsequently, we conducted simulations repeatedly for 5000 iterations. Observing that the means of the gamma priors yield the real parameter values with the given hyper-parameters. The determination of hyper-parameters relies on informative priors, which are derived from the Maximum Likelihood Estimates (MLEs) of $(\lambda_1,\lambda_2,\gamma)$ by aligning the mean and variance of $(\hat{\lambda_1}^j,\hat{\lambda_2}^j,\hat{\gamma}^j)$ with those of specified priors (Gamma priors). Here, $j = 1,2,3,\ldots,k$, where k denotes the number of available samples from the PR distribution. By equating the moments of $(\hat{\lambda_1}^j,\hat{\lambda_2}^j,\hat{\gamma}^j)$ with those of the gamma priors, we derive the following set of equations:

$$\frac{1}{k}\sum_{j=1}^{k}\hat{\gamma}^j=\frac{c_1}{d_1} \quad , \quad \frac{1}{k-1}\sum_{j=1}^{k}\left(\hat{\gamma}^j-\frac{1}{k}\sum_{j=1}^{k}\hat{\gamma}^j\right)^2=\frac{c_1}{d_1^2},$$
$$\frac{1}{k}\sum_{j=1}^{k}\hat{\lambda_1}^j=\frac{c_2}{d_2} \quad , \quad \frac{1}{k-1}\sum_{j=1}^{k}\left(\hat{\lambda_1}^j-\frac{1}{k}\sum_{j=1}^{k}\hat{\lambda_1}^j\right)^2=\frac{c_2}{d_2^2},$$
$$\frac{1}{k}\sum_{j=1}^{k}\hat{\lambda_2}^j=\frac{c_3}{d_3} \quad \text{and} \quad \frac{1}{k-1}\sum_{j=1}^{k}\left(\hat{\lambda_2}^j-\frac{1}{k}\sum_{j=1}^{k}\hat{\lambda_2}^j\right)^2=\frac{c_3}{d_3^2}.$$

By solving the aforementioned system of equations, the estimated hyper-parameters can be expressed as follows:

$$c_2=\frac{\left(\frac{1}{k}\sum_{j=1}^{k}\hat{\lambda_1}^j\right)^2}{\frac{1}{k-1}\sum_{j=1}^{k}\left(\hat{\lambda_1}^j-\frac{1}{k}\sum_{j=1}^{k}\hat{\lambda_1}^j\right)^2}, \quad d_2=\frac{\frac{1}{k}\sum_{j=1}^{k}\hat{\lambda_1}^j}{\frac{1}{k-1}\sum_{j=1}^{k}\left(\hat{\lambda_1}^j-\frac{1}{k}\sum_{j=1}^{k}\hat{\lambda_1}^j\right)^2}$$
$$c_3=\frac{\left(\frac{1}{k}\sum_{j=1}^{k}\hat{\lambda_2}^j\right)^2}{\frac{1}{k-1}\sum_{j=1}^{k}\left(\hat{\lambda_2}^j-\frac{1}{k}\sum_{j=1}^{k}\hat{\lambda_2}^j\right)^2}, \quad d_3=\frac{\frac{1}{k}\sum_{j=1}^{k}\hat{\lambda_2}^j}{\frac{1}{k-1}\sum_{j=1}^{k}\left(\hat{\lambda_2}^j-\frac{1}{k}\sum_{j=1}^{k}\hat{\lambda_2}^j\right)^2} \tag{28}$$

$$c_1 = \frac{\left(\frac{1}{k}\sum_{j=1}^{k}\hat{\gamma}^j\right)^2}{\frac{1}{k-1}\sum_{j=1}^{k}\left(\hat{\gamma}^j - \frac{1}{k}\sum_{j=1}^{k}\hat{\gamma}^j\right)^2}, \qquad d_1 = \frac{\frac{1}{k}\sum_{j=1}^{k}\hat{\gamma}^j}{\frac{1}{k-1}\sum_{j=1}^{k}\left(\hat{\gamma}^j - \frac{1}{k}\sum_{j=1}^{k}\hat{\gamma}^j\right)^2}.$$

We executed the MCMC algorithm 12,000 times for each of the 5000 replications. We then discarded the initial 2000 values during the burn-in period. Given that Markov chains inherently produce samples with autocorrelation, we opted for a thinning strategy, selecting every third variate to achieve uncorrelated samples from the post-burn-in sample pool. As a result, we generated 1000 uncorrelated samples from Markov chains by repeating this thinning process 5000 times.

In the simulation scenario, we present bias values and mean squared errors (MSEs) for the point estimates, along with average lengths (ALs) and corresponding coverage probabilities (CPs) of the approximate confidence intervals. Tables 1–4 display all results from these simulation schemes. The performance of the point and interval estimations can be itemized as follows:

- Our observations consistently show reduced biases, MSEs, and ALs as sample sizes increase.
- The CPs mostly align closely with their anticipated 95% level.
- In general, the informative Bayes estimation method outperforms MLE, with the disparity between the two estimators decreasing as the sample size grows. This highlights the Bayesian methods' advantage for smaller samples.
- In particular, confidence intervals based on the Highest Posterior Density (HPD) method tend to be smaller than those based on the Approximate Confidence Interval (ACI) method, while still providing similar CPs.
- Altering the pre-determined number of failures or stress change time yields comparable performances across all cases, demonstrating the consistent efficiency and productivity of the theoretical findings.
- Increasing the sample size generally leads to improvements in bias, MSE, and the precision of confidence intervals across all methods. This is expected because larger samples provide more information about the population. The number of bootstrap samples m also influences the Bootstrap method's accuracy and precision, with a higher m usually leading to better estimates.
- changing the stress transition time point τ affects the estimation, especially for Bayesian estimation under ELF that adjusts based on the distribution's tail properties. Different τ values can lead to variations in bias and MSE, suggesting the importance of choosing an appropriate τ value for accurate estimation.

Table 1. Some simulation measures from MLE, bootstrap, Bayesian based on SELF, and ELF for parameters of PR distribution based on TRV: $\lambda_1 = 1.5, \gamma = 1.2, \lambda_2 = 1.8, \beta = 0.8$.

				MLE				Bootstrap		SELF			ELF $c = -1.25$			ELF $c = 1.25$		
n	τ	m		Bias	MSE	LACI	CP	LBP	LBT	Bias	MSE	LCCI	Bias	MSE	LCCI	Bias	MSE	LCCI
40	0.6	25	λ_1	−0.2909	0.2916	1.7842	96.1%	0.0561	0.0569	−0.0752	0.0274	0.6008	−0.0508	0.0272	0.5998	−0.0631	0.0298	0.6240
			γ	0.5299	0.6215	2.2892	95.2%	0.0704	0.0691	0.0470	0.0206	0.5408	0.0382	0.0207	0.5390	0.0264	0.0200	0.5441
			λ_2	0.2413	0.7516	3.2657	94.8%	0.0957	0.0962	0.0232	0.0256	0.6184	0.0241	0.0256	0.6195	0.0146	0.0250	0.6224
			β	0.3789	0.5924	2.6275	95.0%	0.0799	0.0795	−0.0552	0.0169	0.4607	−0.0507	0.0167	0.4602	−0.0656	0.0190	0.4694
		35	λ_1	0.2035	0.2412	1.3889	96.3%	0.0726	0.0730	0.0741	0.0210	0.4968	0.0468	0.0211	0.4973	0.0607	0.0198	0.4895
			γ	0.4141	0.4885	2.2082	95.4%	0.0698	0.0699	−0.0460	0.0151	0.3970	−0.0359	0.0150	0.3951	−0.0168	0.0166	0.4061
			λ_2	−0.0342	0.4384	2.5934	95.6%	0.0805	0.0800	0.0157	0.0169	0.5044	0.0063	0.0169	0.5042	−0.0065	0.0170	0.5058
			β	0.2641	0.4174	2.4269	96.6%	0.1088	0.1088	0.0491	0.0147	0.4050	0.0410	0.0152	0.4059	0.0590	0.0179	0.3967
	0.9	25	λ_1	−0.1263	0.1698	1.5383	95.3%	0.0493	0.0494	−0.0609	0.0244	0.5561	−0.0498	0.0242	0.5546	−0.0597	0.0261	0.5569
			γ	0.4162	0.6104	2.1593	95.4%	0.0684	0.0583	0.0461	0.0192	0.5148	0.0355	0.0200	0.5048	0.0235	0.0192	0.5524
			λ_2	0.2353	0.6907	2.5138	95.1%	0.0777	0.0780	0.0176	0.0238	0.5957	0.0185	0.0239	0.5949	0.0094	0.0234	0.5919
			β	0.1743	0.3717	2.2913	94.7%	0.0734	0.0725	−0.0528	0.0152	0.4527	−0.0509	0.0152	0.4523	−0.0609	0.0182	0.5492
		35	λ_1	0.1045	0.1523	1.1130	95.9%	0.0671	0.0673	0.0607	0.0150	0.4670	0.0373	0.0200	0.4675	0.0565	0.0185	0.4511
			γ	0.4030	0.4685	2.0774	95.8%	0.0587	0.0609	−0.0457	0.0146	0.3842	−0.0327	0.0136	0.3419	−0.0163	0.0160	0.4292
			λ_2	0.1219	0.1666	1.4238	95.3%	0.0428	0.0429	0.0113	0.0147	0.4899	0.0051	0.0157	0.4901	0.0057	0.0169	0.4883
			β	0.1530	0.3199	2.1296	95.8%	0.0982	0.1079	0.0489	0.0132	0.3944	0.0393	0.0122	0.3544	0.0561	0.0159	0.4309
100	0.6	75	λ_1	−0.0995	0.1601	1.5198	95.2%	0.0486	0.0484	0.0158	0.0182	0.5072	0.0167	0.0182	0.5077	0.0081	0.0179	0.5052
			γ	0.4028	0.3386	1.6471	94.3%	0.0541	0.0550	0.0348	0.0154	0.4495	0.0346	0.0155	0.4499	0.0236	0.0145	0.4477
			λ_2	−0.0497	0.2607	1.9928	94.8%	0.0630	0.0643	0.0291	0.0212	0.5632	0.0172	0.0213	0.5625	0.0081	0.0206	0.5556
			β	0.4319	0.5261	2.2852	94.5%	0.0714	0.0715	0.0508	0.0152	0.4397	0.0452	0.0145	0.4405	0.0407	0.0139	0.4305
		90	λ_1	0.0882	0.1522	1.1671	95.9%	0.0529	0.0533	0.0148	0.0124	0.4386	0.0087	0.0172	0.4391	0.0071	0.0162	0.4289
			γ	0.3253	0.2541	1.5100	94.5%	0.0508	0.0497	−0.0325	0.0118	0.3807	−0.0315	0.0117	0.3081	−0.0152	0.0124	0.3726
			λ_2	−0.0392	0.2081	1.7523	95.0%	0.0532	0.0533	0.0039	0.0136	0.4509	0.0045	0.0136	0.4514	−0.0041	0.0137	0.4506
			β	0.4057	0.4709	1.9445	95.0%	0.0770	0.0787	0.0480	0.0132	0.3861	0.0349	0.0113	0.3287	0.0388	0.0121	0.3783
	0.9	75	λ_1	0.0911	0.0899	1.0850	94.4%	0.0314	0.0318	0.0128	0.0174	0.4962	0.0153	0.0175	0.4961	0.0079	0.0167	0.4928
			γ	0.2835	0.1965	1.3365	95.2%	0.0422	0.0419	0.0158	0.0129	0.4219	0.0167	0.0130	0.4221	0.0075	0.0125	0.4221
			λ_2	0.0419	0.0923	0.9197	95.2%	0.0295	0.0298	0.0274	0.0171	0.4803	0.0162	0.0172	0.4819	0.0079	0.0163	0.4717
			β	0.2108	0.1767	1.4266	95.2%	0.0444	0.0447	0.0328	0.0149	0.3453	0.0341	0.0141	0.4054	0.0207	0.0134	0.4050
		90	λ_1	0.0810	0.0733	0.9337	94.7%	0.0416	0.0413	0.0126	0.0120	0.4317	0.0081	0.0163	0.4332	0.0071	0.0154	0.4184
			γ	0.2098	0.1822	0.9443	95.6%	0.0471	0.0468	−0.0237	0.0114	0.3597	−0.0274	0.0103	0.2994	−0.0108	0.0120	0.3628
			λ_2	0.0317	0.0827	0.9078	95.4%	0.0279	0.0280	0.0029	0.0120	0.4102	0.0039	0.0112	0.4095	0.0039	0.0118	0.4081
			β	0.1410	0.1238	1.0821	95.9%	0.0606	0.0604	0.0413	0.0127	0.3042	0.0263	0.0102	0.3042	0.0116	0.0115	0.3401

Table 2. Some simulation measures from MLE, bootstrap, Bayesian based on SELF, and ELF for parameters of PR distribution based on TRV: $\lambda_1 = 1.5, \gamma = 1.2, \lambda_2 = 1.8, \beta = 0.3$.

				MLE				Bootstrap		SELF			ELF $c = -1.25$			ELF $c = 1.25$		
n	τ	m		Bias	MSE	LACI	CP	LBP	LBT	Bias	MSE	LCCI	Bias	MSE	LCCI	Bias	MSE	LCCI
40	0.6	25	λ_1	−0.3308	0.2386	1.4870	94.7%	0.0451	0.0459	−0.0420	0.0254	0.6011	−0.0409	0.0252	0.5991	−0.0523	0.0274	0.6088
			γ	0.4901	0.4934	1.9734	95.2%	0.0623	0.0621	0.0215	0.0228	0.5649	0.0228	0.0229	0.5642	0.0101	0.0225	0.5605
			λ_2	0.1578	0.3716	2.3093	93.5%	0.0742	0.0741	0.0267	0.0260	0.6403	0.0276	0.0261	0.6444	0.0184	0.0256	0.6265
			β	0.1282	0.0660	0.8731	94.3%	0.0284	0.0285	−0.0988	0.0034	0.2074	−0.0220	0.0034	0.2078	−0.0320	0.0039	0.2060
		35	λ_1	0.0776	0.2268	1.0065	95.8%	0.0629	0.0632	0.0337	0.0190	0.5233	0.0344	0.0190	0.5232	0.0271	0.0184	0.5246
			γ	0.4732	0.4537	1.8195	95.5%	0.0711	0.0705	−0.0076	0.0151	0.4924	−0.0067	0.0151	0.4915	−0.0101	0.0155	0.4939
			λ_2	−0.1569	0.3127	2.1050	95.6%	0.0707	0.0707	−0.0073	0.0177	0.5031	−0.0066	0.0177	0.5032	−0.0133	0.0181	0.5082
			β	0.1128	0.0520	0.6350	95.4%	0.0411	0.0411	0.0852	0.0021	0.2006	0.0209	0.0031	0.2006	0.0298	0.0036	0.1925
	0.9	25	λ_1	0.3061	0.2147	1.3998	95.4%	0.0409	0.0410	−0.0416	0.0220	0.5343	−0.0384	0.0219	0.5342	−0.0496	0.0234	0.5436
			γ	0.4320	0.4638	1.8635	95.7%	0.0608	0.0606	0.0202	0.0203	0.5583	0.0213	0.0213	0.5183	0.0101	0.0226	0.5476
			λ_2	0.1485	0.3400	1.2760	95.6%	0.0680	0.0678	0.0225	0.0261	0.6143	0.0234	0.0262	0.5061	0.0140	0.0256	0.6064
			β	0.1196	0.0611	0.8163	94.6%	0.0263	0.0262	−0.0208	0.0032	0.2053	−0.0206	0.0031	0.2001	−0.0313	0.0031	0.2048
		35	λ_1	−0.0674	0.1845	1.0016	95.9%	0.0613	0.0628	0.0314	0.0171	0.4842	0.0304	0.0172	0.4832	0.0231	0.0164	0.4777
			γ	0.3999	0.4152	1.7359	95.8%	0.0704	0.0684	−0.0070	0.0146	0.4748	−0.0056	0.0146	0.4742	−0.0091	0.0151	0.4795
			λ_2	0.1474	0.3048	1.1014	96.5%	0.0397	0.0397	−0.0061	0.0152	0.4716	−0.0055	0.0152	0.4713	−0.0125	0.0156	0.4710
			β	0.0734	0.0486	0.5815	96.1%	0.0370	0.0372	0.0201	0.0021	0.2005	0.0201	0.0014	0.1925	0.0283	0.0029	0.1920
100	0.6	75	λ_1	−0.0969	0.1007	1.1855	93.9%	0.0378	0.0378	−0.0105	0.0181	0.5247	−0.0096	0.0181	0.5244	−0.0187	0.0186	0.5230
			γ	0.3355	0.2237	1.3073	95.2%	0.0399	0.0401	0.0195	0.0203	0.5065	0.0206	0.0205	0.5063	0.0094	0.0187	0.4921
			λ_2	−0.0389	0.1561	1.2542	93.5%	0.0481	0.0485	0.0139	0.0224	0.5605	0.0148	0.0225	0.4956	0.0064	0.0222	0.5523
			β	0.1133	0.0494	0.6976	94.5%	0.0217	0.0216	0.0200	0.0017	0.1926	0.0200	0.0029	0.1933	0.0268	0.0031	0.1871
		90	λ_1	0.0614	0.0912	0.9262	95.2%	0.0388	0.0388	0.0308	0.0162	0.4449	0.0277	0.0168	0.4452	0.0092	0.0159	0.4358
			γ	0.3313	0.2126	1.2581	95.9%	0.0407	0.0405	0.0069	0.0130	0.4336	0.0043	0.0131	0.4327	0.0089	0.0123	0.4290
			λ_2	−0.0314	0.1277	1.0928	95.0%	0.0385	0.0384	−0.0060	0.0150	0.4520	−0.0053	0.0153	0.4511	−0.0114	0.0146	0.4598
			β	0.0692	0.0382	0.4354	94.7%	0.0240	0.0240	0.0201	0.0015	0.1872	0.0191	0.0011	0.1877	0.0195	0.0011	0.1126
	0.9	75	λ_1	0.0911	0.0791	1.0200	94.9%	0.0334	0.0334	0.0092	0.0163	0.4957	0.0092	0.0163	0.4963	0.0084	0.0158	0.4896
			γ	0.2783	0.1912	1.1032	94.8%	0.0384	0.0384	0.0184	0.0187	0.4951	0.0192	0.0188	0.4951	0.0083	0.0178	0.4507
			λ_2	0.0199	0.0909	0.9317	95.0%	0.0297	0.0288	0.0124	0.0185	0.5052	0.0137	0.0186	0.4504	0.0053	0.0178	0.5045
			β	0.0787	0.0246	0.5317	95.3%	0.0171	0.0170	0.0201	0.0010	0.1823	0.0193	0.0020	0.1823	0.0189	0.0031	0.1802
		90	λ_1	0.0510	0.0691	0.6013	95.8%	0.0351	0.0344	0.0090	0.0152	0.4206	0.0082	0.0152	0.4223	0.0021	0.0149	0.4148
			γ	0.2090	0.1696	1.0311	95.2%	0.0401	0.0394	0.0024	0.0111	0.4026	0.0032	0.0111	0.4033	−0.0043	0.0109	0.4046
			λ_2	0.0115	0.0632	0.7802	96.5%	0.0264	0.0259	−0.0058	0.0129	0.4485	−0.0051	0.0129	0.4483	−0.0051	0.0091	0.4473
			β	0.0615	0.0230	0.4061	95.7%	0.0187	0.0185	0.0200	0.0010	0.1722	0.0181	0.0010	0.1722	0.0110	0.0010	0.1522

Table 3. Some simulation measures from MLE, bootstrap, Bayesian based on SELF, and ELF for parameters of PR distribution based on TRV: $\lambda_1 = 0.6, \gamma = 2, \lambda_2 = 0.7, \beta = 0.3$.

				MLE				Bootstrap		SELF			ELF $c = -1.25$			ELF $c = 1.25$		
n	τ	m		Bias	MSE	LACI	CP	LBP	LBT	Bias	MSE	LCCI	Bias	MSE	LCCI	Bias	MSE	LCCI
40	0.6	25	λ_1	−0.1998	0.1449	0.7621	96.0%	0.0327	0.0329	−0.0477	0.0182	0.3608	−0.0746	0.0152	0.3615	−0.0619	0.0134	0.3561
			γ	0.2789	0.4008	1.8532	93.8%	0.0814	0.0818	0.0447	0.0279	0.6292	0.0531	0.0279	0.6298	0.0297	0.0276	0.6339
			λ_2	0.1927	0.1262	1.1712	95.4%	0.0514	0.0519	0.0688	0.0197	0.4464	0.0705	0.0201	0.4454	0.0532	0.0166	0.4370
			β	−0.1116	0.0305	0.3889	94.0%	0.0166	0.0164	−0.0692	0.0084	0.2207	−0.0682	0.0084	0.2224	−0.0584	0.0069	0.2084
		35	λ_1	0.1260	0.1380	0.7400	96.4%	0.0442	0.0443	0.0465	0.0128	0.3536	0.0663	0.0130	0.3536	0.0550	0.0111	0.3459
			γ	0.2355	0.3055	1.7076	95.4%	0.0904	0.0910	0.0342	0.0202	0.5307	0.0348	0.0202	0.5309	0.0287	0.0199	0.5275
			λ_2	0.0517	0.0590	0.8553	95.6%	0.0391	0.0388	0.0256	0.0092	0.3572	0.0267	0.0093	0.3585	0.0160	0.0086	0.3555
			β	0.0581	0.0195	0.3499	96.4%	0.0220	0.0225	0.0662	0.0075	0.2163	0.0672	0.0077	0.2162	0.0573	0.0061	0.2019
	0.9	25	λ_1	−0.0919	0.0314	0.5944	96.6%	0.0262	0.0264	−0.0348	0.0048	0.2355	−0.0340	0.0048	0.2361	−0.0422	0.0052	0.2291
			γ	0.1929	0.2398	1.7663	95.2%	0.0754	0.0758	0.0094	0.0205	0.5689	0.0052	0.0205	0.5686	−0.0092	0.0205	0.5646
			λ_2	0.1410	0.0346	0.4753	95.8%	0.0215	0.0213	0.0385	0.0095	0.3348	0.0398	0.0097	0.3357	0.0268	0.0081	0.3274
			β	−0.1529	0.0290	0.2949	94.2%	0.0131	0.0132	−0.0216	0.0022	0.2143	−0.0593	0.0032	0.2042	−0.0301	0.0032	0.1944
		35	λ_1	0.0812	0.0291	0.5708	96.9%	0.0371	0.0366	0.0326	0.0041	0.2031	0.0326	0.0031	0.2131	0.0409	0.0041	0.2193
			γ	−0.1912	0.1355	1.2343	96.6%	0.0557	0.0559	0.0081	0.0147	0.4481	0.0049	0.0147	0.4485	0.0239	0.0147	0.4489
			λ_2	0.0491	0.0325	0.2734	96.5%	0.0126	0.0126	0.0246	0.0085	0.2672	0.0206	0.0087	0.2680	0.0145	0.0072	0.2571
			β	−0.0499	0.0149	0.2827	94.6%	0.0127	0.0127	0.0204	0.0021	0.2032	0.0256	0.0031	0.2033	0.0075	0.0026	0.1831
100	0.6	75	λ_1	−0.0888	0.0181	0.5272	96.6%	0.0238	0.0238	0.0219	0.0046	0.2130	0.0229	0.0047	0.2304	0.0130	0.0051	0.2197
			γ	0.1727	0.1880	1.3465	94.2%	0.0584	0.0584	0.0085	0.0192	0.5551	0.0051	0.0203	0.5559	0.0082	0.0202	0.5564
			λ_2	0.0570	0.0276	0.4613	96.0%	0.0207	0.0207	0.0369	0.0091	0.2033	0.0370	0.0091	0.3304	0.0258	0.0071	0.3163
			β	0.0301	0.0059	0.2908	94.8%	0.0127	0.0127	−0.0028	0.0014	0.1484	−0.0518	0.0021	0.1482	−0.0293	0.0014	0.1462
		90	λ_1	0.0712	0.0126	0.5264	95.4%	0.0294	0.0292	0.0203	0.0040	0.2013	0.0231	0.0030	0.2031	0.0124	0.0041	0.2030
			γ	0.1683	0.1234	1.2051	94.4%	0.0547	0.0507	0.0075	0.0138	0.4283	0.0045	0.0138	0.4383	0.0072	0.0137	0.4380
			λ_2	0.0370	0.0242	0.2592	94.2%	0.0121	0.0122	0.0203	0.0056	0.2047	0.0202	0.0057	0.2570	0.0125	0.0050	0.2470
			β	0.0300	0.0051	0.2338	94.8%	0.0126	0.0116	0.0019	0.0012	0.1317	0.0236	0.0020	0.1349	0.0054	0.0012	0.1410
	0.9	75	λ_1	0.0756	0.0152	0.3813	96.0%	0.0171	0.0170	0.0193	0.0043	0.2013	0.0203	0.0044	0.2015	0.0129	0.0038	0.1981
			γ	−0.0977	0.0595	0.8773	96.0%	0.0383	0.0387	0.0081	0.0164	0.4916	0.0050	0.0164	0.4919	0.0079	0.0164	0.4920
			λ_2	0.0416	0.0274	0.2007	95.2%	0.0092	0.0092	0.0275	0.0080	0.1807	0.0356	0.0081	0.1815	0.0247	0.0067	0.1732
			β	−0.0080	0.0042	0.1561	95.2%	0.0070	0.0067	−0.0018	0.0014	0.1381	−0.0509	0.0018	0.1381	−0.0290	0.0010	0.1372
		90	λ_1	0.0632	0.0120	0.3506	96.2%	0.0249	0.0250	0.0173	0.0040	0.1928	0.0203	0.0028	0.1985	0.0092	0.0032	0.1827
			γ	−0.0817	0.0493	0.7108	96.4%	0.0313	0.0314	0.0072	0.0129	0.4136	0.0042	0.0129	0.4371	0.0061	0.0127	0.4319
			λ_2	0.0230	0.0206	0.1993	96.0%	0.0089	0.0088	0.0188	0.0049	0.1587	0.0189	0.0053	0.1587	0.0118	0.0046	0.1564
			β	−0.0074	0.0031	0.1471	96.2%	0.0076	0.0074	0.0010	0.0012	0.1248	0.0113	0.0013	0.1285	0.0052	0.0009	0.1284

Table 4. Some simulation measures from MLE, bootstrap, Bayesian based on SELF, and ELF for parameters of PR distribution based on TRV: $\lambda_1 = 0.6, \gamma = 2, \lambda_2 = 0.7, \beta = 0.8$.

				MLE				Bootstrap		SELF			ELF $c = -1.25$			ELF $c = 1.25$		
n	τ	m		Bias	MSE	LACI	CP	LBP	LBT	Bias	MSE	LCCI	Bias	MSE	LCCI	Bias	MSE	LCCI
40	0.6	25	λ_1	−0.0942	0.0975	1.8407	94.4%	0.0379	0.0382	−0.0613	0.0136	0.3633	−0.0599	0.0134	0.3638	−0.0732	0.0149	0.3561
			γ	0.3890	0.5133	2.3610	96.2%	0.1002	0.1019	0.0160	0.0285	0.6367	0.0168	0.0285	0.6383	0.0085	0.0284	0.6491
			λ_2	0.2001	0.1825	1.4812	95.0%	0.0679	0.0678	0.0559	0.0153	0.4141	0.0576	0.0156	0.4147	0.0404	0.0127	0.3986
			β	−0.1482	0.2855	1.1319	96.4%	0.0486	0.0491	−0.0486	0.0159	0.4415	−0.0471	0.0157	0.4391	−0.0625	0.0177	0.4456
		35	λ_1	0.0820	0.0967	1.1768	95.0%	0.0523	0.0526	0.0593	0.0128	0.3374	0.0426	0.0128	0.3276	0.0682	0.0135	0.3458
			γ	0.3482	0.5068	2.2636	97.2%	0.1198	0.1191	0.0148	0.0188	0.5247	0.0162	0.0188	0.5251	0.0081	0.0185	0.5277
			λ_2	0.0160	0.0805	1.1115	95.6%	0.0485	0.0488	0.0539	0.0106	0.3311	0.0550	0.0108	0.3326	0.0394	0.0092	0.3227
			β	0.1352	0.2146	1.1057	96.8%	0.0723	0.0729	0.0410	0.0150	0.3672	0.0375	0.0152	0.3695	0.0615	0.0132	0.3601
	0.9	25	λ_1	−0.0822	0.0417	0.7350	88.2%	0.0358	0.0359	−0.0486	0.0056	0.2287	−0.0478	0.0056	0.2278	−0.0557	0.0063	0.2278
			γ	0.1871	0.2469	1.8064	95.4%	0.0837	0.0843	0.0053	0.0219	0.5513	0.0061	0.0219	0.5507	−0.0072	0.0220	0.5530
			λ_2	0.1577	0.0430	0.5280	94.8%	0.0249	0.0249	0.0311	0.0088	0.3443	0.0325	0.0089	0.3446	0.0182	0.0077	0.3279
			β	−0.0414	0.0722	0.8377	95.7%	0.0383	0.0383	−0.0459	0.0133	0.4165	−0.0456	0.0132	0.4165	−0.0609	0.0164	0.3715
		35	λ_1	0.0793	0.0371	0.6949	95.8%	0.0428	0.0423	0.0381	0.0052	0.2133	0.0391	0.0052	0.2033	0.0481	0.0051	0.2070
			γ	−0.1791	0.1439	1.2860	95.6%	0.0567	0.0565	0.0041	0.0169	0.5039	0.0046	0.0169	0.5032	−0.0018	0.0170	0.5084
			λ_2	0.0104	0.0405	0.2861	95.4%	0.0128	0.0126	0.0305	0.0087	0.2515	0.0305	0.0088	0.2516	0.0141	0.0073	0.2444
			β	−0.0392	0.0681	0.7567	95.8%	0.0336	0.0339	0.0092	0.0106	0.3479	0.0106	0.0130	0.3048	−0.0533	0.0122	0.3482
100	0.6	75	λ_1	−0.0155	0.0225	0.5852	95.4%	0.0263	0.0265	0.0223	0.0051	0.2183	0.0382	0.0054	0.2185	0.0144	0.0056	0.2177
			γ	0.1731	0.2335	1.7403	95.4%	0.0771	0.0773	0.0047	0.0214	0.5357	0.0059	0.0204	0.5371	0.0062	0.0203	0.5419
			λ_2	0.0443	0.0370	0.4735	94.6%	0.0232	0.0239	0.0297	0.0081	0.2934	0.0317	0.0081	0.2942	0.0176	0.0071	0.2850
			β	0.0413	0.0556	0.6791	95.8%	0.0340	0.0377	−0.0133	0.0086	0.3675	−0.0412	0.0086	0.3666	−0.0523	0.0090	0.3623
		90	λ_1	0.0147	0.0213	0.5674	95.9%	0.0323	0.0320	0.0213	0.0042	0.2086	0.0315	0.0052	0.2019	0.0141	0.0050	0.1928
			γ	0.1328	0.1320	1.1807	96.4%	0.0379	0.0478	0.0039	0.0147	0.4574	0.0043	0.0148	0.4569	0.0017	0.0144	0.4606
			λ_2	0.0102	0.0345	0.2722	96.0%	0.0099	0.0113	0.0283	0.0079	0.2453	0.0300	0.0081	0.2503	0.0136	0.0071	0.2408
			β	0.0318	0.0510	0.6704	96.4%	0.0305	0.0305	0.0090	0.0079	0.3334	0.0102	0.0082	0.3004	0.0483	0.0081	0.3347
	0.9	75	λ_1	0.0149	0.0184	0.4049	94.0%	0.0188	0.0188	0.0217	0.0042	0.2022	0.0280	0.0043	0.2022	0.0123	0.0038	0.2130
			γ	−0.1066	0.0595	0.8613	95.8%	0.0386	0.0386	0.0046	0.0177	0.4874	0.0054	0.0178	0.4883	0.0053	0.0171	0.4817
			λ_2	0.0416	0.0274	0.1928	95.6%	0.0093	0.0092	0.0260	0.0060	0.1798	0.0306	0.0061	0.1802	0.0153	0.0050	0.1755
			β	−0.0396	0.0414	0.4319	95.2%	0.0189	0.0190	−0.0116	0.0083	0.3528	−0.0401	0.0083	0.3525	−0.0514	0.0074	0.3550
		90	λ_1	0.0132	0.0172	0.3954	95.8%	0.0237	0.0232	0.0202	0.0043	0.1926	0.0250	0.0042	0.1826	0.0114	0.0032	0.1821
			γ	−0.0912	0.0495	0.7302	96.5%	0.0332	0.0323	0.0031	0.0124	0.4160	0.0040	0.0124	0.4165	0.0017	0.0123	0.4177
			λ_2	0.0092	0.0256	0.1896	96.5%	0.0090	0.0089	0.0228	0.0048	0.1562	0.0249	0.0058	0.1565	0.0128	0.0048	0.1494
			β	−0.0265	0.0408	0.4249	95.6%	0.0208	0.0207	0.0066	0.0070	0.3079	0.0041	0.0071	0.3179	−0.0305	0.0069	0.3287

6. The Optimal Stress Change Time and Sensitivity Analysis

In this section, we describe an optimal method based on asymptotic variances in maximum likelihood estimators. The inverse Fisher information matrix's diagonals can be used to compute the parameters' asymptotic variances. In this section, we used the sum of coefficients of variations (SVCs) as the optimal function instead of the sum of parameter variances, as recommended and implemented by Samanta et al. ([29,30]). Samanta et al. [29] proposed a method to calculate an optimal solution by minimizing the predicted value of the SVC. Since the sum of variances can be calculated using the variance of any specific parameter if the parameter values are on a different scale. That is why we employed the expected value of the SVC by maximizing $E(\phi(\tau))$, where

$$\phi(\tau) = \frac{\sqrt{F_{11}^{-1}}}{\hat{\lambda_1}} + \frac{\sqrt{F_{22}^{-1}}}{\hat{\lambda_2}} + \frac{\sqrt{F_{33}^{-1}}}{\hat{\gamma}} + \frac{\sqrt{F_{44}^{-1}}}{\hat{\beta}}, \tag{29}$$

where F_{ii}^{-1} is the element in the main diagonal of the inverse Fisher information matrix that was described by Equation (22). However, the closed forms of the parameters' posterior variances may be imprecisely estimated. Samanta et al. [30] recommend adopting the Gibbs sampling technique for computation.

Step 1: Obtain the samples U_1, U_2 and $U = \min\{U_1, U_2\}$ using given τ , n, r and parameter values.
Step 2: The objective function $\phi(\tau)$ is calculated.
Step 3: For N times, repeat Step 1 to Step 2, and obtain $\phi^1(\tau), \phi^2(\tau), \ldots, \phi^N(\tau)$.
Step 4: The median of the objective functions is obtained and applied to $\phi^m(\tau)$.
Step 5: For all possible values of τ repeat Step 1 to Step 4 .
Step 6: The optimal τ for which $\phi^m(\tau)$ is the minimum is obtained.

Optimal stress change time τ values, indicated by τ^* are determined for given n, r, and ψ_i for $i = 1, \ldots, 4$ and are reported in Table 5.

Table 5. Optimal stress change time τ for different sample sizes and parameter values by SVC $\phi(\tau)$.

n	τ	m	**Table 1**	**Table 2**	**Table 3**	**Table 4**
40	0.6	25	0.4225	0.3565	0.5263	0.5119
		35	0.3850	0.3477	0.3439	0.3934
	0.9	25	0.4215	0.3301	0.2896	0.2695
		35	0.3384	0.3605	0.2680	0.2661
100	0.6	75	0.3600	0.2797	0.1999	0.2476
		90	0.3199	0.2737	0.1683	0.2312
	0.9	75	0.2850	0.2638	0.1686	0.2050
		90	0.2349	0.2352	0.1567	0.2008

From Table 5, it is evident that the optimal stress change times, denoted as τ, fall within the range of 0.6 to 0.9 for the first parameter set. As the range of the generated dataset is not extensive, there is not a significant deviation in the range of τ in this initial case. It is noticeable that the stress change times utilized in the simulations closely align with the optimal stress change times. Hence, the consistency and effectiveness of the simulation outcomes are contingent upon accurately determining the stress change time.

7. Real Data Examples

In this section, two real data sets are examined for the suitability of the PR model with tampered random variables under the Type-II censoring framework.

7.1. HIV Infection to AIDS

This example discusses the application of a real-life dataset focusing on male individuals and their progression from HIV infections to AIDS over nearly 15 years. According to the United States Center for Disease Control and Prevention, 54% of total diagnosed adult AIDS cases in the U.S. up to 1996 were due to intimate contact with a person who was HIV positive, also, an additional 40% of incident cases occurred in that same year. A subset of the 54% who also engaged in injection drug use accounted for an additional 7% of cumulative and 5% of incident cases in 1996. These data were collected during the era of antiretroviral combined therapy in 1996. For further background information, readers are directed to studies by Dukers et al. [31] and Xiridou et al. [32], while Putter et al. [33] and Geskus et al. [34] cite this dataset as an example for competing risk analysis. The dataset encompasses instances where some patients either remained uninfected or their outcomes were censored in the study.

We focused on a pre-determined number of failures, setting r as 150 from a complete dataset of $n = 222$. We also examined stress change times: $\tau = 4.6$. For clarity, we present the competing risk data as follows in Table 6, where the black color is $t_i < \tau$ and the gold color is $t_i > \tau$.

Table 6. Data from HIV Infection to AIDS dataset.

t_i	c	t_i	c	t_i	c	t_i	c	t_i	c	t_i	c	t_i	c	t_i	c	t_i	c	t_i	c
0.112	1	2.048	1	2.798	1	3.373	0	3.8	0	4.389	1	5.018	0	5.566	0	5.982	1	6.461	0
0.137	1	2.053	1	2.814	1	3.439	1	3.817	1	4.394	1	5.021	1	5.574	0	6.018	1	6.511	1
0.474	1	2.155	0	2.866	1	3.477	0	3.819	0	4.4	1	5.082	1	5.582	0	6.042	0	6.516	1
0.824	1	2.177	0	2.875	0	3.477	1	3.88	1	4.52	1	5.106	1	5.618	0	6.042	1	6.579	0
0.884	1	2.234	0	2.891	0	3.486	0	3.94	1	4.523	1	5.12	1	5.667	0	6.045	0	6.733	0
0.969	1	2.283	0	2.982	1	3.513	0	3.953	0	4.583	0	5.224	1	5.678	0	6.054	0	6.801	0
1.013	1	2.322	1	3.039	1	3.535	0	3.975	0	4.608	0	5.251	0	5.7	1	6.177	0	6.82	1
1.101	1	2.513	1	3.064	0	3.584	1	4.033	1	4.69	0	5.314	1	5.703	1	6.195	0	6.85	1
1.205	1	2.533	0	3.064	1	3.592	0	4.079	1	4.734	1	5.336	1	5.723	0	6.199	0	6.866	0
1.44	0	2.565	1	3.195	0	3.639	0	4.099	0	4.811	0	5.374	1	5.73	1	6.218	1	6.943	1
1.462	1	2.571	1	3.214	0	3.647	0	4.219	0	4.854	1	5.454	1	5.736	1	6.224	0	6.955	0
1.503	1	2.631	1	3.22	1	3.663	0	4.219	0	4.909	1	5.478	1	5.886	1	6.267	0	6.979	1
1.593	1	2.672	0	3.242	0	3.707	0	4.23	1	4.966	1	5.525	1	5.889	1	6.311	1	7.006	0
1.837	0	2.683	0	3.258	1	3.724	0	4.334	1	4.981	0	5.555	0	5.908	1	6.412	0	7.17	1
1.889	1	2.705	0	3.315	0	3.797	1	4.375	1	5.013	0	5.563	1	5.938	0	6.439	1	7.302	0

Table 7 showcases the MLE alongside various fit metrics for the HIV Infection to AIDS dataset, utilizing both the baseline model and SSLT as complete datasets. The analysis derived from Table 7 indicates an adequate fit of the model to the data, evidenced by a Kolmogorov–Smirnov P-value (KSPV) exceeding 0.05. Furthermore, the table provides a range of fit indices, including the Consistent Akaike Information Criterion (CAIC), Akaike Information Criterion (AIC), Bayesian Information Criterion (BIC), and Hannan–Quinn Information Criterion (HQIC), all of which serve as measures of goodness-of-fit.

Table 7. MLE and different measures of HIV Infection to AIDS data.

$t_i < \tau$	size				KSD	PVKS	AIC	BIC	CAIC	HQIC
$t_i < max(t)$	220	λ	5.6859	0.9111	0.0890	0.1853	602.8644	608.8856	602.9460	605.3106
		γ	1.3300	0.0913						
$t_i < \tau$	121	λ	3.9018	0.7312	0.1217	0.1813	260.7185	265.5074	260.8724	262.6399
		γ	1.4365	0.1381						
$\tau < t_i < max(t)$	101	λ	44.2060	20.9781	0.0612	0.9585	145.3549	152.0572	145.7241	148.0139
		γ	1.1368	0.1974						
		β	0.0432	0.0353						

Table 8 presents the maximum likelihood and Bayesian point estimation in addition to the interval estimates for the PR parameters derived from step-stress life testing using the Tampered Random Variable model. Table 8 presents a reliability analysis that evaluates the reliability function of various models through maximum likelihood and Bayesian methods for estimating parameters. The models analyzed include those with a risk factor from cause I, from cause II, and both under standard conditions, followed by an examination under an accelerated framework. Additionally, the reliability of the TRV model is analyzed in the context of two competing risk factors. The findings suggest that the TRV model exhibits the greatest reliability among the models assessed, underscoring the robustness of our proposed model. Figure 1 depicts the likelihood profile for the PR parameters based on SSLT under the TRV model which indicates the existence of the maximum value for the log-likelihood function. Figure 2 illustrates the trace plots and marginal posterior probability density functions of the parameters for the PR distribution, employing SSLT under the TRV model, as obtained via Bayesian estimation.

Table 8. MLE and Bayesian estimation for the parameters of PR based on SSLT under TRV.

	MLE				Bayesian			
	Estimates	StEr	Lower	Upper	Estimates	StEr	Lower	Upper
λ_1	6.0925	1.0225	4.0884	8.0966	6.1924	0.7669	4.6570	7.6641
γ	1.1113	0.1052	0.9050	1.3175	1.1176	0.0774	0.9679	1.2601
λ_2	6.6004	1.1181	4.4089	8.7919	6.7538	0.8359	5.3139	8.4690
β	0.5539	0.0991	0.3596	0.7482	0.5801	0.1257	0.3312	0.8086
$1 - F_1(\bar{t};\gamma,\lambda_1)$	0.70965				0.71305			
$1 - F_2(\bar{t};\gamma,\lambda_2)$	0.74660				0.75253			
$1 - F_1(\bar{t};\gamma,\lambda_1,\beta)$	0.74056				0.74077			
$1 - F_2(\bar{t};\gamma,\lambda_2,\beta)$	0.77422				0.77705			
$1 - (1 - F_1(t_r))(1 - F_2(t_r))$	0.97519				0.97895			

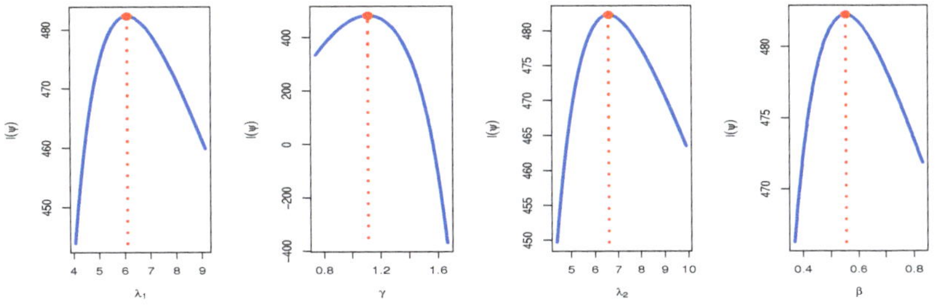

Figure 1. Likelihood profile (blue line) for parameters of PR based on SSLT under TRV model with the maximum likelihood estimation (red dot): HIV infection to AIDS data.

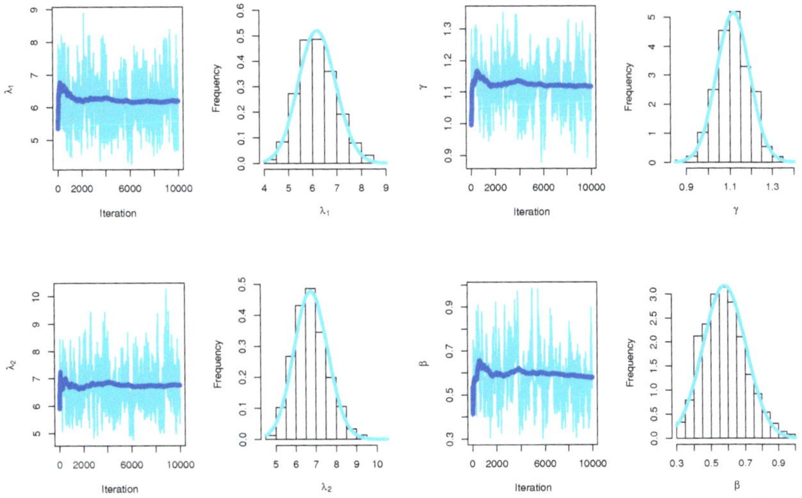

Figure 2. MCMC plots for parameters of PR based on SSLT under TRV model for HIV Infection to AIDS data.(The blue color indicates the convergence line).

7.2. Electrical Appliances Data

The real-world dataset analyzed in reference [35] (p. 441) examines 36 small electronic components subjected to an automated life test, where failures are categorized into 18 types. However, out of the 33 identified failures, only seven modes were observed, with modes 6–9 recurring more than twice. Mode 9 failure is particularly significant. Consequently, the dataset is categorized into two failure causes, c = 0 (mode 9 failure) and c = 1 (all other modes). The provided data presents the failure times in sequence along with the respective cause of each failure, the stress change time is selected to be $\tau = 2500$ as detailed in Table 9.

Table 9. Electrical appliances data.

t_i	c	t_i	c	t_i	c	t_i	c	t_i	c
11	1	381	1	1594	1	2400	0	2694	0
35	1	708	1	1925	0	2451	1	2702	1
49	1	958	1	1990	0	2471	0	2761	1
170	1	1062	1	2223	0	2551	0	2831	1
329	1	1167	0	2327	1	2568	0	3034	0

Table 10 discusses MLE and different measures used for the electrical appliances data in baseline model and SSLT model as complete data. From the results in Table 10, we note that the data are fitting of this model where the KSPV is greater than 0.05. Also, some different measures have been obtained as CAIC, AIC, BIC goodness-of-ft measures, and HQIC.

Table 10. MLE and different measures for electrical appliances data.

$x_i < \tau$	Size		Estimate	StEr	KSD	PVKS	AIC	BIC	CAIC	HQIC
$x_i < max(x_i)$	25	λ	53.3420	43.2098	0.2403	0.0938	423.9574	426.3951	424.5028	424.6335
		γ	0.5803	0.1044						
$x_i < \tau$	18	λ	22.3780	16.9145	0.1772	0.5644	296.2920	298.0727	297.0920	296.5375
		γ	0.4859	0.1002						
$\tau < x_i < max(x_i)$	7	λ	2814.8644	5262.1741	0.2066	0.8722	94.8192	94.6569	102.8192	92.8135
		γ	0.7401	0.1678						
		β	0.0036	0.0012						

Table 11 presents the maximum likelihood and Bayesian point estimation in addition to the interval estimates for the PR parameters derived from step-stress life testing using the Tampered Random Variable model for the electrical appliances data. Similar to the discussion of reliability analysis in the first data set in Table 8, the reliability analysis presented in Table 11 indicates that the TRV model outperformed the other models. Figure 3 depicts the likelihood profile of PR parameters based on SSLT under the TRV model for electrical appliances data. From Figure 3, we can conclude that the parameters of PR distribution based on SSLT under TRV have maximum value for the log-likelihood function for electrical appliances data. Figure 4 shows the trace plots and marginal posterior probability density functions of the parameters for the PR distribution based on SSLT under the TRV model, derived through Bayesian estimation for the electrical appliances data.

Table 11. MLE and Bayesian estimation for the parameters of PR based on SSLT under TRV: electrical appliances data.

	MLE				Bayesian			
	Estimates	StEr	Lower	Upper	Estimates	StEr	Lower	Upper
λ_1	30.436	11.750	7.407	73.065	30.475	2.933	24.635	36.016
γ	0.456	0.091	0.277	0.635	0.453	0.022	0.412	0.495
λ_2	37.277	16.856	4.240	89.915	37.313	3.475	31.022	44.277
β	0.086	0.043	0.0015	0.179	0.096	0.042	0.020	0.171
$1 - F_1(t_r; \gamma, \lambda_1)$	0.44465				0.46349			
$1 - F_2(t_r; \gamma, \lambda_2)$	0.58258				0.59872			
$1 - F_1(t_r; \gamma, \lambda_1, \beta)$	0.12041				0.15590			
$1 - F_2(t_r; \gamma, \lambda_2, \beta)$	0.24386				0.28945			
$1 - (1 - F_1(t_r))(1 - F_2(t_r))$	0.97064				0.97549			

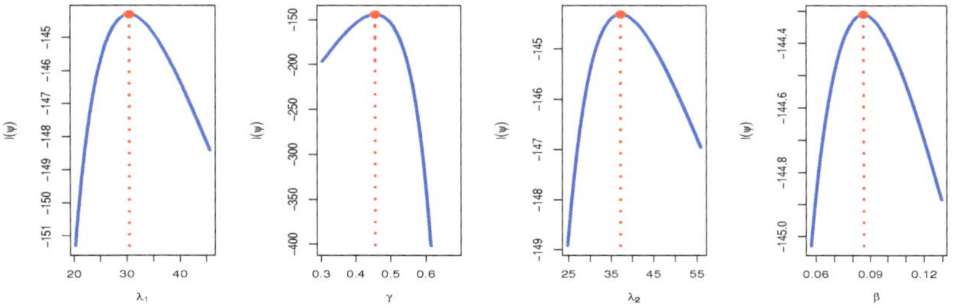

Figure 3. Likelihood profile (blue line) for parameters of PR based on SSLT under TRV model with the maximum likelihood estimation (red dot): electrical appliances data.

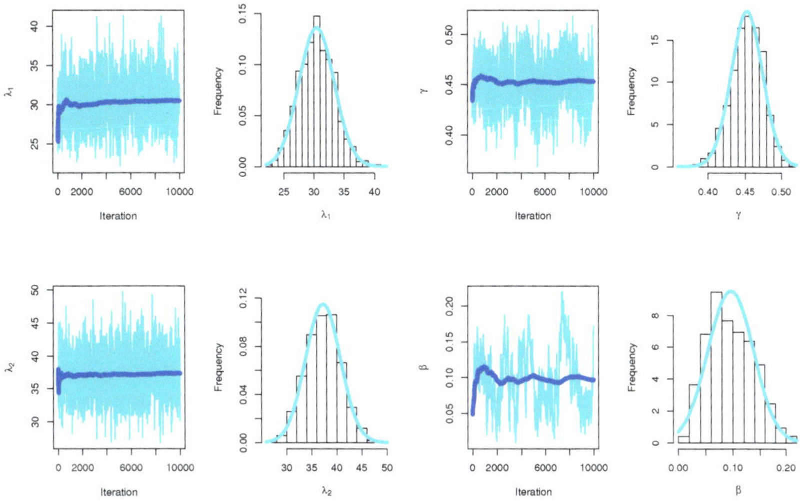

Figure 4. MCMC plots for parameters of PR based on SSLT under TRV model: electrical appliances (The blue color indicates the convergence line).

8. Conclusions

In conclusion, this work has significantly contributed to the field of reliability engineering through the application of the Tampered Random Variable (TRV) model within the step-stress life testing (SSLT) framework, particularly focusing on the Power Rayleigh distribution in the context of competing risks. By integrating TRV with SSLT under such complex scenarios, the study has addressed critical gaps in current research, particularly the various applications of TRV modeling in competing risk analyses.

The methodological advancements presented in this paper, including the use of maximum likelihood estimation and the Bayesian methods for inferential analysis, as well as Monte Carlo simulations for estimator performance evaluation, represent a robust approach to understanding and improving product reliability under varied stress conditions. These techniques have been validated through empirical analysis of real-world datasets from the medical sector, regarding AIDS infection, and the electrical engineering domain, focusing on electronic component failures. The reliability evaluations underscore the model's empirical suitability and the potential for broader application.

Furthermore, the study's exploration of Type-II censoring schemes as a solution to information shortage in lifetime experiments highlights the practical value of the research, offering other options for more cost-effective and efficient testing methodologies. The comparison of TRV modeling with other established models (CEM and TFR) within a competing risks framework not only clarifies the conditions under which these models converge but also showcases the unique advantages of TRV in handling complex, multi-step-stress situations and discrete or multivariate lifetime data.

The comprehensive analysis and the resulting insights into model precision, reliability, and risk management presented in this study provide a solid foundation for future research in this area. It opens up new ways for the development of more accurate and dependable models, enhancing the decision making process and risk management strategies in the medical, industrial, and mechanical domains.

Author Contributions: Conceptualization, H.H.A.; Methodology, H.H.A. and D.A.R.; Software, E.M.A.; Validation, D.A.R.; Formal analysis, H.H.A., E.M.A., and D.A.R.; Investigation, H.H.A. and D.A.R.; Resources, E.M.A.; Data curation, E.M.A.; Writing—original draft, H.H.A. and D.A.R.; Writing—review and editing, D.A.R. and H.H.A.; Funding acquisition, H.H.A. All authors have read and agreed to the published version of the manuscript.

Funding: This work was supported by the Deanship of Scientific Research, Vice Presidency for Graduate Studies and Scientific Research, King Faisal University, Saudi Arabia, Grant No. [GrantA071].

Data Availability Statement: Data are contained within the article.

Conflicts of Interest: The authors declare no conflicts of interest.

References

1. Xu, A.; Wang, B.; Zhu, D.; Pang, J.; Lian, X. Bayesian reliability assessment of permanent magnet brake under small sample size. *IEEE Trans. Reliab.* 2024, *early access*. [CrossRef]
2. Wang, W.; Cui, Z.; Chen, R.; Wang, Y.; Zhao, X. Regression analysis of clustered panel count data with additive mean models. *Stat. Pap.* **2023**, 1–22. [CrossRef]
3. Zhou, S.; Xu, A.; Tang, Y.; Shen, L. Fast Bayesian inference of reparameterized gamma process with random effects. *IEEE Trans. Reliab.* **2024**, *73*, 399–412. [CrossRef]
4. Sedyakin, N. On one physical principle in reliability theory. *Techn. Cybern.* **1966**, *3*, 80–87.
5. Nelson, W. Accelerated life testing-step-stress models and data analyses. *IEEE Trans. Reliab.* **1980**, *29*, 103–108. [CrossRef]
6. Bhattacharyya, G.; Soejoeti, Z. A tampered failure rate model for step-stress accelerated life test. *Commun. Stat.-Theory Methods* **1989**, *18*, 1627–1643. [CrossRef]
7. Madi, M.T. Multiple step-stress accelerated life test: The tampered failure rate model. *Commun. Stat.-Theory Methods* **1993**, *22*, 295–306. [CrossRef]
8. Goel, P.K. *Some Estimation Problems in the Study of Tampered Random Variables*; Technical Report No. 50; Department of Statistics, Carnegie-Mellon University: Pittsburgh, PA, USA, 1971; pp. 2436–2436.
9. DeGroot, M.H.; Goel, P.K. Bayesian estimation and optimal designs in partially accelerated life testing. *Nav. Res. Logist. Q.* **1979**, *26*, 223–235. [CrossRef]
10. Khan, M.A.; Aslam, M. Statistical inferences under step stress partially accelerated life testing based on multiple censoring approaches using simulated and real-life engineering data. *Sci. Rep.* **2020**, *10*, 1–18. [CrossRef]
11. Sultana, F.; Dewanji, A. Tampered random variable modeling for multiple step-stress life test. *Commun. Stat.-Theory Methods* **2021**, *52*, 5387–5406. [CrossRef]
12. Sultana, F.; Çetinkaya, Ç; Kundu, D. Step-stress Life-testing under Tampered Random Variable Modeling for Weibull Distribution in Presence of Competing Risk Data. *Qual. Reliab. Eng.* 2023, *accepted paper*. [CrossRef]
13. Ramadan, D.A.; Almetwally, E.M.; Tolba, A.H. Statistical inference to the parameter of the Akshaya distribution under competing risks data with application HIV infection to aids. *Ann. Data Sci.* **2023**, *10*, 1499–1525. [CrossRef]
14. Tolba, A.H.; Almetwally, E.M.; Sayed-Ahmed, N.; Jawa, T.M.; Yehia, N.; Ramadan, D.A. Bayesian and non-Bayesian estimation methods to independent competing risks models with type II half logistic weibull sub-distributions with application to an automatic life test. *Therm. Sci.* **2022**, *26*, 285–302. [CrossRef]
15. Rosaiah, K.; Kantam, R.R.L. Acceptance sampling based on the inverse Rayleigh distribution. *Econ. Qual. Control* **2005**, *20*, 277–286. [CrossRef]
16. Merovci, F. Transmuted Rayleigh distribution. *Aust. J. Stat.* **2013**, *42*, 21–31. [CrossRef]
17. Cordeiro, G.M.; Cristino, C.T.; Hashimoto, E.M.; Ortega, E.M. The beta generalized Rayleigh distribution with applications to lifetime data. *Stat. Pap.* **2013**, *54*, 133–161. [CrossRef]
18. Ahmad, A.; Ahmad, S.P.; Ahmed, A. Transmuted inverse Rayleigh distribution: A generalization of the inverse Rayleigh distribution. *Math. Theory Model* **2014**, *4*, 90–98.

19. Gomes, A.E.; da-Silva, C.Q.; Cordeiro, G.M.; Ortega, E.M. A new lifetime model: The Kumaraswamy generalized Rayleigh distribution. *J. Stat. Comput. Simul.* **2014**, *84*, 290–309. [CrossRef]
20. Nofal, Z.M.; Abd El Hadi, N.E. Exponentiated transmuted generalized Raleigh distribution: A new four-parameter Rayleigh distribution. *Pak. J. Stat. Oper. Res.* **2015**, *11*, 115–134.
21. Iriarte, Y.A.; Vilca, F.; Varela, H.; Gomez, H.W. Slashed generalized Rayleigh distribution. *Commun. Stat.-Theory Methods* **2017**, *46*, 4686–4699. [CrossRef]
22. El-Sherpieny, E.S.A.; Almetwally, E.M.; Muse, A.H.; Hussam, E. Data analysis for COVID-19 deaths using a novel statistical model: Simulation and fuzzy application. *PLoS ONE* **2023**, *18*, e0283618. [CrossRef]
23. Haj Ahmad, H.; Bdair, O.M.; Naser, M.F.M.; Asgharzadeh, A. The rayleigh lindley distribution: A new generalization of rayleigh distribution with physical applications. *Rev. Investig. Oper.* **2023**, *44*, 205–222.
24. Shen, Z.; Alrumayh, A.; Ahmad, Z.; Abu-Shanab, R.; Al-Mutairi, M.; Aldallal, R. A new generalized Rayleigh distribution with analysis to big data of an online community. *Alex. Eng. J.* **2022**, *61*, 11523–11535. [CrossRef]
25. Mahmoud, M.; Kilany, N.; El-Refai, L. Inference of the lifetime performance index with power Rayleigh distribution based on progressive first-failure–censored data. *Qual. Reliab. Eng. Int.* **2020**, *36*, 1528–1536. [CrossRef]
26. Tierney, L. Markov chains for exploring posterior distributions. *Ann. Stat.* **1994**, *22*, 1701–1728. [CrossRef]
27. Balakrishnan, N.; Sandhu, R.A. Best linear unbiased and maximum likelihood estimation for exponential distributions under general progressive Type-ll censored samples. *Indian J. Stat. Ser. B* **1996**, *58*, 1–9.
28. R Core Team, R. *R: A Language and Environment for Statistical Computing*; R Core Team, R: Vienna, Austria, 2021.
29. Samanta, D.; Kundu, D.; Ganguly, A. Order restricted bayesian analysis of a simple step stress model. *Sankhya B* **2018**, *80*, 195–221. [CrossRef]
30. Samanta, D.; Gupta, A.; ; Kundu, D. Analysis of weibull step-stress model in presence of competing risk. *IEEE Trans. Reliab.* **2019**, *68*, 420–438. [CrossRef]
31. Dukers, N.H.; Renwick, N.; Prins, M.; Geskus, R.B.; Schulz, T.F.; Weverling, G.J.; Coutinho, R.A.; Goudsmit, J. Risk factors for human herpesvirus 8 seropositivity and seroconversion in a cohort of homosexula men. *Am. J. Epidemiol.* **2000**, *151*, 213–224. [CrossRef]
32. Xiridou, M.; Geskus, R.; De Wit, J.; Coutinho, R.; Kretzschmar, M. The contribution of steady and casual partnerships to the incidence of HIV infection among homosexual men in Amsterdam. *Aids* **2003**, *17*, 1029–1038. [CrossRef]
33. Putter, H.; Fiocco, M.; Geskus, R.B. Tutorial in biostatistics: Competing risks and multi-state models. *Stat. Med.* **2007**, *26*, 2389–2430. [CrossRef]
34. Geskus, R.B.; González, C.; Torres, M.; Del Romero, J.; Viciana, P.; Masiá, M.; Blanco, J.R.; Iribarren, M.; De Sanjosé, S.; Hernández-Novoa, B.; et al. Incidence and clearance of anal high-risk human papillomavirus in HIV-positive men who have sex with men: Estimates and risk factors. *Aids* **2016**, *30*, 37. [CrossRef] [PubMed]
35. Lawless, J.F. *Statistical Models and Methods for Lifetime Data*; John Wiley and Sons: New York, NY, USA, 2011.

mathematics

MDPI

Article

Different Statistical Inference Algorithms for the New Pareto Distribution Based on Type-II Progressively Censored Competing Risk Data with Applications

Essam A. Ahmed [1,2], Tariq S. Alshammari [3] and Mohamed S. Eliwa [4,5,*]

1 College of Business Administration, Taibah University, Al-Madina 41411, Saudi Arabia
2 Department of Mathematics, Sohag University, Sohag 82524, Egypt
3 Department of Mathematics, College of Science, University of Ha'il, Hail, Saudi Arabia
4 Department of Statistics and Operations Research, College of Science, Qassim University, Saudi Arabia
5 Department of Mathematics, Faculty of Science, Mansoura University, Mansoura 35516, Egypt
* Correspondence: m.eliwa@qu.edu.sa

Abstract: In this research, the statistical inference of unknown lifetime parameters is proposed in the presence of independent competing risks using a progressive Type-II censored dataset. The lifetime distribution associated with a failure mode is assumed to follow the new Pareto distribution, with consideration given to two distinct competing failure reasons. Maximum likelihood estimators (MLEs) for the unknown model parameters, as well as reliability and hazard functions, are derived, noting that they are not expressible in closed form. The Newton–Raphson, expectation maximization (EM), and stochastic expectation maximization (SEM) methods are employed to generate maximum likelihood (ML) estimations. Approximate confidence intervals for the unknown parameters, reliability, and hazard rate functions are constructed using the normal approximation of the MLEs and the normal approximation of the log-transformed MLEs. Additionally, the missing information principle is utilized to derive the closed form of the Fisher information matrix, which, in turn, is used with the delta approach to calculate confidence intervals for reliability and hazards. Bayes estimators are derived under both symmetric and asymmetric loss functions, with informative and non-informative priors considered, including independent gamma distributions for informative priors. The Monte Carlo Markov Chain sampling approach is employed to obtain the highest posterior density credible intervals and Bayesian point estimates for unknown parameters and reliability characteristics. A Monte Carlo simulation is conducted to assess the effectiveness of the proposed techniques, with the performances of the Bayes and maximum likelihood estimations examined using average values and mean squared errors as benchmarks. Interval estimations are compared in terms of average lengths and coverage probabilities. Real datasets are considered and examined for each topic to provide illustrative examples.

Keywords: progressive type-II censored data; competing risk model; expectation maximization algorithm; stochastic expectation maximization algorithm; Bayes theorem; Metropolis–Hastings algorithm; observed Fisher information matrix; computer simulation; statistics and numerical data

MSC: 62F10; 62F12; 62F15; 62F40

Citation: Ahmed, E.A.; Alshammari, T.S.; Eliwa, M.S. Different Statistical Inference Algorithms for the New Pareto Distribution Based on Type-II Progressively Censored Competing Risk Data with Applications. *Mathematics* **2024**, *12*, 2136. https://doi.org/10.3390/math12132136

Academic Editor: Diana Mindrila

Received: 16 May 2024
Revised: 1 July 2024
Accepted: 4 July 2024
Published: 7 July 2024

1. Introduction

Lifetime studies offer a valuable approach across engineering sciences, medical fields, and economics for exploring the survival distribution of entities or individuals. Analyzing data from such studies necessitates the consideration of the anticipated lifetime distribution, with typical distributions including the exponential, Weibull, Burr, Pareto, and Gamma distributions. Among these, the Pareto distribution family is widely acknowledged in the literature for its ability to model data with heavy tails. One significant outcome of advancements in Pareto modeling is the spotlight it casts on the advantageous properties for measuring income inequality. The Pareto distribution has found applications in various disciplines such as actuarial science, economics, finance, life testing, reliability, survival analysis, and engineering, owing to its tractable nature as a lifespan model. However, it becomes evident that modeling data with heavy-tailed distributions necessitates considerations beyond the capabilities of the Pareto distribution alone. Hence, there arises a need for alternatives. Several distributions based on the Pareto framework have been proposed to address this need. The new Pareto (NP) distribution, introduced by [1], is one such recently discovered distribution that merits further investigation due to its significance across multiple domains. The probability density function (PDF) and cumulative distribution function (CDF) of the NP distribution, denoted $NP(\alpha, \beta)$, are as follows:

$$f(x,\alpha,\beta) = \frac{2\alpha\beta^{\alpha}x^{\alpha-1}}{(x^{\alpha}+\beta^{\alpha})^2}, x \geq \beta, \alpha > 0, \beta > 0, \tag{1}$$

and

$$F(x,\alpha,\beta) = 1 - \frac{2\beta^{\alpha}}{x^{\alpha}+\beta^{\alpha}}, x \geq \beta, \alpha > 0, \beta > 0, \tag{2}$$

where α and β denote the shape and scale parameters of the distribution, respectively. The survival function (SF) for the NP distribution is derived from Equation (2) as follows:

$$S(x,\alpha,\beta) = \frac{2\beta^{\alpha}}{x^{\alpha}+\beta^{\alpha}}, x \geq \beta, \alpha > 0, \beta > 0. \tag{3}$$

The hazard rate function (HRF) can be formulated as

$$H(x,\alpha,\beta) = \frac{f(x,\alpha,\beta)}{1-F(x,\alpha,\beta)} = \frac{\alpha x^{\alpha-1}}{(x^{\alpha}+\beta^{\alpha})}, x \geq \beta, \alpha > 0, \beta > 0. \tag{4}$$

To highlight certain unique characteristics of this function, its first derivative with respect to x can be expressed as follows:

$$H^{'}(x,\alpha,\beta) = \frac{\alpha(\alpha-1)x^{\alpha-2}(x^{\alpha}+\beta^{\alpha}) - \alpha^2 x^{2\alpha-2}}{(x^{\alpha}+\beta^{\alpha})^2} = \frac{\alpha x^{\alpha-2}\Delta(x)}{(x^{\alpha}+\beta^{\alpha})^2}. \tag{5}$$

As mentioned in some of the literature, ref. [2] reported that the NP hazard rate function can be unimodal shaped or decreasing depending on its parameters. For $x > \beta > 0$, it becomes evident that $\Delta(x) = (\alpha-1)\beta^{\alpha} - x^{\alpha}$ and $\Delta(\beta) = \beta^{\alpha}(\alpha-2) < 0$ if $\alpha < 2$. Consequently, for $\alpha < 2$, $H^{'}(x,\alpha,\beta) < 0$, indicating that $H(x,\alpha,\beta)$ is decreasing concerning x. In the case of $\alpha > 2$, the function $H(x,\alpha,\beta)$ exhibits a distinct mode at $x = x_0$, where $H(x,\alpha,\beta)$ increases for all $x < x_0$ and decreases for all $x > x_0$, with $x_0 = \beta\{\alpha-1\}^{1/\alpha}$. Figure 1 shows how the behavior of $H(x,\alpha,\beta)$ is changed when the significant increasing of the parameters α and β happened, as we can see this effect very clearly from the graph. It is clear to us from this figure that the hazard rate of the NP distribution may exhibit either an upside-down bathtub (unimodal) shape or a decreasing trend, contingent upon the values of various parameters.

The NP distribution is often employed for modeling a wide range of real-world datasets, such as insurance, reliability, engineering, and economics. For example, ref. [1] delved into the mathematical properties of the NP distribution, demonstrating its utility in modeling income and reliability data through the analysis of seven real datasets. Ref. [3] used the new Pareto distribution to explain the income of the upper-class group using the Malaysian household incomes dataset from 2012–2019. Further insights into this distribution were uncovered in [2], where the authors derived simpler formulas for relevant risk measures and inequality indices. Ref. [4] conducted parameter estimation for the NP distribution using two distinct methodologies: classical estimation and Bayesian inference, with the latter utilizing importance sampling. Additionally, ref. [5] explored parameter estimation across various phases of progressively censored samples derived from the NP distribution and made predictions regarding failure times for removed units using a Bayesian approach and Markov chain Monte Carlo techniques. Furthermore, ref. [6] applied this distribution to data on rainfall and COVID-19. Ref. [7] utilized Type-II censored data and full samples to estimate NP distribution parameters employing the mixed Gibbs sampling method. They used the NP distribution to model the data of the air conditioning system in an aircraft, as well as some data for mechanical units. Recently, ref. [8] applied the NP distribution to some different data, including the net worth of the affluent in Singapore and China, the daily increment of NASDAQ-100, and the daily new-case rate of COVID-19.

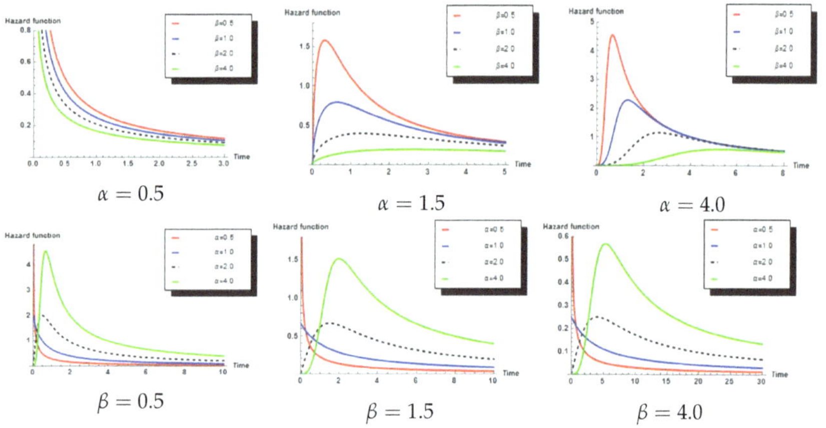

Figure 1. The effects of parameters α and β on the hazard function of $NP(\alpha, \beta)$.

Recognizing the presence or absence of competing risks is pivotal when considering the distribution of units' lifetimes in research. In realistic testing scenarios, it is well understood that failures often stem from multiple factors, competing within the lifespan continuum. This phenomenon is referred to as a competing risk model in the statistical literature and finds widespread application across various scientific domains, including economics, medical sciences, electronic engineering, and social sciences. In the medical field, for instance, reducing mortality rates from specific diseases is a key objective of public health initiatives. However, assessing the impact on overall mortality and life expectancy is complicated by the presence of competing causes of death. Individuals face risks from various ailments like heart disease, diabetes, hypertension, cancer, AIDS, and tuberculosis, all vying for their longevity, even though death is typically attributed to a single cause. Similarly, in assessing the reliability of car tires, factors such as sidewall damage, punctures, or tread wear can lead to failure, highlighting the competition among different failure modes. In industrial reliability contexts, the focus on specific system components prone to failure may overlook potential risks posed by other components, thus introducing a competing risk problem. Competition among failure causes exists in each scenario, with only one mode typically leading to failure. Therefore, the precise inference of each failure mode in the presence of others becomes necessary. Moreover, in medical survival analysis and industrial experimentation, the item of interest often becomes inadvertently lost or removed before failure, leading to data censorship as a result of these investigations.

Due to the challenge of obtaining comprehensive lifespan data for every component in life experiments, researchers often categorize control systems into two types: Type-I and Type-II. Type-I censorship occurs when the experimental duration is fixed, and the number of failures becomes random. Conversely, Type-II censorship arises when the experimental time is unpredictable, but the number of failures is predetermined. Both types entail that any remaining items cannot be removed before the test concludes. Therefore, a progressive

Type-II censoring approach, introduced in [9], aims to streamline resources and reduce costs by implementing multiple censoring stages according to a predefined scheme. The practical implementation of progressive Type-II censorship involves initiating the experiment with n units to be tested. The experiment concludes at the failure time of the mth (where $m < n$) unit. Upon the occurrence of the first failure, R_1 units are randomly selected from the remaining $n - 1$ units, and the failure time, $X_{1:m:n}$, is recorded. Subsequently, R_2 units are randomly removed after noting the second failure time, $X_{2:m:n}$, and this process continues until the mth product fails. At this point, all surviving units are removed, and the failure time, $X_{m:m:n}$, is noted. Here, R_m represents the number of units removed for the mth time. Thus, $X_{1:m:n}, X_{2:m:n}, \ldots, X_{m:m:n}$ constitute the progressive Type-II censored sample. It is essential to note that the total number of units, n, is the sum of $R_1, R_2, \ldots, R_m$ and m, with $R_1, R_2, \ldots, R_m$ being predefined constants. Further insights into censorship-related studies can be found in [9], which serves as a valuable resource in this area. Building upon the progressive Type-II censored sample drawn from an NP distribution, ref. [6] investigated issues related to estimating unknown characteristics and predicting the failure periods of the eliminated units.

In recent years, there has been a growing interest among researchers in statistical inference, particularly concerning the presence of competing risks and various forms of censoring in available data. A multitude of recent studies have addressed this topic, spanning different censoring schemes and statistical models. For instance, ref. [10] explored competing risk analysis under Type-II hybrid censoring for the two-parameter exponential distribution. Ref. [11] investigated statistical inference for competing hazard models using progressive interval censored Weibull data. Ref. [12] analyzed a competing risk model with Weibull distributions under a unified hybrid censoring scheme, deriving maximum likelihood estimators and approximate confidence intervals for distributional parameters. Similarly, ref. [13] focused on statistical inference for competing risk models with inverted exponentiated distributions under a unified hybrid censoring scheme. Additionally, researchers have examined various Bayesian and maximum likelihood estimation techniques under different censoring scenarios and model assumptions. Notably, there is a noticeable gap in the literature regarding the application of the new Pareto lifespan distribution within competing risk models under any form of censorship. Hence, our study aimed to address this gap by studying the maximum likelihood and Bayesian estimation of the unknown parameters, reliability, and hazard rate functions of the NP distribution under progressively censored competing risks models. Moreover, we introduce innovative methodologies, such as employing expectation maximization (EM) and stochastic EM (SEM) algorithms for deriving maximum likelihood estimators for the NP distribution, a novel approach not previously explored. To achieve these objectives, we employ a variety of methods, including the Newton–Raphson method, EM, SEM algorithms, and Bayesian estimation techniques, utilizing different loss functions and prior distributions. Furthermore, we conducted a Monte Carlo simulation analysis to evaluate the efficacy of the proposed estimators using metrics such as absolute bias, mean squared error, average width, and coverage probability. Finally, we illustrate our findings using a real-world dataset, emphasizing the practical relevance of our research.

This paper's structure is organized as follows: Section 2 provides descriptions of the models under the competing risks scenario. In Section 3, maximum likelihood estimators (MLEs) and corresponding approximate confidence intervals (ACIs) for the NP distribution's unknown parameters in the progressively censored competing risk setting are derived using the Newton–Raphson (NR) method. The existence and uniqueness of MLEs are also discussed in this section. Section 4 applies the EM and SEM approximation methods to obtain the MLEs of the unknown parameters. Additionally, a Fisher information matrix is constructed to facilitate approximate interval estimations in this section. Section 5 utilizes the Markov chain Monte Carlo (MCMC) technique to generate Bayes estimates for the unknown parameters, reliability, and hazard functions, assuming independent gamma priors for the unknown parameters. HPD credible intervals for the unknown parameters are also established based on MCMC samples. Section 6 presents a Monte Carlo simulation investigation to compare the suggested estimates in terms of average bias, mean squared error (MSE), average width (AW), and coverage probability (CP). In Section 7, real-world datasets and a simulation example are examined for illustrative purposes. Finally, Section 8 concludes the paper with some closing remarks.

2. Model and Data Overview

Consider a life testing experiment commencing with a set of $n \in N$ identical units, where the failure times of these units are denoted by $X_1, X_2, \ldots, X_n$. To simplify, let us assume that there are only two distinct causes of failure. Thus, for each unit $i = 1, \ldots, n$, we have $X_i = min\{X_{i1}, X_{i2}\}$, where $X_{ik}, (k = 1, 2)$ represents the latent failure time of the ith unit under the kth cause of failure. We make the assumptions that the latent failure times X_{i1} and X_{i2} are statistically independent and the pairs (X_{i1}, X_{i2}) are identically and independently distributed (i.i.d).

As is clear from Figure 1, the effect of the β parameter is not significant on the hazard function, unlike the α parameter. Therefore, here, we assume that the failure times follow the NP distribution, with a common scale parameter β and distinct shape parameters (α_k, $k = 1, 2$). The CDF $F_k(x)$ and the PDF of the jth failure cause of a random variable X_{ij} are provided in [1] as follows:

$$f_k(x, \Theta) = \frac{2\alpha_k \beta^{\alpha_k} x^{\alpha_k - 1}}{(x^{\alpha_k} + \beta^{\alpha_k})^2}, \text{ and } F_k(x, \Theta) = 1 - \frac{2\beta^{\alpha_k}}{x^{\alpha_k} + \beta^{\alpha_k}}, x \geq \beta, \alpha_k > 0, \beta > 0. \tag{6}$$

Given the observation $X_i = \min\{X_{i1}, X_{i2}\}$, where X_{i1} and X_{i2} represent two distinct failure times for a test unit, we solely consider the smaller of the two, indicating the overall failure time. Subsequently, the CDF of this overall failure time is readily derived as follows:

$$\begin{aligned} F(x, \Theta) &= 1 - (1 - F_1(x, \Theta))(1 - F_2(x, \Theta)) = 1 - \left(\frac{2\beta^{\alpha_1}}{x^{\alpha_1} + \beta^{\alpha_1}}\right)\left(\frac{2\beta^{\alpha_2}}{x^{\alpha_2} + \beta^{\alpha_2}}\right) \\ &= 1 - 4\prod_{j=1}^{2} \frac{\beta^{\alpha_k}}{(x^{\alpha_k} + \beta^{\alpha_k})}, x \geq \beta, \alpha_k > 0, \beta > 0. \end{aligned} \tag{7}$$

Consequently, the PDF can be represented as

$$\begin{aligned} f(x, \Theta) &= \frac{4\beta^{(\alpha_1+\alpha_2)}\left[(\alpha_1 + \alpha_2)x^{(\alpha_1+\alpha_2)} + \alpha_2\beta^{\alpha_1}x^{\alpha_2} + \alpha_1\beta^{\alpha_2}x^{\alpha_1}\right]}{x(x^{\alpha_1} + \beta^{\alpha_1})^2(x^{\alpha_2} + \beta^{\alpha_2})^2} \\ &= 4\beta^{(\alpha_1+\alpha_2)} \sum_{k=1}^{2} \frac{\alpha_k x^{\alpha_k - 1}}{(x^{\alpha_k} + \beta^{\alpha_k})^2(x^{\alpha_{3-k}} + \beta^{\alpha_{3-k}})} \\ &= \sum_{k=1}^{2} f_k(x, \Theta)\bar{F}(x, \Theta), x \geq \beta, \alpha_k > 0, \beta > 0. \end{aligned} \tag{8}$$

Subsequently, the survival and hazard functions manifest in the subsequent form:

$$S(t, \Theta) = 4\prod_{j=1}^{2} \frac{\beta^{\alpha_k}}{(t^{\alpha_k} + \beta^{\alpha_k})}, t \geq \beta, \alpha_k > 0, \beta > 0, \tag{9}$$

and

$$H(t, \Theta) = \sum_{k=1}^{2} \frac{\alpha_k x^{\alpha_k - 1}}{(x^{\alpha_k} + \beta^{\alpha_k})}, t \geq \beta, \alpha_k > 0, \beta > 0. \tag{10}$$

Let

$$I(\delta_i = k) = \begin{cases} 1, & \delta_i = k \\ 0, & \text{otherwise :} \end{cases}, i = 1, 2, \ldots, m,$$

then

$$m_k = \sum_{i=1}^{m} I(\delta_i = k),$$

signifies the total count of units that failed because of cause $k(k = 1, 2)$ and $m_1 + m_2 = m$. In exploiting the independence of the latent failure times X_{i1} and X_{i2} for $i = 1, 2, \ldots, n$, the relative risk rate attributed to a specific cause (let us say, cause 1) can be derived as follows:

$$\begin{aligned} p = P(X_1 < X_2) &= \int_{\beta}^{\infty} F_1(x, \alpha_1, \beta) f_2(x, \alpha_2, \beta) dx \\ &= \int_{\beta}^{\infty} \frac{2\alpha_2 \beta^{\alpha_2} x^{\alpha_2 - 1}}{(x^{\alpha_2} + \beta^{\alpha_2})^2} \left(1 - \frac{2\beta^{\alpha_1}}{x^{\alpha_1} + \beta^{\alpha_1}}\right) dx \\ &= 1 - \int_{\beta}^{\infty} \frac{4\alpha_2 \beta^{\alpha_1 + \alpha_2} x^{\alpha_2 - 1}}{(x^{\alpha_2} + \beta^{\alpha_2})^2 (x^{\alpha_1} + \beta^{\alpha_1})} dx. \end{aligned} \tag{11}$$

A numerical approach is necessary to solve the integral on the right side of Equation (11) since it lacks an analytical solution. Once $P(X_1 < X_2)$ is determined, $P(X_2 < X_1)$ can be computed using the relationship $P(X_2 < X_1) = 1 - P(X_2 < X_1)$. Hence, if $m_1 \sim$ Binomial (m, p), then $m_2 \sim$ Binomial$(m, 1 - p)$. In progressive Type-II censoring, the total number of failed individuals m and the predefined censored scheme $(R_1, R_2, \ldots, R_m)$ are specified in advance, where $R_m = n - m - \sum_{i=1}^{m} R_i$. Consequently, the observed set of progressive Type-II censored data with competing risks can be expressed as

$$(X_{1:m:n}, \delta_1, R_1), (X_{J:m:n}, \delta_2, R_m), \ldots, (X_m, \delta_m, R_m).$$

To simplify the notation, we denote $X_{i:m:n}$ as X_i. Then, the likelihood equation for the progressive Type-II censored data, based on the competing risk model, is expressed as

$$L = C(R) \prod_{i=1}^{m} [f_1(x_i) \bar{F}_2(x_i)]^{I(\delta_i = 1)} [f_2(x_i)(\bar{F}_1(x_i))]^{I(\delta_i = 2)} [\bar{F}_1(x_i) \bar{F}_2(x_i)]^{R_i}, \tag{12}$$

where

$$C(R) = n(n - R_1 - 1)(n - R_1 - R_2 - m + 1), \text{and } \bar{F}_k(x_i) = 1 - F_k(x_i), k = 1, 2. \tag{13}$$

3. Estimation of Maximum Likelihood Using Different Algorithms

A widely employed and highly regarded statistical estimation technique is MLE. MLE stands out as a preferred and efficient choice for parametric estimation procedures, owing to its consistency, efficiency, and asymptotic normality properties. This section is dedicated to utilizing maximum likelihood to derive both point and interval estimates for the unknown parameters α_1, α_2, and β, as well as for the reliability $S(t, \alpha_1, \alpha_2, \beta)$ and hazard $H(t, \alpha_1, \alpha_2, \beta)$ functions, within the framework of progressive Type-II censoring with competing risk data.

3.1. MLEs via Newton–Raphson Algorithm

A widely utilized numerical technique for determining MLEs is the NR algorithm. In this context, we outline the NR algorithm for computing the MLEs of α_1, α_2, β, $S(t)$, and $H(t)$. Let $x_{1:m:n} < x_{2:m:n} < \cdots < x_{m:m:n}$ denote the order statistics of Type-II progressively censored data with competing risks from independent NP distributions $NP(\alpha_1, \beta)$ and $NP(\alpha_2, \beta)$, respectively, and a pre-fixed censoring scheme $(R_1, R_2, \ldots, R_m)$. Henceforth, we will use $\mathbf{x} = (x_1, x_2, \ldots, x_m)$ in lieu of $(x_{1:m:n}, x_{2:m:n}, \ldots, x_{m:m:n})$. The likelihood function utilizing Equations (6) and (12) is presented as follows:

$$L \propto \prod_{i=1}^{m} \prod_{k=1}^{2} \left\{ \frac{\alpha_k \beta^{\alpha_k} \beta^{\alpha_{3-k}} x_i^{\alpha_k - 1}}{\left(x_i^{\alpha_k} + \beta^{\alpha_k}\right)^2 \left(x_i^{\alpha_{3-k}} + \beta^{\alpha_{3-k}}\right)} \right\}^{I(\delta_i = k)} \prod_{i=1}^{m} \prod_{k=1}^{2} \left\{ \frac{\beta^{\alpha_k}}{\left(x_i^{\alpha_k} + \beta^{\alpha_k}\right)} \right\}^{R_i}.$$

In disregarding the additive constant, the log-likelihood function is expressed as

$$\mathcal{L} = \ln L = \sum_{k=1}^{2} m_k \log \alpha_k + n \log \beta \sum_{k=1}^{2} \alpha_k + \sum_{k=1}^{2}\sum_{i=1}^{m} (\alpha_k - 1) I(\delta_i = k) \log x_i$$
$$- \sum_{k=1}^{2}\sum_{i=1}^{m} \Delta_k(\delta_i, R_i) \log\left(x_i^{\alpha_k} + \beta^{\alpha_k}\right), \tag{14}$$

where

$$\Delta_k(\delta_i, R_i) = I(\delta_i = k) + R_i + 1, ; i = 1, 2, \cdots ; \text{and } k = 1, 2. \tag{15}$$

Taking the derivative of Equation (15) with respect to α_k and β, where $k = 1, 2$, yields the partial derivatives of the likelihood as follows:

$$\frac{\partial \mathcal{L}}{\partial \alpha_k} \propto \frac{m_k}{\alpha_k} + n \log \beta + \sum_{i=1}^{m} I(\delta_i = k) \log x_i - \sum_{i=1}^{m} \Delta_k(\delta_i, R_i) \left[\frac{x_i^{\alpha_k} \log x_i + \beta^{\alpha_k} \log \beta}{(x_i^{\alpha_k} + \beta^{\alpha_k})} \right], \tag{16}$$

and

$$\frac{\partial \mathcal{L}}{\partial \beta} \propto \sum_{k=1}^{2} \left\{ \frac{n\alpha_k}{\beta} - \sum_{i=1}^{m} \frac{\Delta_k(\delta_i, R_i) \alpha_k \beta^{\alpha_k - 1}}{(x_i^{\alpha_k} + \beta^{\alpha_k})} \right\}. \tag{17}$$

Concurrently solving the intricate nonlinear equations $\frac{\partial \mathcal{L}}{\partial \alpha_k} = 0$ for $k = 1, 2$ and $\frac{\partial \mathcal{L}}{\partial \beta} = 0$ allows for the determination of the MLEs of α_k and β. However, obtaining closed forms for Equations (16) and (17) proves challenging. Hence, various numerical techniques, such as NR, are employed to compute the MLEs for the unknown parameters. Notably, given $x \geq \beta$, the MLE for β is straightforwardly derived as $\beta = x_1$. Furthermore, it is evident that the MLE for α_k can be obtained as a fixed-point solution of the following equation:

$$\Phi(\alpha_k) = 1/\alpha_k, k = 1, 2, \tag{18}$$

where

$$\Phi(\alpha_k) = m_k^{-1} \left\{ n \log x_1 + \sum_{i=1}^{m} I(\delta_i = k) \log x_i - \sum_{i=1}^{m} \frac{\Delta_k(\delta_i, R_i)\left(x_i^{\alpha_k} \log x_i + x_1^{\alpha_k} \log x_1\right)}{(x_i^{\alpha_k} + x_1^{\alpha_k})} \right\}, k = 1, 2. \tag{19}$$

The prevalent approach for numerically solving Equation (19) is through the straightforward iterative method, $\Phi\left(\alpha_k^{(j)}\right) = \alpha_k^{(j)}$, where (j) is the value obtained in the jth iteration. The solutions to the likelihood equations are denoted as $\hat{\alpha}_k$ $(k = 1, 2)$ and $\hat{\beta}$. The following steps illustrate the specific processes involved in this iteration method:

Step 1: Initial values for $\Theta = (\alpha_1, \alpha_2, \beta)$ should be given with $j = 0$; that is, $\Theta^{(0)} = (\alpha_1^{(0)}, \alpha_2^{(0)}, \beta^{(0)})$.

Step 2: In the jth iteration, calculate $\left(\frac{\partial \mathcal{L}}{\partial \alpha_1}, \frac{\partial \mathcal{L}}{\partial \alpha_2}, \frac{\partial \mathcal{L}}{\partial \beta}\right)^{\mathrm{T}}\Big|_{\alpha_1 = \alpha_1^{(j)}, \alpha_2 = \alpha_2^{(j)}, \beta = \beta^{(j)}}$ and $I = I(\alpha_1^{(j)}, \alpha_2^{(j)}, \beta^{(j)})$, where

$$I = I(\alpha_1^{(j)}, \alpha_2^{(j)}, \beta^{(j)}) = \begin{bmatrix} -\mathcal{L}_{11} & -\mathcal{L}_{12} & -\mathcal{L}_{13} \\ -\mathcal{L}_{21} & -\mathcal{L}_{22} & -\mathcal{L}_{23} \\ -\mathcal{L}_{31} & -\mathcal{L}_{32} & -\mathcal{L}_{33} \end{bmatrix}_{\alpha_1 = \alpha_1^{(j)}, \alpha_2 = \alpha_2^{(j)}, \beta = \beta^{(j)}}. \tag{20}$$

The observed information matrix of the parameters α_1, α_2, and β is denoted as I. The elements of the matrix I are as follows:

$$\mathcal{L}_{kk} = \frac{\partial^2 \mathcal{L}}{\partial \alpha_k^2} \propto -\frac{m_k}{\alpha_k^2} - \sum_{i=1}^{m} \left[\frac{\Delta_k(\delta_i, R_i) \beta^{\alpha_k} x_i^{\alpha_k} \log^2(x_i/\beta)}{(x_i^{\alpha_k} + \beta^{\alpha_k})^2} \right], k = 1, 2, \tag{21}$$

$$\mathcal{L}_{12} = \mathcal{L}_{21} = \frac{\partial^2 \mathcal{L}}{\partial \alpha_k \partial \alpha_{3-k}} = \frac{\partial^2 \mathcal{L}}{\partial \alpha_{3-k} \partial \alpha_k} = 0 \tag{22}$$

$$\mathcal{L}_{k3} = \mathcal{L}_{3k} = \frac{\partial^2 \mathcal{L}}{\partial \alpha_k \partial \beta} = \frac{\partial^2 \mathcal{L}}{\partial \beta \partial \alpha_k} = \frac{n}{\beta} - \sum_{i=1}^{m} \frac{\Delta_k(\delta_i, R_i)\beta^{\alpha_k - 1}\left[x_i^{\alpha_k}(1 - \alpha_k \log(x_i/\beta)) + \beta^{\alpha_k}\right]}{\left(x_i^{\alpha_k} + \beta^{\alpha_k}\right)^2}, \tag{23}$$

and

$$\begin{aligned}\mathcal{L}_{33} = \frac{\partial^2 \mathcal{L}}{\partial \beta^2} \propto & -\frac{n}{\beta^2}\sum_{k=1}^{2}\alpha_k - 2\sum_{k=1}^{2}\sum_{i=1}^{m}\frac{\alpha_k \beta^{\alpha_k - 2} I(\delta_i = k)\left[\beta^{\alpha_k} + (\alpha_k - 1)x_i^{\alpha_k}\right]}{\left(x_i^{\alpha_k} + \beta^{\alpha_k}\right)^2} \\ & - \sum_{k=1}^{2}\sum_{i=1}^{m}\frac{[I(\delta_i = 3-k) + R_i]\alpha_k \beta^{\alpha_k - 2}\left[\beta^{\alpha_k} + (\alpha_k - 1)x_i^{\alpha_k}\right]}{\left(x_i^{\alpha_k} + \beta^{\alpha_k}\right)^2}, k = 1, 2.\end{aligned} \tag{24}$$

Step 3: Set

$$(\alpha_1^{(j+1)}, \alpha_2^{(j+1)}, \beta^{(j+1)})^{\mathrm{T}} = A \times \left(\frac{\partial \mathcal{L}}{\partial \alpha_1}, \frac{\partial \mathcal{L}}{\partial \alpha_2}, \frac{\partial \mathcal{L}}{\partial \beta}\right)^{\mathrm{T}}\Bigg|_{\alpha_1 = \alpha_1^{(j)}, \alpha_2 = \alpha_2^{(j)}, \beta = \beta^{(j)}},$$

where $A = (\alpha_1^{(j)}, \alpha_2^{(j)}, \beta^{(j)})^{\mathrm{T}} + I^{-1}(\alpha_1^{(j)}, \alpha_2^{(j)}, \beta^{(j)})$, $(\alpha_1, \alpha_2, \beta)^{\mathrm{T}}$ is the transpose of vector $(\alpha_1, \alpha_2, \beta)$, and $I^{-1}(\alpha_1^{(j)}, \alpha_2^{(j)}, \beta^{(j)})$ represents the inverse of the matrix $I(\alpha_1^{(j)}, \alpha_2^{(j)}, \beta^{(j)})$.

Step 4: In setting $j = j + 1$, the MLEs of the parameters (denoted by $\hat{\alpha}_1, \hat{\alpha}_2, \hat{\beta}$) can be obtained by repeating Steps 2 and 3 until $\left|\left(\alpha_1^{(j+1)}, \alpha_2^{(j+1)}, \beta^{(j+1)}\right)^{\mathrm{T}} - \left(\alpha_1^{(j)}, \alpha_2^{(j)}, \beta^{(j)}\right)^{\mathrm{T}}\right| < \varepsilon$, where ε is a threshold value that is fixed in advance.

Utilizing the acquired point estimators and the invariance property of MLEs, we can derive the estimates of the reliability and hazard functions from Equations (9) and (10), which are expressed as follows:

$$\hat{S}(t) = 4\beta^{\hat{\alpha}_1 + \hat{\alpha}_2}\left[\prod_{k=1}^{2}(t^{\hat{\alpha}_k} + \hat{\beta}^{\hat{\alpha}_k})\right]^{-1}, \text{ and } \hat{H}(t) = \sum_{k=1}^{2}\frac{\hat{\alpha}_k x^{\hat{\alpha}_k - 1}}{(t^{\hat{\alpha}_k} + \hat{\beta}^{\hat{\alpha}_k})}, t > \beta > 0. \tag{25}$$

3.2. Existence and Uniqueness of MLEs

The existence and uniqueness of the MLEs are fundamental aspects to consider in statistical inference.

Theorem 1. *The maximum likelihood estimator of α_k, where $k = 1, 2$, exists and is unique for the lifetimes of objects subject to competing risks and following the NP distribution with parameters (α_1, β) and (α_2, β) when $m_k > 0$ for $k = 1, 2$.*

Proof. Since $x \geq \beta$, we determine the MLE $\hat{\beta} = x_1$, and the MLE $\hat{\alpha}_k$ of α_k, $k = 1, 2$, can be reported from the solution of the following equation:

$$\frac{\partial \mathcal{L}}{\partial \alpha_k} \propto \frac{m_k}{\alpha_k} + n \log x_1 + \sum_{i=1}^{m} I(\delta_i = k)\log x_i - \sum_{i=1}^{m}\Delta_k(\delta_i, R_i)\left[\frac{x_i^{\alpha_k}\log x_i + x_1^{\alpha_k}\log x_1}{(x_i^{\alpha_k} + x_1^{\alpha_k})}\right], k = 1, 2. \tag{26}$$

The left-hand side of Equation (26) is a continuous function. As α_k approaches 0, the left-hand side of Equation (26) tends to infinity, and as $\alpha_k \to \infty$, the left-hand side of Equation (26) tends to ϕ, where ϕ represents

$$\begin{aligned}\phi = \lim_{\alpha_k \to \infty} \frac{\partial \mathcal{L}}{\partial \alpha_k} &= n \log x_1 - \sum_{i=1}^{m} I(\delta_i = k) \log x_i - \sum_{i=1}^{m} [I(\delta_i = 3-k) + R_i] \log x_i \\ &= n \log x_1 - \sum_{i=1}^{m} \log x_i = -\sum_{i=2}^{m} \log x_i < 0. \end{aligned} \tag{27}$$

Hence, the solution of Equation (26) exists. Additionally, we find that the second derivative of $\alpha_k (k = 1, 2)$ takes the following form:

$$\frac{\partial^2 \mathcal{L}}{\partial \alpha_k^2} \propto -\frac{m_k}{\alpha_k^2} - \sum_{i=1}^{m} \left[\frac{\Delta_k(\delta_i, R_i) \beta^{\alpha_k} x_i^{\alpha_k} \log^2(x_i/\beta)}{\left(x_i^{\alpha_k} + \beta^{\alpha_k}\right)^2} \right] < 0. \tag{28}$$

As Equation (28) is always negative, Equation (26) has a unique solution, and this solution represents the maximum likelihood estimator (MLE) of $\alpha_k (k = 1, 2)$. Consequently, we deduce that $\frac{\partial \mathcal{L}}{\partial \alpha_k}$ is a continuous function on $(0, \infty)$, and it monotonically decreases from ∞ to negative values. This demonstrates the existence and uniqueness of the MLE of $\alpha_k (k = 1, 2)$. □

3.3. Approximate Confidence Intervals

In this section, we derive the approximate confidence intervals (ACIs) for the unknown parameters α_1, α_2, and β, as well as for the reliability and hazard functions, utilizing the asymptotic normality of the MLEs. This approach is based on the asymptotic properties of MLEs. Based on the regularity conditions, the MLEs $(\hat{\alpha}_1, \hat{\alpha}_2, \hat{\beta})$ are approximately normally distributed with mean $(\alpha_1, \alpha_2, \beta)$ and variance–covariance matrix $I^{-1}(\hat{\alpha}_1, \hat{\alpha}_2, \hat{\beta})$. Or equivalently,

$$(\hat{\alpha}_1, \hat{\alpha}_2, \hat{\beta}) - (\alpha_1, \alpha_2, \beta) \sim N\left(0, I^{-1}(\hat{\alpha}_1, \hat{\alpha}_2, \hat{\beta})\right), \tag{29}$$

where from (20),

$$I^{-1}(\hat{\alpha}_1, \hat{\alpha}_2, \hat{\beta}) = \begin{bmatrix} -\mathcal{L}_{11} & -\mathcal{L}_{12} & -\mathcal{L}_{13} \\ -\mathcal{L}_{21} & -\mathcal{L}_{22} & -\mathcal{L}_{23} \\ -\mathcal{L}_{31} & -\mathcal{L}_{32} & -\mathcal{L}_{33} \end{bmatrix}^{-1}_{\Theta = \hat{\Theta}} = \begin{bmatrix} Var(\hat{\alpha}_1) & Cav(\hat{\alpha}_1, \hat{\alpha}_2) & Cav(\hat{\alpha}_1, \hat{\beta}) \\ Cav(\hat{\alpha}_2, \hat{\alpha}_1) & Var(\hat{\alpha}_2) & Cav(\hat{\alpha}_2, \hat{\beta}) \\ Cav(\hat{\beta}, \hat{\alpha}_2) & Cav(\hat{\beta}, \hat{\alpha}_2) & Var(\hat{\beta}) \end{bmatrix}, \tag{30}$$

is the inverse of the observed Fisher information matrix with $\Theta = (\alpha_1, \alpha_2, \beta)$. Furthermore, the expression $\mathcal{L}_{ij} = \frac{\partial^2 \mathcal{L}}{\partial \Theta_i \partial \Theta_j}$, for $i, j = 1, 2, 3$, is given by Equations (21)–(24). Thus, the $100(1-\gamma)\%$ ACIs for α_1, α_2, and β are given by

$$\left(\hat{\alpha}_1 \pm Z_{\gamma/2}\sqrt{Var(\hat{\alpha}_1)}\right), \left(\hat{\alpha}_2 \pm Z_{\gamma/2}\sqrt{Var(\hat{\alpha}_2)}\right), \text{and } \left(\hat{\beta} \pm Z_{\gamma/2}\sqrt{Var(\hat{\beta})}\right), \tag{31}$$

where $Z_{\gamma/2}$ is the upper $(\gamma/2)$-th point of the standard normal distribution $N(0, 1)$. Alternatively, the ACIs for $S(t)$ and the hazard function $H(t)$ can be constructed using the asymptotic normality of the MLE. To derive these intervals, we employ the delta method (Greene [14]) to estimate the variances of their estimators. The delta method is a statistical technique used to approximate the probability distribution of a function of an asymptotically normal estimator by employing the Taylor series approximation. In utilizing this method, the approximate variances of $\hat{S}(t)$ and $\hat{H}(t)$ are determined through the following steps:

Step 1: Let Ω_1 and Ω_2 be two quantities that take the following forms:

$$\Omega_1 = \left(\frac{\partial S(t)}{\partial \alpha_1}, \frac{\partial S(t)}{\partial \alpha_2}, \frac{\partial S(t)}{\partial \beta}\right) \text{ and } \Omega_2 = \left(\frac{\partial H(t)}{\partial \alpha_1}, \frac{\partial H(t)}{\partial \alpha_2}, \frac{\partial H(t)}{\partial \beta}\right),$$

where from Equations (9) and (10),

$$\frac{\partial S(t)}{\partial \alpha_k} = \frac{4\beta^{\alpha_k+\alpha_{3-k}} t^{\alpha_k}\{\log t - \log \beta\}}{(t^{\alpha_k}+\beta^{\alpha_k})^2 (t^{\alpha_{3-k}}+\beta^{\alpha_{3-k}})}, k = 1, 2,$$

$$\frac{\partial S(t)}{\partial \beta} = -\frac{4\beta^{\alpha_1+\alpha_2-1}\{(\alpha_1+\alpha_2)t^{\alpha_1+\alpha_2} + \alpha_1\beta^{\alpha_2}t^{\alpha_1} + \alpha_2\beta^{\alpha_1}t^{\alpha_2}\}}{(t^{\alpha_1}+\beta^{\alpha_1})^2(t^{\alpha_2}+\beta^{\alpha_2})^2},$$

$$\frac{\partial H(t)}{\partial \alpha_k} = \frac{t^{\alpha_k-1}\{t^{\alpha_k} + \beta^{\alpha_k}(1+\alpha_k(\log t - \log \beta))\}}{(t^{\alpha_k}+\beta^{\alpha_k})^2}, k = 1, 2,$$

and

$$\frac{\partial H(t)}{\partial \beta} = -\frac{1}{\beta t}\sum_{k=1}^{2}\frac{t^{\alpha_k}\alpha_k^2\beta^{\alpha_k}}{(t^{\alpha_k}+\beta^{\alpha_k})^2}, k = 1, 2.$$

Step 2: Using the following formulas, determine the approximate variances of $S(t)$ and $H(t)$:

$$Var(\hat{S}(t)) \backsimeq \left[\Omega_1 Var(\hat{\Theta})\Omega_1^T\right]_{\Theta=\hat{\Theta}}, \text{ and } Var(\hat{H}(t)) \backsimeq \left[\Omega_2 Var(\hat{\Theta})\Omega_2^T\right]_{\Theta=\hat{\Theta}}, \quad (32)$$

where $Var(\hat{\Theta})$ is obtained from (30) for $\hat{\Theta} = (\hat{\alpha}_1, \hat{\alpha}_2, \hat{\beta})$.

Step 3: Determine the $100(1-\gamma)\%$ asymptotic confidence intervals for $S(t)$ and $H(t)$ using the following formula:

$$\left(\hat{S}(t) \pm Z_{\gamma/2}\sqrt{Var(\hat{S}(t))}\right), \text{ and } \left(\hat{H}(t) \pm Z_{\gamma/2}\sqrt{Var(\hat{H}(t))}\right) \quad (33)$$

3.4. Log-Normal Approximation Confidence Intervals (LACIs)

In instances where the lower bound of the asymptotic confidence intervals may fall below 0, conflicting with the prerequisite $\Theta > 0$, the issue can be circumvented through log transformation and the delta method. We have $\ln\hat{\Theta} \sim N(\ln\Theta, Var(\ln\hat{\Theta}))$, where $Var(\ln\hat{\Theta}) = \frac{Var(\hat{\Theta})}{\hat{\Theta}^2}$. The log-transformed MLEs of $\Theta = (\alpha_1, \alpha_2, \beta)$ obtain their asymptotic confidence intervals at $100(1-\gamma)\%$ as follows:

$$\left[\hat{\Theta}\exp\left(-Z_{(\gamma/2)}\sqrt{\frac{Var(\hat{\Theta})}{\hat{\Theta}^2}}\right), \hat{\Theta}\exp\left(Z_{(\gamma/2)}\sqrt{\frac{Var(\hat{\Theta})}{\hat{\Theta}^2}}\right)\right], \hat{\Theta} = (\hat{\alpha}_1, \hat{\alpha}_2, \hat{\beta}). \quad (34)$$

Similarly, the $100(1-\gamma)\%$ LACIs of $S(t)$ and $H(t)$ can be listed as

$$\left[\hat{S}(t)\exp\left(-Z_{(\gamma/2)}\sqrt{\frac{Var(\hat{S}(t))}{\hat{S}(t)^2}}\right), \hat{S}(t)\exp\left(Z_{(\gamma/2)}\sqrt{\frac{Var(\hat{S}(t))}{\hat{S}(t)^2}}\right)\right], \quad (35)$$

and

$$\left[\hat{H}(t)\exp\left(-Z_{(\gamma/2)}\sqrt{\frac{Var(\hat{H}(t))}{\hat{H}(t)^2}}\right), \hat{H}(t)\exp\left(Z_{(\gamma/2)}\sqrt{\frac{Var(\hat{H}(t))}{\hat{H}(t)^2}}\right)\right]. \quad (36)$$

4. Expectation Maximization (EM) Algorithm

Given that the ML estimators are not straightforwardly derived and that initial values can impact the maximum likelihood estimates obtained from the NR method, we resort to the EM algorithm to compute the estimations of the unknown parameters (α_1, α_2, β), alongside the reliability function $S(t)$ and hazard function $H(t)$ of the NP models. Ref. [15] introduced a generic iterative approach for computing the MLEs of unknown parameters when observed and censored data are available, known as the EM algorithm. This method offers several advantages, such as its capability to address complex problems, the assured increase in log-likelihood with each iteration, the straightforward yet meticulous computations, and the elimination of the necessity for second- and higher-order derivatives. The EM algorithm comprises two key steps: the expectation step (E-step) and the maximization step (M-step). In the E-step, the conditional expectation of the missing data is determined based on the observed data and the current parameter estimates, and the pseudo-log-likelihood function is calculated. Subsequently, in the M-step, the likelihood function under the observed, and censored data are maximized. This method is particularly useful for analyzing partially or censored datasets, such as those encountered in the Type-II progressive censoring with competing risks sample model. Thus, we employed the EM algorithm to obtain the MLEs of α_1, α_2, β, $S(t)$, and $H(t)$.

4.1. Point Estimation via EM Algorithm

Consider the observed sample denoted as $X = (x_{1:m:n}, x_{2:m:n}, \ldots, x_{m:m:n})$ and the censored data as $Z = (Z_1, Z_2, .., Z_m)$, where $Z_i = (z_{i1}, z_{i2}, , z_{iR_i})$ for $i = 1, 2, \ldots, m$. Here, the censored data represent the missing data. It is important to note that the complete sample comprises both the observed sample and the censored data. Thus, the complete sample is represented as $W = (X, Z)$. For the parameter set $\Theta = (\alpha_1, \alpha_2, \beta)$, the likelihood function of the complete sample data W is expressed as

$$L_c(\Theta) \propto \prod_{i=1}^{m}\prod_{k=1}^{2}\left\{ [f_k(x_{i:m:n})\bar{F}_{3-k}(x_{i:m:n})] \prod_{j=1}^{R_i} [f_k(z_{ij})\bar{F}_{3-k}(z_{ij})\right\}^{I(\delta_i = k)}. \tag{37}$$

Hence, the log-likelihood function for α_1, α_2, and β derived from a complete sample is given by

$$\mathcal{L}_c = \sum_{k=1}^{2} n_k \ln \alpha_k + n \ln \beta \sum_{k=1}^{2} \alpha_k + \sum_{k=1}^{2} (\alpha_k - 1)\left\{ \sum_{i=1}^{m} I(\delta_i = k)\left(\log x_i + \sum_{j=1}^{R_i} \log z_{ij} \right)\right\}$$
$$- \sum_{k=1}^{2}\left\{ \sum_{i=1}^{m} \Psi(\delta_i)\left[\log(x_i^{\alpha_k} + \beta^{\alpha_k}) + \sum_{j=1}^{R_i} \log(z_{ij}^{\alpha_k} + \beta^{\alpha_k}) \right]\right\}, \tag{38}$$

where $n_k = \sum_{i=1}^{m} I(\delta_i = k)(R_i + 1)$ represents the total number of failures for each cause in the complete dataset with size n, $n = n_1 + n_2$, and $\Psi(d) = 2I(d = k) + I(d = 3 - k) = I(d = k) + 1$. After obtaining the log-likelihood function in Equation (38) with respect to α and β and setting the normal equations to zero, the maximum likelihood estimators (MLEs) of the parameters α_1, α_2, and β for the entire sample W can be determined as follows:

$$\frac{\partial \mathcal{L}_c}{\partial \alpha_k} = \frac{n_k}{\alpha_k} + n \ln \beta + \sum_{i=1}^{m} I(\delta_i = k)\left[\log x_i + \sum_{j=1}^{R_i} \log z_{ij} \right]$$
$$- \sum_{i=1}^{m} \Psi(\delta_i)\left[\frac{(x_i^{\alpha_k} \log x_i + \beta^{\alpha_k} \log \beta)}{(x_i^{\alpha_k} + \beta^{\alpha_k})} + \sum_{j=1}^{R_i} \frac{z_{ij}^{\alpha_k} \log z_{ij} + \beta^{\alpha_k} \log \beta}{(z_{ij}^{\alpha_k} + \beta^{\alpha_k})} \right] = 0, k = 1, 2, \tag{39}$$

and

$$\frac{\partial \mathcal{L}_c}{\partial \beta} = \frac{n \sum_{k=1}^{2} \alpha_k}{\beta} - \sum_{k=1}^{2}\sum_{i=1}^{m} \Psi(\delta_i)\left[\frac{\alpha_k \beta^{\alpha_k - 1}}{x_i^{\alpha_k} + \beta^{\alpha_k}} + \sum_{j=1}^{R_i} \frac{\alpha_k \beta^{\alpha_k - 1}}{z_{ij}^{\alpha_k} + \beta^{\alpha_k}} \right] = 0. \tag{40}$$

The pseudo-log-likelihood function changes during the E-step to become

$$\frac{n_k}{\alpha_k} + n \ln \beta + \sum_{i=1}^{m} I(\delta_i = k) \log x_i + \sum_{i=1}^{m} I(\delta_i = k) \sum_{j=1}^{R_i} E\left[\log z_{ij} | z_{ij} > x_i\right]$$
$$- \sum_{i=1}^{m} \Psi(\delta_i)\left[\frac{x_i^{\alpha_k} \log x_i + \beta^{\alpha_k} \log \beta}{x_i^{\alpha_k} + \beta^{\alpha_k}} + E\left(\frac{z_{ij}^{\alpha_k} \log z_{ij} + \beta^{\alpha_k} \log \beta}{z_{ij}^{\alpha_k} + \beta^{\alpha_k}} | z_{ij} > x_i \right)\right] = 0, k = 1, 2, \tag{41}$$

and

$$\frac{n\sum_{k=1}^{2}\alpha_k}{\beta}-\sum_{k=1}^{2}\sum_{i=1}^{m}\Psi(\delta_i)\left[\frac{\alpha_k\beta^{\alpha_k-1}}{x_i^{\alpha_k}+\beta^{\alpha_k}}+\sum_{j=1}^{R_i}E\left[\frac{\alpha_k\beta^{\alpha_k-1}}{z_{ij}^{\alpha_k}+\beta^{\alpha_k}}|z_{ij}>x_i\right]\right]=0. \tag{42}$$

In the following steps, we need to derive the following result.

Theorem 2. *Given $X_1 = x_1, \ldots, X_i = x_i$, the conditional distribution of z_{ij}, $j = 1, \ldots, R_i$, which has the left-truncated NP distribution at x_i, is given by*

$$f_{Z|x}(z_{ij}|X_i = x_i = x_i, \alpha_1, \alpha_2, \beta) = \frac{f_Z(z_{ij}|\alpha_1, \alpha_2, \beta)}{1 - F_X(x_i|\alpha_1, \alpha_2, \beta)}, z_{ij} > x_i, \tag{43}$$

where $F_X(x_i; \alpha_1, \alpha_2, \beta)$ and $f_Z(z_{ij}; \alpha_1, \alpha_2, \beta)$ can be found in Equations (7) and (8), respectively.

Proof. The proof is comprehensible and can be explored in detail in [16]. Hence, based on Equation (43), the conditional expectations described in Equations (39) and (42) are generated, as illustrated below:

$$\begin{aligned}E_1(x_i,\alpha_1,\alpha_2,\beta) &= E[\log z_i|z_i>x_i]\\ &= \frac{4\beta^{(\alpha_1+\alpha_2)}}{1-F_X(x_i;\alpha_1,\alpha_2,\beta)}\sum_{k=1}^{2}\int_{x_i}^{\infty}\frac{\alpha_k y^{\alpha_k-1}\log y}{(y^{\alpha_k}+\beta^{\alpha_k})^2(y^{\alpha_{3-k}}+\beta^{\alpha_{3-k}})}dy,\end{aligned} \tag{44}$$

$$\begin{aligned}E_2(x_i,\alpha_1,\alpha_2,\beta) &= E\left[\frac{z_{ij}^{\alpha_k}\log z_{ij}+\beta^{\alpha_k}\log\beta}{z_{ij}^{\alpha_k}+\beta^{\alpha_k}}|z_{ij}>x_i\right]\\ &= \frac{4\beta^{(\alpha_1+\alpha_2)}}{1-F_X(x_i;\alpha_1,\alpha_2,\beta)}\sum_{k=1}^{2}\int_{x_i}^{\infty}\frac{\alpha_k y^{\alpha_k-1}\log(y^{\alpha_k}\log y+\beta^{\alpha_k}\log\beta)}{(y^{\alpha_k}+\beta^{\alpha_k})^3(y^{\alpha_{3-k}}+\beta^{\alpha_{3-k}})}dy,\end{aligned} \tag{45}$$

and

$$\begin{aligned}E_3(x_i,\alpha_1,\alpha_2,\beta) &= E\left[\frac{\alpha_k\beta^{\alpha_k-1}}{z_{ij}^{\alpha_k}+\beta^{\alpha_k}}|z_{ij}>x_i\right]\\ &= \frac{4\beta^{(\alpha_1+\alpha_2)}}{1-F_X(x_i;\alpha_1,\alpha_2,\beta)}\sum_{k=1}^{2}\int_{x_i}^{\infty}\frac{\alpha_k^2 y^{\alpha_k-1}\beta^{\alpha_k-1}}{(y^{\alpha_k}+\beta^{\alpha_k})^3(y^{\alpha_{3-k}}+\beta^{\alpha_{3-k}})}dy.\end{aligned} \tag{46}$$

In the M-step of the EM algorithm, at the $(l+1)$ iteration, the value of $\beta^{(l+1)}$ is first obtained by solving the following equation:

$$\frac{n\sum_{k=1}^{2}\alpha_k^{(l)}}{\beta^{(l+1)}}-\sum_{k=1}^{2}\sum_{i=1}^{m}\Psi(\delta_i)\left[\frac{\alpha_k^{(l)}\left(\beta^{(l+1)}\right)^{\alpha_k^{(l)}-1}}{x_i^{\alpha_k^{(l)}}+(\beta^{(l+1)})^{\alpha_k^{(l)}-1}}+\sum_{j=1}^{R_i}E_3\left(x_i,\alpha_1^{(l)},\alpha_2^{(l)},\beta^{(l)}\right)\right]=0.$$

The estimate of β might then be obtained by

$$\hat{\beta}=\left[n\sum_{k=1}^{2}\alpha_k^{(l)}\right]\left\{\sum_{k=1}^{2}\sum_{i=1}^{m}\Psi(\delta_i)\left[\frac{\alpha_k^{(l)}\left(\beta^{(l+1)}\right)^{\alpha_k^{(l)}-1}}{x_i^{\alpha_k^{(l)}}+(\beta^{(l+1)})^{\alpha_k^{(l)}-1}}+\sum_{j=1}^{R_i}E_3(x_i,\alpha_1^{(l)},\alpha_2^{(l)},\beta^{(l)})\right]\right\}^{-1}. \tag{47}$$

After obtaining $\beta^{(l+1)}$, solve the following to obtain $\alpha_1^{(l+1)}$ and $\alpha_2^{(l+1)}$:

$$\frac{n_k}{\alpha_k^{(l+1)}} + n\ln\beta^{(l+1)} + \sum_{i=1}^{m} I(\delta_i = k)\log x_i - \sum_{i=1}^{m}\Psi(\delta_i)\left[\frac{x_i^{\alpha_k^{(l+1)}}\log x_i + \beta^{(l+1)\alpha_k^{(l+1)}}\log\beta^{(l+1)}}{x_i^{\alpha_k^{(l+1)}} + \beta^{(l+1)\alpha_k^{(l)}}}\right]$$
$$+\sum_{i=1}^{m} I(\delta_i = k)\sum_{j=1}^{R_i} E_1(x_i, \alpha_1^{(l)}, \alpha_2^{(l)}, \beta^{(l+1)}) - \sum_{i=1}^{m}\Psi(\delta_i)\sum_{j=1}^{R_i} E_2(x_i, \alpha_1^{(l)}, \alpha_2^{(l)}, \beta^{(l+1)}) = 0, k = 1, 2.$$

The estimate of α_k might then be obtained by

$$\hat{\alpha}_k = \frac{n_k}{\left\{\sum_{i=1}^{m}\Psi(\delta_i)\left[\frac{x_i^{\alpha_k^{(l+1)}}\log x_i+\beta^{(l+1)\alpha_k^{(l+1)}}\log\beta^{(l+1)}}{x_i^{\alpha_k^{(l+1)}}+\left(\beta^{(l+1)}\right)^{\alpha_k^{(l)}}}\right] - \sum_{i=1}^{m} I(\delta_i = k)\log x_i\right\} - n\ln\beta^{(l+1)}}$$
$$+\frac{n_k}{\sum_{i=1}^{m}\Psi(\delta_i)\sum_{j=1}^{R_i} E_2(x_i, \alpha_1^{(l)}, \alpha_2^{(l)}, \beta^{(l+1)}) - \sum_{i=1}^{m} I(\delta_i = k)\sum_{j=1}^{R_i} E_1(x_i, \alpha_1^{(l)}, \alpha_2^{(l)}, \beta^{(l+1)})}. \quad (48)$$

The next iteration uses $(\alpha_1^{(l+1)}, \alpha_2^{(l+1)}, \beta^{(l+1)})$ as the new value of $(\alpha_1, \alpha_2, \beta)$. Now, an iterative process can be employed to obtain the necessary maximum likelihood estimates of α_1, α_2 , and β. This iterative procedure continues until

$$\left|\alpha_1^{(l+1)} - \alpha_1^{(l)}\right| + \left|\alpha_2^{(l+1)} - \alpha_2^{(l)}\right| + \left|\beta^{(l+1)} - \beta^{(l)}\right| < \varepsilon,$$

for a predetermined small value of ε and some l. This suggested approach converges to the local maximum likelihood as the log-likelihood increases with each iteration. We use the maximum likelihood estimates of the parameters based on the entire sample as starting values in the expectation maximization technique. Henceforth, we denote the maximum likelihood estimates of α_1, α_2, and β as their respective MLEs: $\hat{\alpha}_{1EM}$, $\hat{\alpha}_{2EM}$, and $\hat{\beta}_{EM}$. Furthermore, the MLEs of $S(t)$ and $H(t)$ can be straightforwardly derived using the in-variance property of the MLEs. These are given as

$$\hat{S}_{EM}(t) = \frac{4\beta_{EM}^{\hat{\alpha}_{1EM}+\hat{\alpha}_{2EM}}}{\prod_{k=1}^{2}\left(t^{\hat{\alpha}_{kEM}} + \hat{\beta}^{\hat{\alpha}_{kEM}}\right)}, \text{ and } \hat{H}(t) = \sum_{k=1}^{2}\frac{\hat{\alpha}_{kEM}x^{\hat{\alpha}_{kEM}-1}}{\left(t^{\hat{\alpha}_{kEM}} + \hat{\beta}^{\hat{\alpha}_{kEM}}\right)}, t > \beta > 0. \quad (49)$$

□

4.2. Point Estimation via Stochastic Expectation Maximization (SEM) Algorithm

Most EM algorithms follow a simple closed-form approach, iterating through two phases. However, a notable limitation of the EM method is its susceptibility to becoming trapped in a saddle point, particularly when handling high-dimensional or complex data such as censored lifespan models (see [17]). Notably, the EM equations mentioned earlier lack a closed form, necessitating numerical computation. Hence, to obtain maximum likelihood estimators, we employed the SEM method in this context. The SEM algorithm, proposed in [18], replaces the expectation step (E-step) of the EM method with a stochastic step (S-step). This modification involves substituting each missing datum with a randomly generated value from the conditional distribution of the missing data, based on the observed data and current parameter values. By incorporating random values obtained from the conditional distribution of the missing data, the SEM algorithm augments the observed sample. It is often preferred over the EM approach due to its simplicity, avoidance of complex computations, and lack of reliance on calculating conditional expectations. Moreover, studies have demonstrated that the SEM approach is robust to initial values and performs effectively with small sample sizes; see [19]. Let $W = (X, Z)$ represent the complete dataset, where X

= $(x_1, x_2, \ldots, x_m)$ denotes the observed data, and $Z_i = (z_{i1}, z_{i2}, , z_{iR_i})$ for $i = 1, 2, \ldots, m$, and $j = 1, 2, \ldots, R_i$ signifies the censored data when x_i fails. Following the SEM algorithm's principle, we initially generate the missing samples z_{ij}, $i = 1, 2, \ldots, m$ and $j = 1, 2, \ldots, R_i$, from the truncated conditional distribution below:

$$G_Z(z_{ij}; \alpha_1, \alpha, \beta | z_{ij} > x_i) = \frac{F_Z(z_{ij}; \Theta) - F_{X_i}(x_i; \alpha_1, \alpha, \beta)}{1 - F_{X_i}(x_i; \alpha_1, \alpha, \beta)}, z_{ij} > x_i. \tag{50}$$

To obtain a random sample from Equation (50), we start by generating a random value from the uniform distribution $U(0,1)$, denoted as u. Subsequently, the realization of z_{ij} is derived as

$$z_{ij} = F^{-1}(u + (1-u)F_{X_i}(x_i; \Theta)), \tag{51}$$

where $F^{-1}(.)$ represents the inverse function of $F(.)$, as defined in Equation (7). Subsequently, in replacing each value of z_{ij} with the value generated by Equation (53), the ML estimators of α_1, α_2, and β at the $(l+1)$th stage are obtained from Equations (39) and (40) using

$$\beta^{(l+1)} = \left[n \sum_{k=1}^{2} \alpha_k^{(l)} \right] \left\{ \sum_{k=1}^{2} \sum_{i=1}^{m} \Psi(\delta_i) \left[\frac{\alpha_k^{(l)} \left(\beta^{(l)}\right)^{\alpha_k^{(l)}-1}}{x_i^{\alpha_k^{(l)}} + \left(\beta^{(l)}\right)^{\alpha_k^{(l)}-1}} + \sum_{j=1}^{R_i} \frac{\alpha_k^{(l)} \left(\beta^{(l)}\right)^{\alpha_k^{(l)}-1}}{z_{ij}^{\alpha_k^{(l)}} + \left(\beta^{(l)}\right)^{\alpha_k^{(l)}}} \right] \right\}^{-1}, \tag{52}$$

and

$$\alpha_k^{(l+1)} = \frac{n_k}{\left\{ \sum_{i=1}^{m} \Psi(\delta_i) \left[\frac{x_i^{\alpha_k^{(l)}} \log x_i + \left(\beta^{(l+1)}\right)^{\alpha_k^{(l)}} \log \beta^{(l+1)}}{x_i^{\alpha_k^{(l)}} + \left(\beta^{(l+1)}\right)^{\alpha_k^{(l)}}} \right] - \sum_{i=1}^{m} I(\delta_i = k) \log x_i \right\} - n \ln \beta^{(l+1)}}$$
$$+ \frac{n_k}{\sum_{i=1}^{m} \Psi(\delta_i) \sum_{j=1}^{R_i} \frac{z_{ij}^{\alpha_k^{(l)}} \log z_{ij} + \left(\beta^{(l+1)}\right)^{\alpha_k^{(l)}} \log \beta^{(l+1)}}{z_{ij}^{\alpha_k^{(l)}} + \left(\beta^{(l+1)}\right)^{\alpha_k^{(l)}}} - \sum_{i=1}^{m} I(\delta_i = k) \sum_{j=1}^{R_i} \log z_{ij}}. \tag{53}$$

Like in the EM algorithm, the iterations can be terminated when $\left|\alpha_1^{(h+1)} - \alpha_1^{(h)}\right| + \left|\alpha_2^{(h+1)} - \alpha_2^{(h)}\right| + \left|\beta^{(h+1)} - \beta^{(h)}\right| < \varepsilon$.

4.3. Fisher Information Matrix

This section presents the observed Fisher information matrix derived from [20]'s concept of missing values. It is worth noting that this observed Fisher information matrix can be utilized for constructing asymptotic confidence intervals. The missing information principle, referenced in various scholarly articles, can be summarized as follows:

$$\text{Observed information} = \text{Complete information} - \text{Missing information}. \tag{54}$$

Let us employ the notation shown as follows: $\Theta = (\alpha_1, \alpha, \beta)$; X: the observed data; W: the complete data; and $I_X(\Theta)$, $I_W(\Theta)$, and $I_{W|X}(\Theta)$ represent the observed, complete, and missing information, respectively. By the definition in [20], observed information is the difference between complete and missing information; this may be stated in the following manner:

$$I_X(\Theta) = I_W(\Theta) - I_{W|X}(\Theta). \tag{55}$$

The complete information matrix $I_W(\Theta)$ is provided as follows:

$$I_W(\Theta) = -E\left[\frac{\partial^2 \ln L(W,\Theta)}{\partial \Theta^2}\right]_{\Theta=(\alpha_1,\alpha_2,\beta)}$$
$$= \begin{bmatrix} a_{11}(\alpha_1,\alpha_2,\beta) & a_{12}(\alpha_1,\alpha_2,\beta) & a_{13}(\alpha_1,\alpha_2,\beta) \\ a_{21}(\alpha_1,\alpha_2,\beta) & a_{22}(\alpha_1,\alpha_2,\beta) & a_{23}(\alpha_1,\alpha_2,\beta) \\ a_{31}(\alpha_1,\alpha_2,\beta) & a_{32}(\alpha_1,\alpha_2,\beta) & a_{33}(\alpha_1,\alpha_2,\beta) \end{bmatrix}_{|\Theta=\hat{\Theta}}. \quad (56)$$

The symbol $a_{ij}(\alpha_1,\alpha_2,\beta)$ represents the matrices' elements for $I_W(\Theta)$. They are listed in the following order:

$$a_{kk} = -E\left[\frac{\partial^2 \ln L(\Theta)}{\partial \alpha_k^2}\right] =$$
$$2\sum_{i=1}^{n} E\left[\frac{\beta^{\alpha_k}(\log x_i - \log\beta)^2 x_i^{\alpha_k}}{\left(x_i^{\alpha_k}+\beta^{\alpha_k}\right)^2}\right] - \sum_{i=1}^{n} E\left[\frac{U_1(\underline{x},\alpha_k,\alpha_{3-k},\beta)^2}{V(\underline{x},\alpha_k,\alpha_{3-k},\beta)^2}\right] + \sum_{i=1}^{n} E\left[\frac{U_2(\underline{x},\alpha_k,\alpha_{3-k},\beta)}{V(\underline{x},\alpha_k,\alpha_{3-k},\beta)}\right],$$

where

$$U_1(\underline{x},\alpha_k,\alpha_{3-k},\beta) = \left[\begin{array}{c} \beta^{\alpha_{3-k}}(1+\alpha_k \ln x_i)x_i^{\alpha_k} + \alpha_{3-k}\beta^{\alpha_k}x_i^{\alpha_{3-k}}\ln\beta \\ +(1+(\alpha_k+\alpha_{3-k})\ln x_i)x_i^{(\alpha_k+\alpha_{3-k})} \end{array}\right],$$

$$U_2(\underline{x},\alpha_k,\alpha_{3-k},\beta) = \beta^{\alpha_{3-k}}\ln x_i(2+\alpha_k \ln x_i)x_i^{\alpha_k} + \alpha_2\beta^{\alpha_k}x_i^{\alpha_{3-k}}\ln^2\beta$$
$$+\ln x_i[2+(\alpha_k+\alpha_{3-k})\ln x_i]x_i^{\alpha_k+\alpha_{3-k}},$$

and

$$V(\underline{x},\alpha_k,\alpha_{3-k}\beta) = \left[\alpha_k\beta^{\alpha_{3-k}}x_i^{\alpha_k} + \alpha_{3-k}\beta^{\alpha_k}x_i^{\alpha_{3-k}} + (\alpha_k+\alpha_{3-k})x_i^{(\alpha_k+\alpha_{3-k})}\right],$$

$$a_{k(3-k)} = a_{(3-k)k} = -E\left[\frac{\partial^2 \ln L(\Theta)}{\partial\alpha_k\partial\alpha_{3-k}}\right] = -E\left[\frac{\partial^2 \ln L(\Theta)}{\partial\alpha_{3-k}\partial\alpha_k}\right]$$
$$= \sum_{i=1}^{n} E\left[\frac{\prod_{k=1}^{2} U_1(\underline{x},\alpha_k,\alpha_{3-k},\beta)(1+V(\underline{x},\alpha_k,\alpha_{3-k},\beta)\,)}{V(\underline{x},\alpha_k,\alpha_{3-k},\beta)^2}\right],$$

$$a_{k3} = a_{3k} = -E\left[\frac{\partial^2 \ln L(\eta)}{\partial\alpha_k\partial\beta}\right] = -E\left[\frac{\partial^2 \ln L(\eta)}{\partial\beta\partial\alpha_k}\right]$$
$$= \frac{-n}{\beta} + 2\sum_{i=1}^{n} E\left[\frac{\beta^{\alpha_k-1}\left[\beta^{\alpha_k} + (1+\alpha_k\log\beta - \alpha_k\log x_i)x_i^{\alpha_k}\right]}{\left(x_i^{\alpha_k}+\beta^{\alpha_k}\right)^2}\right]$$
$$+\sum_{i=1}^{n}\left[\frac{\alpha_{3-k}x_i^{\alpha_{3-k}}\left[x_i^{\alpha_k}\beta^{(\alpha_k+\alpha_{3-k})}(\alpha_{3-k}+\alpha_k(\alpha_k-\alpha_{3-k})\log\beta + \alpha_k(\alpha_{3-k}-\alpha_k)\log x_i)\right]}{(V(\underline{x},\alpha_k,\alpha_{3-k},\beta))^2}\right]$$
$$+\sum_{i=1}^{n}\left[\frac{\alpha_{3-k}x_i^{\alpha_{3-k}}\left[\alpha_{3-k}\beta^{\alpha_{3-k}}x_i^{2\alpha_k} + \alpha_{3-k}\beta^{2\alpha_k}x_i^{\alpha_{3-k}}\right]}{V(\underline{x},\alpha_k,\alpha_{3-k},\beta)^2}\right]$$
$$+\sum_{i=1}^{n}\frac{\alpha_{3-k}x_i^{\alpha_{3-k}}\left[(\beta^{\alpha_k}(\alpha_{3-k}+\alpha_k(\alpha_k+\alpha_{3-k})\log\beta - \alpha_k(\alpha_k+\alpha_{3-k})\log x_i))x_i^{(\alpha_k+\alpha_{3-k})}\right]}{V(\underline{x},\alpha_k,\alpha_{3-k},\beta)^2},$$

and

$$a_{33} = -E\left[\frac{\partial^2 \ln L(\eta)}{\partial \beta^2}\right] = \frac{n}{\beta^2}\sum_{k=1}^{2}\alpha_k + 2\sum_{k=1}^{2}\sum_{i=1}^{n} E\left[\frac{\alpha_k \beta^{\alpha_k - 2}\left[(\alpha_k - 1)x_i^{\alpha_k} - \beta^{\alpha_k}\right]}{\left(x_i^{\alpha_k} + \beta^{\alpha_k}\right)^2}\right]$$

$$-\sum_{i=1}^{n} E\left[\frac{\alpha_k^2 \alpha_{3-k}^2\left(\beta^{\alpha_k} x_i^{\alpha_{3-k}} + \beta^{\alpha_{3-k}} x_i^{\alpha_k}\right)^2}{\beta^2 [V(\underline{x}, \alpha_k, \alpha_{3-k}\beta)]^2}\right]$$

$$+\sum_{i=1}^{n} E\left[\frac{\alpha_k \alpha_{3-k}\left[(\alpha_{3-k} - 1)\beta^{\alpha_{3-k}} x_i^{\alpha_k} + (\alpha_k - 1)\beta^{\alpha_k} x_i^{\alpha_{3-k}}\right]}{\beta^2 [V(\underline{x}, \alpha_k, \alpha_{3-k}, \beta)]}\right],$$

where for every function $g(y)$, the expected value is provided by

$$E[g(y)] = \int_0^\infty g(y) f(y, \alpha_1, \alpha_2, \beta) dy = 4\beta^{(\alpha_1 + \alpha_2)} \sum_{k=1}^{2} \alpha_k \int_\beta^\infty \frac{g(y) y^{\alpha_k - 1}}{(y^{\alpha_k} + \beta^{\alpha_k})^2 (y^{\alpha_{3-k}} + \beta^{\alpha_{3-k}})} dy. \quad (57)$$

Moreover, when considering a single observation that was censored at the time of the ith failure, the Fisher information matrix is obtained as follows:

$$I^i_{W|X}(\Theta) = -E_{Z_i|X_i}\left[\frac{\partial^2 \ln f_{Z|X}(z_i | z_i > x_i, \alpha_1, \alpha_2, \beta)}{\partial \Theta^2}\right]_{\Theta = (\theta, \beta, \lambda)}$$

$$= \begin{bmatrix} b_{11}(x_i, \alpha_1, \alpha_2, \beta) & b_{12}(x_i, \alpha_1, \alpha_2, \beta) & b_{13}(x_i, \alpha_1, \alpha_2, \beta) \\ b_{21}(x_i, \alpha_1, \alpha_2, \beta) & b_{22}(x_i, \alpha_1, \alpha_2, \beta) & b_{23}(x_i, \alpha_1, \alpha_2, \beta) \\ b_{31}(x_i, \alpha_1, \alpha_2, \beta) & b_{32}(x_i, \alpha_1, \alpha_2, \beta) & b_{33}(x_i, \alpha_1, \alpha_2, \beta) \end{bmatrix}_{|\Theta = \hat{\Theta}}. \quad (58)$$

wherein $f_{Z|X}(z_i | z_i > x_i, \alpha_1, \alpha_2, \beta)$ is given by Equation (43). It is therefore simple to retrieve the expected missing information as

$$I_{W|X}(\Theta) = \sum_{i=1}^{m} R_i I^i_{W|X}(\Theta), \quad (59)$$

where $I^i_{W|X}(\Theta)$ and $I^m_{W|X}(\Theta)$ are the information matrix of a single observation for the truncated NP distribution with left truncation at x.

For brevity and simplicity, assume that $F = f_{Z|X}(z_i | z_i > x_i, \alpha_1, \alpha_2, \beta)$. From Equations (7), (8), and (43), the logarithm of the PDF of the truncated NP distribution with left truncation at x_i can be listed as the PDF of the truncated NP distribution with left truncation at x_i, which can be listed as

$$\ln F = \ln\left[(\alpha_1 + \alpha_2) z_i^{(\alpha_1 + \alpha_2)} + \alpha_2 \beta^{\alpha_1 z_i \alpha_2} + \alpha_1 \beta^{\alpha_2 z_i \alpha_1}\right] - B + \sum_{k=1}^{2} \ln(x_i^{\alpha_k} + \beta^{\alpha_k}), \quad (60)$$

where

$$B = \ln(z_i) + 2\sum_{k=1}^{2} \ln(x_i^{\alpha_k} + \beta^{\alpha_k}).$$

The negative expected value of the second partial derivatives with respect to α_1, α_2, and β are obtained by

$$b_{kk}(.) = -E\left[\frac{\partial^2 \ln F}{\partial \alpha_k^2}\right] = -\frac{x_i^{\alpha_k}\beta^{\alpha_k}(\ln x_i - \ln\beta)^2}{(x_i^{\alpha_k}+\beta^{\alpha_k})^2} + 2\sum_{k=1}^{2} E\left[\frac{z_{ij}^{\alpha_k}\beta^{\alpha_k}(\ln z_{ij} - \ln\beta)^2}{(z_{ij}^{\alpha_k}+\beta^{\alpha_k})^2}\right]$$
$$+E\left[\frac{z_i^{2\alpha_1}(z_{ij}^{\alpha_2}+\beta^{\alpha_2})^2 - z_i^{(\alpha_1+\alpha_2)}\alpha_2\beta^{\alpha_1}(z_{ij}^{\alpha_2}(\alpha_1+\alpha_2)+\alpha_1\beta^{\alpha_2})\ln^2 z_{ij}}{\left[(\alpha_1+\alpha_2)z_i^{(\alpha_1+\alpha_2)} + \alpha_2\beta^{\alpha_1 z_i \alpha_2} + \alpha_1\beta^{\alpha_2 z_i \alpha_1}\right]^2}\right]$$
$$+E\left[\frac{2z_i^{(\alpha_1+\alpha_2)}\alpha_2\beta^{\alpha_1}(z_{ij}^{\alpha_2}+\beta^{\alpha_2})\ln\beta - z_i^{(\alpha_1+\alpha_2)}\alpha_2\beta^{\alpha_1}(z_{ij}^{\alpha_2}(\alpha_1+\alpha_2)+\alpha_1\beta^{\alpha_2})\ln^2 z_{ij}}{\left[(\alpha_1+\alpha_2)z_i^{(\alpha_1+\alpha_2)} + \alpha_2\beta^{\alpha_1 z_i \alpha_2} + \alpha_1\beta^{\alpha_2 z_i \alpha_1}\right]^2}\right]$$
$$+E\left[\frac{2z_i^{(\alpha_1+\alpha_2)}\alpha_2\beta^{\alpha_1}\ln z_{ij}(-z_{ij}^{\alpha_2} - \beta^{\alpha_2} + (z_{ij}^{\alpha_2}(\alpha_1+\alpha_2)+\alpha_1\beta^{\alpha_2})\ln\beta)}{\left[(\alpha_1+\alpha_2)z_i^{(\alpha_1+\alpha_2)} + \alpha_2\beta^{\alpha_1 z_i \alpha_2} + \alpha_1\beta^{\alpha_2 z_i \alpha_1}\right]^2}\right],$$

$$b_{k3}(.) = b_{3k}(.) = -E\left[\frac{\partial^2 \ln F}{\partial \alpha_k \partial \beta}\right] = -\frac{\beta^{\alpha_k-1}\left[x_i^{\alpha_k}+\beta^{\alpha_k}+\alpha_k x_i^{\alpha_k}(\ln z_{ij} - \ln\beta)\right]}{(x_i^{\alpha_k}+\beta^{\alpha_k})^2}$$
$$+E\left[\frac{\beta^{\alpha_k-1}\left[z_{ij}^{\alpha_k}+\beta^{\alpha_k}+\alpha_k z_{ij}^{\alpha_k}(\ln z_{ij} - \ln\beta)\right]}{(z_{ij}^{\alpha_k}+\beta^{\alpha_k})^2}\right], k = 1,2,$$

$$b_{33}(.) = -E\left[\frac{\partial^2 \ln F}{\partial \beta^2}\right] = -\frac{\alpha_1\beta^{\alpha_k-2}\left[(\alpha_1-1)x_i^{\alpha_k} - \beta^{\alpha_k})\right]}{(x_i^{\alpha_k}+\beta^{\alpha_k})^2} + E\left[\frac{\alpha_1\beta^{\alpha_k-2}\left[(\alpha_1-1)z_{ij}^{\alpha_k} - \beta^{\alpha_k})\right]}{(z_{ij}^{\alpha_k}+\beta^{\alpha_k})^2}\right].$$

The inverse of the observed Fisher information matrix $I_X(\Theta)$ at the maximum likelihood estimates $\hat{\Theta}_{EM} = (\hat{\Theta}_{1EM}, \hat{\Theta}_{2EM}, \hat{\Theta}_{EM})$ is the asymptotic variance–covariance matrix of the $\hat{\Theta}_{EM}$.

$$I_X^{-1}(\hat{\Theta}_{EM}) = \left[I_W(\hat{\Theta}_{EM}) - I_{W|X}(\hat{\Theta}_{EM})\right]^{-1}. \tag{61}$$

Accordingly, for $\Theta = (\alpha_1, \alpha_2, \beta)$, the $100(1-\gamma)\%$ asymptotic confidence interval is then given by

$$\left(\hat{\alpha}_{1EM} \pm Z_{\gamma/2}\sqrt{Var(\hat{\alpha}_{1EM})}\right), \left(\hat{\alpha}_{2EM} \pm Z_{\gamma/2}\sqrt{Var(\hat{\alpha}_{2EM})}\right), \text{and } \left(\hat{\beta}_{EM} \pm Z_{\gamma/2}\sqrt{Var(\hat{\beta}_{EM})}\right), \tag{62}$$

and the $100(1-\gamma)\%$ log-transformed MLE confidence intervals of $\Theta = (\alpha_1, \alpha_2, \beta)$ through the EM algorithm are as follows:

$$\left[\hat{\Theta}_{EM}\exp\left(-Z_{(\gamma/2)}\sqrt{\frac{Var(\hat{\Theta}_{EM})}{\hat{\Theta}^2_{EM}}}\right), \hat{\Theta}\exp\left(Z_{(\gamma/2)}\sqrt{\frac{Var(\hat{\Theta}_{EM})}{\hat{\Theta}^2_{EM}}}\right)\right], \hat{\Theta} = (\hat{\alpha}_1, \hat{\alpha}_2, \hat{\beta}). \tag{63}$$

In using the delta method in the same way as presented in Section 3, the estimators of $S(t)$ and $H(t)$ can be obtained.

5. Bayesian Estimation

In this section, Bayesian estimates for the NP distributions are generated based on progressively censored competing risks samples, focusing on the unknown parameters α_1, α_2, and β, alongside the reliability function $S(t)$ and the hazard function $H(t)$. Estimations are derived considering three distinct loss functions: the general entropy loss function (GELF), LINEX loss function (LLF), and squared error loss function (SELF). While the LINEX and GE loss functions exhibit asymmetry, the squared error loss function is symmetric, affording equal weight to both over- and under-estimations. Given its prevalence in the Bayesian literature, the SELF is often chosen for its symmetry; however, in cases where over-estimation or under-estimation holds varying degrees of significance, asymmetric loss functions such as the LINEX come into play. In Bayesian estimation, prior distributions for parameters are essential, drawing from previous experiences and available parameter information. Since the NP distribution lacks a natural conjugate prior, and joint conjugate priors are not feasible for unknown parameters, independent gamma priors with hyper-parameters (a_i, b_i), $i = 1, 2, 3$, are recommended for all positive parameters in the model.

$$\pi_1(\alpha_1) \propto \alpha_1^{a_1-1} e^{-b_1\alpha_1},\ \pi_2(\alpha_2) \propto \alpha_2^{a_2-1} e^{-b_2\alpha_2} \text{ and } \pi_3(\beta) \propto \beta^{a_3-1} e^{-b_3\beta}; \alpha_1, \alpha_2, \beta_2 > 0. \quad (64)$$

In this context, it is presumed that each hyper-parameter, denoted as (a_i, b_i), $i = 1, 2, 3$, is known and non-negative. This assumption leads to the proposition of a specific prior distribution, representing a scenario where non-informative priors for the parameters are available. Consequently, the joint prior distribution of $(\alpha_1, \alpha_2, \beta)$ can be expressed as follows:

$$\pi(\alpha_1, \alpha_2, \beta) \propto \alpha_1^{a_1-1} \alpha_2^{a_2-1} \beta^{a_3-1} e^{-(b_1\alpha_1 + b_2\alpha_2 + b_3\beta)}. \quad (65)$$

In combining the prior information from Equation (65) with the likelihood function presented in Equation (14), the joint posterior distribution can be represented as follows:

$$\pi^*(\alpha_1, \alpha_2, \beta | \underline{x}) \propto \alpha_1^{a_1-1} \alpha_2^{a_2-1} \beta^{a_3-1} e^{-(b_1\alpha_1 + b_2\alpha_2 + b_3\beta)} \prod_{i=1}^{m} \prod_{k=1}^{2} \left\{ \frac{\alpha_k \beta^{\alpha_k} \beta^{\alpha_{3-k}} x_i^{\alpha_k - 1}}{\left(x_i^{\alpha_k} + \beta^{\alpha_k}\right)^2 \left(x_i^{\alpha_{3-k}} + \beta^{\alpha_{3-k}}\right)} \right\}^{I(\delta_i = k)}$$

$$\times \prod_{i=1}^{m} \prod_{k=1}^{2} \left\{ \frac{\beta^{\alpha_k}}{\left(x_i^{\alpha_k} + \beta^{\alpha_k}\right)} \right\}^{R_i}. \quad (66)$$

The Bayes estimator of any function of α_1, α_2 , and β, say $\Phi(\alpha_1, \alpha_2, \beta)$ under the SE, LINEX, and GE loss functions, can be expressed as

$$\hat{\Phi}_{BS} = E(\Phi(\alpha_1, \alpha_2, \beta) | \underline{x}) = \frac{\int_{x_1}^{\infty} \int_0^{\infty} \int_0^{\infty} \Phi(\alpha_1, \alpha_2, \beta) \pi^*(\alpha_1, \alpha_2, \beta | \underline{x}) d\alpha_1 d\alpha_2 d\beta}{\int_{x_1}^{\infty} \int_0^{\infty} \int_0^{\infty} \pi^*(\alpha_1, \alpha_2, \beta | \underline{x}) d\alpha_1 d\alpha_2 d\beta}, \quad (67)$$

$$\hat{\Phi}_{LINEX} = -\frac{1}{c} \log \frac{\int_{x_1}^{\infty} \int_0^{\infty} \int_0^{\infty} e^{-c\Phi(\alpha_1, \alpha_2, \beta)} \pi^*(\alpha_1, \alpha_2, \beta | \mathbf{x}) d\alpha_1 d\alpha_2 d\beta}{\int_{x_1}^{\infty} \int_0^{\infty} \int_0^{\infty} \pi^*(\alpha_1, \alpha_2, \beta | \underline{x}) d\alpha_1 d\alpha_2 d\beta}, \quad (68)$$

and

$$E_{\alpha_1, \alpha_2, \beta | \underline{x}}\left(g(\alpha_1, \alpha_2, \beta)^{-q}\right) = \frac{\int_{x_1}^{\infty} \int_0^{\infty} \int_0^{\infty} \Phi(\alpha_1, \alpha_2, \beta)^{-q} \pi^*(\alpha_1, \alpha_2, \beta | \underline{x}) d\alpha_1 d\alpha_2 d\beta}{\int_{x_1}^{\infty} \int_0^{\infty} \int_0^{\infty} \pi^*(\alpha_1, \alpha_2, \beta | \underline{x}) d\alpha_1 d\alpha_2 d\beta}. \quad (69)$$

Numerically solving the Bayes estimators becomes necessary as they are not explicitly obtained under the SE or LINEX loss functions. Hence, we propose deriving the Bayes estimators of α_1, α_2 , and β using the MCMC method. Initially, the fully conditional posterior distribution $\pi^*(\alpha_k | \underline{x}, \beta)$ of α_k is

$$\pi^*(\alpha_k | \underline{x}, \beta) \propto \alpha_k^{a_k + m_k - 1} \beta^{n\alpha_k} e^{-b_k\alpha_k} \prod_{i=1}^{m} x_i^{(\alpha_k - 1) I(\delta_i = k)} \prod_{i=1}^{m} \left(x_i^{\alpha_k} + \beta^{\alpha_k}\right)^{-(\Delta_k(\delta_i, R_i))}, k = 1, 2, \quad (70)$$

and the fully conditional posterior distribution $\pi^*(\beta | \underline{x}, \alpha_1, \alpha_2)$ of β is

$$\pi^*(\beta|\underline{x},\alpha_1,\alpha_2) \propto \beta^{a_3-1+n(\alpha_k+\alpha_{3-k})}e^{-b_3\beta}\prod_{i=1}^{m}\prod_{k=1}^{2}\left\{\frac{x_i^{\alpha_k-1}}{\left(x_i^{\alpha_k}+\beta^{\alpha_k}\right)^2\left(x_i^{\alpha_{3-k}}+\beta^{\alpha_{3-k}}\right)}\right\}^{I(\delta_i=k)}$$
$$\times\prod_{i=1}^{m}\prod_{k=1}^{2}\left\{\frac{1}{\left(x_i^{\alpha_k}+\beta^{\alpha_k}\right)}\right\}^{R_i}. \quad (71)$$

The absence of closed-form solutions for the three posterior distributions necessitates the utilization of the Metropolis–Hastings algorithm to derive our Bayes estimators from posterior samples. The Metropolis–Hastings (M-H) algorithm is a valuable tool for generating random samples from the posterior distribution by leveraging a proposal density, as detailed in [21,22]. The Algorithm 1 proceeds through the following stages:

Algorithm 1 MCMC algorithm

Step 1: Choose an initial guess of $(\alpha_1,\alpha_2,\beta)$ denoted by $(\alpha_1^{(0)},\alpha_2^{(0)},\beta^{(0)})$, and set $i=1$.

Step 2: Using the following Metropolis–Hustings, generate $\beta^{(l)}$ from π^*, with the normal proposal distribution, $N\left(\beta^{(i-1)},Var(\beta)\right)$, where $(\beta^{(i-1)})$ is the current value of β, and $Var(\beta)$ is a variance of β. Perform the following:

- i Generate a proposal β^* from $N\left(\beta^{(i-1)},Var(\beta)\right)$;
- ii Evaluate the acceptance probability $\rho=\min\left\{1,\frac{\pi^*(\beta^*|\alpha_1^{(i-1)},\alpha_2^{(i-1)},\underline{x})}{\pi^*(\beta^{(j-1)}|\alpha_1^{(i-1)},\alpha_2^{(i-1)},\underline{x})}\right\}$;
- iii Generate a u from a Uniform$(0,1)$ distribution;
- iv If $u\le\rho$, accept the proposal and set $\beta^{(l)}=\beta^*$; else, set $\beta^{(l)}=\beta^{(l-1)}$.

Step 3: In the same way as in the previous step, generate $\alpha_k^{(i)}$ from $\pi^*\left(\alpha_k^{(i-1)}|\underline{x},\beta^{(i)}\right)$ with the normal proposal distribution $N\left(\alpha_k^{(i-1)},Var(\alpha_k)\right)$, where $(\alpha_k^{(i-1)})$ is the current value of α_k, and $Var(\alpha_k)$ is a variance of α_k.

Step 4: For given t, compute the reliability and hazard functions:

$$S(t,\alpha_1^{(i)},\alpha_2^{(i)},\beta^{(i)})=4\prod_{j=1}^{2}\left[\frac{\beta^{(i)\alpha_k^{(i)}}}{t^{\alpha_k^{(i)}}+\beta^{(i)\alpha_k^{(i)}}}\right],$$

and

$$H(t,\Theta)=\sum_{k=1}^{2}\left[\frac{\beta^{(i)\alpha_k^{(i)}}}{t^{\alpha_k^{(i)}}+\beta^{(i)\alpha_k^{(i)}}}\right].$$

Step 5: Set $i=i+1$.

Step 6: Repeat steps $(2-4)$ N times and obtain the desired number of samples. After discarding the first M burn-in samples, the remaining $N-M$ samples are used to obtain the Bayesian estimates.

Step 7: It is now possible to calculate the Bayes estimate of $\Phi=\Phi(\alpha_1,\alpha_2,\beta)$ under the SE, LINEX, and GE loss functions as

$$\hat{\Phi}_S=\frac{1}{N-M}\sum_{i=M+1}^{N}\Phi(\alpha_1^{(1)},\alpha_2^{(1)},\beta^{(1)}),$$

$$\hat{\Phi}_{LINEX}=\frac{-1}{c}\log\left[\frac{1}{N-M}\sum_{i=M+1}^{N}e^{-c\Phi(\alpha_1^{(1)},\alpha_2^{(1)},\beta^{(1)})}\right],$$

and

$$\hat{\beta}_{GE}=\left[\frac{1}{N-M}\sum_{l=M+1}^{N}\Phi(\alpha_1^{(1)},\alpha_2^{(1)},\beta^{(1)})^{-q}\right]^{\frac{-1}{q}},$$

where $\Phi=\Phi(\alpha_1,\alpha_2,\beta)$ denotes parameters α_1, α_2, β, $S(t)$, and $H(t)$.

Step 8: Order $\Phi^{(M+1)},\Phi^{(M+2)},\ldots,\Phi^{(N)}$ as $\Phi_{(1)}<\Phi_{(2)}<\cdots<\Phi_{(N-M)}$. Consequently,

$$\left(\Phi_{[(N-M)\alpha/2]},\Phi_{[(N-M)(1-\alpha/2)]}\right)$$

yields the $100(1-\gamma)\%$ Bayesian credible interval of Φ. In this case, $[q]$ stands for the integer part of q.

6. Actual Data Illustration

The main objective of this part is to illustrate how the recommended techniques may be used in real-world situations. The first example includes the failure times of electrical appliances. The source of this dataset is listed in [23]. This real data collection, which [24] first looked at, contains 18 electrical equipment failure times. Only failure mode 11 is seen to occur more than twice out of all the failure modes found in these data. In this article, failure mode 11 was regarded as reason number one (cause-1), and the other failure modes as cause number two (cause-2)). In this instance, we treat X_1 and X_2 as failing because of two competing risks under cause-1 and cause-2, respectively. It is observed that there are 8 failures attributed to cause-1 and 13 failures attributed to cause-2. Data analysis begins with dividing the original set of data by 100. Since this transformation is only a scale change, it will not have an impact on this study's results. Table 1 displays the electrical appliance survival times results.

Table 1. Electrical appliance survival times.

Cause1	0.98	4.13	4.95	6.92	10.65	11.93	14.67	19.37
Cause2	0.12	4.95	0.16	5.57	0.16	6.16	0.46	11.07
	0.46	14.67	0.52	0.98	2.70			

We first look at whether or not these datasets can be analyzed using the NP distribution before moving on. It is commonly known that the Kolmogorov–Simirnov (K–S) test may be used with both extremely small and large samples. The Kolmogorov–Smirnov distances and the accompanying p-values (with bracket) for cause-1 and -2 are, respectively, 0.3637 (0.2405) and 0.1792 (0.7981). Because both causes' associated p-values are clearly higher than the significance level of $\gamma = 0.05$, we are unable to reject the null hypothesis that the data for electronic applications originate from the NP distribution. Figure 2 displays the plot of the empirical vs. the fitted CDFs. Furthermore, the most widely used graphical approach for model validation is quantile–quantile (Q-Q), which determines if the fitted model agrees with the provided data. Let $\hat{F}(x)$ be the estimated value of $F(x)$ given $x_1, x_2, \ldots, x_n$. Point $\hat{F}(i/n)$ vs. $x_{i:n}$, where $i = 1, 2, \ldots, n$, was plotted as a Q-Q scatter diagram. The estimated and observed quantiles are displayed in a Q-Q plot. The points on the Q-Q plot should demonstrate a 45 straight line if the model matches the data well. Figure 3 displays the appliance failure data together with the quantile–quantile (Q-Q) plot. As can be seen in Figure 3, there is a rough straight-line pattern that suggests a good match for the NP model. Furthermore, one crucial graphical technique for determining whether or not the data may be applied to a particular distribution is the total time test (TTT) plot. The TTT plot's empirical form is provided by $\tau(k/n) = \left[\sum_{i=1}^{k} x_{i:n} - (n-k)x_{k:n}\right] / (\sum_{i=1}^{n} x_i)$, $k = 1, 2, \ldots, n$, where $x_{i:n}$ denotes the sample's order statistics. In a graphic representation of the TTT plot, the HRF is growing (or decreasing) if the TTT plot is concave (or convex), and it is constant if the TTT plot is graphically portrayed as a straight diagonal. When the TTT plot is convex and then concave, the HRF is U-shaped; otherwise, it is unimodal. Figure 4 shows the TTT plot for the two datasets. The failure data's total time on the test (TTT) plot in Figure 4 illustrates the increasing and decreasing hazard rate for cause-1 and cause-2, respectively, which is consistent with the NP distribution.

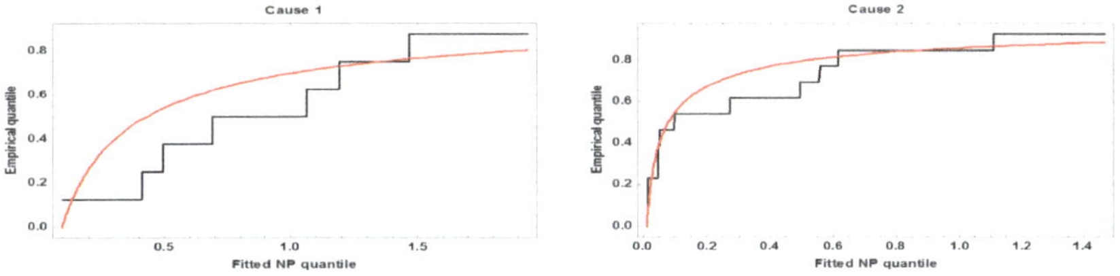

Figure 2. Fitted and empirical CDFs associated with NP distribution for two causes.

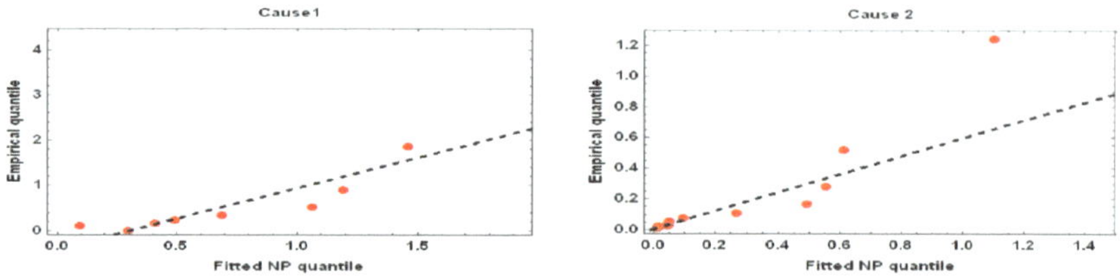

Figure 3. Q-Q plots associated with NP distribution for two causes.

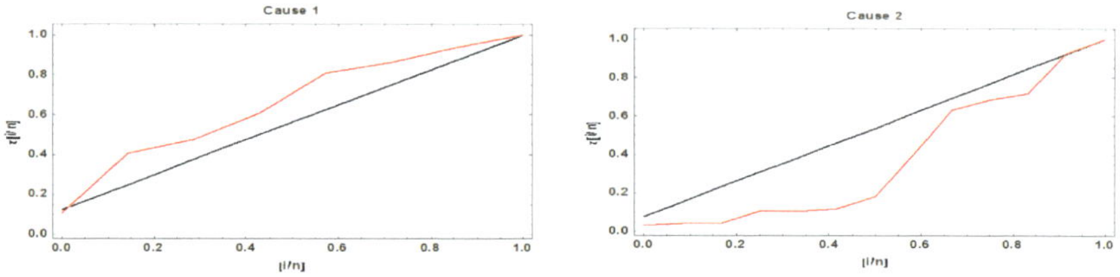

Figure 4. T-T plots associated with NP distribution for two causes.

Three groups of progressive Type-II samples are formed based on the total failure data from Table 1 and are displayed in Table 2. These generated samples are used to estimate the MLEs and Bayes estimates of $\alpha_1, \alpha_2, \beta$, $S(t)$, and $H(t)$ together with their standard errors (in parenthesis).

Table 2. Progressive Type-II data for electrical appliance data, with $n = 21$ and $m = 15$.

Sc. I	$(6, 0^{14})$	$(0.12, 2)$	$(0.16, 2)$	$(0.16, 2)$	$(0.46, 2)$	$(0.46, 2)$
		$(0.52, 2)$	$(0.98, 2)$	$(0.98, 1)$	$(2.70, 2)$	$(4.95, 2)$
		$(55.7, 2)$	$(61.6, 2)$	$(10.65, 1)$	$(11.93, 1)$	$(14.67, 2)$
Sc. II	$(1^6, 0^9)$	$(0.12, 2)$	$(0.16, 2)$	$(0.46, 2)$	$(0.46, 2)$	$(0.52, 2)$
		$(0.98, 2)$	$(0.98, 1)$	$(2.70, 2)$	$(4.13, 1)$	$(4.95, 2)$
		$(4.95, 1)$	$(6.92, 1)$	$(11.93, 1)$	$(14.67, 2)$	$(14.67, 1)$
Sc. III	$(0^{14}, 6)$	$(0.12, 2)$	$(0.16, 2)$	$(0.16, 2)$	$(0.46, 2)$	$(0.46, 2)$
		$(0.52, 2)$	$(0.98, 2)$	$(0.98, 1)$	$(2.70, 2)$	$(4.13, 1)$
		$(4.95, 2)$	$(4.95, 1)$	$(5.57, 2)$	$(6.16, 2)$	$(10.65, 1)$

In real-world data analysis, choosing the starting parameters for an NR algorithm is a challenging task. For Sc. I, the initial guess of β is taken to be the x_1, and a graphical approach presented in [25] is utilized to calculate the MLE shape parameters α_1 and α_2, where, from Equation (19), the curves of $\frac{1}{\alpha_k}$ and $\Phi(\alpha_k)$ are plotted in Figure 5. According to Figure 5, for $k = 1, 2$, the crossing points of the two functions $(\frac{1}{\alpha_k}, \Phi(\alpha_k))$ are around 0.1543 and 0.4605, respectively. To acquire the MLEs of α_1 and α_2, we thus recommend starting the iteration with $\alpha_1 = 0.15$ and $\alpha_2 = 0.46$ as initial values. Moreover, a contour plot of the log-likelihood function of α_1 and α_2 with respect to $\beta = x_1$ is shown in Figure 6.

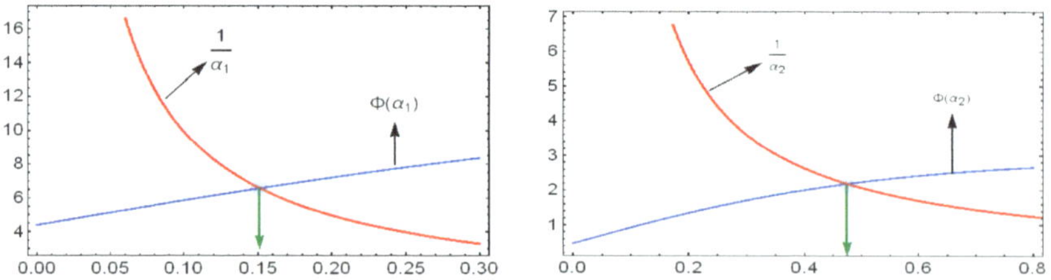

Figure 5. The MLE shape parameters with NP distribution.

Based on these initial values, we first calculated the MLEs of $\alpha_1, \alpha_2, \beta$, $S(t)$, and $H(t)$ using the NR approach. Table 3 presents the values of the MLEs and estimated standard errors of the MLEs of α_1, α_2, β, $S(t)$, and $H(t)$, where $t = 0.4$ is the calculation point for the reliability function $S(t)$ and the hazard rate $H(t)$ function. We also analyzed the dataset (Sc. I) using the EM and SEM methods developed in Section 4. The NR approach is used to set the starting values of α_1, α_2 and β for the EM and SEM algorithms as related MLEs. Table 3 also includes a list of all point estimation results for the MLEs using the EM and SEM methods, along with estimated standard errors. It can be seen from the standard errors that the estimations produced by the NR approach are often higher than those produced by the EM or SEM algorithms. Next, the 95% asymptotic confidence intervals of α_1, α_2, β, $S(t)$, and $H(t)$ were computed using the asymptotic normality of the MLEs and missing information principal techniques.

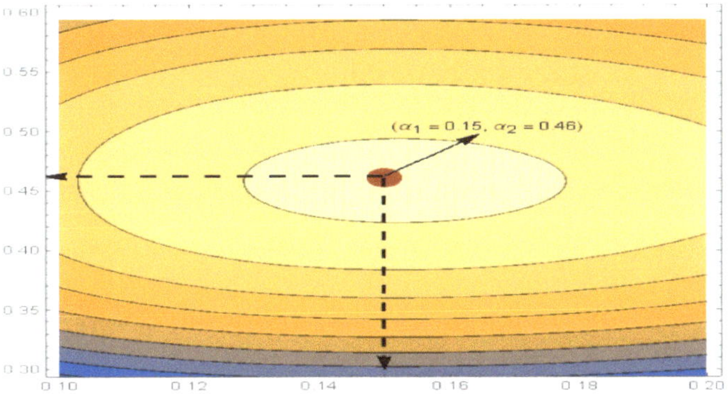

Figure 6. Contour plot of the log-likelihood function of α_1 and α_2.

The next step would be to compute the Bayes estimates of α_1, α_2, β, $S(t)$, and $H(t)$ against the SE, LINEX, and GE loss functions using the MCMC samples. Selecting the hyper-parameters in a real dataset while using Bayesian estimating is a very difficult operation since the true value is not known in advance. Hence, we combine the non-informative prior assumptions ($a_i = b_i = 0, i = 1, 2, 3$) with the MCMC approach to build Bayes estimators. The initial assumptions for executing the MCMC algorithm are thought to be the MLEs estimations of α_1, α_2, and β. In employing 30,000 posterior sample points and ignoring the burn-in of the first 5000 times, the Bayes estimates of α_1, α_2, β, $S(t)$, and $H(t)$ are evaluated using the M-H method. A proposed density function that follows a normal distribution may be built in order to use the MCMC approaches, using the variance–covariance matrix and mean equal to the MLEs of the unknown parameters.

Table 3. ML and Bayesian point estimates (first row) and their standard errors (second row) of α_1, α_2, β, $S(t)$, and $H(t)$ for the real data.

Scheme	Parameter	MLE			Bayes				
					Loss Function				
		NR	EM	SEM	SEL	LINEX		GEL	
						$c=-2$	$c=2$	$q=-2$	$c=2$
Sc. I: $(6,0^{14})$	α_1	0.1523	0.1111	0.1091	0.1501	0.1510	0.1498	0.1513	0.1466
		(0.0218)	(0.0061)	(0.0065)	(0.0048)	(0.0521)	(0.0072)	(0.0014)	(3.1353)
	α_2	0.4638	0.4917	0.4835	0.4567	0.4650	0.4534	0.4603	0.4458
		(0.0356)	(0.0319)	(0.0291)	(0.0146)	(0.7900)	(0.0117)	(0.0137)	(0.3379)
	β	0.1200	0.1200	0.1200	0.1141	0.1142	0.1140	0.1143	0.1139
		(0.0146)	(0.0143)	(0.0133)	(0.0009)	(0.0077)	(0.0014)	(0.0002)	(1.1689)
	$S(t=0.5)$	0.6070	0.6104	0.6162	0.6006	0.6038	0.5994	0.6017	0.5974
		(0.0224)	(0.0237)	(0.0217)	(0.0092)	(0.9349)	(0.0055)	(0.0110)	(0.0874)
	$H(t=0.5)$	0.7807	0.7773	0.7616	0.7742	0.7987	0.7650	0.7802	0.7561
		(0.0611)	(0.0474)	(0.0432)	(0.0247)	(0.7132)	(0.0105)	(0.0392)	(0.1149)

Table 3. *Cont.*

Scheme	Parameter	MLE			Bayes				
					Loss Function				
		NR	EM	SEM	SEL	LINEX		GEL	
						$c=-2$	$c=2$	$q=-2$	$c=2$
Sc. II: $(1^6,0^9)$	α_1	0.2318	0.1722	0.1834	0.1560	0.1569	0.1557	0.1572	0.1527
		(0.0234)	(0.0093)	(0.0118)	(0.0048)	(0.0537)	(0.0071)	(0.0015)	(2.7634)
	α_2	0.3056	0.3561	0.3782	0.4108	0.4181	0.4080	0.4142	0.4005
		(0.0263)	(0.0204)	(0.0241)	(0.0136)	(0.5889)	(0.0119)	(.0115)	(0.4275)
	β	0.1200	0.1200	0.1200	0.1156	0.1156	0.1155	0.1156	0.1154
		(0.0157)	(0.0157)	(0.0137)	(0.0009)	(0.0077)	(0.0014)	(0.0002)	(1.1245)
	$S(t=0.5)$	0.6566	0.6594	0.6406	0.6276	0.6304	0.6265	0.6285	0.6249
		(0.0213)	(0.0246)	(0.0210)	(0.0085)	(0.9958)	(0.0049)	(0.0107)	(0.0722)
	$H(t=0.5)$	0.6411	0.6378	0.6851	0.7066	0.7268	0.6992	0.7119	0.6910
		(0.0492)	(0.0306)	(0.0369)	(0.0223)	(0.9597)	(0.0107)	(0.0323)	(0.1335)
Sc. III: $(0^{14},6)$	α_1	0.1290	0.2161	0.2236	0.1481	0.1489	0.1478	0.1492	0.1447
		(0.0160)	(0.0113)	(0.0158)	(0.0047)	(0.0497)	(0.0069)	(0.0014)	(3.1217)
	α_2	0.2950	0.2182	0.2261	0.3992	0.4054	0.3967	0.4022	0.3900
		(0.0227)	(0.0114)	(0.0165)	(0.0130)	(0.5229)	(0.0116)	(0.0106)	(0.4435)
	β	0.1200	0.1200	0.1200	0.1157	0.1157	0.1157	0.1157	0.1155
		(0.0170)	(0.0181)	(0.0155)	(0.0009)	(0.0076)	(0.0014)	(0.0002)	(1.1088)
	$S(t=0.5)$	0.7198	0.7162	0.7071	0.6389	0.6413	0.6378	0.6396	0.6364
		(0.0206)	(0.0250)	(0.0204)	(0.0083)	(1.0171)	(0.0047)	(0.0105)	(0.0661)
	$H(t=0.5)$	0.4971	0.5010	0.5212	0.6784	0.6953	0.6721	0.6832	0.6644
		(0.0367)	(0.0204)	(0.0296)	(0.0211)	(0.96918)	(0.0107)	(0.0293)	(0.1419)

It is not a secret that, for the LINEX loss function, overestimation has a larger penalty than underestimation when $c > 0$, and the opposite is true when $c < 0$. Additionally, when c approaches 0, the LINEX loss function becomes symmetric and behaves somewhat like the SE loss function. The Bayes estimates were produced under the LINEX loss function discussed here for two different values of $c = -5$ and $+2$. Furthermore, in the general entropy (GE) loss function, the values of parameter q were chosen to be -2 and $+2$. The point Bayes estimates for α_1, α_2, β, $S(t)$, and $H(t)$ were been computed and are presented in Table 3, along with the estimated standard errors and the corresponding 95% credible ranges in Table 4.

Table 4. The 95% interval estimates of α_1, α_2, β, $S(t)$, and, $H(t)$ for the real data.

Scheme	Method	α_1	α_2	β	$S(t=0.5)$	$H(t=0.5)$
Sc. I: $(6, 0^{14})$	ACI	(−0.0127, 0.3172)	(0.1936, 0.7340)	(0.0094, 0.2306)	(0.4370, 0.7768)	(0.3165, 1.2448)
	LACI	(0.0515, 0.4498)	(0.2590, 0.8305)	(0.0477, 0.3016)	(0.4588, 0.8030)	(0.4308, 1.4147)
	EM	(0.0730, 0.1689)	(0.0679, 0.8047)	(0.0486, 0.2963)	(0.4548, 0.8194)	(0.4891, 1.2353)
	SEM	(0.0690, 0.1727)	(0.3062, 0.7636)	(0.0519, 0.2775)	(0.4717, 0.8047)	(0.4949, 1.1718)
	HPD	(0.1149, 0.1892)	(0.3496, 0.5749)	(0.1077, 0.1208)	(0.5305, 0.6692)	(0.5986, 0.9751)
Sc. II: $(1^6, 0^9)$	ACI	(0.0535, 0.4101)	(0.1064, 0.5049)	(0.0010, 0.2390)	(0.4951, 0.8180)	(0.2679, 1.0142)
	LACI	(0.1074, 0.5002)	(0.1592, 0.5866)	(0.0445, 0.3235)	(0.5135, 0.8396)	(0.3582, 1.1474)
	EM	(0.1144, 0.2591)	(0.1114, 0.5502)	(0.0445, 0.3235)	(0.4969, 0.8751)	(0.4430, 0.9185)
	SEM	(0.1124, 0.2993)	(0.2332, 0.6136)	(0.0505, 0.2852)	(0.4995, 0.8216)	(0.4550, 1.0317)
	HPD	(0.1206, 0.1949)	(0.3145, 0.5210)	(0.1090, 0.1221)	(0.5594, 0.6896)	(0.5536, 0.8917)

Table 4. *Cont.*

Scheme	Method	α_1	α_2	β	$S(t=0.5)$	$H(t=0.5)$
Sc. III: $(0^{14}, 6)$	ACI	(0.0068, 0.2511)	(0.12215, 0.4679)	(−0.0089, 0.2489)	(0.5637, 0.8758)	(0.2186, 0.7755)
	LACI	(0.0500, 0.3325)	(0.1642, 0.5301)	(0.0410, 0.3514)	(0.5795, 0.894)	(0.2839, 0.8704)
	EM	(0.1452, 0.3216)	(0.1452, 0.3247)	(0.0382, 0.3770)	(0.5495, 0.9334)	(0.3673, 0.6833)
	SEM	(0.1306, 0.3829)	(0.1299, 0.3934)	(0.0449, 0.3204)	(0.5678, 0.8805)	(0.3389, 0.8016)
	HPD	(0.1144, 0.1860)	(0.3082, 0.5035)	(0.1092, 0.1223)	(0.5743, 0.6981)	(0.535, 0.8502)

To assess the convergence of the Markov chain Monte Carlo (MCMC) method for the studied dataset, we present density plots as well as trace plots of the MCMC outputs for the parameters α_1, α_2, β, $S(t=0.4)$, and $H(t=0.4)$ in Figure 7. These graphs demonstrate the good convergence of all the parameters considered for different chains. Furthermore, it appears that the samples come from the same posterior density because all of the plots display a very strong overlap of the density plots for different chains. Moreover, as Figure 6 illustrates, the estimated estimations of α_1, α_2, β, $S(t=0.4)$, and $H(t=0.4)$ for each sample were typically symmetrical. Thus, the trace and density graphs appear to indicate that the MCMC approach has a high degree of convergence.

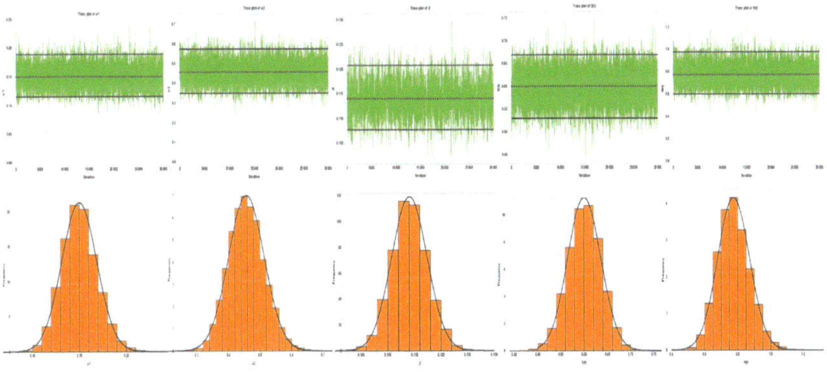

Figure 7. MCMC trace plot (**first row**) and histogram (**second row**) of $\alpha_1\alpha_2, \beta, S(t)$, and $H(t)$ for electrical appliance dataset. Dashed lines (---) represent the posterior means and soled lines (—) represent the lower and upper bounds of the 95% probability interval.

The results are shown in Tables 3 and 4, where the recommended Bayes estimates outperform the most frequent estimates in terms of the lowest standard errors. Moreover, the HPD credible intervals estimates outperform the ACIs, LACIs, EM, and SEM estimates with respect to the shortest ILs. This example also shows how similar the results of all the traditional estimations are to one another. However, it should be mentioned that the MLEs that employed the SEM approach had the lowest standard errors. As a result, machine learning estimates obtained from SEM algorithms frequently performed better than estimates obtained from the NR and EM techniques. Furthermore, it is important to keep in mind that the techniques utilized to compute the point and interval estimators based on Sc. I were also applied to Sc. II and Sc. III; Tables 4 and 5 present the results.

Table 5. The average confidence lengths (AL) and the corresponding coverage percentages (CP) of α_1, α_2, β, $S(t)$, and $H(t)$ based on ML and Bayes estimates for censoring Scheme I.

(n,m)		MLE								MCMC			
		Normal		Log-Normal		EM		SEM		NIP		IP	
		AL	CP	AL	CP	AL	CP	AL	CP	AL	CP	AL	CP
(30, 20)	α_1	1.6237	0.813	2.1950	0.875	0.9999	0.859	0.8271	0.879	0.6282	0.821	0.3709	0.953
	α_2	1.8054	0.829	2.1247	0.894	1.1388	0.894	0.9577	0.893	0.6601	0.893	0.3912	0.962
	β	1.1514	0.939	0.9871	0.907	0.6561	0.993	0.5441	0.965	0.1149	0.955	0.0888	0.962
	$S(t)$	0.6043	0.931	0.4201	0.918	0.3986	0.974	0.3984	0.993	0.2481	0.887	0.1547	0.955
	$H(t)$	1.7221	0.814	2.4435	0.945	1.1505	0.957	1.0194	0.989	0.6702	0.821	0.3983	0.947
	α_1	1.3480	0.958	1.7874	0.958	0.9596	0.942	0.7644	0.930	0.6000	0.921	0.3610	0.951
	α_2	1.4875	0.959	2.0442	0.967	1.0825	0.975	0.9265	0.925	0.6509	0.932	0.3891	0.973
	β	0.9636	0.967	0.9715	0.981	0.6422	0.992	0.5230	0.975	0.1086	0.940	0.0818	0.962
	$S(t)$	0.5822	0.919	0.4166	0.964	0.3851	0.981	0.3891	0.961	0.2344	0.937	0.1510	0.964
	$H(t)$	1.3756	0.934	2.1249	0.972	1.0145	0.946	0.9082	0.958	0.6618	0.939	0.3929	0.947
(50, 35)	α_1	1.2839	0.918	1.4703	0.92	0.7550	0.976	0.6222	0.975	0.5180	0.933	0.3367	0.963
	α_2	1.4564	0.933	1.6739	0.913	0.8297	0.988	0.6838	0.981	0.5379	0.932	0.3562	0.957
	β	0.7670	0.941	0.8492	0.987	0.4937	0.982	0.4152	0.992	0.0658	0.941	0.0537	0.951
	$S(t)$	0.3253	0.939	0.3310	0.933	0.3032	0.979	0.3018	0.998	0.1907	0.954	0.1299	0.959
	$H(t)$	1.5651	0.925	1.6949	0.913	0.8331	0.990	0.7230	0.986	0.5360	0.955	0.3619	0.975
(50, 45)	α_1	0.8591	0.984	0.8793	0.982	0.7458	0.98	0.5786	0.927	0.4609	0.962	0.3255	0.960
	α_2	0.9730	0.991	0.9849	0.973	0.7969	0.982	0.6612	0.960	0.4832	0.947	0.3414	0.933
	β	0.5536	0.956	0.5688	0.988	0.4890	0.963	0.4065	0.971	0.0640	0.956	0.0513	0.960
	$S(t)$	0.3050	0.950	0.3090	0.961	0.3021	0.955	0.3029	0.965	0.1754	0.963	0.1225	0.967
	$H(t)$	1.0225	0.980	0.9891	0.979	0.8227	0.968	0.6432	0.968	0.4920	0.953	0.3474	0.968
(80, 60)	α_1	0.8289	0.982	0.8434	0.981	0.5849	0.967	0.4650	0.945	0.3932	0.923	0.2984	0.955
	α_2	0.9331	0.969	0.9519	0.973	0.6289	0.952	0.5183	0.936	0.4051	0.920	0.3142	0.946
	β	0.5308	0.990	0.5584	0.980	0.3909	0.967	0.3256	0.976	0.0398	0.979	0.0349	0.954
	$S(t)$	0.2513	0.964	0.2535	0.967	0.2408	0.982	0.2397	0.987	0.1462	0.942	0.1120	0.972
	$H(t)$	0.9486	0.965	0.9531	0.972	0.6359	0.980	0.5298	0.973	0.4116	0.963	0.3193	0.969
(80, 75)	α_1	0.6538	0.981	0.6768	0.945	0.5681	0.962	0.4429	0.973	0.3748	0.929	0.2913	0.962
	α_2	0.7093	0.957	0.7327	0.980	0.6158	0.957	0.4813	0.964	0.3895	0.937	0.3038	0.952
	β	0.4266	0.966	0.4394	0.982	0.3827	0.981	0.3171	0.958	0.0362	0.963	0.0337	0.967
	$S(t)$	0.2441	0.947	0.2461	0.971	0.2378	0.964	0.2396	0.972	0.1356	0.965	0.1066	0.960
	$H(t)$	0.7276	0.993	0.7456	0.994	0.6214	0.989	0.4642	0.987	0.3995	0.971	0.3128	0.946

7. Simulation Study

This section compares the performances of suggested estimate methods under progressive Type-II censoring using Monte Carlo simulations. A Monte Carlo simulation analysis was carried out using the Mathematica version 11 statistical program. A comparison is made between point estimators for competing risk lifetimes parameters based on the following perspectives:

(i) Bias = $\frac{1}{Ns}\sum_{i=1}^{Ns}\left|\Theta_i - \hat{\Theta}_i\right|$, where Θ_i and $\hat{\Theta}_i$ stand for the unknown parameters and the associated estimations, and Ns is the number of simulation repeats. The higher agreement of the experimental data with the estimated model is shown by the smaller value of the average bias.

(ii) Mean squared error (MSE) = $\frac{1}{Ns}\sum_{i=1}^{Ns}\left(\Theta_i - \hat{\Theta}_i\right)^2$. An improved estimate performance is shown by a lower MSE value. Each and every result is derived from 1000 replications.

Additionally, average confidence lengths (ACLs) (a better interval estimate performance is correlated with a smaller width) and coverage percentages (CPs) (the probability that the real parameter values lie within the range of the interval estimations) are used to evaluate interval estimators of asymptotic and HPD intervals. If the CP is 95%, then a confidence interval estimator works well. The significance threshold that we employed was $\gamma = 0.05$. We created progressive Type-II censored competing risk data, where the parameters' true values of the NP distribution are arbitrarily assumed to be $(\alpha_1, \alpha_2, \beta) = (0.7, 0.8, 0.5)$. The hazard rate functions in Figure 1 were used to determine the values of α_1, α_2, and β. In order to conduct this simulation research, we chose three distinct censoring schemes, which are detailed below, for $(n, m) = (30, 20)$, $(30, 25)$, $(50,35)$, $(50,45)$, and $(80, 75)$:

Sc. I: $R_1 = R_2 = \cdots = R_{m-1} = 0$ and $R_m = n - m$;

Sc. II: $R_1 = n - m$ and $R_2 = R_3 \cdots = R_{m-1} = 0$;

Sc. III: $R_1 = R_2 = \cdots = R_{m-1} = 1$ and $R_{n-m+1} = \cdots = R_m = 0$.

Here, Sc. I is equivalent to the conventional Type-II censoring technique, and Sc. II is Type-I censoring. After the samples are created, we use the NR, EM, and SEM algorithms to obtain the MLEs of the parameters α_1, α_2, β, $S(t)$, and $H(t)$. The EM and SEM procedures began, in each case, with the true values of the parameters, and the iteration ended when the absolute value of the difference between the two successive iterations for each of the three parameters was less than 10^{-5}. Figures 8–12 display the ABs and MSEs of the MLEs for α_1, α_2, β, $S(t)$, and $H(t)$, respectively. Additionally, we use the missing information principle, the asymptotic distribution of the MLEs, and the log-transformed MLEs to generate the 95% confidence intervals for α_1, α_2, β, $S(t)$, and $H(t)$ (at time $t = 1.0$).

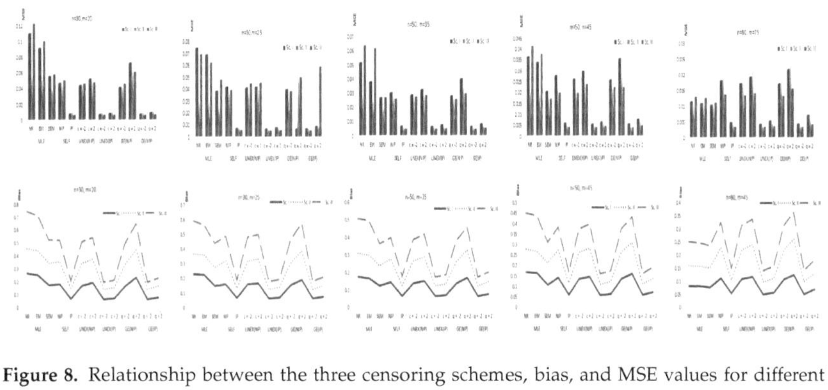

Figure 8. Relationship between the three censoring schemes, bias, and MSE values for different estimates of α_1 at different sample sizes.

In Bayesian calculations, the non-informative prior (NIP) and informative prior (IP) for unknown parameters $(\alpha_1, \alpha_2, \beta, S(t), H(t))$ are taken into account. The hyper-parameters allocated to proper Bayes are $a_1 = 4$, $a_2 = 5$, $a_3 = 15$, and $b_1 = b_1 = b_3 = 10$, for which the corresponding true means and prior means are the same. The hyper-parameters are allocated values of $a_i = b_i = 0$ (NIP) for an improper prior case. Prior I and prior II are used to denote informative and non-informative Bayes estimators, respectively. Additionally, squared error loss (SEL), LINEX loss, and general entropy loss (GEL) are the three distinct loss functions that are taken into consideration when evaluating all Bayes estimators of unknown parameters (α_1, α_2, β, $S(t)$, $H(t)$). It was assumed that the values for the LINEX and GE loss parameters were ± 2 for both c and q. The Bayesian estimates are obtained using the M-H technique described in Section 4. In order to construct α_1, α_2, and β for this procedure, we used normal proposal densities from Equations (70) and (71). The initial values of α_1, α_2, and β in this procedure are the MLEs, or actual parameters values. Following Section 5's instruction, we created $N = 10{,}000$ MCMC samples and discarded the first $M = 1000$ values as the burn-in period. Thus, the SE, LINEX, and GE loss functions were used to construct the Bayes estimates of α_1, α_2, β, $S(t)$, and $H(t)$, based on 9000 M-H sample data points. Using the M-H samples, we obtained the 95% symmetric credible intervals of α_1, α_2, β, $S(t)$, and $H(t)$. Figures 8–12 show the relationships between the progressively censoring schemes (Cs.), bias, and MSE values for different estimates of α_1, α_2, β, $S(t)$, and $H(t)$ at different sample sizes.

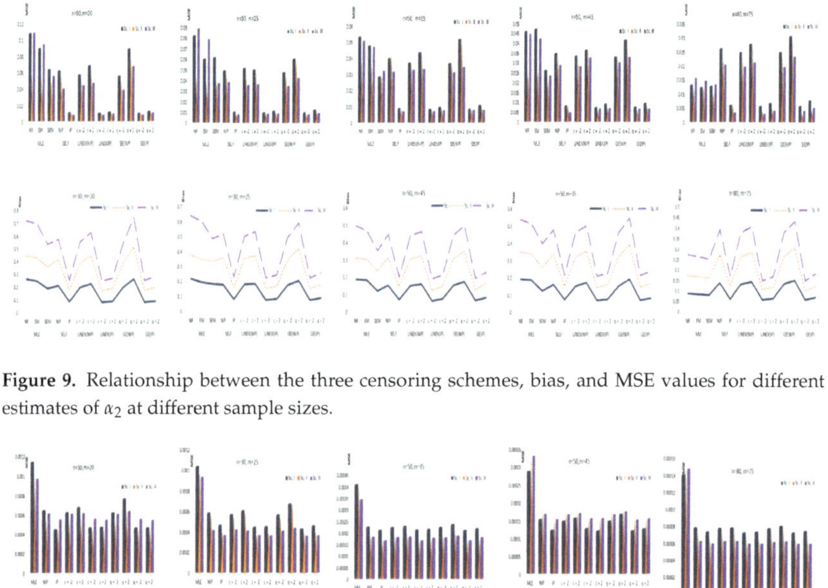

Figure 9. Relationship between the three censoring schemes, bias, and MSE values for different estimates of α_2 at different sample sizes.

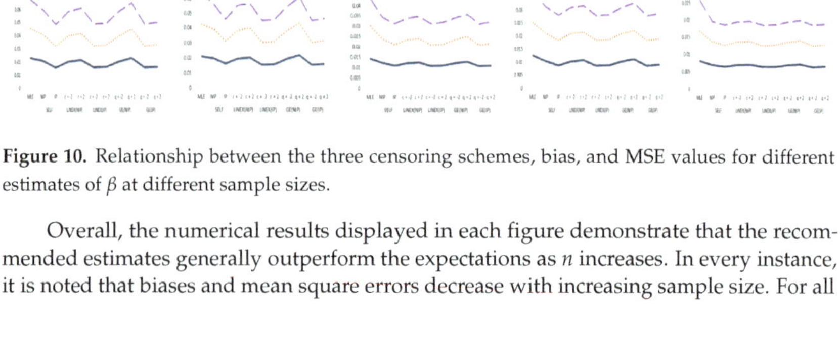

Figure 10. Relationship between the three censoring schemes, bias, and MSE values for different estimates of β at different sample sizes.

Overall, the numerical results displayed in each figure demonstrate that the recommended estimates generally outperform the expectations as n increases. In every instance, it is noted that biases and mean square errors decrease with increasing sample size. For all unknown parameters, the Bayesian estimators outperform the MLE estimators, primarily because the Bayesian approach considers both the data and prior information about the unknown parameters, whereas the MLEs just consider the data. Informative prior Bayesian estimators perform better than non-informative prior estimators for the same reason. When m increases with a fixed n or n increases with a fixed m, the estimator performances of all the unknown parameters improve. Both the MSE and the bias of the Bayesian estimators and MLEs decrease. Therefore, one strategy to improve the outcome estimate is to increase the effective sample size. The best Bayes estimates for the reliability function $S(t)$ are under the GE loss function with $q = +2$, whereas the Bayesian estimates under the LINEX loss function with $c = -2$ are better than those under the SE or GE loss functions for α_1, α_2, β, and $H(t)$. The values of the MSE and bias of all Bayes estimates for α_1, α_2, β, $S(t)$, and $H(t)$, are presented also in Figures 8–12. Finally, the ALs and associated CPs of the estimated confidence intervals for α_1, α_2, β, $S(t)$, and $H(t)$, are also presented in Tables 5–7.

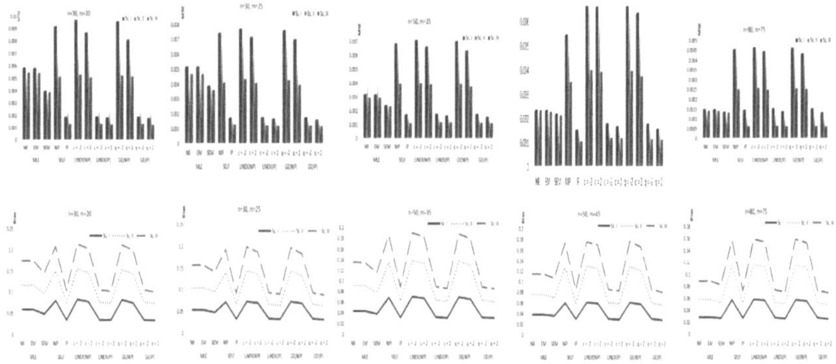

Figure 11. Relationship between the three censoring schemes, bias, and MSE values for different estimates of $S(t)$ at different sample sizes.

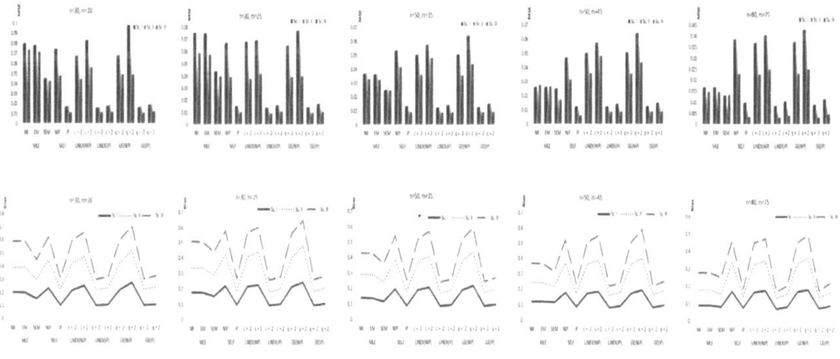

Figure 12. Relationship between the three censoring schemes, bias, and MSE values for different estimates of $H(t)$ at different sample sizes.

We found that the SEM strategy produces better confidence intervals than either the NR or EM strategies in terms of average length. Still, when compared to the five recommended methods (NR, log normal, EM, SEM, and HPD), credible intervals from the HPD perform best. This suggests that their ALs are smaller than at the nominal level, and their CPs are closer to those at the nominal level. Furthermore, informative prior Bayesian credible intervals outperform non-informative prior intervals. Additionally, Bayesian credible intervals with an informative prior perform better than intervals with a non-informative prior. This is true for all parameters α_1, α_2, β, $S(t)$, and $H(t)$.

Finally, it can be seen that for point and interval estimations with various combinations of sample size (m, n) and censoring schemes, the Bayesian technique with an informative prior provides excellent results. Additionally, varied estimates of the parameters exhibit varied performances under varied loss functions. As a result, if it is possible to access previous knowledge about the unknown parameters, the Bayesian approach is preferable. If not, we often select the Bayesian approach with a non-informative prior when the sample size is modest.

Table 6. The average confidence lengths (ALs) and the corresponding coverage percentages (CPs) of α_1, α_2, β, $S(t)$, and $H(t)$ based on ML and Bayes estimates for censoring Scheme II.

(n, m)	SC. II.	MLE								MCMC			
		Normal		Log-Normal		EM		SEM		NIP		IP	
		AL	CP	AL	CP	AL	CP	AL	CP	AL	CP	AL	CP
(30, 20)	α_1	1.5193	0.895	1.6258	0.882	0.9935	0.928	0.7426	0.893	0.6698	0.939	0.3817	0.957
	α_2	1.7420	0.921	1.6777	0.915	1.0981	0.948	0.8834	0.909	0.6671	0.946	0.3898	0.961
	β	0.9044	0.958	1.1778	0.935	0.6592	0.950	0.5508	0.955	0.1093	0.963	0.0793	0.953
	$S(t)$	0.4201	0.975	0.4715	0.960	0.3997	0.970	0.3899	0.973	0.2534	0.971	0.1528	0.958
	$H(t)$	1.8289	0.961	1.8724	0.965	1.0964	0.974	0.9507	0.964	0.7084	0.980	0.4041	0.980
(30, 25)	α_1	1.2936	0.941	1.4567	0.967	0.9519	0.938	0.7178	0.941	0.6087	0.933	0.3673	0.948
	α_2	1.4650	0.965	1.5402	0.955	1.0203	0.925	0.8544	0.953	0.6261	0.972	0.3835	0.936
	β	0.8076	0.989	0.9287	0.971	0.6531	0.996	0.5227	0.962	0.1076	0.981	0.0784	0.968
	$S(t)$	0.4157	0.986	0.4258	0.989	0.3897	0.986	0.3733	0.971	0.2331	0.967	0.1508	0.955
	$H(t)$	1.5254	0.957	1.6910	0.976	1.0679	0.950	0.8992	0.964	0.6516	0.942	0.3949	0.970
(50, 35)	α_1	1.2055	0.948	1.3211	0.957	0.7105	0.965	0.6148	0.893	0.5241	0.963	0.3396	0.965
	α_2	1.3169	0.963	1.4668	0.969	0.8546	0.970	0.6863	0.925	0.5227	0.957	0.3530	0.932
	β	0.7246	0.959	0.8016	0.979	0.5434	0.979	0.4267	0.962	0.0656	0.951	0.0511	0.950
	$S(t)$	0.3171	0.974	0.3375	0.963	0.3165	0.983	0.3110	0.960	0.1921	0.976	0.1285	0.912
	$H(t)$	1.4375	0.968	1.5127	0.982	0.9394	0.974	0.8413	0.907	0.5463	0.980	0.3603	0.956
(50, 45)	α_1	0.9863	0.939	1.1346	0.986	0.6559	0.954	0.5835	0.964	0.4596	0.978	0.3251	0.976
	α_2	1.0898	0.944	1.2108	0.982	0.8202	0.959	0.6485	0.962	0.4750	0.969	0.3395	0.957
	β	0.5891	0.919	0.6406	0.987	0.4996	0.960	0.4037	0.974	0.0545	0.983	0.0442	0.948
	$S(t)$	0.3082	0.941	0.3248	0.965	0.3079	0.952	0.3039	0.978	0.1716	0.982	0.1254	0.969
	$H(t)$	1.0969	0.963	1.2396	0.984	0.8263	0.903	0.6334	0.980	0.4870	0.985	0.3460	0.973
(80, 60)	α_1	0.9161	0.942	1.0169	0.940	0.5611	0.949	0.4416	0.931	0.3910	0.947	0.2987	0.920
	α_2	0.9987	0.950	1.1559	0.936	0.6181	0.957	0.4962	0.935	0.4153	0.952	0.3161	0.927
	β	0.5661	0.923	0.6106	0.967	0.3879	0.964	0.3240	0.941	0.0407	0.962	0.0358	0.940
	$S(t)$	0.2579	0.946	0.2619	0.977	0.2391	0.959	0.2311	0.940	0.1478	0.968	0.1116	0.953
	$H(t)$	1.0316	0.962	1.1103	0.981	0.6150	0.961	0.4751	0.953	0.4163	0.972	0.3184	0.946
(80, 75)	α_1	0.6010	0.961	0.6203	0.948	0.5501	0.959	0.4264	0.973	0.3425	0.958	0.2758	0.949
	α_2	0.6664	0.947	0.6861	0.962	0.6082	0.971	0.4835	0.964	0.3604	0.967	0.2889	0.951
	β	0.4067	0.956	0.4177	0.967	0.3757	0.961	0.3124	0.957	0.0381	0.955	0.0343	0.953
	$S(t)$	0.2405	0.969	0.2424	0.982	0.2294	0.974	0.2206	0.970	0.1303	0.959	0.1046	0.960
	$H(t)$	0.6694	0.965	0.6841	0.980	0.6066	0.98	0.4728	0.952	0.3575	0.969	0.2904	0.957

Table 7. The average confidence lengths (ALs) and the corresponding coverage percentages (CPs) of α_1, α_2, β, $S(t)$, and $H(t)$ based on ML and Bayes estimates for censoring Scheme III.

(n,m)	SC. II.	MLE								MCMC			
		Normal		Log-Normal		EM		SEM		NIP		IP	
		AL	CP	AL	CP	AL	CP	AL	CP	AL	CP	AL	CP
(30, 15)	α_1	1.5068	0.899	1.4165	0.858	0.9673	0.885	0.7862	0.944	0.6975	0.962	0.3918	0.961
	α_2	1.6727	0.901	1.7901	0.874	1.1628	0.909	0.8341	0.958	0.8879	0.954	0.4046	0.958
	β	1.0934	0.912	0.9106	0.906	0.6737	0.924	0.5371	0.966	0.0941	0.927	0.0761	0.974
	$S(t)$	0.4288	0.933	0.4575	0.961	0.4020	0.964	0.4025	0.984	0.2780	0.932	0.1532	0.966
	$H(t)$	1.7791	0.947	1.8364	0.931	1.1477	0.950	0.7730	0.975	0.9280	0.974	0.4366	0.970

Table 7. *Cont.*

(n,m)	SC. II.	MLE								MCMC			
		Normal		Log-Normal		EM		SEM		NIP		IP	
		AL	CP	AL	CP	AL	CP	AL	CP	AL	CP	AL	CP
(30, 25)	α_1	1.3599	0.913	1.3795	0.925	0.9205	0.921	0.7257	0.967	0.6561	0.887	0.3765	0.964
	α_2	1.5155	0.925	1.5960	0.930	1.0907	0.955	0.7793	0.935	0.7043	0.927	0.3968	0.949
	β	0.8601	0.973	0.8983	0.941	0.6328	0.956	0.5146	0.969	0.0871	0.947	0.0660	0.964
	$S(t)$	0.4246	0.924	0.4310	0.966	0.3961	0.971	0.3949	0.978	0.2374	0.927	0.1516	0.973
	$H(t)$	1.5901	0.927	1.6530	0.965	1.0548	0.975	0.7520	0.967	0.7243	0.913	0.4109	0.982
(50, 35)	α_1	1.1810	0.927	1.2832	0.934	0.7542	0.935	0.5438	0.961	0.5691	0.943	0.3529	0.958
	α_2	1.2960	0.932	1.3718	0.939	0.8244	0.960	0.6138	0.972	0.5942	0.937	0.3789	0.952
	β	0.7910	0.965	0.7164	0.951	0.4985	0.957	0.4045	0.954	0.0518	0.985	0.0505	0.961
	$S(t)$	0.3211	0.919	0.3403	0.945	0.3054	0.968	0.3171	0.956	0.2032	0.972	0.1310	0.964
	$H(t)$	1.3949	0.941	1.4298	0.967	0.8321	0.974	0.5927	0.964	0.6323	0.968	0.3892	0.956
(50, 45)	α_1	0.9155	0.929	0.9655	0.947	0.7336	0.951	0.5136	0.960	0.4960	0.939	0.3374	0.971
	α_2	1.0021	0.932	0.8673	0.954	0.8185	0.956	0.6092	0.941	0.5172	0.923	0.3542	0.957
	β	0.5830	0.970	0.6341	0.962	0.4952	0.970	0.3910	0.965	0.0465	0.947	0.0460	0.962
	$S(t)$	0.3160	0.961	0.3228	0.969	0.3026	0.968	0.3116	0.971	0.1774	0.912	0.1278	0.967
	$H(t)$	1.0408	0.968	1.1734	0.959	0.8248	0.973	0.5837	0.959	0.5487	0.913	0.3681	0.960
(80, 60)	α_1	0.7033	0.949	0.7603	0.944	0.5849	0.949	0.4429	0.962	0.4814	0.940	0.3286	0.967
	α_2	0.7783	0.938	0.8430	0.957	0.6654	0.907	0.5021	0.956	0.4995	0.912	0.3537	0.981
	β	0.4680	0.968	0.5810	0.962	0.3909	0.968	0.3082	0.972	0.0362	0.973	0.0336	0.973
	$S(t)$	0.3098	0.977	0.3194	0.977	0.2413	0.983	0.2397	0.966	0.1621	0.867	0.1172	0.965
	$H(t)$	0.9821	0.951	1.0190	0.964	0.6566	0.987	0.4692	0.961	0.5280	0.867	0.3666	0.947
(80, 75)	α_1	0.6538	0.955	0.6768	0.978	0.5726	0.968	0.3996	0.969	0.3748	0.953	0.2913	0.962
	α_2	0.7093	0.943	0.7327	0.980	0.6289	0.965	0.4813	0.973	0.3895	0.987	0.3045	0.957
	β	0.4266	0.976	0.4394	0.981	0.3878	0.971	0.3050	0.968	0.0301	0.960	0.0296	0.967
	$S(t)$	0.2441	0.981	0.2461	0.979	0.2408	0.967	0.2378	0.971	0.1356	0.959	0.1067	0.961
	$H(t)$	0.7276	0.963	0.7456	0.994	0.6359	0.989	0.4640	0.987	0.3995	0.981	0.3126	0.948

8. Conclusions

In this study, we took into consideration progressive Type-II censoring data for the statistical inference of unknown lifespan parameters when competing failure mechanisms are present but independent. The lifespan distribution for each cause of failure is assumed to correspond to a NP distribution. The maximum likelihood and Bayesian estimate techniques are taken into consideration in order to accomplish our goal. This study was conducted on the point and approximate confidence interval estimations for the hazard rate functions, reliability, and unknown parameters. We note that in using both the SEM and EM techniques, the complexity related to the numerical computation of the MLEs may have been significantly reduced. To produce the Bayesian estimates under the squared error, LINEX, and GE loss functions, and the corresponding credible intervals, the Metropolis–Hastings method was used in the Bayesian paradigm. We analyzed a real-world dataset to illustrate the techniques presented in this article. We then carried out an extensive simulation analysis to contrast the effectiveness of different estimators. When there is no subjective information, it was found that the MLEs perform rather well. With subjective information available, the Bayesian estimators perform better than the MLEs, as predicted.

While units failing because of two competing risks are subject to progressive Type-II censoring, it should be noted that the results also apply to cases where there are multiple causes of failure and to other types of failure data cases, including complete data, Type-II censoring, and progressive first-failure censoring. The optimal design and sample plan of progressive censoring under the competing risk model appear to be interesting topics for additional investigation and will be looked into in the future.

Author Contributions: Methodology, E.A.A. and M.S.E.; software, E.A.A. and M.S.E.; validation, T.S.A.; formal analysis, E.A.A. and M.S.E.; investigation, E.A.A. and T.S.A.; resources, T.S.A.; data curation, E.A.A.; writing—review and editing, E.A.A. and M.S.E.; visualization, T.S.A. and M.S.E. All authors have read and agreed to the published version of the manuscript.

Funding: Qassim University under project number (QU-APC-2024-9/1).

Data Availability Statement: All datasets are listed within the article.

Acknowledgments: The researchers would like to thank the Deanship of Graduate Studies and Scientific Research at Qassim University for financial support (QU-APC-2024-9/1).

Conflicts of Interest: The authors declare no conflicts of interest.

References

1. Bourguignon, M.; Saulo, H.; Fernandez, R.N. A new Pareto-type distribution with applications in reliability and income data. *Phys. A Stat. Mech. Appl.* **2016**, *457*, 166–175. [CrossRef]
2. Sarabia, J.M.; Jorda, V.; Prieto, F. On a new Pareto-type distribution with applications in the study of income inequality and risk analysis. *Phys. A Stat. Mech. Appl.* **2019**, *527*, 121277. [CrossRef]
3. Abd Raof, A.S.; Haron, M.A.; Safari, M.A.M.; Siri, Z. Modeling the incomes of the upper-class group in Malaysia using new Pareto-type distribution. *Sains Malays.* **2022**, *51*, 3437–3448. [CrossRef]
4. Nik, A.S.; Asgharzadeh, A.; Nadarajah, S. Comparisons of methods of estimation for a new Pareto-type distribution. *Statistica* **2019**, *79*, 291–319.
5. Nik, A.S.; Asgharzadeh, A.; Raqab, M.Z. Estimation and prediction for a new Pareto-type distribution under progressive type-II censoring. *Math. Comput. Simul.* **2021**, *190*, 508–530.
6. Nik, A.S.; Asgharzadeh, A.; Baklizi, A. Inference based on new Pareto-type records with applications to precipitation and COVID-19 data. *Stat. Optim. Inf. Comput.* **2023**, *11*, 243–257. [CrossRef]
7. Li, F.; Wei, S.; Zhao, M. Bayesian estimation of a new Pareto-type distribution based on mixed Gibbs sampling algorithm. *Mathematics* **2023**, *12*, 18. [CrossRef]
8. Safari, M.A.M.; Masseran, N. Robust estimation techniques for the tail index of the new Pareto-type distribution. *Empir. Econ.* **2024**, *66*, 1161–1189. [CrossRef]
9. Balakrishnan, N.; Aggarwala, R. *Progressive Censoring: Theory, Methods, and Applications*; Springer Science & Business Media: Berlin/Heidelberg, Germany, 2000.
10. Mao, S.; Shi, Y.; Wang, L. Exact inference for two exponential populations with competing risks data. *J. Syst. Eng. Electron.* **2014**, *25*, 711–720.
11. Azizi, F.; Haghighi, F.; Tabibi, G.N. Statistical inference for competing risks model under progressive interval censored Weibull data. *Commun. Stat. Simul. Comput.* **2020**, *49*, 1931–1944. [CrossRef]
12. Dutta, S.; Kayal, S. Bayesian and non-Bayesian inference of Weibull lifetime model based on partially observed competing risks data under unified hybrid censoring scheme. *Qual. Reliab. Eng. Int.* **2022**, *38*, 3867–3891. [CrossRef]
13. Dutta, S.; Ng, H.K.T.; Kayal, S. Inference for a general family of inverted exponentiated distributions under unified hybrid censoring with partially observed competing risks data. *J. Comput. Appl. Math.* **2023**, *422*, 114934. [CrossRef]
14. Greene, W.H. *Econometric Analysis*, 4th ed.; Prentice-Hall: New York, NY, USA, 2000.

15. Dempster, A.P.; Laird, N.M.; Rubin, D.B. Maximum likelihood from incomplete data via the EM algorithm. *J. R. Stat. Soc. B Stat. Methodol.* **1977**, *39*, 1–22. [CrossRef]
16. Ng, H.K.T.; Chan, P.S.; Balakrishnan, N. Estimation of parameters from progressively censored data using EM algorithm. *Comput. Stat. Data Anal.* **2002**, *39*, 371–386. [CrossRef]
17. Zhang, M.; Zhisheng, Y.; Min, X. A stochastic EM algorithm for progressively censored data analysis. *Qual. Reliab. Eng. Int.* **2013**, *30*, 711–722. [CrossRef]
18. Celeux, G.; Diebolt, J. The SEM algorithm: A probabilistic teacher algorithm derived from the EM algorithm for the mixture problem. *Comput. Stat. Quar.* **1985**, *2*, 73–82.
19. Cariou, C.; Chehdi, K. Unsupervised texture segmentation/classification using 2-d autoregressive modeling and the stochastic expectation-maximization algorithm. *Pattern Recognit. Lett.* **2008**, *29*, 905–917. [CrossRef]
20. Louis, T.A. Finding the observed information matrix using the EM algorithm. *J. R. Stat. Soc. B Stat. Methodol.* **1982**, *44*, 226–233. [CrossRef]
21. Metropolis, N.; Rosenbluth, A.; Rosenbluth, M.; Teller, A.; Teller, E. Equations of state calculations by fast computing machines. *J. Chem. Phys.* **1953**, *21*, 1087–1092. [CrossRef]
22. Hastings, W.K. Monte Carlo sampling methods using Markov chains and their applications. *Biometrika* **1970**, *57*, 97–109. [CrossRef]
23. Lawless, J.F. *Statistical Models and Methods for Lifetime Data*, 2nd ed.; John Wiley and Sons: New York, NY, USA, 2003.
24. Park, C.; Kulasekera, K.B. Parametric inference of incomplete data with competing risks among several groups. *IEEE Trans. Reliab.* **2004**, *53*, 11–21. [CrossRef]
25. Balakrishnan, N.; Kateri, M. On the maximum likelihood estimation of parameters of Weibull distribution based on complete and censored data. *Stat. Probab. Lett.* **2008**, *78*, 2971–2975. [CrossRef]

mathematics

Article

The Efficiency of Hazard Rate Preservation Method for Generating Discrete Rayleigh–Lindley Distribution

Hanan Haj Ahmad [1,2]

[1] Department of Basic Science, The General Administration of Preparatory Year, King Faisal University, Hofuf 31982, Al-Ahsa, Saudi Arabia; hhajahmed@kfu.edu.sa
[2] Department of Mathematics and Statistics, College of Science, King Faisal University, Hofuf 31982, Al-Ahsa, Saudi Arabia

Abstract: In this study, we introduce two novel discrete counterparts for the Rayleigh–Lindley mixture, constructed through the application of survival and hazard rate preservation techniques. These two-parameter discrete models demonstrate exceptional adaptability across various data types, including skewed, symmetric, and monotonic datasets. Statistical analyses were conducted using maximum likelihood estimation and Bayesian approaches to assess these models. The Bayesian analysis, in particular, was implemented with the squared error and LINEX loss functions, incorporating a modified Lwin Prior distribution for parameter estimation. Through simulation studies and numerical methods, we evaluated the estimators' performance and compared the effectiveness of the two discrete adaptations of the Rayleigh–Lindley distribution. The simulations reveal that Bayesian methods are especially effective in this setting due to their flexibility and adaptability. They provide more precise and dependable estimates for the discrete Rayleigh–Lindley model, especially when using the hazard rate preservation method. This method is a compelling alternative to the traditional survival discretization approach, showcasing its significant potential in enhancing model accuracy and applicability. Furthermore, two real data sets are analyzed to assess the performance of each analog.

Keywords: discretization methods; hazard rate; maximum likelihood; Bayesian inference; simulation; Monte Carlo Markov Chain

MSC: 62E10; 62F15; 62N05; 60E05; 62P30

Citation: Ahmad, H.H. The Efficiency of Hazard Rate Preservation Method for Generating Discrete Rayleigh–Lindley Distribution. *Mathematics* **2024**, *12*, 1261. https://doi.org/10.3390/math12081261

Academic Editor: Manuel Alberto M. Ferreira

Received: 21 March 2024
Revised: 16 April 2024
Accepted: 20 April 2024
Published: 22 April 2024

1. Introduction

With every passing day, the data available in our world are growing rapidly, requiring the development of new statistical distributions to create more accurate representations of various phenomena and experiments being examined. Although most lifetime data appear continuous, the reality is that these are discrete observations, promoting the search for more suitable techniques to convert continuous distribution into discrete forms that more closely align with the data of interest. There are multiple motivations for frequently employing discrete distributions in statistical modeling.

Discrete distributions model data that assume a countable or finite set of numbers, like the number of units being tested, the tally of people in a queue, the occurrence of tails in flipping a coin, or the count of failed products in the manufacturing process. These distributions are particularly straightforward and interpretable because they represent data that adopt a specific range of values. The probability mass function (pmf)and the probability generating function (pgf) associated with discrete distribution are basic functions specifying the likelihood of each potential result. Moreover, discrete distributions often come with closed-form formulas for their pmf or pgf, facilitating their mathematical handling and enabling efficient calculation of probabilities and statistical moments without needing to resort to integration. Additionally, discrete distributions are versatile in modeling numerous real-life scenarios, such as species distribution within ecosystems, genetic variation within populations, or traffic flow across networks. Hence, they are computationally efficient, as their pmf or pgf can be used to compute probabilities and moments without having to integrate over an interval.

Recently, different discrete models have been created, mainly in medical, engineering, reliability, and survival analysis, among others. For detailed insights and employment for discrete distributions, one might consult references [1,2] in addition to other sources. Consequently, numerous researchers have extensively contributed to the creation and enhancement of discrete reliability theory from diverse perspectives. The analysis of continuous random variables commonly employs a range of techniques, including the probability density function (PDF), cumulative distribution function (CDF), moments, and hazard rate functions, along with other methods. To convert these continuous models into discrete ones, several discretization techniques have been proposed and documented in scholarly articles, aiming to establish a suitable discrete distribution that mirrors the continuous model. Different discretization methods appeared in the literature, see for example, Refs. [3–5] that provide a review of several discretization methods.

Commonly, researchers adopt a widely recognized discretization method based on the survival function. In references [6,7], the discrete analogs of normal and the Rayleigh distributions were presented, respectively, with the authors employing the survival technique for the discretization method. Following this methodology, the discrete version of Burr Type-II distribution was explored in [8]. Also, Ref. [9] discussed the discrete additive analog of the Weibull distribution. Ref. [10] discretized the half-logistic distribution and employed it in a reliability and risk analysis. One may refer also to [11–17] for more examples of discrete versions of the distributions.

Haj Ahmad and Almetwally [18] used three different discretization methods for the generalized Pareto distribution, the results look stimulating and there is motivation to continue using them in this field. Still, there is an enduring necessity to refine existing discrete models and develop new ones for better representation and fitting of big data that appear and spread constantly in everyday human life.

In this paper, we apply two distinct discretization techniques to convert the continuous Rayleigh–Lindley distribution (RLD) into a discrete form.

- The survival discretization method: The advantage of employing the survival discretization technique lies in its ability to preserve the statistical characteristics of the basic distribution, such as the median and percentiles, alongside the distribution's general shape. However, a limitation of this approach is its computational demand, often necessitating the use of numerical methods to handle complicated distributions.
- The hazard rate preservation method: This technique is designed to maintain the hazard function's structure when transitioning from a continuous to a discrete setting. The hazard function, which represents the instant rate of failure at any given time, is crucial for understanding the likelihood of an event occurring at a specific moment, provided the event has not yet occurred. One of the primary benefits of this method is its ability to closely replicate the behavior of the original continuous distribution in a discrete framework. This is particularly valuable in reliability engineering and survival analysis, where the timing of events is critical. A limitation of this approach is that it may require substantial computational resources, especially for complex distributions or when high precision is needed. This may limit its applicability in real-time or resource-constrained environments.

Efficiency in discretization methods is defined by how effectively these methods can transform continuous data into discrete forms while ensuring accuracy and retaining the usefulness of the data with minimal information loss. There are several ways to measure the efficiency of discretization methods, depending on the specific application and the type of data being discretized. Some common measures of efficiency include information loss, number of intervals, discretization error, and robustness, among others. The efficiency in this work is examined through the idea of minimizing the bias and mean squared error of the estimated parameters; this is established for each discrete analog of the RLD.

This study aims to achieve several key objectives. Firstly, to present two new discrete versions of the continuous RLD and explore their characteristics. Secondly, to conduct inferential statistics for the parameters of these newly generated distributions and assess the estimation performance. Thirdly, to evaluate the effectiveness of the new discrete distributions by examining the bias and mean square error (MSE) of the estimators through simulations and numerical methods, including the Monte Carlo technique, and finally, to use real data examples as illustrative examples of the applications of the discrete analog of RLD.

The novelty of this research stems from the fact that the hazard preservation discretization method has been virtually unused by researchers. Hence, we will explore the two analogs, examining how the frequentist and Bayesian estimation techniques perform when determining the point estimation for the parameters of the proposed discrete distributions. Ultimately, our goal is to identify which analog demonstrates greater efficiency in reducing bias and mean squared error (MSE) within the estimation framework

The remainder of this paper is structured as follows: Model descriptions and the discretization methods are detailed in Section 2. Section 3 is dedicated to evaluating some statistical functions for both analogs. In Section 4, the maximum likelihood estimation is carried out, while Section 5 delves into Bayesian inference. Simulation analysis, results, and discussions are provided in Section 6. Section 7 showcases real data analysis, and finally, concluding remarks are offered in Section 8.

2. Rayleigh–Lindley Distribution and Methods of Discretization

The Rayleigh–Lindley Distribution (RLD) is a continuous distribution that builds upon the foundations of both the Rayleigh and Lindley distributions. The statistical characteristics, inferential statistics, and reliability analysis of the RLD have been thoroughly investigated by Haj Ahmad et al. [19]. This distribution offers several advantages over the original distributions and numerous others. Additionally, previous research has shown that the Rayleigh–Lindley Distribution (RLD) is more effective at handling datasets with smaller values than larger ones. Moreover, as the hazard rate increases, the Rayleigh–Lindley model demonstrates superior performance in fitting data sets from the engineering field when compared to the Weibull, Lindley, Rayleigh, Burr X, and Power Lindley distributions. However, its continuous nature restricts its applicability for originally discrete datasets. By discretizing the RLD, we obtain a new distribution that can handle count data while retaining the RLD's capacity for tail modeling. This paper introduces two discrete analogs of the RLD.

The probability density function (pdf) for the continuous RLD is given as follows

$$f(x;\alpha,\theta)=\frac{\theta^2}{\alpha^2}(\theta+1)(x+1)e^{\theta x}\left[\frac{e^{\theta x}(\theta+1)-(1+\theta+\theta x)}{(1+\theta+\theta x)^3}\right]\exp\left[\frac{-1}{2\alpha^2}\left(\frac{(1+\theta)e^{\theta x}}{(1+\theta+\theta x)}-1\right)^2\right],x>0, \tag{1}$$

and the survival function (S) is given by

$$S(x;\alpha,\theta)=exp\left[\frac{-1}{2\alpha^2}\left(\frac{(1+\theta)e^{\theta x}}{(1+\theta+\theta x)}-1\right)^2\right], \tag{2}$$

in which $\theta,\alpha>0$ are the scale parameters.

The hazard rate function for the RLD is

$$h(x;\alpha,\theta)=\frac{\theta^2}{\alpha^2}(\theta+1)(x+1)e^{\theta x}\left[\frac{e^{\theta x}(\theta+1)-(1+\theta+\theta x)}{(1+\theta+\theta x)^3}\right]. \tag{3}$$

In this study, our objective is to establish a new discrete version of the RLD, leading to the creation of two discrete analogs. The first analog is derived using the survival discretization method and is referred to as DRLD1. The second discrete analog is derived using the hazard preservation method and is denoted by DLRD2. The pmf and the CDF of each distribution and their properties are presented in the following subsections.

2.1. The Method of Survival Discretization

Roy [6,7] introduced the pmf for a discrete distribution, utilizing the survival function for its definition, and expressed it in the following manner:

$$P(X=k)=S(k)-S(k+1), \qquad k=0,1,2,\ldots \tag{4}$$

with $S(x)$ denoting the survival function given by Equation (2); therefore, the pmf for the first Discrete Rayleigh–Lindley distribution analog (DRLD1) is given by:

$$P(k) = e^{-\omega(\alpha,\theta,k)} - e^{-\omega(\alpha,\theta,k+1)} \tag{5}$$

where $\omega(\alpha,\theta,i) = \frac{1}{2\alpha^2}\left[\frac{(1+\theta)e^{\theta i}}{1+\theta+\theta i} - 1\right]^2$.

The DRLD1 distribution under the survival discretization method has a CDF

$$F(k)=1-e^{-\omega(\alpha,\theta,k+1)} \tag{6}$$

The hazard rate function for the DRLD1 is given by

$$h_{DRLD1}(k) = exp[-\omega(\alpha,\theta,k) + \omega(\alpha,\theta,k+1)] - 1$$

Figures 1 and 2 illustrate the behavior of the pmf and the hazard rate function, respectively, for DRLD1 with different parameter values.

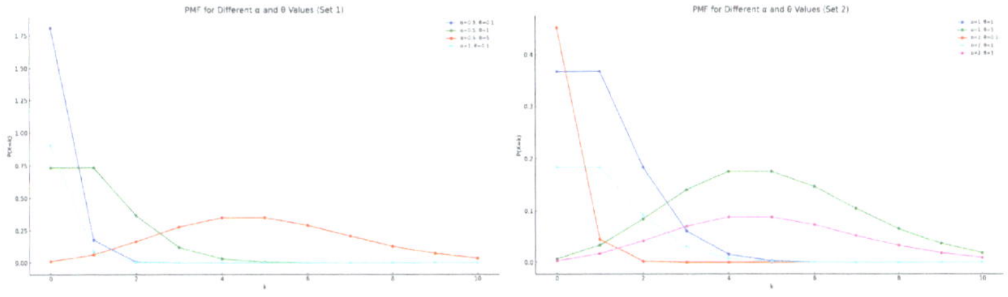

Figure 1. Graphs for the pmf of the DRLD1 distribution with various parameter values of α and θ.

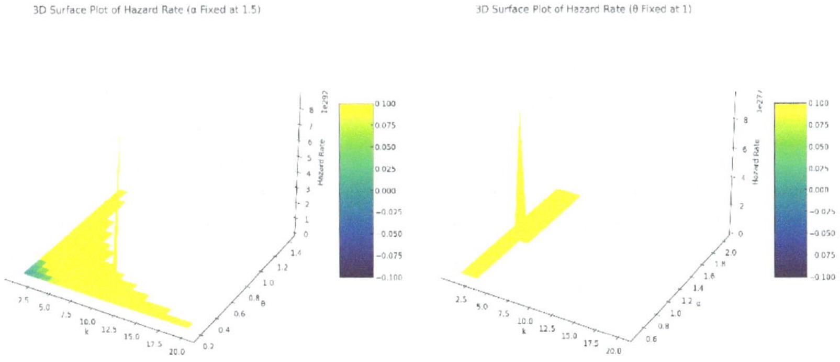

(**a**) (**b**)

Figure 2. 3D surface plot of hazard rate of DRLD1 distribution. (**a**) $\alpha = 1.5$, (**b**) $\theta = 1$.

The limiting behavior of DRLD1 at the boundary points are:

$\lim_{k \to \infty} P(k) = 0$
$\lim_{\alpha \to \infty} P(k) = 0$
$\lim_{k \to 0} P(k) = 1 - e^{-\omega(\alpha,\theta,1)}$
$\lim_{\theta \to \infty} P(k) = 0$

From the limiting behavior and Figure 1, we can summarize the effect of the parameters on the pmf of DRLD1 as follows:

- Effect of α: As the value of α increases, the pmf curves tend to flatten, indicating a broader spread of the probabilities across different values of k. This suggests that a larger α parameter will reduce the rate at which probabilities decay, leading to a more uniform distribution of the probability mass over the range of k. It highlights α's role in controlling the dispersion of the distribution.
- Effect of θ: The parameter θ affects the shape and skewness of the pmf curves. For a fixed α, varying θ alters how quickly the probabilities decrease as k increases. Higher values of θ tend to produce curves that drop off more sharply. This effect might be due to the exponential terms involving θ in the ω function, affecting the exponential decay rate of the pmf .

In summary, the parameter α primarily influences the spread or variation of the distribution, with higher values leading to a flatter pmf curve. On the other hand, θ plays a crucial role in determining the distribution's shape and how the probability mass is concentrated across different k values, with higher values leading to a more pronounced decay in probabilities as k increases.

This analysis illustrates the importance of these parameters in shaping the behavior of the distribution and highlights the flexibility of the DRLD1 model in accommodating various probability distributions based on the choice of parameter values.

In Figure 2, the hazard rate function is plotted with different parameter values for DRLD1, from which we can illustrate the behavior of the hazard function. The effect of the parameters on the hazard rate is presented as follows:

- Hazard rate with $\theta = 1$: With a fixed θ, the increasing values of α tend to modulate the hazard rate's sensitivity to changes in k. Specifically, lower values of α yield steeper curves, indicating a higher hazard rate change rate over k. Conversely, higher α values result in more gradual curves, suggesting a slower change in the hazard rate over k. This style highlights α's influence on spreading the risk over time, with higher values smoothing the rate of change in the hazard rate.
- Hazard rate with $\alpha = 1$: Keeping α constant, the variation in θ values reveals distinct trends in the hazard rate's evolution. Lower θ values produce relatively flat curves, indicating a more uniform hazard rate across k. As θ increases, the curves show a sharper descent, underscoring a rapid decrease in the hazard rate after an initial peak. This behavior showcases θ's role in determining the hazard rate's peak and subsequent decline, with higher values accelerating the peak's onset.

These findings illuminate the effects of α and θ on the DRLD1's hazard rate. α acts as a dispersion control, affecting the pace at which the hazard rate changes over time. θ influences the distribution's skewness and the rapidity of the hazard rate's peak, affecting how quickly the probability of an event occurring decreases after reaching a certain point. The utilization of distinct colors for each parameter combination in the plots not only aids in visual discrimination but also in comprehensively understanding the accurate impact of α and θ on the hazard rate's behavior, providing valuable insights for modeling and interpreting the dynamics of events described by the DRLD1.

2.2. Hazard Preservation Method

This method maintains the integrity of the hazard rate function through a two-step process. Initially, assume X is a continuous random variable with CDF $F(x)$, ranges over the interval $[0, \infty)$, and is utilized to create X_1, a new continuous random variable. This new variable X_1 is characterized by a hazard rate function $h_{X_1}(x) = e^{-F(x)}$, $(x \geq 0)$. A comprehensive understanding of this methodology is referred to in [5], which serves as an excellent resource. The discrete analog Y has a survival function that is described as follows:

$$P(Y \geq k) = (1 - h_{X_1}(1))(1 - h_{X_1}(2)) \ldots (1 - h_{X_1}(k-1)),\ k = 1, 2, \ldots, m \tag{7}$$

The pmf is written by

$$P(Y = k) = \begin{cases} h_{X_1}(0), & k = 0, \\ (1 - h_{X_1}(1))(1 - h_{X_1}(2)) \ldots (1 - h_{X_1}(k-1)) h_{X_1}(k), & k = 1, 2, \ldots, m, \\ 0, & otherwise. \end{cases} \tag{8}$$

It is important to highlight that the domain of Y corresponds to the value of m (which may not necessarily be finite), so it is chosen to ensure the condition $0 \leq h(y) \leq 1$ is fulfilled.

For X_1 we define its hazard rate as

$$h_{X_1}(y) = \exp\left[e^{\left[\frac{-1}{2\alpha^2} \left(\frac{(1+\theta)e^{\theta y}}{(1+\theta+\theta y)} - 1 \right)^2 \right]} - 1 \right],$$

It is obvious that the condition $0 \leq h(y) \leq 1$ holds. The survival function in Equation (7) for the DRLD2 is written as

$$P(Y \geq k) = \prod_{i=1}^{k-1} \left(1 - exp\left(e^{-\omega(\alpha,\theta,i)} - 1 \right) \right). \tag{9}$$

Hence, the CDF is

$$P(Y < k) = 1 - \prod_{i=1}^{k-1} \left(1 - exp\left(e^{-\omega(\alpha,\theta,i)} - 1 \right) \right) \tag{10}$$

The pmf is written as

$$P(Y = k) = \begin{cases} 1, & k = 0 \\ \exp\left[e^{-\omega(\alpha,\theta,k)} - 1 \right] \prod_{i=1}^{k-1} (1 - \exp\left[e^{-\omega(\alpha,\theta,i)} - 1 \right]), & k = 1, 2, \ldots, m \end{cases} \tag{11}$$

The hazard rate for the DRLD2 is given by

$$h_{DRLD2}(k) = \exp\left[e^{-\omega(\alpha,\theta,k)} - 1 \right], \quad k = 1, 2, \ldots, m \tag{12}$$

It is clear from Equation (12) that the hazard rate is a decreasing function for k.

Figures 3 and 4 show the behavior of the pmf and the hazard rate function respectively of the DRLD2 for different parameter values.

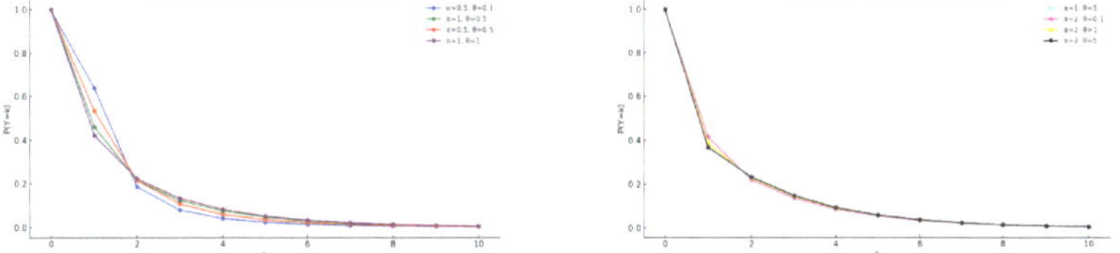

Figure 3. Graphs for the pmf of the DRLD2 with various parameter values of α and θ.

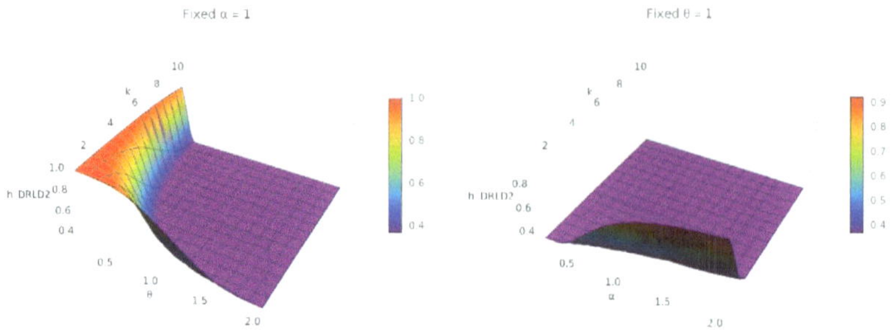

(**a**) (**b**)

Figure 4. 3D surface plot of hazard rate of DRLD2 distribution. (**a**) $\alpha = 1$, (**b**) $\theta = 1$.

For the pmf of DRLD2 it can be realized from Figure 3 that for different values of the parameters, the pmf decreases, The influence of α and θ is summarized as follows:

- As k increases from 1 onwards, the probability $P(Y = k)$ shows a decreasing trend for all combinations of α and θ. However, the rate of decrease and the pattern of the probabilities vary significantly with different values of α and θ. This variation illustrates how these parameters modulate the distribution, affecting both the likelihood of higher k values and the distribution's tail.
- The decay pattern of $P(Y = k)$ as k increases suggests that the distribution's tail becomes thinner or heavier depending on the values of α and θ. For some parameter combinations, the probability decreases more sharply, indicating a thinner tail. In contrast, other combinations show a more gradual decrease, suggesting a heavier tail and hence a higher probability of larger k values.
- Comparing curves of different colors (each representing a unique combination of α and θ) indicates that higher values of α and/or θ generally result in a quicker drop-off in the probability as k increases. This suggests that larger values of these parameters make higher k values less likely, potentially due to the increased spread or dispersion introduced by α and the rate of decrease in probability mass with k influenced by θ.
- The product term in the pmf for $k > 0$ accumulates the effect of all previous k values, introducing a dependence that shapes the overall distribution. The gradual decrease for $k > 1$ highlights the cumulative impact of preceding probabilities, emphasizing the distribution's memory of past values. This effect is particularly noticeable in distributions where the probabilities do not drop to near-zero immediately, illustrating the balance between the likelihood of consecutive events.

In summary, the plot and the behavior of $P(Y = k)$ underscore the critical roles of α and θ in determining the distribution's characteristics. The parameters not only influence the initial probabilities but also significantly impact the distribution's long-tail behavior, with implications for how likely higher k values are under different conditions.

The unimodality property of the DRLD1 and DRLD2 distribution, as well as the decreasing hazard rate curve, are consistent with the characteristics of the continuous RL distribution, see [19].

Figure 4 represents the hazard rate behavior for the DRLD2 under different parameter values. In plot (a) where α is fixed at 1, the hazard rate is plotted against θ and k. The color gradient represents the magnitude of the hazard rate, with red being higher and blue being lower. This plot shows a more pronounced curve in the surface as θ increases, indicating that the hazard rate is more sensitive to changes in θ than to changes in α from the next plot (b). As θ increases, for a given value of k, the hazard rate decreases, suggesting that the parameter θ has an inverse relationship with the hazard rate.

In plot (b) where θ is fixed at 1, the hazard rate is shown as a function of α and k. Again, the color gradient from red to blue indicates a decrease in the hazard rate value.

The surface plot shows that as k increases, the hazard rate decreases smoothly without any abrupt changes. For a fixed k, as α increases, the hazard rate decreases as well, which can be observed from the gradient of the surface.

The behavior of the hazard rate in these plots can point out the reliability of a system, where a lower hazard rate suggests a lesser likelihood of failure over time. The exact interpretation would depend on the context of the parameters α, θ, and k, which could represent physical properties or design parameters in an engineering system.

Understanding these relationships can help in designing systems with desired reliability characteristics or in making predictions about system longevity or failure rates. For a more detailed analysis, it would be necessary to know the specific context and definitions of these parameters. Consequently, finding the estimated values of these parameters will lead to a better understanding and prediction of the system's reliability and failure times. Hence, our next step is to use statistical inference to observe classical and Bayesian estimations for the model parameters.

3. Statistical Functions

In this section, statistical functions such as Quantile, moments, skewness, kurtosis, and ordered statistics are discussed for both discrete analogs of RLD.

3.1. Quantile Function

Due to the complexity of the CDF of DRLD1 and DRLD2, isolating k in the expression of $\omega(\alpha, \theta, k)$ analytically is non-trivial and likely not possible to be exact due to the nature of the expression involving both exponential and rational terms in k. Instead of an exact analytical expression, one can use approximations or numerical methods for practical applications. See Table 1 for some quantile values.

Table 1. Statistics for DLRD1 samples.

(α, θ)	Q_1	Median	Q_3	Mean	Variance	Skewness	Kurtosis	Range
(0.30, 0.80)	0.00	0.00	0.00	0.1906	0.1543	1.5755	3.4821	0–1
(1.50, 0.50)	2.00	3.00	4.00	3.0135	1.1724	−0.3520	2.8274	0–6
(1.00, 0.50)	2.00	2.00	3.00	2.3541	0.8954	−0.2713	2.7926	0–5
(0.20, 0.40)	0.00	1.00	1.00	0.8416	0.3669	0.0978	2.6259	0–3
(0.60, 0.60)	1.00	1.00	2.00	1.2112	0.4552	−0.0750	2.5224	0–3

3.2. Moments

Moments are important statistical functions. They provide comprehensive information about the shape and characteristics of a probability distribution and have many applications in quality control, risk management, and environmental studies among others. To find the moments for the DRLD1, assume two non-negative random variables $k \sim DRLD1(\alpha, \theta)$, and $l \sim DRLD2(\alpha, \theta)$. The sth moment, say μ'_s and $\mu^{*\prime}_s$ for DRLD1 and DRLD2 can be expressed, respectively, as follows:

$$\mu'_s = \sum_{k=0}^{\infty} k^s \left[e^{-\omega(\alpha,\theta,k)} - e^{-\omega(\alpha,\theta,k+1)} \right]. \tag{13}$$

and

$$\mu^{*\prime}_s = \sum_{l=0}^{\infty} l^s P(Y = l). \tag{14}$$

where $P(Y = l)$ is defined by Equation (11). An exact expression for the sth moment cannot be derived, therefore the Matlab (R2023a) software is useful for numerically evaluating the moment. Tables 1 and 2 explore some functions like the mean, variance, skewness (SK), and kurtosis (Kt) for different values of α and θ for DRLD1 and DRLD2, respectively. It can be noticed that the DRLD1 distribution is appropriate for modeling under-dispersed data since in this model the variance is smaller than the mean, which is the case with some standard classical discrete distributions. In addition, the positive and negative skewness values show that this distribution can be skewed to the right or left. Also, a minimal skew value that tends to zero indicates a possible symmetry curve for the pmf. The statistics for the DRLD2 indicate the suitability of this distribution to model both over and under-dispersed data since the variance can be greater and less than the mean. For different parameter values, the skewness can be positive and negative and some values are small enough to ensure a symmetric pattern of the pmf. A higher kurtosis is an indicator of substantial tail risk and are potential outliers compared to a normal distribution. One can realize the distribution changes by varying θ and α.

Table 2. Statistics for DLRD2 samples.

(α, θ)	Q_1	Median	Q_3	Mean	Variance	Skewness	Kurtosis	Range
(0.30, 0.80)	0.00	1.00	2.00	1.25	3.58	2.29	9.67	0–15
(1.50, 0.50)	0.00	1.00	1.00	0.51	0.27	0.16	1.59	0–3
(1.00, 0.50)	0.00	0.00	1.00	0.51	0.29	0.76	7.56	0–7
(0.20, 0.40)	0.00	1.00	1.00	0.82	1.80	3.58	21.75	0–16
(2.00, 3.00)	0.00	1.00	2.00	1.36	3.91	2.14	9.18	0–17
(3.50, 0.50)	0.00	1.00	1.00	0.51	0.25	−0.01	1.07	0–2
(4.00, 0.50)	0.00	1.00	1.00	0.50	0.25	0.01	1.06	0–2

3.3. Order Statistics

Let $Z_1, Z_2, \ldots, Z_n$ be a random sample with the DRLD1 and $Z_{1:n}, Z_{2:n}, \ldots, Z_{n:n}$ denote the corresponding order statistics. Then, the *CDF* of *ith* order statistics at the value z can be written as follows

$$F_{i:n}(z;\alpha,\theta) = \sum_{i=1}^{n} \binom{n}{m} [F_i(z;\alpha,\theta)]^m [1 - F_i(z;\alpha,\theta)]^{n-m}. \tag{15}$$

By using the negative Binomial theorem, we have

$$F_{i:n}(z;\alpha,\theta) = \sum_{i=1}^{n} \sum_{j=1}^{n-m} \binom{n}{m} \binom{n-m}{j} (-1)^j [F_i(z;\alpha,\theta)]^{m+j}. \tag{16}$$

Therefore,

$$F_{i:n}(z;\alpha,\theta) = \sum_{i=1}^{n} \sum_{j=1}^{n-m} \binom{n}{m} \binom{n-m}{j} (-1)^j \left[1 - e^{-\omega(\alpha,\theta,k+1)}\right]^{m+j}. \tag{17}$$

Consequently, the pmf of the ith order statistic under the DRLD1 can be derived and expressed as follows

$$f_{i:n}(z;\alpha,\theta) = \sum_{i=1}^{n} \sum_{j=1}^{n-m} \binom{n}{m} \binom{n-m}{j} (-1)^j \left[e^{-\omega(\alpha,\theta,k)} - e^{-\omega(\alpha,\theta,k+1)}\right]^{m+j}.$$

So, the *rth* moments of $z_{i:n}$ can be written as follows

$$E(Z_{i:n}^r) = \sum_{z=0}^{\infty} \sum_{i=1}^{n} \sum_{j=1}^{n-m} \binom{n}{m} \binom{n-m}{j} (-1)^j z^r \left[e^{-\omega(\alpha,\theta,k)} - e^{-\omega(\alpha,\theta,k+1)}\right]^{m+j}.$$

In a similar argument, the order statistics under the DRLD2 can be obtained by using Equations (10) and (11).

4. Maximum Likelihood Estimation

In this part of the study, we calculate the undetermined parameters for both versions of the DRLD distribution by applying the Maximum Likelihood Estimation (MLE) approach. To determine the required estimators, we use numerical methods, specifically adopting the well-known Newton–Raphson method for the numerical computation.

Let $x_1, \ldots, x_n$ represent a random sampling from DRLD1. From the pmf in Equation (5), the log likelihood function is given by:

$$\ell(\alpha,\theta) = \sum_{k=1}^{n} \log(e^{-\omega(\alpha,\theta,k)} - e^{-\omega(\alpha,\theta,k+1)})$$

The Maximum Likelihood Estimators (MLEs) for the parameters α and θ are derived by calculating the partial derivatives of the likelihood function $\ell(\alpha,\theta)$ for α and θ, respectively. These equations are then set to zero, and the resulting system of equations is solved numerically to obtain the estimations.

$$\frac{\partial \ell(\alpha,\theta)}{\partial \alpha} = \sum_{k=1}^{n} \frac{-\omega_\alpha(\alpha,\theta,k)e^{-\omega(\alpha,\theta,k)} + \omega_\alpha(\alpha,\theta,k+1)e^{-\omega(\alpha,\theta,k+1)}}{e^{-\omega(\alpha,\theta,k)} - e^{-\omega(\alpha,\theta,k+1)}} = 0 \tag{18}$$

$$\frac{\partial \ell(\alpha,\theta)}{\partial \theta} = \sum_{k=1}^{n} \frac{-\omega_\theta(\alpha,\theta,k)e^{-\omega(\alpha,\theta,k)} + \omega_\theta(\alpha,\theta,k+1)e^{-\omega(\alpha,\theta,k+1)}}{e^{-\omega(\alpha,\theta,k)} - e^{-\omega(\alpha,\theta,k+1)}} = 0,$$

Such that $\omega_\alpha(\alpha,\theta,k) = \frac{\partial \omega(\alpha,\theta,k)}{\partial \alpha} = -\frac{2}{\alpha}\omega(\alpha,\theta,k)$ and

$$\omega_\theta(\alpha,\theta,k) = \frac{\partial \omega(\alpha,\theta,k)}{\partial \theta} = \frac{\theta k e^{\theta k}}{\alpha^2}\left[\frac{(1+\theta)e^{\theta k}}{1+\theta+\theta k} - 1\right]\left[\frac{1}{1+\theta+\theta k} + \frac{1+k}{(1+\theta+\theta k)^2}\right].$$

Similarly, the MLEs of α and θ can be evaluated under DRLD2, in this case, the log-likelihood function can be written depending on the pmf in Equation (11) as follows:

$$\mathcal{L}(\alpha,\theta) = \sum_{k=1}^{n} log\left(\exp\left[e^{-\omega(\alpha,\theta,k)} - 1\right]\prod_{i=1}^{k-1}(1-\exp\left[e^{-\omega(\alpha,\theta,i)} - 1\right])\right)$$

$$\mathcal{L}(\alpha,\theta) = \sum_{k=1}^{n}\left(e^{-\omega(\alpha,\theta,k)} - 1\right) + \sum_{k=1}^{n}\sum_{i=1}^{k-1} log\left[1 - \exp\left(e^{-\omega(\alpha,\theta,i)} - 1\right)\right]$$

Therefore, the partial derivatives of $\mathcal{L}(\alpha,\theta)$ to α and θ are

$$\frac{\partial \mathcal{L}(\alpha,\theta)}{\partial \alpha} = \sum_{k=1}^{n} -\omega_\alpha(\alpha,\theta,k)e^{-\omega(\alpha,\theta,k)} + \sum_{k=1}^{n}\sum_{i=1}^{k-1} \frac{\omega_\alpha(\alpha,\theta,i)\exp\left(e^{-\omega(\alpha,\theta,i)} - 1\right)e^{-\omega(\alpha,\theta,i)}}{1 - \exp(e^{-\omega(\alpha,\theta,i)} - 1)} = 0, \tag{19}$$

$$\frac{\partial \mathcal{L}(\alpha,\theta)}{\partial \theta} = \sum_{k=1}^{n} -\omega_\theta(\alpha,\theta,k)e^{-\omega(\alpha,\theta,k)} + \sum_{k=1}^{n}\sum_{i=1}^{k-1} \frac{\omega_\theta(\alpha,\theta,i)\exp\left(e^{-\omega(\alpha,\theta,i)} - 1\right)e^{-\omega(\alpha,\theta,i)}}{1 - \exp(e^{-\omega(\alpha,\theta,i)} - 1)} = 0.$$

Solving the system of two nonlinear Equations (18) and (19) can only be done numerically. Numerous numerical methods have been employed in the literature; in this case, we are utilizing the Newton–Raphson method. The discussion results are presented in Section 6.

5. Bayesian Inference

In this section, we employ the Bayesian approach to determine the unknown parameters of the two discrete RL distributions. The Bayesian technique assumes that the parameters of the model are random variables adhering to a distribution known as the prior distribution. Often, prior information is not readily available, necessitating the selection of an appropriate prior. In this study, we opt for a joint conjugate prior distribution for the parameters α and θ, referred to as the modified Lwin Prior. This prior is specified by assigning a Gamma distribution to α and a Pareto (I) distribution to θ. Consequently,

$$\alpha \sim \text{Gamma}(a_1, b_1) \text{ and } \theta|\alpha \sim \text{Pareto(I)}(\alpha a_2, b_2)$$

where a_1, a_2, b_1, and b_2 are non-negative hyperparameters of the assumed distributions. Reference [20] highlighted that expressing θ conditional on α holds more significance than the reverse. Furthermore, they advocate that it is more pertinent to assume the prior distributions for α and θ as independent.

Thus, the prior distributions for α and θ are presented as follows:

$$\pi_1(\alpha) = \frac{{b_1}^{a_1}}{\Gamma(a_1)} \alpha^{a_1 - 1} e^{-b_1 \alpha},$$

$$\pi_2(\theta|\alpha) = \frac{\alpha a_2}{\theta b_2} \left(\frac{\theta}{b_2}\right)^{-a_2 \alpha}.$$

Therefore, the joint prior function for α and θ is

$$\pi(\alpha, \theta) \propto \frac{\alpha^{a_1}}{\theta} e^{-b_1 \alpha} \left(\frac{\theta}{b_2}\right)^{-a_2 \alpha} \tag{20}$$

The joint posterior of α and θ under condition of data availability is given as

$$p(\alpha, \theta|\underline{x}) = \frac{1}{K} L(\underline{x}/\alpha, \theta) \pi(\alpha, \theta)$$

where $L(\underline{x}/\alpha, \theta)$ is the likelihood function of the DRLD, $\pi(\alpha, \theta)$ is the joint prior given by Equation (20), and $K = \iint L(\underline{x}/\alpha, \theta)\pi(\alpha, \theta) d\alpha d\theta$.

The process of estimating the parameters for the DRLD distribution has been examined through the use of both symmetric squared error (SE) and asymmetric LINEX loss functions. An evaluation of how well the estimators perform under these loss functions was conducted via a simulation study. Criteria such as the bias and mean square error (MSE) are utilized to determine the effectiveness of the estimation techniques.

The following loss functions are employed for estimating posterior functions.

(i) Squared error (SE) loss function: assuming SE loss function, Bayes estimation for the parameters α and θ are defined as the mean or expected value for the joint posterior

$$\hat{\alpha}_{SE} = \frac{1}{K} \iint \alpha L(\underline{x}/\alpha, \theta) \pi(\alpha, \theta) d\alpha d\theta$$

$$\hat{\theta}_{SE} = \frac{1}{K} \iint \theta L(\underline{x}/\alpha, \theta) \pi(\alpha, \theta) d\alpha d\theta \tag{21}$$

(ii) LINEX loss function: under LINEX loss function, estimating parameters with Bayesian method is written as

$$\hat{\alpha}_{LIN} = -\frac{1}{h} \ln\left[\frac{1}{K} \iint e^{-h\alpha} L(\underline{x}/\alpha, \theta) \pi(\alpha, \theta) d\alpha d\theta\right]$$

$$\hat{\theta}_{LIN} = -\frac{1}{h} \ln\left[\frac{1}{K} \iint e^{-h\theta} L(\underline{x}/\alpha, \theta) \pi(\alpha, \theta) d\alpha d\theta\right], \tag{22}$$

where h is the value of the shape factor and it represents the orientation of asymmetry; hence, in our study we select the values of h to be 1.5 and -1.5 in the simulation analysis.

To calculate the expected values and perform the double integration required in Equations (21) and (22), it is necessary to employ numerical approaches. We have chosen to apply the Markov Chain Monte Carlo (MCMC) strategy, specifically utilizing the Gibbs sampling method, hence, we developed an appropriate R code to facilitate this process. For additional information on this technique, interested readers can consult the reference [21].

We have to discuss two cases listed below as we developed two different discretization methods on the continuous RLD.

Case 1

Utilizing the survival discretization method results in the derivation of DRLD1, whose pmf is provided by Equation (5). The corresponding joint posterior density is as follows:

$$p_1(\alpha,\ \theta/\underline{x}) = \frac{1}{K}\prod_{i=1}^{n}\left[e^{-\omega(\alpha,\theta,i)} - e^{-\omega(\alpha,\theta,i+1)}\right]\frac{\alpha^{a_1}}{\theta}e^{-b_1\alpha}\left(\frac{\theta}{b_2}\right)^{-a_2\alpha} \tag{23}$$

$$= G_\alpha(a_1+1,b_1)\Lambda(\alpha,\theta),$$

where, $\Lambda(\alpha,\theta) = \frac{1}{K}\prod_{i=1}^{n}\left[e^{-\omega(\alpha,\theta,i)} - e^{-\omega(\alpha,\theta,i+1)}\right]\frac{\theta^{-\alpha a_2-1}}{b_2^{-\alpha a_2}}$, and $G(.,.)$ denoting the Gamma distribution.

The Bayes inference for the parameters α and θ under SE loss function is obtained using Equation (21) and the posterior density is obtained using Equation (23)

$$\widehat{\alpha}_{SE} = \frac{1}{K}\iint\prod_{i=1}^{n}\left[e^{-\omega(\alpha,\theta,i)} - e^{-\omega(\alpha,\theta,i+1)}\right]\frac{\alpha^{a_1+1}}{\theta}e^{-b_1\alpha}\left(\frac{\theta}{b_2}\right)^{-a_2\alpha}d\alpha d\theta$$

$$\widehat{\theta}_{SE} = \frac{1}{K}\iint\prod_{i=1}^{n}\left[e^{-\omega(\alpha,\theta,i)} - e^{-\omega(\alpha,\theta,i+1)}\right]\theta^{-a_2\alpha}\,\alpha^{a_1}e^{-b_1\alpha}(b_2)^{a_2\alpha}d\alpha d\theta$$

Using the LINEX loss function, Bayesian estimation is derived from Equation (22) in conjunction with the posterior density detailed in Equation (23)

$$\widehat{\alpha}_{LIN} = -\frac{1}{h}\ln\left[\frac{1}{K}\iint\prod_{i=1}^{n}\left[e^{-\omega(\alpha,\theta,i)} - e^{-\omega(\alpha,\theta,i+1)}\right]\frac{\alpha^{a_1}}{\theta}e^{-(b_1+h)\alpha}\left(\frac{\theta}{b_2}\right)^{-a_2\alpha}d\alpha d\theta\right]$$

$$\widehat{\theta}_{LIN} = -\frac{1}{h}\ln\left[\frac{1}{K}\iint\prod_{i=1}^{n}\left[e^{-\omega(\alpha,\theta,i)} - e^{-\omega(\alpha,\theta,i+1)}\right]\frac{\alpha^{a_1}}{\theta}e^{-b_1\alpha-h\theta}\left(\frac{\theta}{b_2}\right)^{-a_2\alpha}d\alpha d\theta\right]$$

Case 2

The second discretization method of RL produces DRLD2 with the pmf presented in Equation (11), so the joint posterior density is given by

$$p_2(\alpha,\theta/\underline{x}) = \frac{1}{K}\prod_{j=1}^{n}\exp\left[e^{-\omega(\alpha,\theta,j)}-1\right]\prod_{i=1}^{j-1}\left(1-\exp\left[e^{-\omega(\alpha,\theta,i)}-1\right]\right)\frac{\alpha^{a_1}}{\theta}e^{-b_1\alpha}\left(\frac{\theta}{b_2}\right)^{-a_2\alpha}$$

$$= \frac{1}{K}G_\alpha(a_1+1,b_1)\Psi(\alpha,\theta)$$

where $\Psi(\alpha,\theta) = \prod_{j=1}^{n}\exp\left[e^{-\omega(\alpha,\theta,j)}-1\right]\prod_{i=1}^{j-1}\left(1-\exp\left[e^{-\omega(\alpha,\theta,i)}-1\right]\right)\frac{\theta^{-a_2\alpha-1}}{b_2^{-a_2\alpha}}$

Bayes estimation for the parameters α and θ under SE loss function is given as

$$\widehat{\alpha}_{SE} = \frac{1}{K}\iint\prod_{j=1}^{n}\exp\left[e^{-\omega(\alpha,\theta,j)}-1\right]\prod_{i=1}^{j-1}\left(1-\exp\left[e^{-\omega(\alpha,\theta,i)}-1\right]\right)\frac{\alpha^{a_1+1}}{\theta}e^{-b_1\alpha}\left(\frac{\theta}{b_2}\right)^{-a_2\alpha}d\alpha d\theta$$

$$\widehat{\theta}_{SE} = \frac{1}{K}\iint\prod_{j=1}^{n}\exp\left[e^{-\omega(\alpha,\theta,j)}-1\right]\prod_{i=1}^{j-1}\left(1-\exp\left[e^{-\omega(\alpha,\theta,i)}-1\right]\right)\alpha^{a_1}e^{-b_1\alpha}\left(\frac{\theta}{b_2}\right)^{-a_2\alpha}d\alpha d\theta$$

With the LINEX loss function, Bayesian estimation for the parameters is obtained by:

$$\widehat{\alpha}_{LIN} = -\frac{1}{h}\ln[\frac{1}{K} \iint \prod_{j=1}^{n} \exp\left[e^{-\omega(\alpha,\theta,j)} - 1\right] \prod_{i=1}^{j-1} (1-\exp\left[e^{-\omega(\alpha,\theta,i)} - 1\right]) \frac{\alpha^{a_1}}{\theta} e^{-(b_1+h)\alpha} \left(\frac{\theta}{b_2}\right)^{-a_2\alpha} d\alpha d\theta]$$

$$\widehat{\theta}_{LIN} = -\frac{1}{h}\ln[\frac{1}{K} \iint \prod_{j=1}^{n} \exp\left[e^{-\omega(\alpha,\theta,j)} - 1\right] \prod_{i=1}^{j-1} (1-\exp\left[e^{-\omega(\alpha,\theta,j)} - 1\right]) \frac{\alpha^{a_1}}{\theta} e^{-b_1\alpha-h\theta} (\frac{\theta}{b_2})^{-a_2\alpha} d\alpha d\theta]$$

6. Simulation Analysis

Through this section, our goal is to assess how well the two discrete variants of the continuous RL distribution perform by examining the point estimation accuracy of the unknown parameters in terms of bias and MSE. Furthermore, we will compare their performance using various loss functions outlined in Section 5. We will present some noteworthy findings and outcomes after this section.

In the simulation scenario, 10,000 iterations of random samples are generated using suitable R-code. Some predetermined parameters values for α and θ are $\{0.5, 2\}$, with a sample size $n = \{50, 100, 150\}$ being considered.

The simulation analysis for estimating the parameters of the two discrete analogs of RL distribution is presented in Tables 3 and 4. Primary findings from the simulation study are summarized as follows:

Table 3. The MLE and the Bayesian inference for DRLD1 with estimation bias and MSE with various values of parameters.

				MLE		Bayes (SE)		Bayes (LINEX-1.5)		Bayes (LINEX 1.5)	
α	θ	n		Bias	MSE	Bias	MSE	Bias	MSE	Bias	MSE
0.5	0.5	50	α	0.2767	0.2983	0.3212	0.2058	0.4068	0.3078	0.2397	0.1283
			θ	0.0082	0.0132	0.0124	0.0089	0.0207	0.0093	0.0392	0.0086
		100	α	0.2356	0.2307	0.1285	0.0419	0.1434	0.0487	0.1136	0.0359
			θ	0.0363	0.0107	−0.0113	0.0045	−0.0183	0.0044	−0.0338	0.0046
		150	α	0.3009	0.0997	0.0808	0.0182	0.0866	0.0197	0.0751	0.0167
			θ	0.0337	0.0020	−0.0104	0.0038	−0.0144	0.0037	−0.0246	0.0038
	2	50	α	0.5566	0.3608	0.5221	0.5461	0.6359	0.7618	0.4130	0.3687
			θ	−0.3494	0.1234	−0.4743	0.3857	−0.4029	0.3057	−0.5439	0.4715
		100	α	0.5084	0.3463	0.4526	0.2419	0.4738	0.2654	0.4059	0.2168
			θ	−0.3299	0.1133	−0.3927	0.1791	−0.3753	0.1629	−0.4083	0.1943
		150	α	0.4642	0.3350	0.4105	0.1794	0.4233	0.1915	0.3955	0.1657
			θ	−0.3048	0.1024	−0.3731	0.1465	−0.3596	0.1355	−0.3847	0.1562
2	0.5	50	α	0.0949	0.0189	0.1408	0.2011	0.2307	0.2592	0.0498	0.1630
			θ	−0.0359	0.0015	−0.0340	0.0024	−0.0332	0.0024	−0.0349	0.0025
		100	α	0.0936	0.0175	0.0366	0.0458	0.0540	0.0493	0.0193	0.0432
			θ	−0.0325	0.0012	−0.0328	0.0018	−0.0308	0.0018	−0.0328	0.0018
		150	α	0.0828	0.0162	0.0199	0.0170	0.0263	0.0177	0.0135	0.0165
			θ	−0.0276	0.0011	−0.0309	0.0017	−0.0302	0.0017	−0.0309	0.0017
	2	50	α	0.4530	0.3211	0.3866	0.4378	0.5541	0.6896	0.2218	0.2658
			θ	−0.1637	0.0295	−0.1801	0.0399	−0.1737	0.0376	−0.1866	0.0422
		100	α	0.3739	0.3020	0.1408	0.0698	0.1632	0.0806	0.1182	0.0603
			θ	−0.1179	0.0215	−0.1206	0.0344	−0.1520	0.0335	−0.1721	0.0415
		150	α	0.2530	0.2695	0.0899	0.0274	0.0982	0.0298	0.0815	0.0251
			θ	−0.1195	0.0176	−0.1209	0.0314	−0.1421	0.0308	−0.1621	0.0405

Table 4. The MLE and the Bayesian inference for DRLD2 with estimation bias and MSE with various values of parameters.

				MLE		Bayes (SE)		Bayes (LINEX-1.5)		Bayes (LINEX 1.5)	
α	θ	n		Bias	MSE	Bias	MSE	Bias	MSE	Bias	MSE
0.5	0.5	50	α	0.2661	0.2635	0.3323	0.2177	0.4237	0.3308	0.2455	0.1332
			θ	0.0026	0.0145	0.0151	0.0089	0.0236	0.0094	0.0063	0.0085
		100	α	0.2373	0.2410	0.1272	0.0432	0.1426	0.0501	0.1119	0.0371
			θ	0.0377	0.0125	−0.0133	0.0048	−0.0131	0.0047	−0.0054	0.0050
		150	α	0.3020	0.0999	0.0801	0.0185	0.0860	0.0201	0.0742	0.0169
			θ	0.0339	0.0020	−0.0125	0.0020	−0.0114	0.0038	−0.0046	0.0040
	2	50	α	0.5574	0.3163	0.4552	0.2611	0.4168	0.1873	0.2431	0.1400
			θ	−0.3524	0.1254	−0.3046	0.1185	−0.2390	0.1031	−0.2154	0.0947
		100	α	0.2587	0.2349	0.2426	0.2183	0.2045	0.1892	0.2040	0.1297
			θ	−0.3340	0.1162	−0.3042	0.1020	−0.2402	0.1018	−0.2044	0.0922
		150	α	0.1574	0.2035	0.1421	0.1750	0.1417	0.1486	0.1390	0.1136
			θ	−0.3046	0.1023	−0.2380	0.0915	−0.1307	0.0901	−0.0939	0.0816
2	0.5	50	α	0.0541	0.0109	−0.0601	0.0101	0.0914	0.0103	−0.0666	0.0101
			θ	−0.0363	0.0015	−0.0391	0.0013	−0.0395	0.0013	−0.0393	0.0013
		100	α	0.0496	0.0102	0.0479	0.0033	−0.0536	0.0035	−0.0564	0.0031
			θ	−0.0330	0.0012	−0.0361	0.0009	−0.0390	0.0009	−0.0385	0.0009
		150	α	0.0317	0.0092	−0.0398	0.0014	−0.0231	0.0014	−0.0449	0.0013
			θ	−0.0288	0.0012	−0.0358	0.0008	−0.0380	0.0008	−0.0369	0.0009
	2	50	α	0.0541	0.0109	0.0431	0.0070	0.0555	0.0114	0.0305	0.0037
			θ	−0.0363	0.0015	−0.0215	0.0013	−0.0206	0.0012	−0.0223	0.0014
		100	α	0.0496	0.0102	0.0322	0.0014	0.0352	0.0017	0.0291	0.0012
			θ	−0.0330	0.0012	−0.0224	0.0012	−0.0218	0.0011	−0.0229	0.0012
		150	α	0.0317	0.0092	0.0214	0.0013	0.0225	0.0014	0.0201	0.0011
			θ	−0.0288	0.0012	−0.0244	0.0012	−0.0234	0.0011	−0.0252	0.0013

- It is evident that the estimated parameter values approach the true values as the sample size increases. This is indicated by the reduction in both MSE and bias with larger sample sizes, demonstrating the consistency of the proposed estimators.
- When working with small sample sizes, Bayesian estimation with LINEX loss function yields the lowest MSE and bias for estimating the parameter θ. In contrast, the SE loss function produces the smallest MSE and bias for estimating α.
- For big sample sizes, LINEX loss function consistently achieves the lowest MSE and bias for the two parameters α and θ.
- For both parameters α and θ, the Bayesian methods generally show a different bias and MSE pattern compared to MLE. Specifically, the Bayesian SE method tends to have lower MSE than MLE in many cases, suggesting that it might provide more accurate and reliable estimates under certain conditions. For nearly all scenarios, both the LINEX and SE loss functions result in the lowest bias and MSE values across various sample sizes.
- The LINEX penalties introduce variability in the performance, with LINEX-1.5 generally resulting in higher bias and MSE for α, especially when $\alpha = 2$, suggesting a sensitivity to the loss function's shape.
- The performance of the estimation methods varies significantly between the two parameter settings α and θ = 0.5 vs. 2. For instance when α and θ are both set to 2, the bias and MSE are generally higher compared to when they are set to 0.5. This suggests that the true values of the parameters can significantly affect the difficulty of the estimation problem.

When comparing the performance of the estimation methods for the parameters of the two DGPD analogs, several insights emerge regarding the performance of these methods across different conditions. Below are some general observations.

- Across both distributions, the Bayesian methods, particularly with the Standard Error (SE) approach, often show a lower MSE compared to the MLE, suggesting that in the context of these simulations, Bayesian methods might offer a more robust approach under certain conditions.
- The bias for parameter α in DRLD2 seems to have less variability across different methods and conditions compared to DRLD1. For example, in DRLD2, the bias values for α are generally closer to zero, especially in the Bayesian SE and LINEX (-1.5) scenarios, indicating potentially more accurate estimations. For parameter θ, the bias is also generally lower in DRLD2, suggesting that the estimation methods may perform better on this distribution for θ.
- The MSE values for both α and θ tend to be lower in DRLD2 across most methods and conditions, indicating a more precise estimation. This is particularly evident in the Bayesian SE and LINEX (-1.5) methods, where the improvement in MSE is clear.
- The impact of increasing sample size on improving bias and MSE appears to be more consistent in DRLD2 than in DRLD1, especially for the Bayesian methods. This suggests that DRLD2 may be more amenable to these estimation techniques as the sample size increases.
- The Bayesian methods, especially with SE and LINEX (-1.5), show a notable improvement in DRLD2 over DRLD1 in terms of both bias and MSE. This could be indicative of the Bayesian methods being particularly well-suited for the characteristics of DRLD2.

The comparison between the two tables highlights that DRLD2 generally allows for more accurate parameter estimation than DRLD1, as evidenced by lower bias and MSE across various methods and conditions. The improvement is particularly noticeable with Bayesian estimation methods, suggesting that these methods may be more effective for distributions with characteristics similar to DRLD2. This could be due to differences in the underlying properties of the two distributions, such as their sensitivity to sample size and the specific challenges they present for parameter estimation.

7. Real Data Examples

This section presents the analysis using a real dataset. The main goal of this section is to examine the usefulness and applicability of the proposed discrete analogs to real phenomena. The first dataset consists of the number of fires that occurred in Greece between 1 July and 31 August 1998. We only take into account fires in forest districts. We considered a sample of data with a size of 24. The minimum value is 1, the first quartile is 4, the median value is 7.5, the mean value is 6.88, the third quartile is 9, the maximum value is 12, and the variance value is 8.9. The data are as follows:

Dataset I: 4, 3, 10, 5, 4, 5, 12, 3, 8, 10, 11, 6, 1, 8, 9, 9, 4, 8, 11, 8, 6, 4, 7, 9.

These data hae been discussed by [22]. We apply the Chi-square goodness of fit measure for testing DRLD1 with the above set of data, the results are explored in Table 5. The observed p-value indicates the suitability of DRLD1 to fit these data. Additionally, Figure 5 illustrates the connection between observed probability distribution with the expected one, as well as the empirical CDF with the expected CDF plot, and finally the Q-Q plot.

Table 5. MLE estimate and chi-square measure for dataset I.

	Parameters	MLE	p-Value	Test Statistics
DRLD1	α	0.1684	0.8366	0
	θ	0.0849		

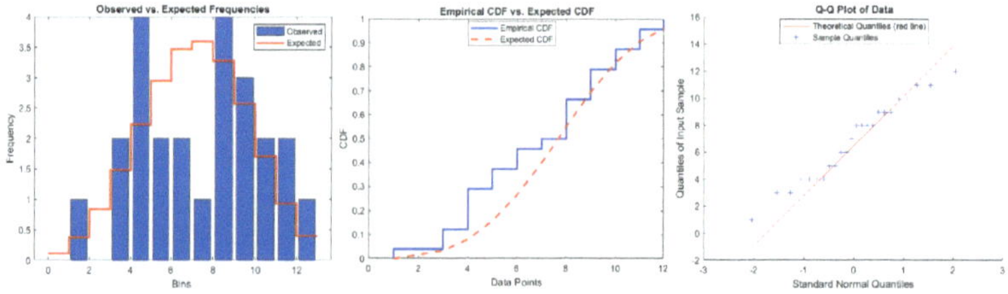

Figure 5. P-P, Empirical CDF VS estimated CDF, and Q-Q plots for dataset I with DRLD1.

The second dataset represents a 25-day COVID-19 data set from the United States Virgin Islands, recorded in May 2021. These data comprise daily new deaths. The data are as follows:

Dataset II: 4, 12, 3, 5, 12, 6, 9, 13, 10, 26, 32, 13, 10, 20, 18, 2, 18, 14, 24, 7, 30, 16, 26, 17, 23. The data are available on the Worldometer website at [23]. Applying the Chi-square goodness of fit test to assess the appropriateness of the DRLD2 distribution for this dataset indicates that this model is relatively well-suited for analyzing these data. The results are detailed in Table 6. Additionally, Figure 6 displays the P-P plot, the empirical and estimated cumulative distribution functions, and the Q-Q plot.

Table 6. MLE estimate and chi-square measure for dataset II.

	Parameters	MLE	p-Value	Test Statistics
DRLD2	α θ	0.1098 0.0305	0.0853	0

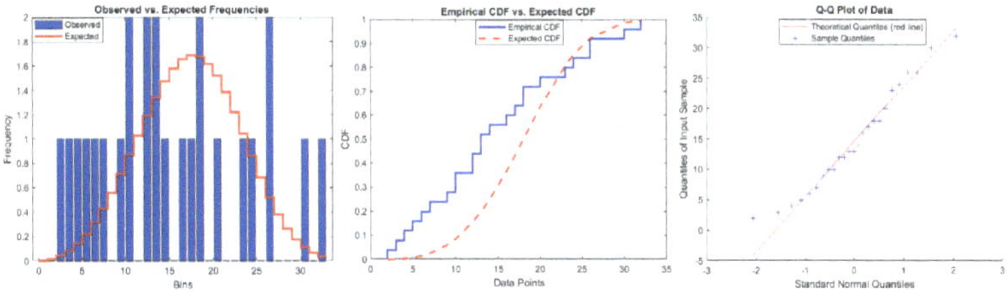

Figure 6. P-P, Empirical CDF VS estimated CDF and Q-Q plots for dataset II with DRLD2.

8. Conclusions

Discrete distributions are a natural choice for modeling data that are limited to a finite or countably infinite set of values, due to their simplicity, closed-form expressions, and ability to model real-world phenomena. They are also computationally efficient and can be used to model categorical data. In this study, the author developed two new discrete analogs of the Raleigh–Lindley distribution. Their statistical properties are discussed, then estimation methods are applied to assess the performance of estimation methods for the two analogs. The simulation study illustrates the performance of MLE and Bayesian methods in estimating DRLD parameters. The choice of estimation method and the specification of the Bayesian loss function can significantly impact the bias and MSE of the estimates. These findings underscore the importance of considering the specific context of the parameter estimation problem, including the sample size and the true parameter values, when selecting an estimation approach. It was obtained that the new hazard preservation method enhances the performance of estimation methods, this is especially evident in the Bayesian estimation approaches, indicating that these techniques may be better suited for distributions that share characteristics with DRLD2. This distinction could stem from the unique attributes of the two distributions, including how they respond to changes in sample size and the particular obstacles they pose for estimating parameters. Finally, two real data examples from the environment and health fields are examined to assess the performance of each analog.

Funding: This work was supported by the Deanship of Scientific Research, Vice Presidency for Graduate Studies and Scientific Research, King Faisal University, Saudi Arabi, Grant No. [GrantA096].

Data Availability Statement: Data are contained within the article

Conflicts of Interest: The author declares no conflicts of interest.

References

1. Xekalaki, E. Hazard function and life distributions in discrete time. *Commun. Stat. Theory Methods* **1983**, *12*, 2503–2509. [CrossRef]
2. Roy, D.; Ghosh, T. A new discretization approach with application in reliability estimation. *IEEE. Trans. Reliab.* **2009**, *58*, 456–461. [CrossRef]
3. Bracquemond, C.; Gaudoin, O. A survey on discrete life time distributions. *Int. J. Reliabil. Qual. Saf. Eng.* **2003**, *10*, 69–98. [CrossRef]
4. Lai, C.D. Issues concerning constructions of discrete lifetime models. *Qual. Technol. Quant. Manag.* **2013**, *10*, 251–262. [CrossRef]
5. Chakraborty, S. Generating discrete analogues of continuous probability distributions—A survey of methods and constructions. *J. Stat. Distrib. Appl.* **2015**, *2*, 6. [CrossRef]
6. Roy, D. The discrete normal distribution. *Commun. Stat. Theor. Methods* **2003**, *32*, 1871–1883. [CrossRef]
7. Roy, D. Discrete Rayleigh distribution. *IEEE. Trans. Reliab.* **2004**, *53*, 255–260. [CrossRef]

8. Al-Huniti, A.A.; Al-Dayjan, G.R. Discrete Burr type III distribution. *Am. J. Math. Stat.* **2012**, *2*, 145–152. [CrossRef]
9. Bebbington, M.; Lai, C.D.; Wellington, M.; Zitikis, R. The discrete additive Weibull distribution: A bathtub-shaped hazard for discontinuous failure data. *Reliab. Eng. Syst. Saf.* **2012**, *106*, 37–44. [CrossRef]
10. Barbiero, A.; Hitaj, A. Discrete half-logistic distributions with applications in reliability and risk analysis. *Ann. Oper. Res.* **2024**, 1–31. [CrossRef]
11. Sarhan, A.M. A two-parameter discrete distribution with a bathtub hazard shape. *Commun. Stat. Appl. Methods* **2017**, *24*, 15–27. [CrossRef]
12. Yari, G.; Tondpour, Z. Discrete Burr XII-Gamma Distributions: Properties and Parameter Estimations. *Iran. J. Sci. Technol. Trans. Sci.* **2017**, *42*, 2237–2249. [CrossRef]
13. Almetwally, E.M.; Ibrahim, G.M. Discrete Alpha Power Inverse Lomax Distribution with Application of COVID-19 Data. *Int. J. Appl. Math.* **2020**, *9*, 11–22.
14. Eliwa, M.S.; Altun, E.; El-Dawoody, M.; El-Morshedy, M. A new three-parameter discrete distribution with associated INAR process and applications. *IEEE Access* **2020**, *8*, 91150–91162. [CrossRef]
15. Al-Babtain, A.; Hadi, A.; Ahmed, N.; Afify, A.Z. A New Discrete Analog of the Continuous Lindley Distribution, with Reliability Applications. *Entropy* **2020**, *22*, 603. [CrossRef] [PubMed]
16. Eldeeb, A.S.; Ahsan-Ul-Haq, M.; Babar, A. A Discrete Analog of Inverted Topp-Leone Distribution: Properties, Estimation and Applications. *Int. J. Anal. Appl.* **2021**, *19*, 695–708.
17. Haj Ahmad, H.; Ramadan, D.A.; Almetwally, E.M. Evaluating the Discrete Generalized Rayleigh Distribution: Statistical Inferences and Applications to Real Data Analysis. *Mathematics* **2024**, *12*, 183. [CrossRef]
18. Ahmad, H.H.; Almetwally, E.M. Generating optimal discrete analogue of the generalized Pareto distribution under Bayesian inference with application. *Symmetry* **2022**, *14*, 1457. [CrossRef]
19. Haj Ahmad, H.; Bdair, O.M.; Naser, M.F.M.; Asgharzadeh, A. The rayleigh lindley distribution: A new generalization of rayleigh distribution with physical applications. *Rev. Investig. Oper.* **2023**, *44*, 1–18.
20. Arnold, B.C.; Press, S.J. Compatible Conditional Distributions. *J. Am. Stat. Assoc.* **1989**, *84*, 152. [CrossRef]
21. Karandikar, R.L. On the markov chain monte carlo (MCMC) method. *Sadhana* **2006**, *31*, 81–104. [CrossRef]
22. Karlis, D.; Xekalaki, E.; Lipitakis, E.A. On some discrete valued time series models based on mixtures and thinning. In Proceedings of the Fifth Hellenic-European Conference on Computer Mathematics and Its Applications, Athens, Greece, 20–22 September 2001; pp. 872–877.
23. Worldometers. Available online: https://www.worldometers.info/coronavirus (accessed on 1 June 2021).

Σ mathematics

Article

Revisited Bayesian Sequential Indicator Simulation: Using a Log-Linear Pooling Approach

Nasser Madani

School of Mining and Geosciences, Nazarbayev University, Astana 010000, Kazakhstan; nasser.madani@nu.edu.kz; Tel.: +7-7471755327

Abstract: It has been more than a decade since sequential indicator simulation was proposed to model geological features. Due to its simplicity and easiness of implementation, the algorithm attracts the practitioner's attention and is rapidly becoming available through commercial software programs for modeling mineral deposits, oil reservoirs, and groundwater resources. However, when the algorithm only uses hard conditioning data, its inadequacy to model the long-range geological features has always been a research debate in geostatistical contexts. To circumvent this difficulty, one or several pieces of soft information can be introduced into the simulation process to assist in reproducing such large-scale settings. An alternative format of Bayesian sequential indicator simulation is developed in this work that integrates a log-linear pooling approach by using the aggregation of probabilities that are reported by two sources of information, hard and soft data. The novelty of this revisited Bayesian technique is that it allows the incorporation of several influences of hard and soft data in the simulation process by assigning the weights to their probabilities. In this procedure, the conditional probability of soft data can be directly estimated from hard conditioning data and then be employed with its corresponding weight of influence to update the weighted conditional portability that is simulated from the same hard conditioning and previously simulated data in a sequential manner. To test the algorithm, a 2D synthetic case study is presented. The findings showed that the resulting maps obtained from the proposed revisited Bayesian sequential indicator simulation approach outperform other techniques in terms of reproduction of long-range geological features while keeping its consistency with other expected local and global statistical measures.

Keywords: Bayesian updating; sequential indicator simulation; kriging; probability aggregation

MSC: 62C10

Citation: Madani, N. Revisited Bayesian Sequential Indicator Simulation: Using a Log-Linear Pooling Approach. *Mathematics* **2022**, *10*, 4669. https://doi.org/10.3390/math10244669

Academic Editor: Diana Mindrila

Received: 28 October 2022
Accepted: 3 December 2022
Published: 9 December 2022

1. Introduction

High-quality probabilistic modeling of categorical variables (e.g., geological domains based on lithological description) is a vital demand in many geoscience disciplines, as it provides the basement for modeling the continuous variables (e.g., grade of elements and properties of rock) throughout the domain of study. This modeling process offers several applications in mineral resource estimation [1,2], statistic reservoir modeling [3,4], and modeling of aquifers in hydrogeology [5], to name a few. To model the categorical variables, among others, truncated Gaussian simulation [6,7], plurigaussian simulation [8,9], multiple-point statistics [5], and sequential indicator simulation [10–12] received significant attention for building such categorical models. The latter, modeling of a continuous variable, is more popular due to its simplicity and straightforwardness, and the algorithm itself is available in several commercial software programs. The resulting models are acceptable where there are no large-scale geological patterns. However, in the case of having curvilinear or long-range geological features, the algorithm produces very patchy and unstructured results, which is a legitimate criticism of this method [13,14]. The reason is that sequential indicator simulation only takes into account the two-point statistical measures of the geological domains that are informed by hard data. To solve this issue, several variations of sequential indicator simulation have been developed that consider the soft information as secondary data to inform the large-scale geological variability at the target simulation nodes, which substantially improves the applicability of the method. In this context, hard and soft data refer to measured values at the borehole via the sampling points and interpretive geological models at target grid nodes, respectively.

A possible solution to include secondary/soft information in the process of simulation is to use the probability aggregation concept [15]. A proper evaluation of different probability aggregation methods with special attention to their applicability using geoscientific data is conducted by Allard et al. [16]. They showed that the log-linear pooling outperforms the linear pooling when integrating them with the truncated Gaussian and the Boolean models in geostatistics. The integration of log-linear pooling in several geostatistical algorithms has already been developed for various applications: interpolation of satellite images [17], 3D multiple-point statistics using 2D training images [18], multiple-point geostatistical simulations using secondary exhaustive information [19], and Bayesian sequential Gaussian simulation for modeling the continuous hydrogeophysical data [20].

Probability aggregation is also employed specifically in sequential indicator simulation. Among others, Bayesian sequential indicator simulation [21] was introduced to integrate a simple Bayesian updating rule to construct a local posterior distribution of the geological domains at the target grid node. This work was developed to model the lithofacies in static reservoirs where the soft data can be informed by seismic attributes for the different lithofacies. Another form of Bayes' law was introduced by Ortiz and Deutsch [22], where the indicator formalism in this sequential simulation technique is updated by using the multiple-point configuration of the soft data. The method is proposed to simulate the continuous variable in the context of mineral resource estimation, where the soft conditional probability is advised by using the production data such as blasthole. However, in these methods, the hard and soft data use equal constant weights, meaning identical influence on the final simulation results. To overcome this limitation, a revised version of Bayesian sequential indicator simulation is proposed in this study for the purpose of modeling the categorical variables (e.g., geological domains) that are integrated with a log-linear pooling approach. The soft information in this enhanced process of simulation is inferred from an interpolation technique using only the hard conditioning data at the sampling point.

The outline of the paper is as follows. In Section 2, the methodology is explained with its main mathematical properties. Then, the proposed approach is tested and validated through a synthetic case study in Section 3. Finally, Section 4 provides the discussion and conclusions.

2. Sequential Indicator Simulation

2.1. Conventional Sequential Indicator Simulation

Sequential indicator simulation [10–12] is a stochastic methodology for modeling M categories, for which they are exhaustive and mutually exclusive at all data locations. This means that one of the categories must predominate at each individual sample point.

In order to perform a sequential indicator simulation algorithm, first, the categorical variable T_h through the hard conditioning data h at sample locations χ is transformed into a matrix of M hard indicator variables, characterized as:

$$I(\chi;\mu) = \begin{cases} 1, \textit{ if category M predominates at sample location } \chi \\ 0, \textit{ otherwise} \end{cases} \qquad \mu = 1,\ldots,\mathrm{M} \quad (1)$$

In this equation, predominating category M at location χ that is converted to indicators necessitates that $\sum_{\mu=1}^{m} I(\chi;\mu) = 1$.

In the second step, a random path is identified to visit each node of the target grid only once. The next step is to estimate the corresponding category at randomly selected target node grid node χ', using the indicators that are obtained through Equation (1) at hard conditioning data. There are different methodologies in geostatistics that can be utilized for this purpose. Among others, the simple kriging method provides promising results. This geostatistical interpolation paradigm is built based on three constraints [23]: (a) the estimator is a weighted linear combination of the available hard data (linearity constraint), (b) the estimator is unbiased, that is, the expectation of the error is 0 (unbiasedness constraint), and (c) the error variance is minimum (optimality constraint). The first restriction necessitates writing the estimator as a linear combination (weighted average) of the neighboring hard data to estimate the variable of interest $T_h(\chi)$ at target grid node χ':

$$T_h^*(\chi') = \sum_{\beta=1}^{v} \omega_\beta^{SK}(\chi')T(\chi_\beta) + \left[1 - \sum_{\beta=1}^{v} \omega_\beta^{SK}(\chi')\right] \times m \quad (2)$$

where data v are composed of neighboring hard conditional data; m is the global declustered mean value of the corresponding variable $T_h(\chi)$, $\omega_\beta^{SK}(\chi';\mu)$ is the weights assigned to the variable T at the location $\chi_\beta(\beta = 1,\ldots,v)$. The weights needed in Equation (2) are achievable by solving a covariance matrix for each χ' as:

$$\sum_{\beta=1}^{v} \omega_\beta^{SK} C(\chi_\alpha - \chi_\beta) = C(\chi_\alpha - \chi') \quad \alpha = 1,\ldots,v \quad (3)$$

where C is the covariance of variable T. Solving the kriging system in Equation (3) to obtain the weights ω_β^{SK} requires the knowledge of covariance of variable T. In this respect, the lineal model of regionalization is widely used to fit such covariances, owing to its mathematical simplicity and tractability [9,23]. In this model, the covariance C is defined as weighted sum of L basic covariances, also called basic nested structures:

$$C(h) = \sum_{l=0}^{L} b^l C_l(h) \quad (4)$$

where for each structure $(l = 1,\ldots,L)$, b^l is the positive sill of basic permissible covariance model $C_l(h)$.

Therefore, one can use the simple kriging paradigm to estimate the corresponding category μ at the target grid node. Since the initial data is already converted to indicators, then the aim of simple kriging in this step is to establish the conditional probability of occurrence of each indicator $I(\chi;\mu)$ at the corresponding target grid node χ':

$$I_h^*(\chi';\mu) = \sum_{\beta=1}^{v} \omega_\beta^{SIK}(\chi';\mu)I(\chi_\beta;\mu) + \left[1 - \sum_{\beta=1}^{v} \omega_\beta^{SIK}(\chi';\mu)\right] \times \pi_\mu \quad (5)$$

where data v are composed of neighboring hard conditional indicators and previously simulated indicator values; the π_μ is the global declustered proportion of each category (i.e., prior probability or prior proportion), which equivalently can be defined as $E\{I(\chi;\mu)\}$; $\omega_\beta^{SIK}(\chi';\mu)$ is the weights assigned to the indicator variable $I(\chi_\beta;\mu)$ at the $\chi_\beta(\beta = 1,\ldots,v)$ of this indicator variable. The weights needed in Equation (5) are achievable by solving a covariance matrix for each χ' using Equation (3). Then, the permissible model of $C(h)$ can be inferred from the direct calculation of covariance over each indicator variable $I(\chi;\mu)$ or it can be deduced from the calculation of variogram over indicator variable $I(\chi;\mu)$. In this case, the variogram can then be converted to covariance to be embedded into Equation (3) for computing the corresponding weights $\omega_\beta^{SIK}(\chi';\mu)$.

Since each M category is independently estimated, an order relation deviation is expected for the estimated conditional probabilities. This signifies that the probabilities do not always sum to one, and the existence of negative probabilities is expected. The reason is that some of the weights obtained by solving the simple kriging system via Equation (3) receive negative or high values. These particular weights lead to producing some negative values or very high values for estimated indicators, which are more than 1 [24–26]. Then, in the fourth step, the deviations in probabilities for order relation should be rectified [26]. In this respect, one can utilize the bounds correction, signifying that all negative estimated values are set to 0, and all high values (more than 1) are set to 1. Then, a normalization of probabilities is needed to force the summation of all estimated conditional probabilities to reach 1 at all target grid nodes $\sum_{\mu=1}^{m} I(\chi'; \mu) = 1$ [23]. Once the probabilities of each indicator are estimated and their order related problems are corrected, then one can define any ordering of the M categories and build a cumulative distribution function (cdf-type function). Afterward, the fifth step includes drawing a random number U from a uniform distribution in $[0,1]$ by Monte Carlo simulation. Therefore, the simulated category $T_h^{(1)}$ at location χ' can be obtained with an inverse of the quantile associated with that generated random number in $[0,1]$. In the sixth step, the simulated value is added to the hard conditioning data, and then the algorithm proceeds to the next target node, following the identified random path, repeating steps from one to five. In order to generate another realization $T_h^{(i)}$, $i = 2, \ldots, r$, with r total number of realizations, one needs to repeat the entire algorithm with a different random path. These simulated values are obtained by only hard conditioning data- h. However, the current methodology is poor in reproduction of connectivity or large-scale geological features between sample points. In the following, we show how soft information can be incorporated to simulate the values at target nodes, enforcing the connectivity in the outcomes.

2.2. Revisited Bayesian Sequential Indicator Simulation

If the secondary evidence, known as soft information, is available at the target node χ', the conditional probability of occurrence of each category μ obtained from Equation (5) can be updated accordingly. Let us denote event Ψ, a category that needs to be simulated at target grid node, and $\Delta_i (i = 1, \ldots, n)$ represent n information of that underlying category, for which they are inferred by using the hard conditioning data, soft data, or any other source of information. Indeed, we aim at approximating the probability $P(\Psi|\Delta_1, \ldots, \Delta_n)$ based upon the coexisting information of the n conditional probabilities. Probability aggregation [16] is used to construct an approximation of the true conditional probability by utilizing an aggregation operator Ω, which is also known as the pooling operator or pooling formula:

$$P(\Psi|\Delta_1, \ldots, \Delta_n) \approx \Omega(P(\Psi|\Delta_1), \ldots, P(\Psi|\Delta_n)) \tag{6}$$

In the case of having a prior probability Δ_0 for the category sought, then Equation (6) can be generalized to:

$$P(\Psi|\Delta_0, \ldots, \Delta_n) \approx \Omega(P(\Psi|\Delta_0), \ldots, P(\Psi|\Delta_n)) \tag{7}$$

Among others, aggregation operators based on the multiplication of probabilities appear to be more suited in geoscience than those formulated based on an addition operator (Allard et al. 2012). These product-based pooling operators are linear operators of the logarithms of the probabilities [16]:

$$\ln P(\Psi|\Delta) = \ln Z + \sum_{i=1}^{n} \lambda_i \ln P(\Psi|\Delta_i) \tag{8}$$

where Z is a normalizing constant. This expression equivalently can be converted to the multiplication of probabilities:

$$P(\Psi|\Delta_0, \ldots, \Delta_n) \propto P(\Psi|\Delta_0)^{1-\sum_{i=1}^{n} \lambda_i} \prod_{i=1}^{n} P(\Psi|\Delta_i)^{\lambda_i} \tag{9}$$

where λ_i are positive weights with restriction $\sum_{i=1}^{n} \lambda_i = 1$ to verify external Bayesianity. In Equation (9), there is no restriction on the weights.

In the context of sequential indicator simulation using the secondary data, subject to having one hard conditioning data Δ_h and one soft information Δ_s, log-linear pooling can be simplified to combine the information provided by these two sources of information:

$$P(\Psi|\Delta_0, \Delta_h, \Delta_s) \propto P(\Psi|\Delta_0)^{1-\lambda_h-\lambda_s} \cdot P(\Psi|\Delta_h)^{\lambda_h} \cdot P(\Psi|\Delta_s)^{\lambda_s} \tag{10}$$

where $P(\Psi|\Delta_0)$ is prior probability of each category π_μ; λ_h and λ_s are the weights that can be arbitrary assigned to the hard and soft data, and $\sum_{i=1}^{n} \lambda_i = 1$ is always verified.

Therefore, the conditional probability of occurrence of each category μ obtained from Equation (5) can be updated by using Equation (10) as:

$$I_U^*(\chi';\mu) = \pi_\mu^{1-\lambda_h-\lambda_s} \times I_h^*(\chi';\mu)^{\lambda_h} \times I_s^*(\chi';\mu)^{\lambda_s} \tag{11}$$

where I_S^* is soft information, for which its conditional probability of occurrence of category μ can be reported from an interpretive geological block model at the target grid node. For instance, wireframing or hand contouring can provide such a geological block model [2]. However, it is somehow tedious to produce such a probabilistic information from a deterministic interpretive model of geological domains in the region of study. There is an alternative way to convert this deterministic model to conditional probabilities, as thoroughly explained in [27]. In this study, we propose to obtain I_S^* values from simple kriging of the indicators at hard conditioning data. These estimated values provide probabilistic information of the occurrence of that category μ at the corresponding target grid node $I_s^*(\chi';\mu)$. However, to obtain $I_h^*(\chi';\mu)^{\lambda_h}$, we take both the neighboring hard conditional indicator data and also the previously simulated indicator values as a typical practice in conventional sequential indicator simulation as explained in Section 2.1. In practice, the soft information I_S^* in this probability aggregation method helps improve the conventional sequential indicator simulation for better reproduction of long-range geological structures. The most important issue in this approach is to properly assign the weights of hard λ_h and soft data λ_s. In the case of $\lambda_h = 1$ and $\lambda_s = 1$, Equation (11) simplifies to the traditional Bayesian sequential indicator simulation as proposed in [21].

2.3. *Optimal Evaluation of Weighting Mechanism*

The proposed log-linear pooling approach in Equation (10) allows one to identify different or equal weighting schemes for a specific category that are coming from different sources of information. These are prior proportion, primary (hard) and secondary (soft) data, where their weights can dictate the intensity of their influence in the final probability aggregation resulting through Equation (11). The proposed methodology in this work highly depends on the proper assignation of these weights so that their improper designation yields completely different results. These weights can be either assigned equally or ranked differently, for instance, based on the opinion of an expert. Allard et al. [16] suggested using a log-likelihood approach to optimally identify these weights. Nassbaumer et al. [20] proposed a multi-step sophisticated approach to change the weights multiple times during the simulation process following a Monte Carlo-type search. In this study, we used an empirical technique to assign the weights that gradually change in the interval of [0, 2], allowing the testing of our algorithm conveniently. The algorithm for identification of optimal weights is as follows:

1. Testing the algorithm with 24 different experiments of λ_h and λ_s as provided in Table 1.
2. Evaluating the performance of the experiments by calculating the error in the reproduction of the proportion of indicators.
3. Evaluating the performance of the experiments by calculating the error in matching percentage with the reference map.
4. Evaluating the performance of the experiments by calculating the error in the reproduction of connectivity measures of the categories.
5. Evaluating the performance of the experiments by calculating the error in the reproduction of spatial continuity of the categories.
6. Fitting polynomial function to interpolate the errors.
7. Obtaining the optimal weights based on minimum error.

Table 1. Twenty-four experiments based on different weighting schemes of λ_h and λ_s; $1-\lambda_h-\lambda_s$ introduces the weight of prior proportion for the category sought.

	λ_h	λ_s	$1-\lambda_h-\lambda_s$	Note
SIS-0–0.5	0	0.5	0.5	Revisited Bayesian simulation (Pure soft simulation)
SIS-0–2	0	1	0	Revisited Bayesian simulation (Pure soft simulation)
SIS-0–1.5	0	1.5	−0.5	Revisited Bayesian simulation (Pure soft simulation)
SIS-0–2	0	2	−1	Revisited Bayesian simulation (Pure soft simulation)
SIS-0.5–0	0.5	0	0.5	Revisited Bayesian simulation (Pure hard simulation)
SIS-0.5–0.5	0.5	0.5	0	Revisited Bayesian simulation
SIS-0.5–1	0.5	1	−0.5	Revisited Bayesian simulation
SIS-0.5–1.5	0.5	1.5	−1	Revisited Bayesian simulation
SIS-0.5–2	0.5	2	−1.5	Revisited Bayesian simulation
SIS-1–0	1	0	0	Revisited Bayesian simulation (Pure hard simulation/Traditional simulation)
SIS-1–0.5	1	0.5	−0.5	Revisited Bayesian simulation
SIS-1–1	1	1	−1	Revisited Bayesian simulation (Doyan's simulation)
SIS-1–1.5	1	1.5	−1.5	Revisited Bayesian simulation
SIS-1–2	1	2	−2	Revisited Bayesian simulation
SIS-1.5–0	1.5	0	−0.5	Revisited Bayesian simulation (Pure hard simulation)
SIS-1.5–0.5	1.5	0.5	−1	Revisited Bayesian simulation
SIS-1.5–1	1.5	1	−1.5	Revisited Bayesian simulation
SIS-1.5–1.5	1.5	1.5	−2	Revisited Bayesian simulation
SIS-1.5–2	1.5	2	−2.5	Revisited Bayesian simulation
SIS-2–0	2	0	−1	Revisited Bayesian simulation (Pure hard simulation)
SIS-2–0.5	2	0.5	−1.5	Revisited Bayesian simulation
SIS-2–1	2	1	−2	Revisited Bayesian simulation
SIS-2–1.5	2	1.5	−2.5	Revisited Bayesian simulation
SIS-2–2	2	2	−3	Revisited Bayesian simulation

In the above criteria, reproduction of proportions refers to the examination of reproduction of π_μ in the simulation results. Connectivity measures $\tau(d)$ is defined as the probability that two target grid nodes belonging to category μ are connected [28]:

$$\tau(d) = P(\chi' \Leftrightarrow \chi' + d | \chi', \chi' + d \in \mu) \tag{12}$$

where $\tau(d)$ is a non-decreasing function as d increases. The connectivity function can be estimated by:

$$\hat{\tau}(d) = \frac{\#N(\chi' \Leftrightarrow \chi' + d | \chi', \chi' + d \in \mu)}{\#N(\chi', \chi' + d \in \mu)} \tag{13}$$

where $\hat{\tau}(d)$ is the estimate of connectivity function for distance d, $\#N(\chi' \Leftrightarrow \chi' + d | \chi', \chi' + d \in \mu)$ the number of target grid nodes, separated by a vector of distance d, that belong to category μ and are connected and $\#N(\chi', \chi' + d \in \mu)$ the number of target grid nodes, deparated by a vector of distance d, that belong to category μ and that may or may not be connected.

Concerning the spatial continuity evaluation, a variogram can be a satisfying measurement. To do so, experimental variogram $\gamma(d)$ computes the average dissimilarity between data separated by vector d. It is calculated as half the average difference between components of every data pair [9,29]:

$$\gamma(d) = \frac{1}{2N(d)} \sum_{\beta=1}^{N(d)} \left[T(\chi_\beta) - T(\chi_\beta + d)\right]^2 \tag{14}$$

where $\left[T(\chi_\beta) - T(\chi_\beta + d)\right]^2$ is a d-increment of the indicator variable T and $N(d)$ is the number of pair.

Through the synthetic case study, we show how the method proposed does work and how these weights can be optimally identified.

3. Results

3.1. Synthetic Case Study

To test the proposed methodology and evaluate the optimal weighting parameters, a synthetic case study is considered. For this purpose, one categorical variable, including two indicators/categories, is non-conditionally simulated on a 2D 300 m × 300 m domain consisting of 300 × 300 nodes by using plurigaussian simulation [8] associated with an anisotropic spatial continuity. Ten realizations are produced, and the one (reference map hereafter) is selected in such a way that it shows long connectivity along easting and relatively short connectivity along elevation coordinates (Figure 1A). In order to mimic the vertical sampling pattern like the ones in the actual case studies, 100 points are randomly sampled from the reference map along elevation to constitute four synthetic boreholes throughout the domain. However, in order to evaluate the algorithm properly, two target areas are of paramount importance. These are identified by two ellipses (I and II) in Figure 1A. In fact, we are interested in producing the realizations that honor the connectivity of the categories through these two critical areas when there are few hard conditioning data points. Therefore, it was intended to sample only one category per each of these two-target areas (Figure 1B). The lack of data in these two areas can be an excellent signature for evaluating the proposed method. Consequently, this synthetic dataset is used in the proposed algorithm to compare the simulation results with a reference map to provide evaluation debates.

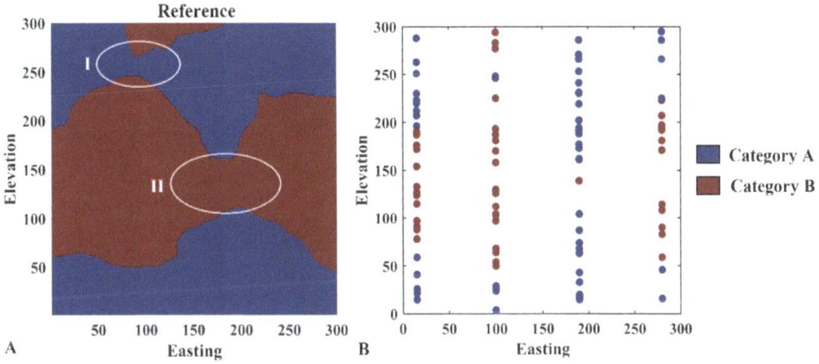

Figure 1. Reference dataset, (**A**) reference map, (**B**) synthetic sampled dataset; only one sample is preserved in the target areas I and II to better evaluate the revisited Bayesian sequential indicator simulation in this study.

The connectivity on the selected map, to be considered a reference map, is also verified by computing the connectivity functions [23] over this realization along the specified coordinates (Equation (13)).

As can be observed from Figure 2, there is approximately 50% and 0.0% probability that categories A and B are connected from west to east and from south to north, respectively. This interpretation is obtained by looking at the connectivity measures at large lag distances when they reach a steady range around the lag of 240. This is also compatible with the visual inspection of the map illustrated in Figure 1A. Therefore, this is an interesting reference map for further analysis in this study.

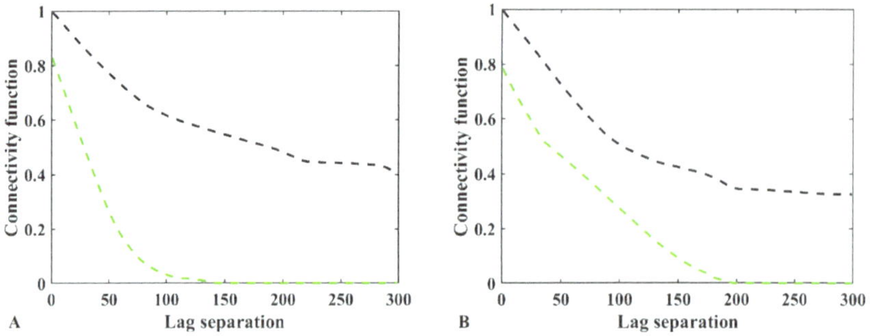

Figure 2. Connectivity measures along Easting (back dashed line) and Elevation (green dashed line) for (**A**) category A, and (**B**) category B.

Once the synthetic sampling points are identified, the next step in the proposed algorithm is to produce a map that reports the soft data at target nodes to be used subsequently for log-linear pooling probability aggregation in revisited Bayesian sequential indicator simulation through Equation (6). In this study, simple kriging is used to map the soft information for each indicator. Therefore, after variogram analysis, the categories/indicators are estimated on the target grid nodes using up to 40 conditioning data in a moving neighborhood configuration (Figure 3). Due to the existence of indicator value at synthetic sample points, these maps are deemed as the probability of occurrence of the corresponding category $P(\Psi|\Delta_s)^{\lambda_s}$ at the target node, for which they can be used as soft information I_S^* in Equation (7). One advantage of this map is that one can recognize the probability of connectivity between the boreholes, which is favorable information for modeling the categories with long connectivity.

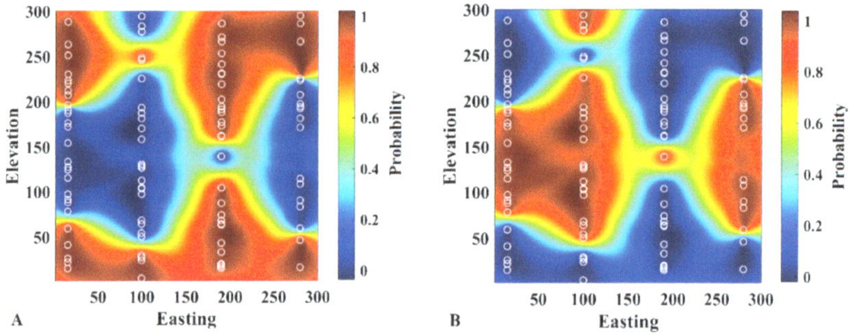

Figure 3. Probability of occurrence of (**A**) category A and (**B**) category B obtained by estimating the conditioning indicator data on the target grid nodes. These maps are used as soft information in the proposed algorithm.

The next step in the revisited Bayesian sequential indicator simulation in this study is to implement the simulation algorithm but using updating scheme of Equation (11) to update the probability of occurrence of the corresponding category $I_h^*(\chi';\mu)^{\lambda_h}$. As mentioned earlier, the updated probability $I_U^*(\chi';\mu)$ highly depends on the assignation of weights λ_h and λ_s. A method is used to test different weighting schemes in Equation (11) to evaluate the optimum weighting values in the revisited Bayesian sequential indicator simulation. In this technique, $\lambda_h \in \{0,2\}$ and $\lambda_s \in \{0.5,2\}$, and the prior proportion ranging with a weight of $1-\lambda_h-\lambda_s$. These 24 trials are shown in Table 1.

In this examination, the trials with $\lambda_h = 0$ and $\lambda_s \in \{0.5,2\}$, signify that the revisited Bayesian sequential indicator simulation is only based on the soft information (pure soft simulation hereafter). Those trials with $\lambda_s = 0$ and $\lambda_h \in \{0.5, 2\}$, denote that the resulting revisited Bayesian sequential indicator simulation is solely yielded on the hard conditioning data (pure hard simulation hereafter). The trial represents the conventional sequential indicator simulation with $\lambda_h = 1$ and $\lambda_s = 0$. The trial providing the simulation results with $\lambda_h = 1$ and $\lambda_s = 1$ presents the traditional Bayesian Updating approach as proposed in Doyen et al. [21] (Doyan's simulation hereafter). There are some links between Doyan's simulation and sequential indicator simulation that uses a collocated cokriging by different implementation mechanisms [13]. All these subgroups are some special cases of the revisited Bayesian sequential indicator simulation as proposed in this study. However, in order to be clearer about these subgroups, another subgroup is defined, which is called Revisited Bayesian simulation hereafter, for which λ_h and λ_s vary between {0, 2} and {0.5, 2}, respectively, but excluding the weights belonging to pure soft simulation, pure hard simulation, traditional simulation, and Doyan's simulation. A summary of these techniques is provided in Table 2. A note is also provided in Table 1 to link the trials in Table 1 to the subgroups in Table 2.

Table 2. Comparison of some key properties of the log-linear pooling aggregation approach in this study; weight of the prior proportion for the category sought is $1 - \lambda_h - \lambda_s$.

	Aggregation Equation for $P(\Psi \mid \Delta_0, \Delta_h, \Delta_s)$	λ_h	λ_s
Pure soft simulation	$P(\Psi\|\Delta_0)^{1-\lambda_s} \cdot P(\Psi\|\Delta_s)^{\lambda_s}$	0	$\{0.5, 2\}$
Pure hard simulation	$P(\Psi\|\Delta_0)^{1-\lambda_h} \cdot P(\Psi\|\Delta_h)^{\lambda_h}$	$\{0.5, 2\}$	0
Traditional simulation	$P(\Psi\|\Delta_0)^{1-\lambda_h-\lambda_s} \cdot P(\Psi\|\Delta_h)^{\lambda_h} \cdot P(\Psi\|\Delta_s)^{\lambda_s}$	1	0
Doyan's simulation	$P(\Psi\|\Delta_0)^{1-\lambda_h-\lambda_s} \cdot P(\Psi\|\Delta_h)^{\lambda_h} \cdot P(\Psi\|\Delta_s)^{\lambda_s}$	1	1
Revisited Bayesian simulation	$P(\Psi\|\Delta_0)^{1-\lambda_h-\lambda_s} \cdot P(\Psi\|\Delta_h)^{\lambda_h} \cdot P(\Psi\|\Delta_s)^{\lambda_s}$	$\{0, 2\}$	$\{0.5, 2\}$

Implementation of simulation for all the trials was conducted under a moving neighborhood scheme with up to 40 hard conditioning data and 40 previously simulated data. The sequential paradigm is followed a random sequence while using a multiple-grid simulation procedure. The number of realizations is 100. Some realizations from each subgroup (presented in Table 2) are selected for visualization (Figure 4). The rest of the realizations for the other trials are provided in Appendix A (Figure A1). As can be seen, the results obtained by using only pure soft information dramatically failed in reproducing the shape of both categories A and B, which do not show the underlying structure. In pure hard simulation, when $\lambda_h \leq 0.5$, the resulting maps are quite patchy (Figure 4B). However, by increasing the weight of hard data, the shape of category B, to some extent, is regenerating. It can be seen that the reproduction of connectivity is highly controlled by the weight assigned to soft data (Figure 4D,F). One of its particular cases is the one with $\lambda_h = 1$ and $\lambda_s = 0$ (Figure 4F), where it shows the result of traditional sequential indicator simulation. The shape of category B on the left and on the right sides is more or less reproduced, but the continuity of category B is disconnected around area II, shown by a white ellipse in Figure 4A. This is a typical problem of traditional sequential indicator simulation, in which the geological features with long connectivity cannot be produced properly. Doyan's simulation, which represents conventional Bayesian updating in Equation (11), could produce the general shape of category B in the places where enough conditioning data are available, but again it failed to produce the connectivity of interest in area II (Figure 4E). In these figures, the revisited Bayesian simulation with $\lambda_h = 1$ and $\lambda_s = 2$ seems to be the most promising one among others.

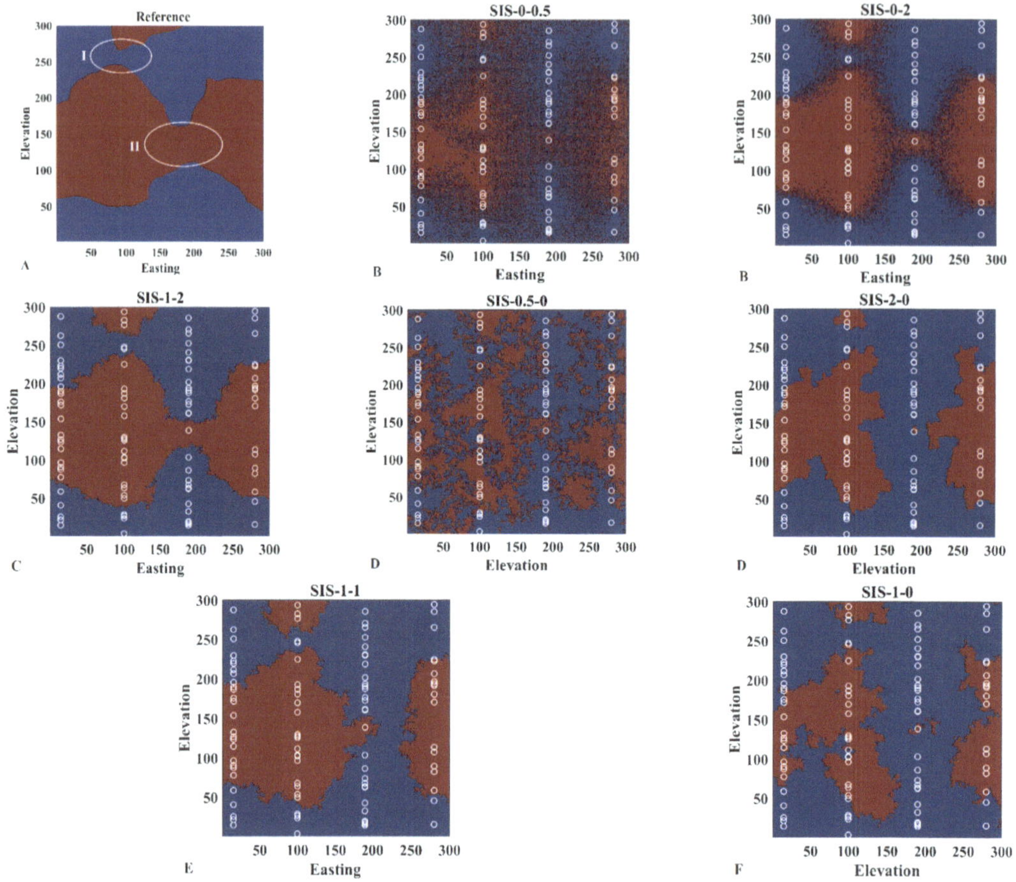

Figure 4. Comparison of (**A**) reference map with resulting simulation maps obtained from (**B**) pure soft simulation, (**C**) revisited Bayesian simulation, (**D**) pure hard simulation, (**E**) Doyan's simulation, and (**F**) traditional simulation.

3.2. Validation

However, the results depicted in Figure 4 are only one single realization of each trial, and one might be interested in examining the probability maps obtained from 100 realizations that assess the uncertainty in the categories at a local (node-by-node) scale for better validation. The maps are constructed by calculating, for each grid node, the frequency of occurrence of category B over 100 conditional realizations (Figure 5). They constitute a complement to the reference map insofar as they show the risk of finding a category different from the one that has been expected. The sectors with little uncertainty are those associated with a high probability for a given category (painted in red in Figure 5), indicating that there is little risk of not finding category B, or those associated with a very low probability (painted in dark blue in Figure 5), indicating that one is pretty sure of not finding this category, while the other sectors (painted in light blue, green or yellow in Figure 5) are more uncertain. As can be observed, the revisited Bayesian simulation method with $\lambda_h = 1$ and $\lambda_s = 2$ provides the most prominent results to give the high probability of connectivity close to areas I and II, as we expect from the reference map. The rest of the probability maps are presented in Appendix B (Figure A2).

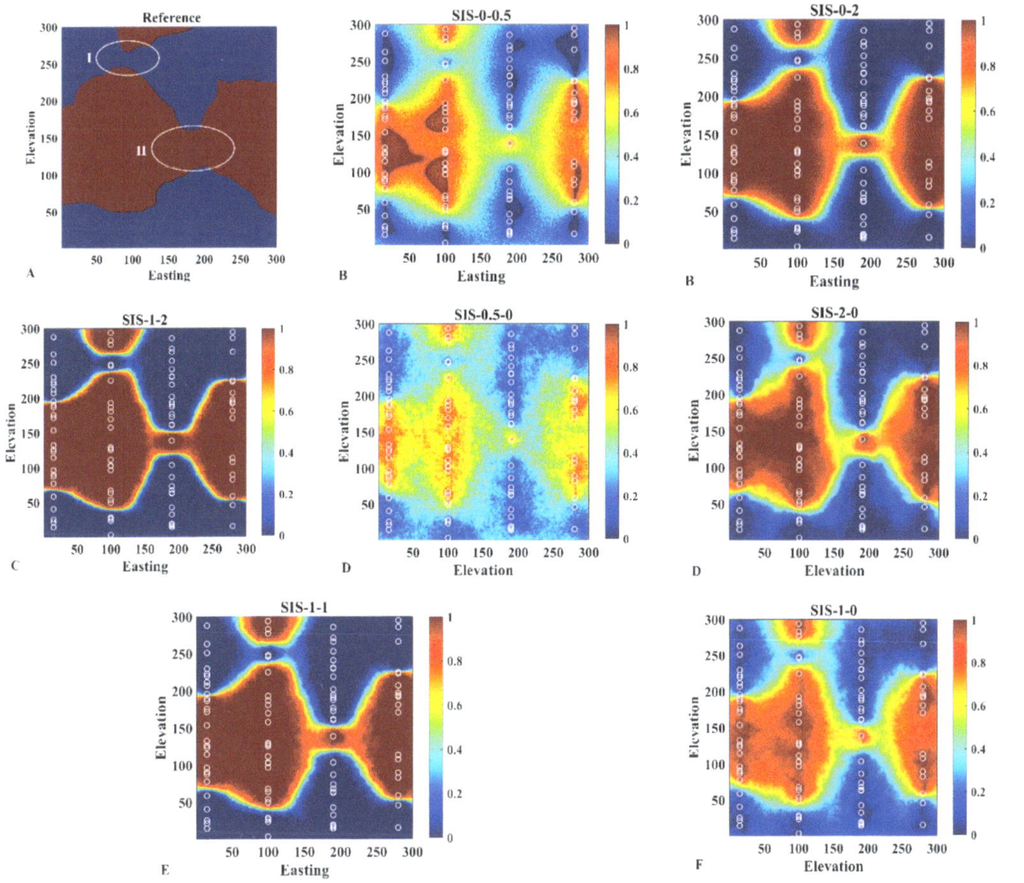

Figure 5. Comparison of (**A**) reference map with resulting simulation maps obtained from (**B**) pure soft simulation, (**C**) revisited Bayesian simulation, (**D**) pure hard simulation, (**E**) Doyan's simulation, and (**F**) traditional simulation.

From the probability maps, one can build a categorical map by selecting, for each grid node, the most probable category. This model can then be compared to the reference map in order to identify the grid nodes for which the category matches the most probable one and also the grid nodes for which there is a mismatch (Figure 6). The latter grid nodes are mostly located near the border of the two categories. As can be observed, revisited Bayesian simulation (Figure 6C) and traditional simulation (Figure 6E) provided better results compared to other experiments.

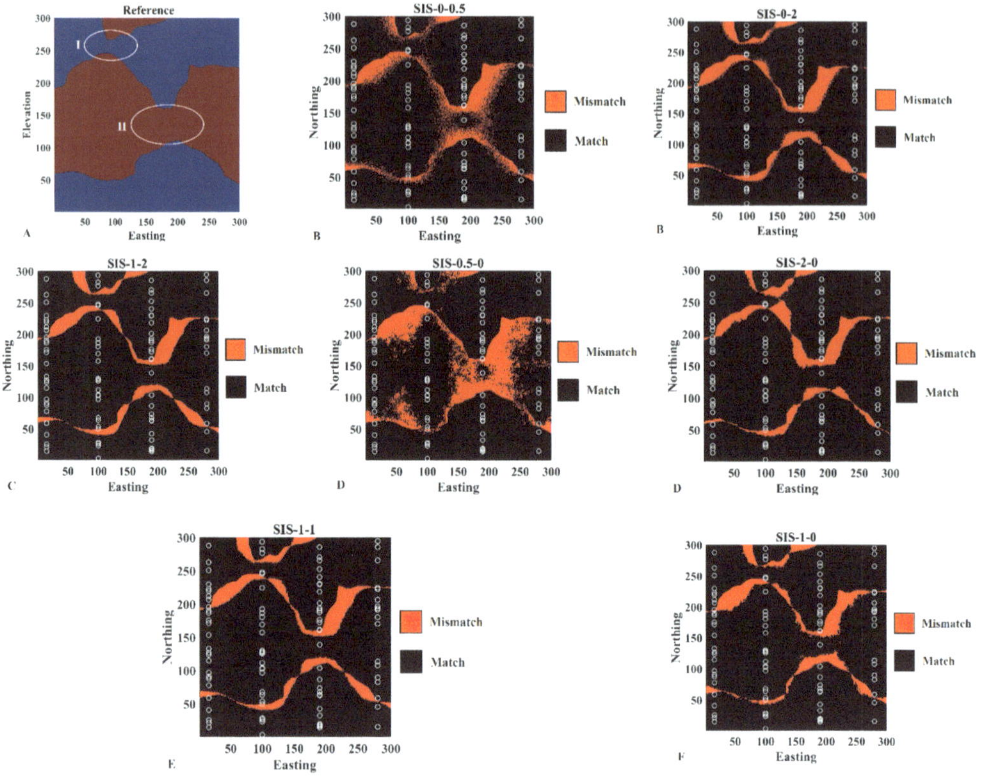

Figure 6. The matches and mismatches with respect to (**A**) reference map for resulting simulation maps obtained from (**B**) pure soft simulation, (**C**) revisited Bayesian simulation, (**D**) pure hard simulation, (**E**) Doyan's simulation, and (**F**) traditional simulation.

3.3. *Assessment of Optimal Weights*

It is shown that the revisited Bayesian simulation method with weights $\lambda_h = 1$ and $\lambda_s = 2$, by visual inspection, can deliver the most promising results compared to the reference dataset; this has been achieved qualitatively by using only the realizations and probability maps. To show this also quantitatively, a validation step is performed to assess the accuracy of the obtained weights. To secure the optimum weights of λ_h and λ_s, four criteria were assessed by calculating the error deducing from:

(A) The difference between the reproduced proportion of indicators through the realizations and the proportion of categories/indicators in the reference map.
(B) The difference in the percentage of the match [27] between the most probable categorical map and the reference map.
(C) The difference between the connectivity function of each indicator and the connectivity function in the reference map.
(D) The difference between the variogram reproduction of the indicators and the original variogram of the indicators in the reference map.

In all these criteria, the underlying parameter is calculated over individual realizations, and then it is averaged. The average value is compared with the original parameter obtained from the reference map. The difference (error) is then taken into account as the main criterion for evaluating the optimal weights of λ_h and λ_s. For each criterion, 24 experiments are available, where their errors are calculated. Different polynomial functions are tested to interpolate the errors within the trials. The one is selected for each that offers the highest R^2.

As can be seen from Figure 7, there are two optimal weight candidates that worth to be discussed. It seems that when the λ_h is equal to 1 and 1.5, and λ_s is equal to 2, the errors in variogram reproduction along Easting (Figure 7G), connectivity for both categories (Figure 7C,E), and slightly variogram reproduction along Northing (Figure 7H) are quite low and almost equivalent. However, this is not true when one considers the errors in proportion reproduction (Figure 7A), match percentage (Figure 6B), and northing connectivity for both categories (Figure 7D,F). The individual realizations and the probability maps obtained from these two weights (Appendix A) also corroborated these quantitative findings. Therefore, these numerical examinations also verified that $\lambda_h = 1$ and $\lambda_s = 2$ are two optimal weights for the corresponding probability aggregation in the proposed revisited sequential indicator simulation approach in this case study.

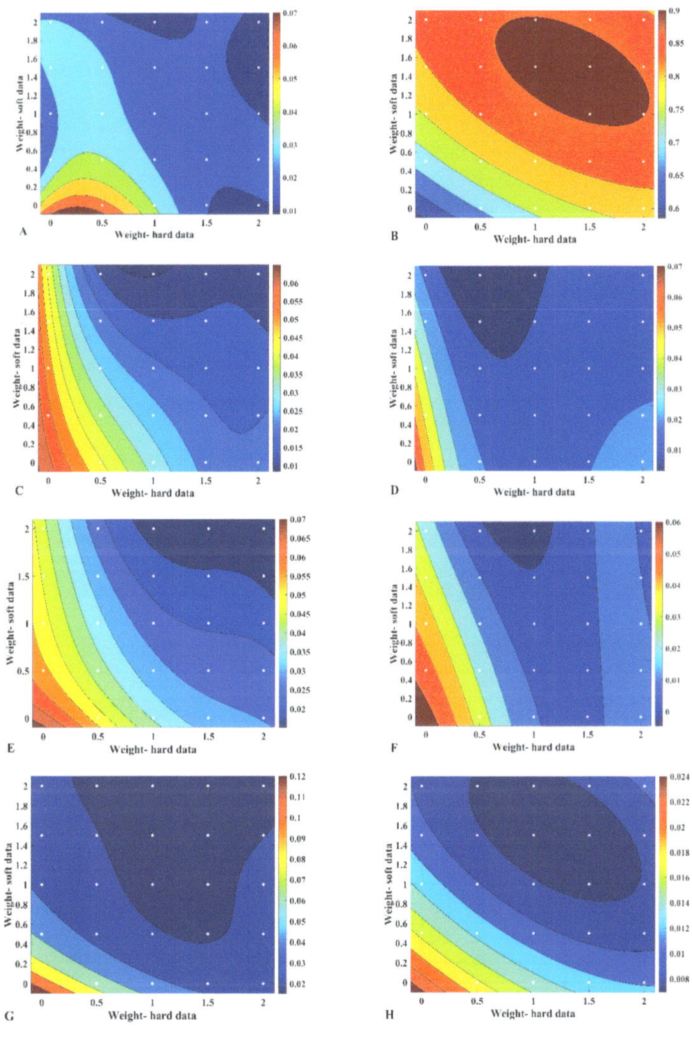

Figure 7. Validation by computing the error for (**A**) Proportion reproduction, (**B**) Match percentage, (**C**) Easting connectivity-R1, (**D**) Northing connectivity-R1, (**E**) Easting connectivity-R2, (**F**) Northing connectivity-R2, (**G**) Variogram-Easting, (**H**) Variogram-Northing; weight-hard data (λ_h) and weight-soft data (λ_s).

4. Discussion and Conclusions

A methodology proposed in this study revisits the traditional Bayesian sequential indicator simulation. This versatile model uses a log-linear pooling probability aggregation approach to integrate the probabilities that are coming from different sources. The algorithm makes the job easier to find the soft information by using only an available geostatistical interpolation technique to inform the soft data at the target data location. Different weighting options were also tested, and through a numerical examination, it was revealed that different weights that are assigned to each source of information produce different results. Indeed, the incorporation of sources of information with different influences was overlooked through all previously Bayesian sequential indicator simulation approaches. The results of this study compared with traditional sequential indicator simulation algorithms, and it was shown that the long-range structures could be better produced. Nevertheless, the method proposed is not restricted to only two sources of information. Equation (5) can be generalized to include more sources of information, and the weights can simply be tuned. However, the methodology needs some numerical experiments to assess the optimal weights. Further research can focus on developing a sophisticated technique to infer the weights automatically. The proposed algorithm is tested in a synthetic case study, but it can also be tested in an actual case study. To set the optimal weights, one solution is to use the production dataset as a benchmark. This information may only be available partially in a domain. The obtained weights can be taken into account for resource reconciliation in other parts of the domain, where only exploratory data are available and one does not have access to production data. Another avenue of research can further develop the method to model the non-stationary domains.

Funding: This research was funded by Nazarbayev University via Faculty Development Competitive Research Grants for 2021–2023, grant number 021220FD4951.

Data Availability Statement: Not applicable.

Acknowledgments: The author acknowledges Nazarbayev University, particularly School of Mining and Geosciences, and Aizhan Nurakysheva for providing the financial open-access support of this work.

Conflicts of Interest: The author declares no conflict of interest.

Appendix A

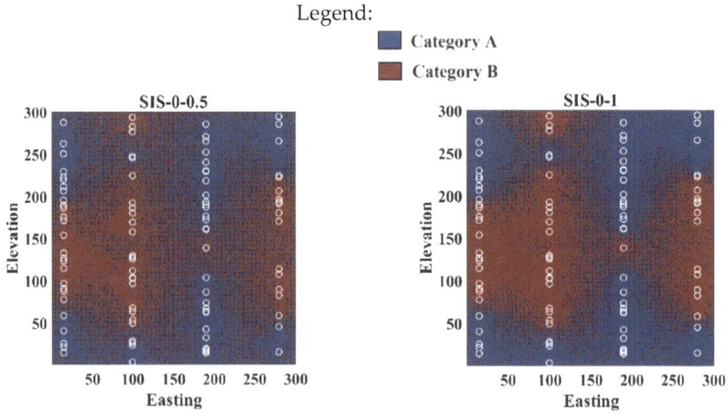

Figure A1. *Cont.*

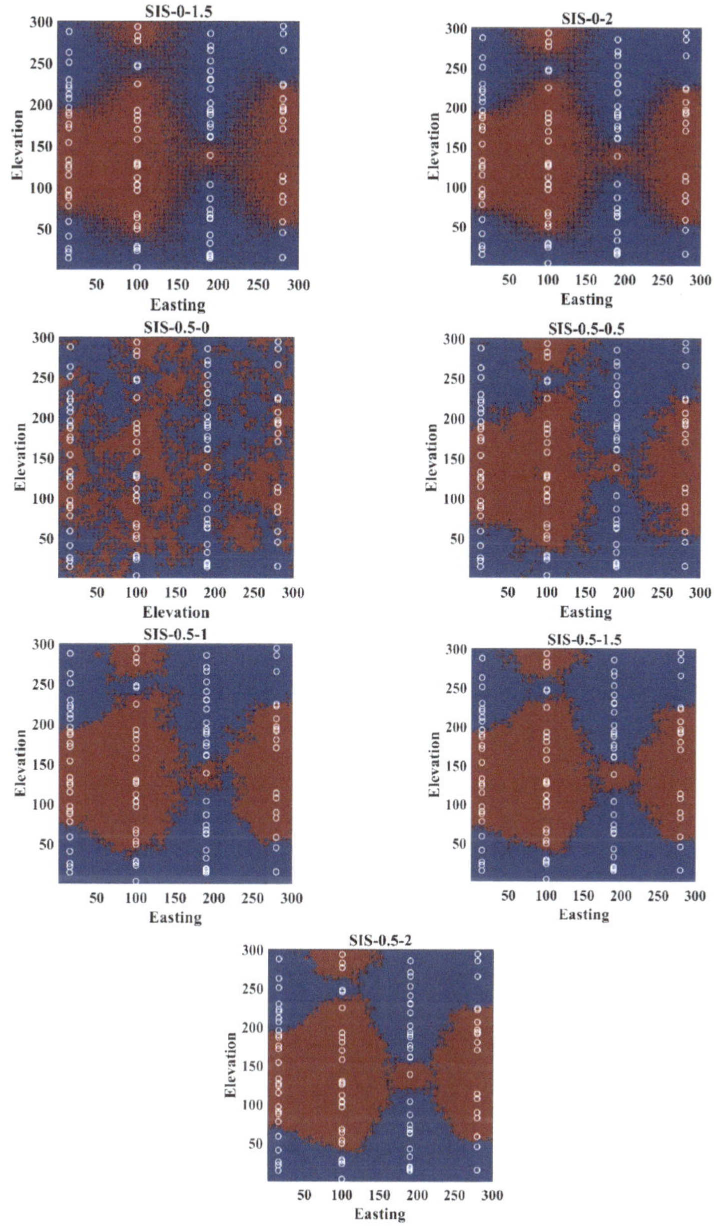

Figure A1. *Cont.*

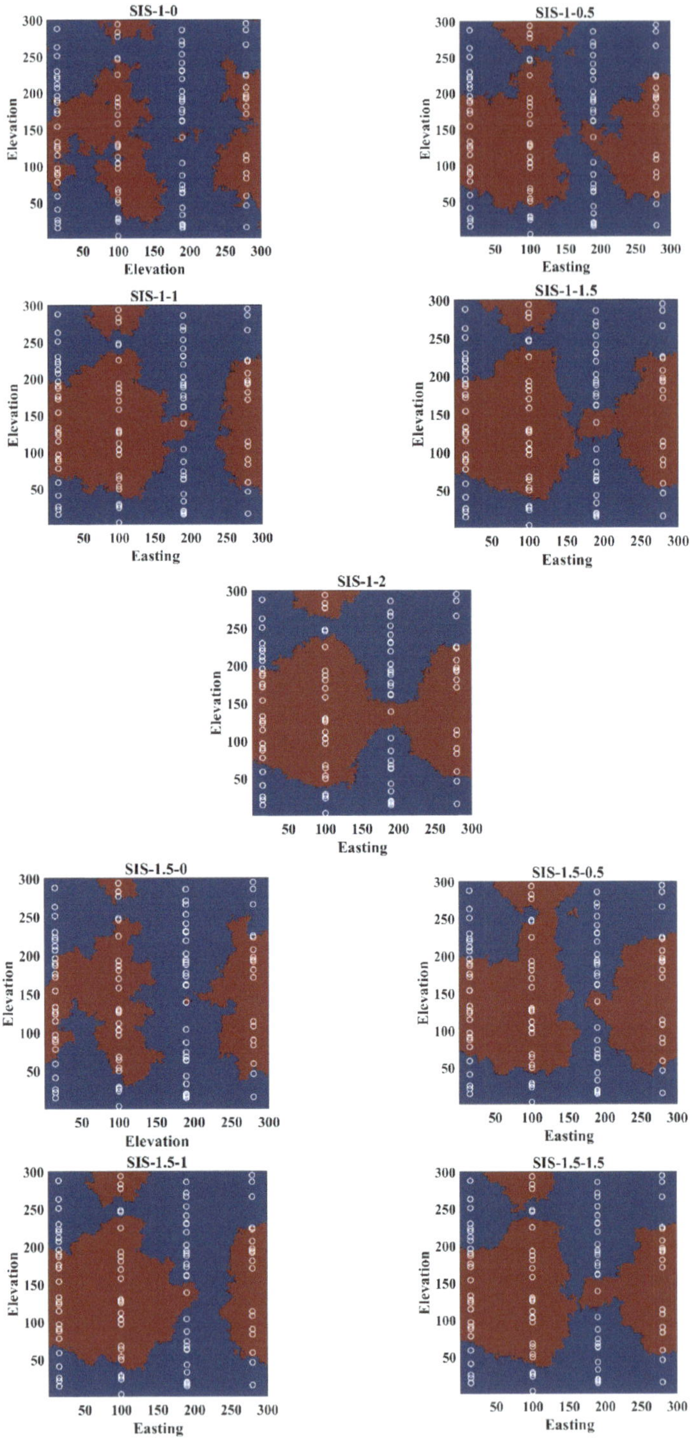

Figure A1. *Cont.*

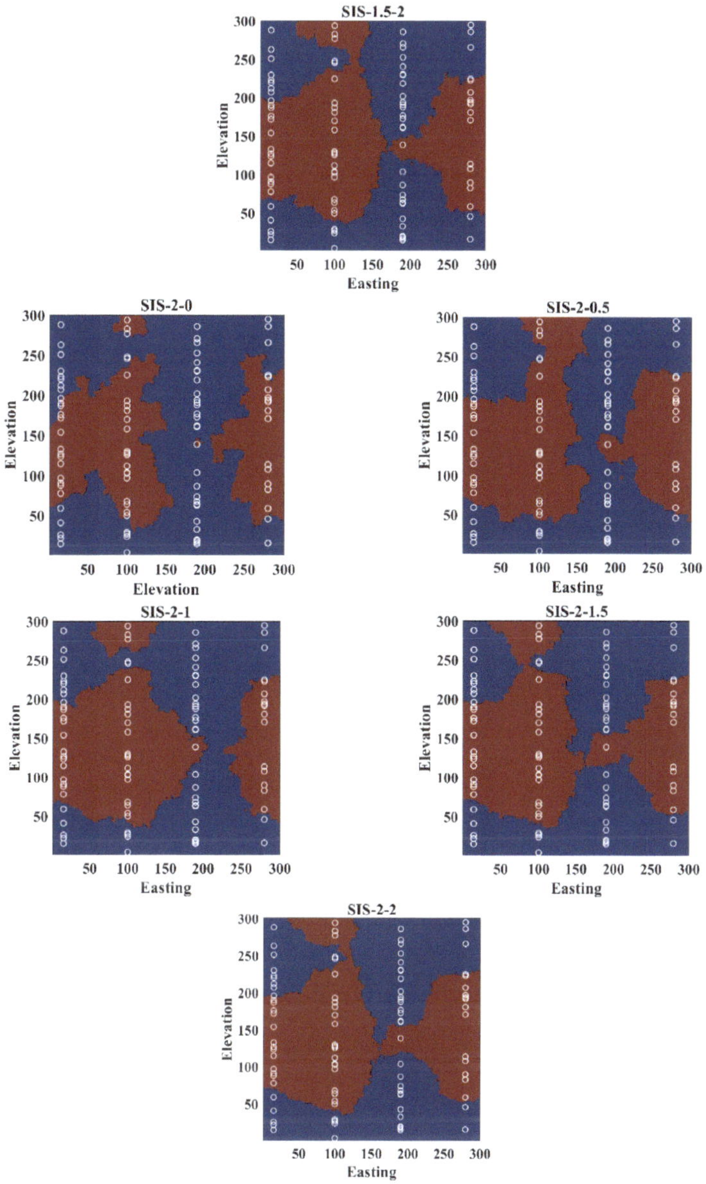

Figure A1. Realizations obtained from revisited Bayesian sequential indicator simulation by different assignations of weights.

Appendix B

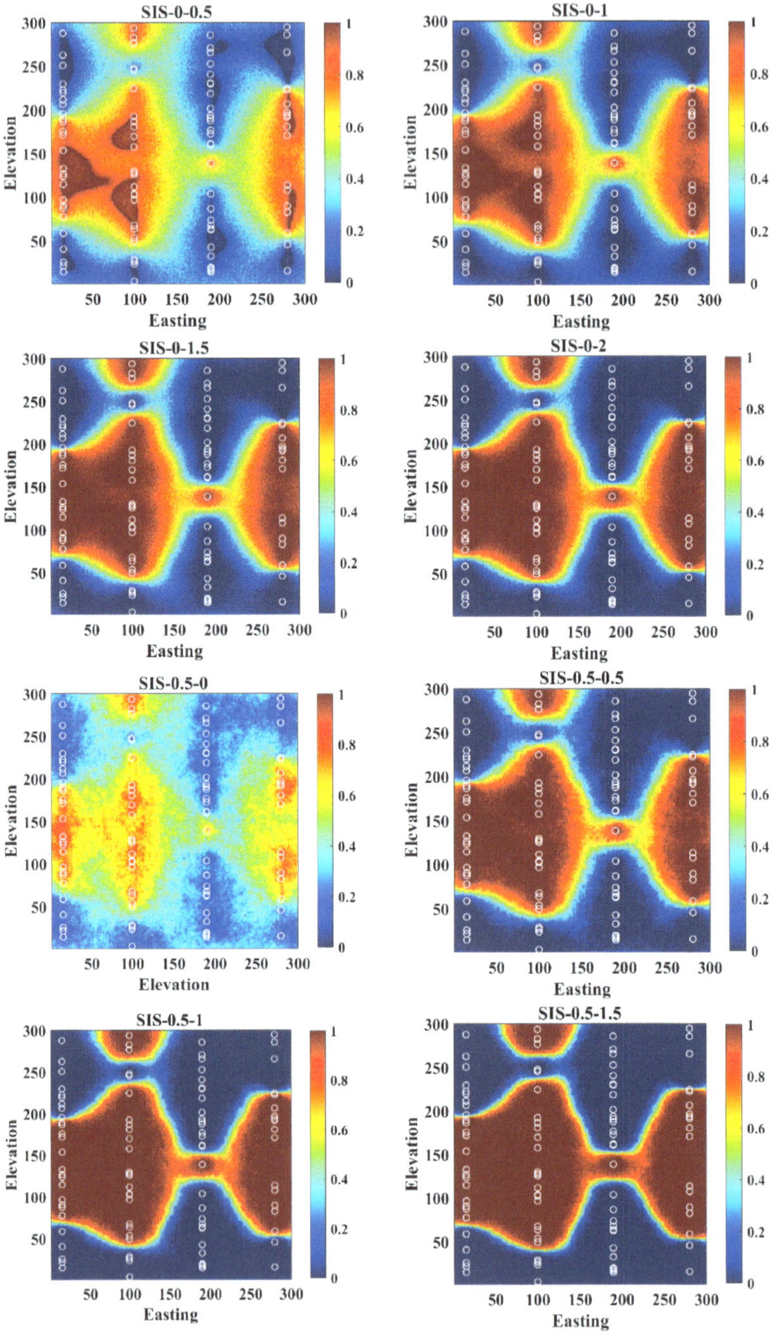

Figure A2. *Cont.*

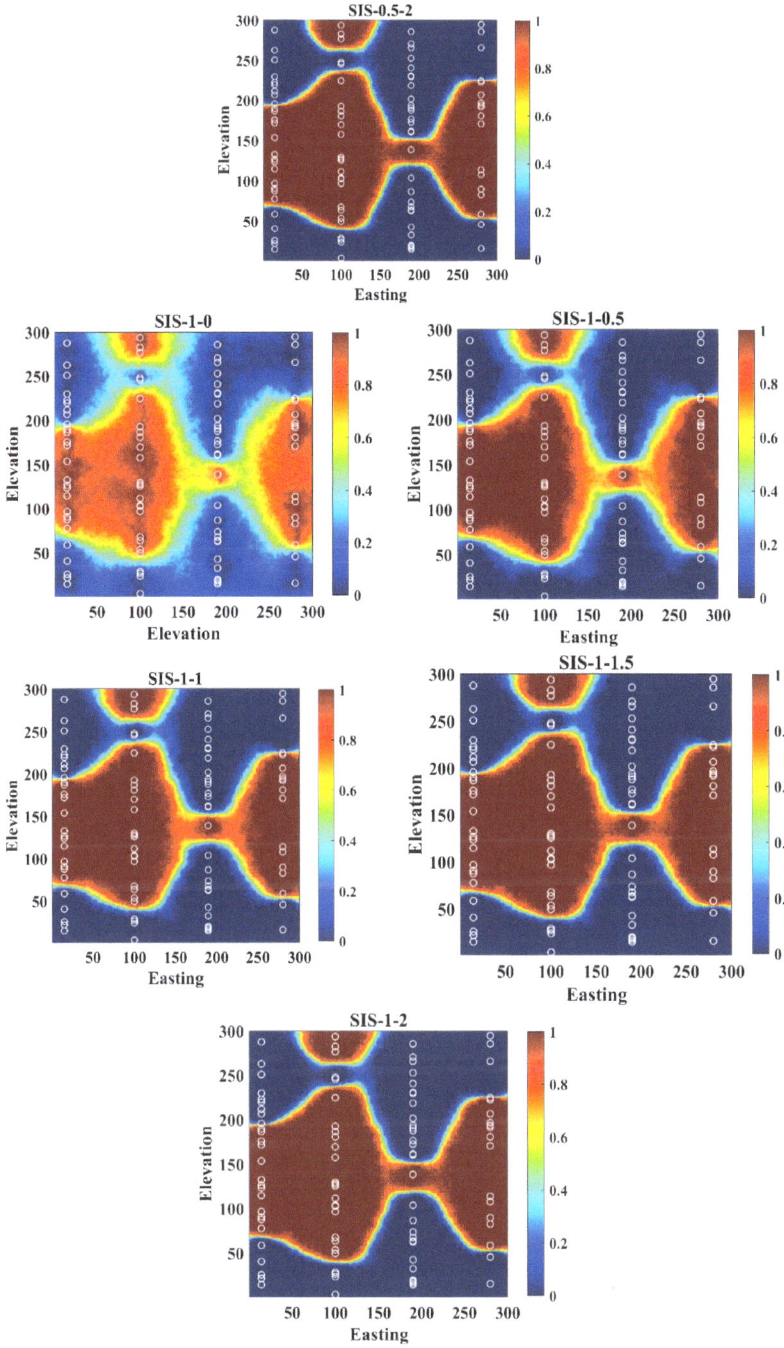

Figure A2. *Cont.*

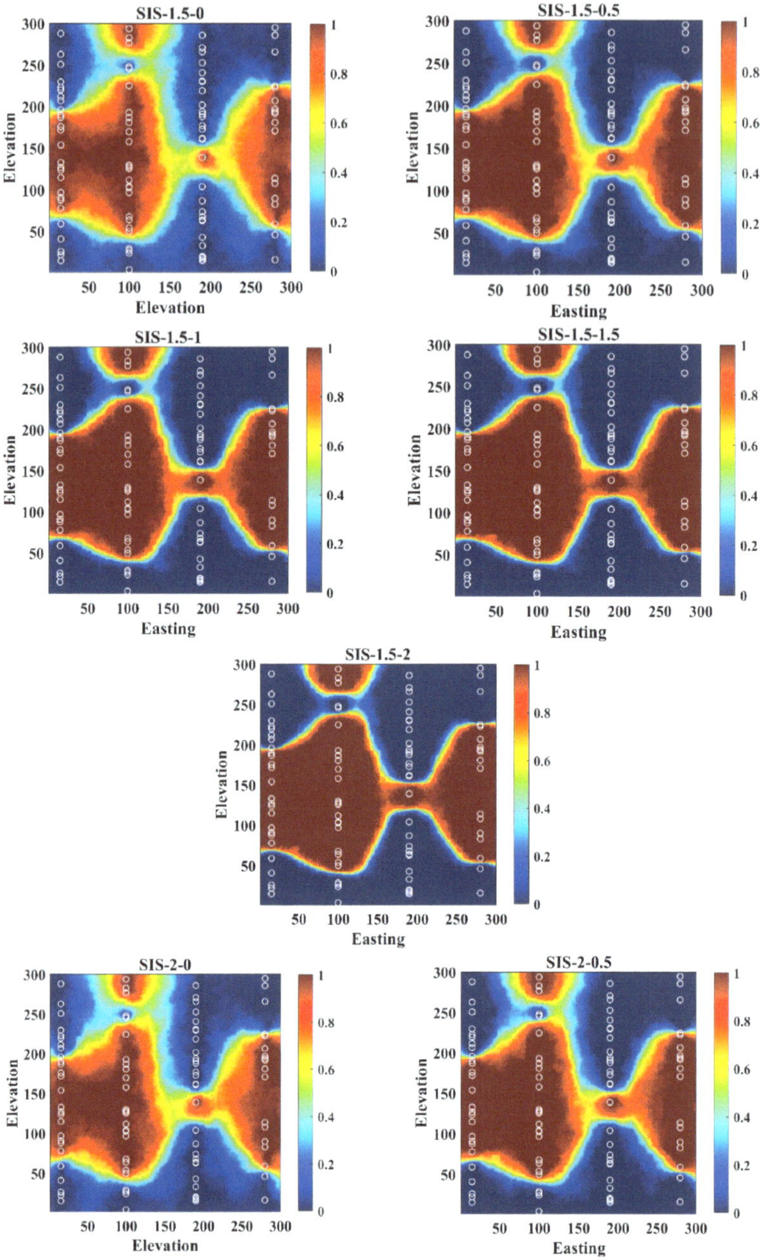

Figure A2. *Cont.*

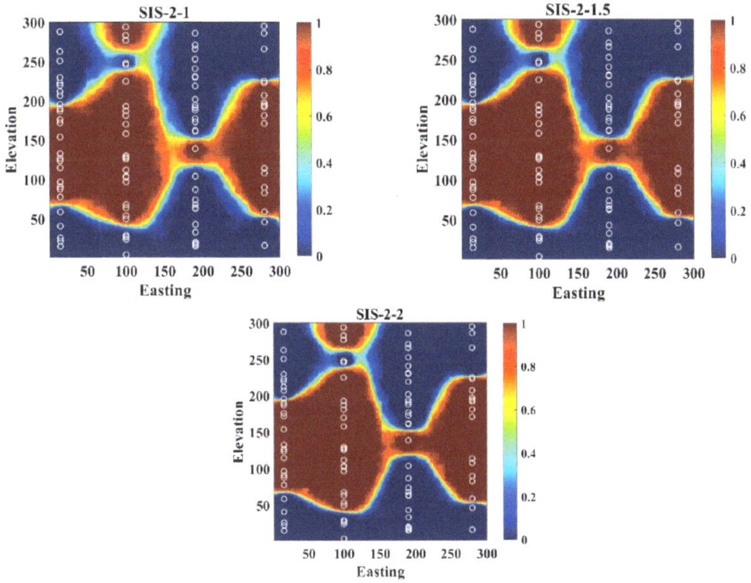

Figure A2. Probability maps obtained from revisited Bayesian sequential indicator simulation by different assignations of weights.

References

1. Madani, N.; Maleki, M.; Emery, X. Nonparametric Geostatistical Simulation of Subsurface Facies: Tools for Validating the Reproduction of, and Uncertainty in, Facies Geometry. *Nat. Resour. Res.* **2019**, *28*, 1163–1182. [CrossRef]
2. Rossi, M.E.; Deutsch, C.V. *Mineral Resource Estimation*; Springer: New York, NY, USA, 2014; p. 332.
3. Sadeghi, M.; Madani, N.; Falahat, R.; Sabeti, H.; Amini, N. Hierarchical reservoir lithofacies and acoustic impedance simulation: Application to an oil field in SW of Iran. *J. Pet. Sci. Eng.* **2022**, *208*, 109552. [CrossRef]
4. Pyrcz, M.J.; Deutsch, C.V. *Geostatistical Reservoir Modeling*; Oxford University Press: Oxford, UK, 2014.
5. Mariethoz, G.; Caers, J. *Multiple-Point Geostatistics: Stochastic Modeling with Training Images*; Wiley: New York, NY, USA, 2014; p. 376.
6. Matheron, G.; Beucher, H.; Galli, A.; Guérillot, D.; Ravenne, C. Conditional simulation of the geometry of fluvio-deltaic reservoirs. In Proceedings of the 62nd Annual Technical Conference and Exhibition of the Society of Petroleum Engineers, Dallas, TX, USA, 27–30 September 1987; pp. 591–599.
7. Galli, A.; Beucher, H.; Le Loc'h, G.; Doligez, B. The pros and cons of the truncated Gaussian method. In *Geostatistical Simulations*; Armstrong, M., Dowd, P.A., Eds.; Kluwer: Dordrecht, The Netherlands, 1984; pp. 217–233.
8. Madani, N. Plurigaussian Simulations. In *Encyclopedia of Mathematical Geosciences*; Encyclopedia of Earth Sciences Series; Daya Sagar, B., Cheng, Q., McKinley, J., Agterberg, F., Eds.; Springer: Cham, Switzerland, 2021. [CrossRef]
9. Chilès, J.P.; Delfiner, P. *Geostatistics: Modeling Spatial Uncertainty*; Wiley: New York, NY, USA, 2012; p. 699.
10. Alabert, F. Stochastic Imaging of Spatial Distributions Using Hard and Soft Information. Master's Thesis, Department of Applied Earth Sciences, Stanford University, Stanford, CA, USA, 1987; p. 332.
11. Journel, A.G.; Alabert, F.G. New method for reservoir mapping. *J. Pet. Technol.* **1990**, *42*, 212–218. [CrossRef]
12. Journel, A.G.; Gómez-Hernández, J.J. Stochastic imaging of the Wilmington clastic sequence. *SPE Form. Eval.* **1993**, *8*, 33–40. [CrossRef]
13. Deutsch, C.V. A sequential indicator simulation program for categorical variables with point and block data: BlockSIS. *Comput. Geosci.* **2006**, *32*, 1669–1681. [CrossRef]
14. Emery, X. Properties and limitations of sequential indicator simulation. *Stoch. Environ. Res. Risk Assess.* **2004**, *18*, 414–424. [CrossRef]
15. Genest, C.; Zidek, J.V. Combining probability distributions: A critique and an annotated bibliography. *Stat. Sci.* **1986**, *1*, 147–148.
16. Allard, D.; Comunian, A.; Renard, P. Probability aggregation methods in geoscience. *Math. Geosci.* **2012**, *44*, 545–581. [CrossRef]
17. Mariethoz, G.; Renard, P.; Froidevaux, R. Integrating collocated auxiliary parameters in geostatistical simulations using joint probability distributions and probability aggregation. *Water Resour. Res.* **2009**, *45*, W08421. [CrossRef]
18. Comunian, A.; Renard, P.; Straubhaar, J. 3D multiple-point statistics simulation using 2D training images. *Comput. Geosci.* **2012**, *40*, 49–65. [CrossRef]

19. Hoffimann, J.; Scheidt, C.; Barfod, A.; Caers, J. Stochastic simulation by image quilting of process-based geological models. *Comput. Geosci.* **2017**, *106*, 18–32. [CrossRef]
20. Nussbaumer, R.; Mariethoz, G.; Gloaguen, E.; Holliger, K. Hydrogeophysical data integration through Bayesian Sequential Simulation with log-linear pooling. *Geophys. J. Int.* **2020**, *221*, 2184–2200. [CrossRef]
21. Doyen, P.M.; Psaila, D.E.; Strandenes, S. Bayesian sequential indicator simulation of channel sands from 3-D seismic data in the Oseberg Field, Norwegian North Sea. In Proceedings of the 69th Annual Technical conference and Exhibition, New Orleans, LA, USA, 25–28 September 1994; pp. 197–211.
22. Ortiz, J.M.; Deutsch, C.V. Indicator simulation accounting for multiple-point statistics. *Math. Geol.* **2004**, *36*, 545–565. [CrossRef]
23. Goovaerts, P. *Geostatistics for Natural Resources Evaluation*; Oxford University Press: New York, NY, USA, 1997.
24. Deutsch, C.V.; Journel, A. *GSLIB: Geostatistical Software and User's Guide*, 2nd ed.; Oxford University Press: New York, NY, USA, 1998.
25. Emery, X.; Ortiz, J. Estimation of mineral resources using grade domains: Critical analysis and a suggested methodology. *J. S. Afr. Inst. Min. Metall.* **2005**, *105*, 247–255.
26. Madani, N.; Maleki, M.; Sepidbar, F. Integration of Dual Border Effects in Resource Estimation: A Cokriging Practice on a Copper Porphyry Deposit. *Minerals* **2021**, *11*, 660. [CrossRef]
27. Madani, N.; Emery, X. Simulation of geo-domains accounting for chronology and contact relationships: Application to the Río Blanco copper deposit. *Stoch. Environ. Res. Risk Assess.* **2015**, *29*, 2173–2191. [CrossRef]
28. Renard, P.; Allard, D. Connectivity metrics for subsurface flow and transport. *Adv. Water Resour.* **2013**, *51*, 168–196. [CrossRef]
29. Maleki, M.; Emery, X.; Mery, N. Indicator Variograms as an Aid for Geological Interpretation and Modeling of Ore Deposits. *Minerals* **2017**, *7*, 241. [CrossRef]

mathematics

Article

Variational Bayesian Variable Selection for High-Dimensional Hidden Markov Models

Yao Zhai †, Wei Liu †, Yunzhi Jin * and Yanqing Zhang

Key Lab of Statistical Modeling and Data Analysis of Yunnan Province, Department of Statistics, Yunnan University, Kunming 650091, China; zhaiyao@itc.ynu.edu.cn (Y.Z.); liuwei_clsa@stu.ynu.edu.cn (W.L.); zhangyanqing@ynu.edu.cn (Y.Z.)
* Correspondence: yzjin@ynu.edu.cn
† These authors contributed equally to this work.

Abstract: The Hidden Markov Model (HMM) is a crucial probabilistic modeling technique for sequence data processing and statistical learning that has been extensively utilized in various engineering applications. Traditionally, the EM algorithm is employed to fit HMMs, but currently, academics and professionals exhibit augmenting enthusiasm in Bayesian inference. In the Bayesian context, Markov Chain Monte Carlo (MCMC) methods are commonly used for inferring HMMs, but they can be computationally demanding for high-dimensional covariate data. As a rapid substitute, variational approximation has become a noteworthy and effective approximate inference approach, particularly in recent years, for representation learning in deep generative models. However, there has been limited exploration of variational inference for HMMs with high-dimensional covariates. In this article, we develop a mean-field Variational Bayesian method with the double-exponential shrinkage prior to fit high-dimensional HMMs whose hidden states are of discrete types. The proposed method offers the advantage of fitting the model and investigating specific factors that impact the response variable changes simultaneously. In addition, since the proposed method is based on the Variational Bayesian framework, the proposed method can avoid huge memory and intensive computational cost typical of traditional Bayesian methods. In the simulation studies, we demonstrate that the proposed method can quickly and accurately estimate the posterior distributions of the parameters with good performance. We analyzed the Beijing Multi-Site Air-Quality data and predicted the PM2.5 values via the fitted HMMs.

Keywords: Hidden Markov Models; high-dimensional data; shrinkage prior; variational inference

MSC: 62F15; 65K10; 62M05

Citation: Zhai, Y.; Liu, W.; Jin, Y.; Zhang, Y. Variational Bayesian Variable Selection for High-Dimensional Hidden Markov Models. *Mathematics* **2024**, *12*, 995. https://doi.org/10.3390/math12070995

Academic Editor: Diana Mindrila

Received: 15 February 2024
Revised: 22 March 2024
Accepted: 24 March 2024
Published: 27 March 2024

1. Introduction

Hidden Markov Models (HMMs) are a statistical model used to describe the evolution of observable events that depend on internal factors or states, which are not directly observable and called hidden states. Each hidden state can transition to another hidden state, including itself, with a certain probability, while we cannot observe them directly, we infer their presence and transitions between them based on observable outputs. HMMs have been widely used in various applications, including speech recognition, bioinformatics, natural language processing, and financial markets. In practice, HMMs often face high-dimensional issues, that is, a large number of covariates (or high-dimensional covariates) and multiple states result in high-dimensional parameters existing in the HMMs. The high-dimensional issue may result in the overfitting for the HMMs. The challenge then becomes identifying important variables or parameters in different hidden states. Thus, efficient parameter estimation and hidden Markov chain recovering are significant for the high-dimensional HMMs.

Currently, there have been many methods for estimating parameter estimation and recovering hidden Markov chains, including recursive algorithms [1–4] and traditional Bayesian methods [5,6]. Bayesian inference [7,8] is a versatile framework that utilizes sophisticated hierarchical data models for learning and consistently quantifies uncertainty in unknown parameters through the posterior distribution [9]. However, computing the posterior is demanding even for moderately intricate models and frequently necessitates approximation. Moreover, traditional Bayesian methods (e.g., Markov chain Monte Carlo (MCMC)) for the HMMs is often considered a black-box method by many statisticians due to its reliance on simulations to produce computable results, which is generally inefficient and unnecessary. Traditional Markov Chain Monte Carlo (MCMC) methods may also exhibit slow convergence and extended running times, as documented in prior studies [10–12].

The Variational Bayesian approach is an alternative to traditional MCMC algorithm in high-dimensional issue. Variational inference (VI) based on Bayesian method [13–16] can approximate posterior distributions quickly [17], since VI uses the Kullback–Leibler (KL) divergence to measure the difference between the variational posterior and the true posterior, and transforms the statistical inference problem into a mathematical optimization problem by minimizing the KL divergence. Therefore, the variational approach is as close as possible to the true posterior distribution according to the KL divergence. Wang and Blei [17] have proved that the variational posterior is consistent to the true posterior distribution. Moreover, there have been many efficient optimization algorithms to approximate complex probability densities such as coordinate-ascent (CAVI) [18] and gradient-based methods [15,19]. Currently, there exist many VI research studies for the HMMs [20–24]. For example, MacKay [20] was the pioneering proponent of the application of variational methods to HMMs, with a focus only on cases with discrete observations. Despite the limited comprehension of the state-removal phenomenon, which is that removing certain states from an HMM for simplifying the HMM while preserving its essential statistical properties does not significantly affect the the ability of the HMM to represent the underly stochastic process, variational methods are gaining popularity for HMMs within the machine learning community. C. A. McGrory [21] extended the deviance information criterion for Bayesian model selection within the Hidden Markov Model framework, utilizing a variational approximation. Nicholas J. Foti [22] devised an SVI algorithm to learn HMM parameters in settings of time-dependent data. Since VI can be seen as a special instance of the EM algorithm [23], Gruhl [23] integrates both approaches and uses the multivariate Gaussian output distribution of VI to train the HMM. Ding [24] employed variational inference techniques to investigate nonparametric Bayesian Hidden Markov Models built on Dirichlet processes. These processes enable an unbounded number of hidden states and adopt an infinite number of Gaussian components to handle continuous observations. However, variational inference has not been fully explored in HMMs, especially in HMMs with a high-dimensional covariate. It is important to note that research frequently entails using high-dimensional covariate datasets in real-world applications. When high-dimensional HMMs contain a large number of parameters, there is a possibility of overfitting the given data set. The challenge then becomes identifying which covariates have a substantial impact on the interpretation of observations and state shifts in each hidden state of the model, and which covariates have a negligible effect. This process is important for improving the predictive accuracy of the model, reducing the risk of overfitting, and improving the interpretability of the model.

In this article, we develop a Variational Bayesian method for variable selection. We utilize the double-exponential shrinkage prior [25–28] as the prior of coefficients in each hidden state models, to screen vital variables that affect each hidden state and obtain hidden Markov regression models. We use mean-field variational inference to identify variational densities for approximating complex posterior densities, via minimizing the difference between the approximate probability density and the actual posterior probability density. Moreover, we adopt the Monte Carlo Co-ordinate Ascent VI (MC-CAVI) [29] algorithm to compute the necessary expectations within the CAVI. Since the variational inference is a fast alternative to the MCMC method and can avoid large memory and intensive computational cost compared to traditional Bayesian methods, the proposed approach inherits the good properties of variational inference, and can quickly and accurately estimate the posterior distributions and the unknown parameters. In the simulation studies and real data analysis, the proposed method outperforms the common methods in term of variable selection and prediction.

The main contributions of this article are as follows: First, the proposed method can perform variable selection for high-dimensional HMMs, and offer the advantage of fitting the model and investigating specific factors that impact the response variable changes simultaneously. Since the proposed method uses double-exponential shrinkage prior, which has the feature of being able to select important variables, the proposed method can simultaneously select important variables to the response variable and estimate the corresponding parameters. Second, since the proposed method is based on the Variational Bayesian framework, the proposed method can avoid huge memory and intensive computational cost of the traditional Bayesian methods, especially for the high-dimensional issue. Finally, we demonstrate that the proposed method can quickly and accurately estimate the posterior distributions of the parameters with good performance in the simulation studies. Moreover, we analyze Beijing Multi-Site Air-Quality data and predict the PM2.5 values well via the fitted HMMs.

The rest of the article is organized as follows: Section 2 introduce the Hidden Markov Model with high-dimensional covariate and shrinkage priors in the Bayesian inference. In Section 3, we propose an efficient Variational Bayesian estimation method with the double-exponential shrinkage prior for variable selection of the high-dimensional HMMs (HDVBHMM). In Section 4, we conduct simulation studies to investigate the finite sample performances of the proposed method. In Section 5, Beijing Multi-Site Air-Quality data are analyzed and the efficiency of the proposed method is verified. Section 6 concludes our work. Technical details are presented in the Appendix A.

2. Model and Notation

2.1. Hidden Markov Model

In this section, we first introduce Hidden Markov Models (HMMs). The HMMs are a type of doubly stochastic process that occurs over discrete time intervals and includes observations y_t and latent states z_t. In a traditional Hidden Markov Model without covariates, the observation y_t depends only on the current potential state z_t. The conditional distribution of the observation y_t when given the potential state $z_t = k$ can be expressed as:

$$y_t \mid z_t = k \sim F_k(\theta_k),$$

where $F_k(\theta_k)$ denotes a certain family of distributions, such as the normal distribution $N(\mu_k, \sigma_k^2)$. Extended HMM models can include covariates $x_t \in R^p$. That is, the set of observations is $y = (y_1, \ldots, y_T)$ and $x = (x_1, \ldots, x_T)$. Specifically, for $y = (y_1, \ldots, y_T)$, $z = (z_1, \ldots, z_T)$ and $x = (x_1, \ldots, x_T)$, the model expression is as follows:

$$y_t \mid x_t, z_t = k, \beta, \sigma^2 \sim N\left(x_t^\top \beta_k, \sigma^2\right) \quad \text{for} \quad t = 1, \ldots, T,$$

where the symbol N represents the normal distribution, σ^2 denotes the variance of y_t, and $\beta = (\beta_1, \ldots, \beta_K)^\top$ is the coefficient of the covariate at all hidden states. In the article, we consider the high-dimensional issue of the covariate. We denote the dummy variable corresponding to z_t as the $v_t = (v_{t1}, \ldots, v_{tK})^\top$, where $v_{tk} = 1$ and other elements being zero if $z_t = k$. Thus,

$$P\left(y_t \mid x_t, \beta, \sigma^2\right) = \prod_{k=1}^{K} P\left(y_t \mid x_t, z_t = k, \beta, \sigma^2\right)^{v_{tk}} = \prod_{k=1}^{K} P\left(y_t \mid x_t, \beta_k, \sigma^2\right)^{v_{tk}}.$$

In hidden Markov chains, each hidden state z_t is independent from $z_1, \ldots, z_{t-2}$ and $z_{t+1}, \ldots, z_T$ conditionally on z_{t-1}. Therefore, we can assume that the probability distribution of z_1 is given by $z_1 \sim P(z_1|\pi) = (\pi_1, \pi_2, \ldots, \pi_K)^\top$, where $\sum_{k=1}^K \pi_k = 1$ and $\pi_k > 0$. The conditional probability of z_t given z_{t-1} is assumed as:

$$P(z_t \mid z_{t-1}, A) = \prod_{k=1}^{K}\prod_{j=1}^{K} A_{jk}^{v_{t-1,j}\, v_{tk}},$$

where A is the transition matrix with elements A_{ij} for $i, j = 1, \ldots, K$, $\sum_{j=1}^K A_{ij} = 1$ and $A_{ij} > 0$, and A_{ij} represents the probability of transitioning from state i to state j. Thus, the joint distribution is as follows:

$$\begin{aligned} & P(\boldsymbol{y}, \boldsymbol{z} \mid \boldsymbol{x}, \pi, A, \beta, \sigma^2) \\ = \; & P(z_1|\pi) \prod_{t=2}^T P(z_t \mid z_{t-1}, A) \prod_{t=1}^T P(y_t \mid x_t, \beta, \sigma^2) \\ = \; & \left(\prod_{k=1}^K \pi_k^{v_{1k}}\right)\left(\prod_{t=2}^T \prod_{k=1}^K \prod_{j=1}^K A_{jk}^{v_{t-1,j}\, v_{tk}}\right)\left(\prod_{t=1}^T \prod_{k=1}^K P(y_t \mid x_t, \beta, \sigma^2, z_t = k)^{v_{tk}}\right). \end{aligned} \quad (1)$$

2.2. Prior Selection in the HMMs

To make Variational Bayesian inference, we require specifying the prior of the parameters π, A, β and σ^2. Based on the characteristics of $\pi = (\pi_1, \pi_2, \ldots, \pi_K)$, $\sum_{k=1}^K \pi_k = 1$ and $\pi_k > 0$, Dirichlet distribution is applied to the prior distribution of π as follows:

$$\pi \sim \mathrm{Dir}\left(\alpha^{(\pi)}\right), \quad (2)$$

where $\alpha^{(\pi)} = \left(\alpha_1^{(\pi)}, \ldots, \alpha_K^{(\pi)}\right)^\top$, $\sum_{k=1}^K \alpha_k^{(\pi)} = 1$, and $\alpha_k^{(\pi)} > 0$. In the model, A denotes the transition matrix of the hidden state $\boldsymbol{z}$ and can be expressed as follows:

$$A = \begin{pmatrix} A_{11} & \ldots & A_{1K} \\ \vdots & \ddots & \vdots \\ A_{K1} & \ldots & A_{KK} \end{pmatrix}.$$

Like Nicholas [22], we specify the prior of the jth row of the transition matrix A as:

$$A_j \sim \mathrm{Dir}\left(\alpha_j^{(A)}\right) \quad \text{for} \quad j = 1, \ldots, K, \quad (3)$$

where $\alpha_j^{(A)} = \left(\alpha_{j1}^{(A)}, \alpha_{j2}^{(A)}, \ldots, \alpha_{jK}^{(A)}\right)^\top$, $\sum_{k=1}^K \alpha_{jk}^{(A)} = 1$, and $\alpha_{jk}^{(A)} > 0$. Since σ^2 is variance of the y, we specify the prior of the σ^2 as

$$\sigma^2 \sim f\left(\sigma^2\right) = \frac{1}{\sigma^2}. \quad (4)$$

In a high-dimensional and sparse issue, we consider the double-exponential shrinkage prior [25,27] as the prior of β, defined as follows:

$$\begin{aligned} & \beta_k \mid \sigma^2, \tau_1, \ldots, \tau_p^2 \sim N_p\left(0, \sigma^2 D_\tau\right), \\ & D_\tau = \mathrm{diag}\left(\tau_1^2, \ldots, \tau_p^2\right), \\ & \tau_m^2 \sim \mathrm{Exp}\left(\frac{\lambda^2}{2}\right) \quad \text{for} \quad m = 1, 2, \ldots, p, \\ & \lambda^2 \sim \Gamma(r, \delta), \end{aligned} \quad (5)$$

where Γ represents gamma distribution and $\text{Exp}(\cdot)$ represents the exponential distribution. The above prior can select important variables of the HMMs in each hidden state. Bayesian approaches can be used to solve the parameter estimation question with the above prior information. However, in high-dimensional data, the traditional Bayesian methods (e.g., MCMC) require huge memory and intensive computational cost. The Variational Bayesian approach is an alternative to the traditional MCMC algorithm in high-dimensional issue. Next, we introduce the proposed Variational Bayesian inference for high-dimensional HMMs.

3. Variational Bayesian Inference for the HMMS

3.1. Mean Field Variational

Mean-field Variational Bayesian inference is a prevalent approach in variational inference, and aims to identify an approximate density by minimizing the difference between the approximate probability density and the actual posterior probability density, while being bounded by the Kullback–Leibler divergence. In this subsection, we proposed the mean-field variational inference for HMMS with the high-dimensional covariates.

Let $\boldsymbol{D}$ be an observed data set, $\boldsymbol{D} = \{\boldsymbol{y}, \boldsymbol{x}\}$ with response set $\boldsymbol{y} = \{y_i \mid i = 1, \ldots, n\}$ and covariate set $\boldsymbol{x} = \{x_i \mid i = 1, \ldots, n\}$, and $\boldsymbol{\theta} = \{\pi, A, \beta, \sigma^2, \tau_m^2, \lambda^2\}$. The $\boldsymbol{\theta}$ and $\boldsymbol{z}$ include all parameters in the HMMs. We focus on the posterior distribution of parameters $\boldsymbol{\theta}$ and the hidden state z_t. Assume that there is an approximate density family $\boldsymbol{Q}$ containing possible densities over the parameters $\boldsymbol{\theta}, \boldsymbol{z}$. Minimizing the Kullback–Leibler (KL) divergence between the member of the family $q(\boldsymbol{\theta}, \boldsymbol{z})$ and the true posterior $P(\boldsymbol{\theta}, z \mid \boldsymbol{D})$ is to obtain the optimal density approximation of the true posterior, with variational inference prioritizing optimization rather than sampling. That is,

$$q^*(\boldsymbol{\theta}, z) = \arg \min_{q(\boldsymbol{\theta}, z) \in \boldsymbol{Q}} \text{KL}(q(\boldsymbol{\theta}, z) \| P(\boldsymbol{\theta}, z \mid \boldsymbol{D})),$$

where the KL-divergence is:

$$\text{KL}(q(\boldsymbol{\theta}, \boldsymbol{z}) \| P(\boldsymbol{\theta}, z \mid \boldsymbol{D})) = \int q(\boldsymbol{\theta}, \boldsymbol{z}) \log\left\{ \frac{q(\boldsymbol{\theta}, z)}{P(\boldsymbol{\theta}, z|\boldsymbol{D})} \right\} d(\boldsymbol{\theta}, z).$$

The KL-divergence can be further written as:

$$\begin{aligned} &\text{KL}(q(\boldsymbol{\theta}, \boldsymbol{z}) \| P(\boldsymbol{\theta}, z \mid \boldsymbol{D})) \\ =& E_q[\log q(\boldsymbol{\theta}, \boldsymbol{z})] - E_q[\log P(\boldsymbol{\theta}, z \mid \boldsymbol{D})] \\ =& E_q[\log q(\boldsymbol{\theta}, \boldsymbol{z})] - E_q[\log P(\boldsymbol{\theta}, z, \boldsymbol{D})] + \log P(\boldsymbol{D}), \end{aligned}$$

where $\log P(\boldsymbol{D})$ is a constant, E_q denotes the expected value of $\boldsymbol{\theta}$ and z drawn from the distribution q. Thus, minimizing the KL divergence is equivalent to maximizing the following evidence lower bound (ELBO):

$$\text{ELBO}(q) = E_q[\log P(\boldsymbol{\theta}, z, \boldsymbol{D})] - E_q[\log q(\boldsymbol{\theta}, z)]. \tag{6}$$

From another perspective, the ELBO comprises the negative KL divergence and $\log P(\boldsymbol{D})$.

According to the mean-field variational framework [30,31], the parameters are assumed to be posterior independent of each other and to be controlled by a separate factor in the variational density. In the HMMs, $q(\boldsymbol{\theta}, \boldsymbol{z})$ is decomposed as:

$$q(\boldsymbol{\theta}, \boldsymbol{z}) = q(\pi) q(A) q\left(\sigma^2\right) q\left(\tau_m^2\right) q\left(\lambda^2\right) \prod_{t=1}^{T} q(z_t). \tag{7}$$

Each parameter $\boldsymbol{\theta}_i$ and latent state z is governed by its own variational factor. The forms of $q(\boldsymbol{\theta}_i)$ and $q(z)$ are unknown, but the form of the hypothesized factorization is determined. In the optimization process, the optimal solutions of these variational factors $q(\boldsymbol{\theta}_i)$ and $q(z)$ are obtained by maximizing the ELBO of Equation (6) by the coordinate ascent method. Based on the consistency of the Variational Bayesian [17], the variational densities over the mean-field family are still consistent to the posterior densities, even though the mean field approximating family can be a brutal approximation. More generally, one can consider structured variational distributions involving partial factorizations that correspond to tractable substructures of parameters [32]. In this article, we only consider the mean field framework. To express the variational posterior formula concisely, we define $\phi = \{\boldsymbol{\theta}, z\}$ and rewrite $q(\boldsymbol{\theta}, z)$ as $q(\phi)$.

3.2. The Coordinate Ascent Algorithm for Optimizing the ELBO

Based on the variational density decomposition, we can obtain each factor of the variational density via maximizing the ELBO. Let $q_i(\phi_i)$ for $i = 1, 2, \ldots, b$ be the ith factor of the variational density in .The common approaches to maximize the ELBO mainly include a Coordinate Ascent Variational Inference (CAVI) and a gradient-based approach [33]. The CAVI approach sequentially optimizes each factor of the variational density of the mean field to obtain a local maximizer for the ELBO, while keeping the others fixed. Based on the CAVI approach, we can obtain the optimal variational density $q_i^*(\phi_i)$ as follows:

$$q_i^*(\phi_i) \propto \exp\{E_{-i}[\log P(\phi_{i_-}, \phi_i, \phi_{i_+}, \boldsymbol{D})]\}, \tag{8}$$

where i_- (or i_+) refers to the ordered indexes that are less than (or greater than) i. Let $\phi_{-i} := (\phi_{i_-}, \phi_{i_+})$. The vector ϕ_{-i} represents the vector ϕ with the ith component ϕ_i removed. The E_{-i} denotes the expectation with respect to ϕ_{-i}.

Based on the joint distribution (1), the priors (2)–(5) and Formula (8), we can derive all variational posteriors (see Appendix A for details). The variational posterior of the π is:

$$q^*(\pi) \sim \mathrm{Dir}\left(\alpha_{(\pi)}\right), \tag{9}$$

where $\alpha_{(\pi)} = E(z_1) + \alpha^{(\pi)}$. The variational posterior of the A_j is:

$$q^*(A_j) \sim \mathrm{Dir}\left(\alpha_{(A_j)}\right) \text{ for } j = 1, \ldots, K, \tag{10}$$

where $\alpha_{(A_j)} = \sum_{t=2}^{T} E(v_{t-1,j} v_{tk}) + \alpha_{jk}^{(A)}$. The variational posterior of the β_k is:

$$q^*(\beta_k) \sim N_p(\beta_k; \mu_k, \Sigma_k), \tag{11}$$

where

$$\begin{aligned} \Sigma_k &= \left(E\left(\tfrac{1}{\sigma_2}\right) \textstyle\sum_{t=1}^{T} E(v_{tk}) x_t x_t^\top + E\left(\tfrac{1}{\sigma^2}\right) E(D_\tau^{-1})\right)^{-1}, \\ \mu_k &= \Sigma_k \left(E\left(\tfrac{1}{\sigma^2}\right) \textstyle\sum_{t=1}^{T} y_t \cdot E(v_{tk}) \cdot x_t\right). \end{aligned}$$

The variational posterior of the σ^2 is:

$$q^*\left(\sigma^2\right) \sim \text{Inverse-Gamma}\left(\alpha_{(\sigma^2)}, \beta_{(\sigma^2)}\right), \tag{12}$$

where $\alpha_{(\sigma^2)} = \frac{T}{2}$ and $\beta_{(\sigma^2)} = \frac{1}{2}\sum_{t=1}^{T}\sum_{k=1}^{K} E(v_{tk})\left[\left(y_t - x_t^\top \mu_k\right)^2 + x_t^\top \Sigma_k x_t\right]$. The variational posterior of the τ_m^2 is:

$$q^*\left(\tau_m^2\right) \sim \text{Generalized-Inverse–Gaussian}\left(C_{\tau_m}, a_{\tau_m}, b_{\tau_m}\right), \tag{13}$$

where $a_{\tau_m} = E(\lambda^2)$, $b_{\tau_m} = E(1/\sigma_2)\sum_{k=1}^{K} E(\beta_{km}^2)$, and $C_{\tau_m} = 1 - K/2$. The variational posterior of the λ^2 is:

$$q^*\left(\lambda^2\right) \sim \Gamma\left(\alpha_{(\lambda^2)}, \beta_{(\lambda^2)}\right), \tag{14}$$

where $\alpha_{(\lambda^2)} = p + r$ and $\beta_{(\lambda^2)} = \delta + \frac{1}{2}\sum_{m=1}^{p} E(\tau_m^2)$.

Based on the dependencies of hidden states, we divide the posterior of z into three parts. The variational posterior of the z_1 is:

$$q^*(z_1) \sim \text{Mult}\left(P_{(z_1)}\right), \tag{15}$$

where the Mult represents multinomial distribution, $P_{(z_1)} = (P_{(z_1)1}, \ldots, P_{(z_1)K})^\top$ and

$$P_{(z_1)k} = \exp\{E[\log \pi_k]\} \exp\left\{E\left[\log P\left(y_1 \mid x_1, \beta_k, \sigma^2\right)\right]\right\} \prod_{j=1}^{K} \exp\left\{E\left[v_{2j}\right] E\left[\log A_{kj}\right]\right\}.$$

The variational posterior of the z_t for $t = 2, \ldots, T-1$ is:

$$q^*(z_t) \sim \text{Mult}\left(P_{(z_t)}\right) \quad \text{for} \quad t = 2, \ldots, T-1, \tag{16}$$

where $P_{(z_t)} = (P_{(z_t)1}, \ldots, P_{(z_t)K})^\top$ and

$$\begin{aligned} P_{(z_t)k} \quad = \quad & \exp\left(E\left[\log P(y_t \mid x_t, \beta_k, \sigma^2)\right]\right) \cdot \textstyle\prod_{j=1}^{K} \exp\left\{E\left[\log A_{jk}\right] E(v_{t-1,j})\right\} \\ & \cdot \textstyle\prod_{j=1}^{K} \exp\left\{E\left[\log A_{kj}\right] E(v_{t+1,j})\right\}. \end{aligned}$$

The variational posterior of the z_T is:

$$q^*(z_T) \sim \text{Mult}\left(P_{(z_T)}\right), \tag{17}$$

where $P_{(z_T)} = (P_{(z_T)1}, \ldots, P_{(z_T)K})^\top$ and

$$P_{(z_T)k} = \exp\left\{E\left[\log P\left(y_T \mid x_T, \beta_k, \sigma^2\right)\right]\right\} \cdot \prod_{j=1}^{K} \exp\left\{E\left[\log A_{jk}\right] E(v_{T-1}, j)\right\}.$$

The expectation $E\left[\log P(y_t \mid x_t, \beta_k, \sigma^2)\right]$ in the above variational posteriors (15)–(17) is expressed as follows:

$$E\left[\log P\left(y_t \mid x_t, \beta_k, \sigma^2\right)\right] = -\frac{1}{2}\log(2\pi) - \frac{1}{2}E[\log(\sigma^2)] - \frac{1}{2}E\left(\frac{1}{\sigma^2}\right)\left[(y_t - x_t^\top \mu_k)^2 + x_t^\top \Sigma_k x_t\right].$$

Note that the expectation part of some parameter posterior formulas is difficult to derive analytically. One feasible method is to use Monte Carlo (MC) sampling to approximate the expectation part that cannot be derived analytically, that is, the Monte Carlo Coordinate Ascent VI (MC-CAVI) [29] algorithm. The MC-CAVI recursion approaches have been proved to be convergent to the maximizer of the ELBO with arbitrarily high probability under regularity conditions. In the article, we also use MC-CAVI to obtain the intractable expectations.

3.3. Implementation

Assume that the expectations $E_{-i}[\log P(\phi_{i-}, \phi_i, \phi_{i+}, \boldsymbol{D})]$ for $i \in I$ within an index set I can be analytically obtained across all updates of the variational density $q^*(\phi)$, and cannot be analytically obtained for $i \notin I$. For the MC-CAVI method, intractable integrals can be approximated using the MC methods if $i \notin I$. Specifically, for $i \notin I$, the samples with the sample size $N \geq 1$ are drawn from the current $q^*_{-i}(\phi_{-i})$ to obtain the expectation estimations as follows:

$$\widehat{E}_{-i}[\log P(\phi_{i-}, \phi_i, \phi_{i+}, \boldsymbol{D})] = \frac{\sum_{n=1}^{N} \log P\left(\phi_{i\_}^{(n)}, \phi_i^{(n)}, \phi_{i_+}^{(n)}, \boldsymbol{D}\right)}{N}.$$

The Algorithm 1 summarizes the implementation of MC-CAVI, where the $q_{i,k}(\phi_i)$ denotes the density of the ith density factor after it has undergone the kth updates, and $q_{-i,k}(\phi_{-i})$ refers to the density of all density factors except the ith factors after the kth updates to the factors preceding the ith factor and the $k-1$ updates to the blocks following it.

Algorithm 1 Main iteration steps of MC-CAVI

Necessary: Number of iteration cycles T.
Necessary: Quantity of Monte Carlo samples denoted as N.
Necessary: $\mathbb{E}_{-i}\left[\log P(\phi_{i\_}, \phi_i, \phi_{i_+}, \boldsymbol{D})\right]$ in closed form for $i \in \mathcal{I}$.
1. Initialize $q_{i,0}(\phi_i)$ for $i = 1, \ldots, b$.
2. for $k = 1 \ldots T$:
3. for $i = 1 \ldots b$:
4. If $i \in \mathcal{I}$:
5. Set $q_{i,k}(\phi_i) \propto \exp\{\mathbb{E}_{-i,k}\left[\log P(\phi_{i\_}, \phi_i, \phi_{i_+}, x, y)\right]\}$;
6. If $i \notin \mathcal{I}$:
7. Obtain N samples $\left(\phi_{i\_,k}^{(n)}, \phi_{i_+,k-1}^{(n)}\right)$ from $q_{-i,k}(\theta_{-i})$ for $n = 1, 2, \ldots, N$;
8. Set $q_{i,k}(\phi_i) \propto \exp\left\{\frac{\sum_{n=1}^{N} \log p\left(\phi_{i\_,k}^{(n)}, \phi_i, \phi_{i_+,k-1}^{(n)}, D\right)}{N}\right\}$;
9. end
10. end.

Combining with the MC-CAVI algorithm, we can summarize the implementation algorithm for variational posteriors for all parameters as follows in Algorithm 2. Based on the Algorithm 2, we can adopt the variational posterior means of the parameters as the estimators.

Algorithm 2 Variational Bayesian Algorithm for the high-dimensional HMMs

Data Input: $\{(x_t, y_t)\}, t = 1, \ldots, T$;
Hyperparameter Input: $\alpha^{(\pi)}, r > 0, \delta > 0$, and $\alpha_{jk}^{(A)}$ for $k, j = 1, \ldots K$;
Initialize: $\alpha_{(\pi)}, \alpha_{(A_j)}, \alpha_{(\sigma^2)}, \beta_{(\sigma^2)}, \beta_{(\lambda^2)}, \Sigma_k$ and μ_k for $k = 1, \ldots K$,
  a_{τ_m} and b_{τ_m} for $m = 1, \ldots, p$, iteration-index $\ell = 1$, a sufficiently small $\epsilon = 10^{-6}$
  and a maximum iteration times $M = 1000$;
While the absolute change of the iterated ELBO $|L^{\ell} - L^{\ell-1}| > \epsilon$ and $\ell < M$ **do**:
  Update $\alpha_{(\pi)}$ and $q^*(\pi)$ according to Equation (9);
  Estimate $E[\log \pi_k]$ by the MC method;
  for $j = 1, \ldots, K$ **:**
    Update $\alpha_{(A_j)}$ and $q^*(A_j)$ according to Equation (10);
    Estimate $E\left[\log A_{jk}\right]$ by the MC method;
  end
  for $k = 1, \ldots, K$ **:**
    Update Σ_k, μ_k and $q^*(\beta_k)$ according to Equation (11);
  end
  Update $\alpha_{(\sigma^2)}, \beta(\sigma^2)$ and $q^*(\sigma^2)$ according to Equation (12);
  Estimate $E\left[\log(\sigma^2)\right]$ by the MC method;
  for $m = 1, \ldots, p$ **:**
    Update a_{τ_m}, b_{τ_m} and $q^*(\tau_m^2)$ according to Equation (13);
  end
  Update $\beta_{(\lambda^2)}$ and $q^*(\lambda^2)$ according to Equation (14);
  Update $P_{(z_1)}$ and $q^*(z_1)$ according to Equation (15);
  Update $P_{(z_t)}$ and $q^*(z_t)$ according to Equation (16);
  Update $P_{(z_T)}$ and $q^*(z_T)$ according to Equation (17);
  Compute the ELBO using the formula (6), denoted as L^{ℓ},
  and the absolute change of the iterated ELBO $|L^{\ell} - L^{\ell-1}|$;
  $\ell \to \ell + 1$;
Output: the variational densities $q^*(\pi)$, $q^*(A_j)$ for $j = 1, \ldots, K$, $q^*(\beta_k)$ for $k = 1, \ldots, K$,
  $q^*(\sigma^2), q^*(\lambda^2), q^*(\tau_m^2)$ for $m = 1, \ldots, p$, and $q^*(z_t)$ for $t = 1, \ldots, T$;
  and the posterior modes of parameters β_k for $k = 1, \ldots, K$.

4. Simulation Studies

In this section, we carry out simulation studies to investigate the finite sample performances of the proposed method, denoted as HDVBHMM. To evaluate the prediction performance, we compare the proposed method with some commonly used and popular methods, including Back Propagation Neural Network (BP), Long Short-Term Memory (LSTM), and Random Forest. The experimental code can be found via the github link (https://github.com/LiuWei-hub/VBHDHMM, accessed on 23 March 2024).

We consider the dataset $\{x_t, y_t : t = 1, \ldots, T\}$, where T is the number of the discrete time intervals, the covariate x_t is generated from the Gaussian distribution $N_p(0, 2I_p)$, and $y_t = x_t^{\top}\beta_{z_t} + \varepsilon_t$, in which the random error $\varepsilon_t \sim N(0, \sigma^2)$, and z_t is hidden state. Here, the initial hidden state z_1 is generated from Mult(π), where $\pi = (\pi_1, \pi_2, \ldots, \pi_K)$. For $t = 2, \ldots, T$, the hidden state z_t is generated from Mult(A_j), where $A_j = (A_{1i}, A_{2i}, \ldots A_{Ki})$ and $A_{jk} = P(z_t = k \mid z_{t-1} = j)$. We set the number of hidden states $K = 3$, $\sigma = 0.4$, and $(\pi_1, \pi_2, \ldots, \pi_K) = (0.6, 0.3, 0.1)^{\top}$.

To assess the predictive performance, we use the samples in the last m time intervals as the testing set and the samples in the first $T - m$ time intervals as the training set. In addition, we use four criteria: (1) the mean absolute percentage error MAPE $= \frac{100\%}{m}\sum_{t=1}^{m}\left|\frac{\hat{y}_t - y_t}{y_t}\right|$, where y_t is the true value and $\hat{y}_t$ represents the predicted value; (2) the root mean square error RMSE $= \sqrt{\frac{1}{m}\sum_{t=1}^{m}(y_t - \hat{y}_t)^2}$; (3) the mean absolute error MAE $= \frac{1}{m}\sum_{t=1}^{m}|y_t - \hat{y}_t|$;

and (4) $R^2 = 1 - \frac{\sum_{t=1}^{m}(\hat{y}_t - y_t)^2}{\sum_{t=1}^{m}(\bar{y} - y_t)^2}$, where $\bar{y}$ represents the sample mean, $\sum_{t=1}^{m}(\hat{y}_t - y_t)^2$ is the error caused by the prediction, and $\sum_{t=1}^{m}(\bar{y} - y_t)^2$ is the error caused by the mean. The smaller the MAPE, RMSE and MAE values are, the better the performance of the method is. The larger the R^2 is, the better the performance of the method is. To evaluate the performance of the parameter estimation, we use two criteria: (1) the root mean square error loss RMSE $= \sqrt{\frac{1}{n}\sum_{i=1}^{n}(\hat{\theta}_i - \theta)^2}$, where n is the number of repeated experiments, $\hat{\theta}_i$ is the estimated value of the parameter obtained in the ith experiment, and θ is the true parameter value; and (2) Bias$(\hat{\theta}) = \frac{1}{n}\sum_{i=1}^{n}\hat{\theta}_i - \theta$. The RMSE and Bias values closer to zero imply better performance for the method. We repeat 10 simulation examples and calculate the average values of the above metrics for each method.

4.1. Experiment 1

In experiment 1 , we consider different dimensions p = 20, 30 and 40. In addition, the state transition matrix A is set as follows:

$$A = \begin{pmatrix} 0.2 & 0.3 & 0.5 \\ 0.1 & 0.6 & 0.3 \\ 0.5 & 0.4 & 0.1 \end{pmatrix}.$$

Due to $K = 3$, we have three regression coefficients $\beta_1, \beta_2, \beta_3$. We set the coefficient as follows:

$$\beta = (\beta_1, \beta_2, \beta_3)^\top = \begin{pmatrix} 0.5 & 1 & 1.5 \\ -2 & -2 & -1.5 \\ 2 & 1.5 & 1 \\ -1 & -1.5 & -2 \\ 0 & 0 & 0 \\ \vdots & \vdots & \vdots \\ 0 & 0 & 0 \end{pmatrix}_{p\times 3}$$

where the first four rows are nonzero and other elements are zero. We set the number of the discrete time intervals $T = 300$ and the sample size in the testing set $m = 10$. In addition, the hyperparameters r, δ in the HDVBHMM method are set to 1. The results are shown in Tables 1 and 2.

In Table 1, the smaller MAPE, RMSE, and MAE index values, the better the algorithm performance. The larger the R^2 index, the better the algorithm performance. Bold indicates the optimal result in each scenario. It is clear that our method is optimal in all cases (bold), especially for $p = 20$, $p = 30$, and $p = 40$. In the small sample case, the prediction performance of the LSTM method decreases significantly as the dimensionality of the covariates increases. The prediction performance of the Random Forest and BP methods is not stable with increasing covariate dimensions. Although the performance of our method decreases as the covariate dimension increases, it is still significantly better than the other methods. Table 2 shows the RMSE and Bias of the estimated values of β and A. From Table 2, we can see that the proposed method performs well. Two metrics are small when the covariate dimension is 20 and 30. When the dimension is increased to 40, the value of the RMSE index increases, but it is still within the acceptable range.

Table 1. Average values of four metrics of all approaches with standard deviation in each parenthesis based on 10 simulations under $T = 300$.

p	Method	Estimate Performance			
		MAPE	RMSE	MAE	R^2
$p = 20$	LSTM	0.957 (1.109)	1.183 (0.132)	1.416 (0.316)	0.871 (0.057)
	BP	0.948 (1.018)	1.210 (0.175)	1.492 (0.436)	0.826 (0.117)
	Random Forest	1.215 (1.844)	1.430 (0.194)	2.081 (0.531)	0.741 (0.119)
	HDVBHMM	**0.467** (0.325)	**1.008** (0.152)	**1.038** (0.329)	**0.887** (0.082)
$p = 30$	LSTM	0.949 (0.485)	1.354 (0.266)	1.898 (0.740)	0.789 (0.144)
	BP	1.312 (0.685)	1.524 (0.198)	2.358 (0.615)	0.659 (0.169)
	Random Forest	1.081 (0.641)	1.568 (0.223)	2.505 (0.717)	0.595 (0.252)
	HDVBHMM	**0.876** (0.919)	**1.186** (0.244)	**1.461** (0.629)	**0.861** (0.073)
$p = 40$	LSTM	1.471 (1.121)	1.404 (2.471)	2.026 (0.716)	0.763 (0.131)
	BP	1.555 (1.218)	1.427 (0.156)	2.060 (0.412)	0.772 (0.117)
	Random Forest	1.210 (0.718)	1.457 (0.231)	2.173 (0.736)	0.754 (0.115)
	HDVBHMM	**1.023** (0.759)	**1.155** (0.264)	**1.398** (0.619)	**0.822** (0.202)

In the Long-Term and Short-Term Memory methods, the learning rate is lr = 0.001, the number of training cycles (Epochs) is set to 50, and the size of the hidden layer is set to 10. The hidden layer of the BP method consists of 20 neurons, and the maximum number of iterations is set to 10,000. In the random forest regression model, the number of trees is set to 100. The bold results are the optimal ones among four methods.

Table 2. Average values of the RMSE and Bias of A and β based on 10 replications in Experiment 1.

p	Parameter	Estimate Performance	
		RMSE	Bias
$p = 20$	β	0.001	0.001
	A	0.002	0.001
$p = 30$	β	0.002	0.001
	A	0.005	0.001
$p = 40$	β	0.004	0.001
	A	0.011	0.001

To better illustrate the performance of parameter estimation, Figure 1 shows box plots of the estimator values of A, β_1, β_2, β_3 under $p = 30$, where the horizontal coordinate is the index of the variables and the vertical coordinate is the values of estimators. The corresponding figures on $p = 20$ and $p = 40$ are shown in Appendix A.2. For the estimators β_1, β_2 and β_3, we can see the first four elements are estimated close to the true value, and the remaining values are estimated clear to zero; This implies that the proposed method can achieve good variable selection performance. In addition, all elements of the state increment matrix A are estimated close to the true values, which also confirms the good performance of our method.

In addition, to further verify that the algorithm is sensitive to the choice of hyperparameters r, δ, we conduct experiments on data with a covariate dimension of 30. Consider the following three experiments, the first with $r = 0.5, \delta = 0.5$; the second with $r = 1.0, \delta = 1.0$; and the third with $r = 1.5, \delta = 1.5$. The experimental results show that the estimation results are not sensitive to the choice of the two hyperparameters r and δ. The images of the Gamma distributions for the three different hyperparameter settings are very similar in shape. This similarity may contribute to the reason why, for a certain range of variations in r and δ values, the model's performance may not show sensitivity to these hyperparameters. We show the results in Appendix A.4.

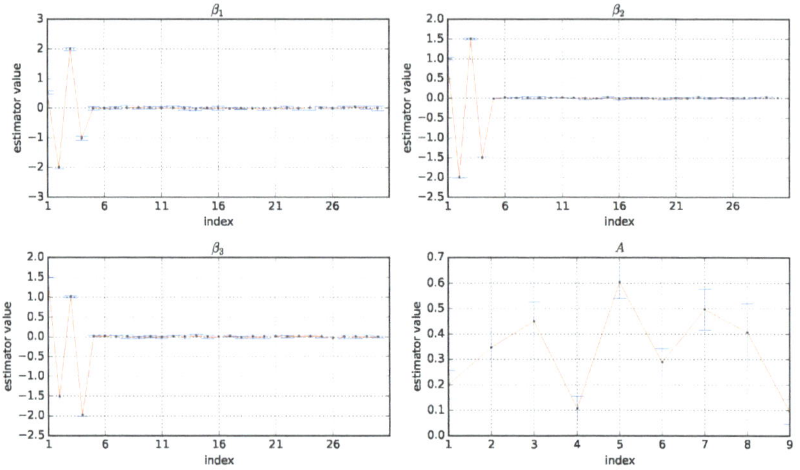

Figure 1. Box plots of the estimator values of A, β_1, β_2, β_3 based on 10 experiments under $p = 30$ and $T = 300$. The horizontal coordinate is the index of the variables and the vertical coordinate is the value of the estimators.

4.2. *Experiment 2*

In experiment 2, we consider the higher dimension cases: p = 60, 90, 120. We set the same A as experiment 1 and the coefficient as follows:

$$\beta = (\beta_1, \beta_2, \beta_3)^\top = \begin{pmatrix} 0.5 & 1 & 1.5 \\ -2 & -2 & -1.5 \\ 2 & 1.5 & 1 \\ -1 & -1.5 & -2 \\ \vdots & \vdots & \vdots \\ 0 & 0 & 0 \end{pmatrix}_{p\times 3}$$

where the first four rows are nonzero and other elements are zero. We set the number of discrete time intervals $T = 600$ and the sample size in the testing set $m = 10$. In addition, the hyperparameters r and δ in the HDVBHMM method are set to 1. The results are shown in Tables 3 and 4.

As can be seen in Table 4, when the covariate dimensions are increased and the sample size reaches 600, our method still performs well among the four methods. It should be noted that when the covariate dimension is 90 and 120, the MAPE metric of the random forest method is slightly smaller than our method. In addition, as the covariate dimension increases from p = 60 to P = 120, the performance of the LSTM method decreases significantly, which is the worst performance among the four methods. This shows that LSTM does not perform well on such small-sample high-dimensional datasets. As the dimensionality of the covariates increases, although the BP and Random Forest methods show better prediction performance than the LSTM method, they are also poorer than the prediction performance of the HDVBHMM method. Overall, our method outperforms the other three methods in terms of prediction performance as the dimensionality increases, suggesting that our method performs better on small-sample high-dimensional datasets.

Figure 2 shows box plots of the estimator values of A, β_1, β_2, β_3 under $p = 90$. The corresponding figures on $p = 60$ and $p = 120$ are shown in Appendix A.3. From Figure 2, we can see that the regression coefficients β_1, β_2, and β_3 are accurately estimated. the first four elements are estimated close to the true value, and the remaining values are estimated clear to zero. It implies that the proposed method can successfully achieve variable screening even as the covariate dimension increases. In addition, all elements of the state increment matrix A are estimated close to the true values, which also confirms the good performance of the proposed method.

Table 3. Average values of four metrics of all approaches with standard deviation in each parenthesis based on 10 simulations under $T = 600$.

p	Method	Estimate Performance			
		MAPE	RMSE	MAE	R^2
$p = 60$	LSTM	1.608 (1.826)	1.524 (0.104)	2.332 (0.304)	0.722 (0.126)
	BP	1.511 (1.326)	1.513 (0.209)	2.328 (0.662)	0.725 (0.143)
	Random Forest	1.236 (1.748)	1.459 (0.175)	2.159 (0.525)	0.728 (0.151)
	HDVBHMM	**0.851** (0.871)	**1.091** (0.218)	**1.235** (0.489)	**0.884** (0.078)
$p = 90$	LSTM	1.690 (2.046)	1.725 (0.152)	2.998 (0.542)	0.523 (0.282)
	BP	2.780 (5.010)	1.606 (0.225)	2.626 (0.706)	0.636 (0.239)
	RF	**0.718** (0.365)	1.374 (0.246)	1.942 (0.694)	0.797 (0.091)
	HDVBHMM	0.878 (0.887)	**1.156** (0.249)	**1.392** (0.617)	**0.862** (0.135)
$p = 120$	LSTM	1.941 (1.839)	1.884 (0.372)	3.677 (1.389)	0.463 (0.323)
	BP	1.235 (1.023)	1.651 (0.274)	2.792 (0.854)	0.684 (0.181)
	Random Forest	**0.832** (0.679)	1.571 (0.193)	2.502 (0.621)	0.718 (0.154)
	HDVBHMM	0.910 (0.717)	**1.321** (0.253)	**1.804** (0.718)	**0.763** (0.393)

In the Long-Term and Short-Term Memory methods, the learning rate is lr = 0.001, the number of training cycles (Epochs) is set to 50, and the size of the hidden layer is set to 10. The hidden layer of the BP method consists of 20 neurons, and the maximum number of iterations is set to 10,000. In the random forest regression model, the number of trees is set to 100. The bold results are the optimal ones among four methods.

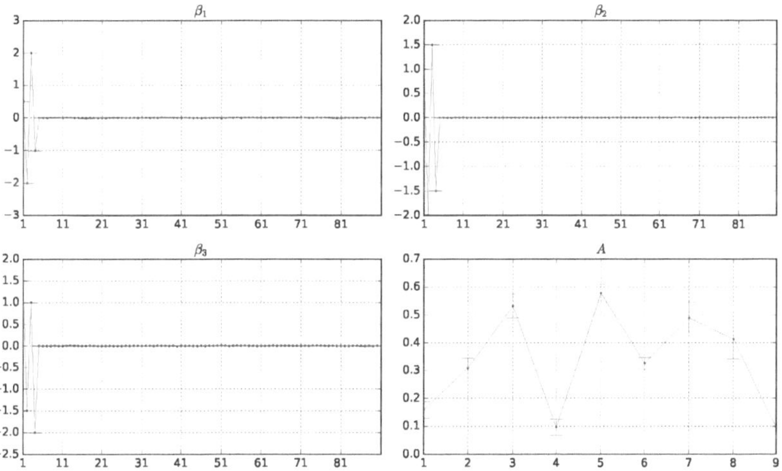

Figure 2. Box plots of the estimator values of A, β_1, β_2, β_3 based on 10 experiments under $p = 90$ and $T = 600$. The horizontal coordinate is the index of the variables and the vertical coordinate is the value of the estimators.

Table 4. Average values of the RMSE and Bias of A and β based on 10 replications in Experiment 2.

p	Parameter	Estimate Performance	
		RMSE	Bias
60	β	0.001	0.001
	A	0.002	0.001
90	β	0.001	0.001
	A	0.005	0.001
120	β	0.015	0.007
	A	0.034	0.009

5. Application to Real Datasets

In this section, we focus on Beijing Multi-Site Air-Quality data, which include 6 major air pollutants and 6 related meteorological variables at multiple locations in Beijing. These air-quality measurements are created by the Beijing Municipal Environmental Monitoring Center. In addition, meteorological data at each air quality location are paired with the nearest weather station provided by the China Meteorological Administration. The data span from 1 March 2013 to 28 February 2017. In our study, we consider PM2.5 concentration as response variable, and PM10 concentration, SO_2 concentration, NO_2 concentration, CO concentration, O_3 concentration, Temperature (TEMP), Pressure (PRES), Dew point temperature (DEWP), Precipitation (RAIN), and Wind speed (WSPM) as covariates; that is, $p = 10$. In order to study the performance on small sample datasets, we delete the missing values in the data and select the data samples in the first 200 time intervals from the Shunyi observation point in Beijing in 2017 for analysis. To assess the predictive performance, we use the first 140 samples as the training set, and the remaining 60 samples as the testing set. We compare the proposed method with the BP neural network, LSTM and Random Forest method similar to Section 4.

One of the main challenges in implementing the HMM is to determine the optimal number of hidden states. The Akaike Information Criterion (AIC) and the Bayesian Information Criterion (BIC) are two common model selection techniques, which select the best model by balancing the fitting accuracy and complexity of the model. In selecting the number of hidden states for a Hidden Markov Model, both AIC and BIC evaluate multiple models containing different numbers of states and select an optimal model that balances fitting accuracy and complexity. Multiple HMMs are trained separately using different numbers of hidden states, then the AIC or BIC values are calculated for each model, and finally the model with the smallest AIC or BIC value is selected [34]. Similar to the work of Dofadar et al. [34], we use AIC and BIC to select the number of the hidden states. The AIC equation used in this study is given by $AIC = 2k - 2L$, where k is the number of free parameters in the model and L is the log probability value. The formula for k used in this research is $k = n^2 + 2n - 1$, where n is the current value of the hidden state. The BIC equation used in this study is expressed as $BIC = \ln(T)k - 2L$, where T is the total number of observations. To find the best number of hidden states, we calculate AIC and BIC values based on the different numbers of hidden states: 2, 3, 4, and 5. The results are shown in Figure 3. Figure 3 shows that when the number of hidden states is 3, the AIC and BIC values are the smallest, indicating that choosing the number of hidden states as 3 is the closest to the real model. Therefore, we set the the number of hidden states $K = 3$.

Similar to Section 4, we calculate MAPE, RMSE, MAE and R^2 to evaluate the predictive performance. Since the time series data are positively skewed, MAE and MASE are the best evaluation metrics for evaluating the model performance [35]. The results are shown in Table 5. From Table 5, we can see that the MAPE and MAE of the proposed HDVBHMM method are smaller than ones of other methods, and the R^2 value of the proposed method is larger than one of other methods, indicating that the performance of the HDVBHMM method is better than other methods. Among the other three competing methods, the MAPE and MAE values of the Random Forest method are lowest among those of the three competing methods, but its MAPE and MAE are still much larger than ones of the proposed method. The BP method is the worst performing among four methods with MAE = 27.570 and MAPE = 1.025.

To better illustrate the predictive performance, Figure 4 shows the true data and predicted values via four methods on the testing set. From Figure 4, we can see that in the first 30 time points, the proposed method fits the true values very well. In the second set of 30 time points, as the prediction time period increases, the predicted values exhibit a slight error, but they are still better than those of other methods. Overall, the prediction accuracy of the proposed method is much better than ones of other methods in term of both short and long time periods.

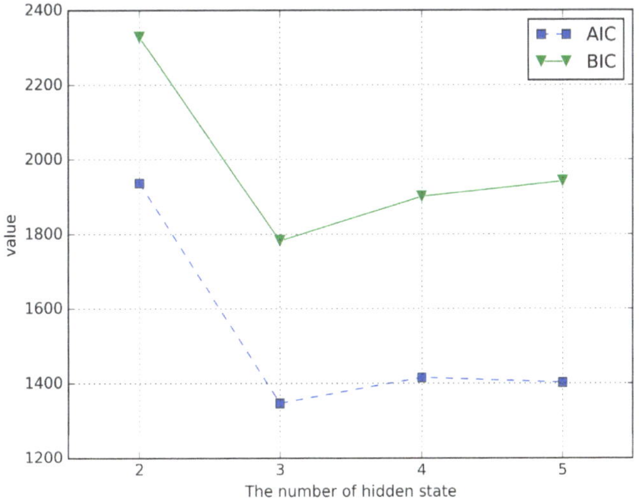

Figure 3. AIC and BIC values when the number of hidden states is 2, 3, 4, and 5 on the real dataset.

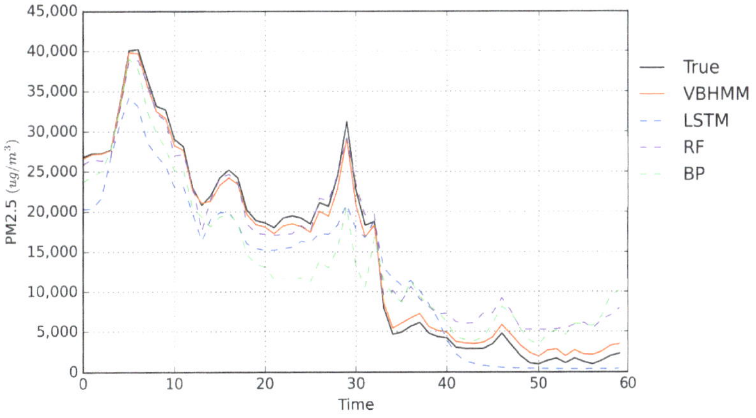

Figure 4. Comparison of observed hourly PM2.5 emissions (test set) with PM2.5 emissions predicted by four methods.

Table 5. Prediction Performance of Four Methods on the testing data.

Method	MAPE	RMSE	MAE	R^2
Random Forest	1.043	4.039	16.318	0.941
BP	1.025	5.250	27.570	0.852
LSTM	0.462	4.924	24.249	0.876
HDVBHMM	**0.317**	**2.249**	**5.058**	**0.993**

In the Long-Term and Short-Term Memory methods, learning rate is lr = 0.001, the number of training cycles (Epochs) is set to 50, and the size of the hidden layer is set to 40. the BP method contains 12 neuron hidden layers and the maximum number of iterations is set to 10,000. The hyperparameters hyperparameters r, δ in the HDVBHMM method were set to 1.0. The bold results are the optimal ones among four methods.

The estimated values of β corresponding to the three states are shown in Table 6. From Table 6, we can see that PM10, SO_2, TEMP (temperature), DEWP (dew point temperature), RAIN (precipitation), and WSPM (wind speed) have the greatest influence on PM2.5 emissions in state 1. PM10, SO_2, TEMP, and DEWP are the four factors that have a negative effect on the presence of PM2.5 emissions in the area, and as these four factors increase, PM2.5 emissions will decrease; meanwhile RAIN and WSPM have a positive effect on the presence of PM2.5 in the area. Rainfall and high wind speed may have increased PM2.5 concentrations through physical effects (such as windblown dust). The prediction formula of the PM2.5 in State 1 is as follows:

$$\begin{aligned} \text{PM2.5}_1 = & -0.833\text{PM10} - 0.349\text{SO}_2 + 0.003\text{CO} - 0.03\text{NO}_2 - 0.09\text{O3} - 2.625\text{TEMP} \\ & - 0.017\text{PRES} - 3.552\text{DEWP} + 1.524\text{RAIN} + 3.547\text{WSPM}. \end{aligned}$$

Table 6. Estimates of the regression coefficients β for each hidden state.

State	PM10	SO_2	NO_2	CO	O_3	TEMP	PRES	DEWP	RAIN	WSPM
State 1	−0.833	−0.349	−0.030	0.003	−0.090	−2.625	−0.017	−3.552	1.524	3.547
State 2	0.965	−0.230	−0.189	−0.003	0.004	−2.154	−0.010	0.745	−4.227	3.891
State 3	−0.303	1.080	1.462	0.000	−0.279	18.55	−0.123	−0.127	9.057	19.217

In addition, PM10, Sulfur Dioxide, Nitrogen Dioxide, TEMP (temperature), DEWP (dew point temperature), RAIN (precipitation), and WSPM (wind speed) have the largest effect on PM2.5 in State 2. The results showed that in state 2, some chemical reactions led to the depletion of gases such as SO_2 and NO_2, which reduced the production of PM2.5, and rainfall also reduced the production of PM2.5. The high wind speed led to an increase in PM2.5 concentration, probably because the wind speed increased the diffusion and transport of particulate matter. The prediction formula of the PM2.5 in this State is as follows:

$$\begin{aligned} \text{PM2.5}_2 = & 0.965\text{PM10} - 0.23\text{SO}_2 - 0.003\text{CO} - 0.189\text{NO}_2 - 0.04\text{O3} - 2.154\text{TEMP} \\ & - 0.01\text{PRES} + 0.745\text{DEWP} - 4.227\text{RAIN} + 3.891\text{WSPM}. \end{aligned}$$

PM10, SO_2, NO_2, O_3, TEMP (temperature), PRES (pressure), DEWP (dew point temperature), RAIN (precipitation), and WSPM (wind speed) have the greatest impact on PM2.5 in state 3. It is worth noting that the increase in variables such as SO_2 and NO_2 leads to an increase in PM2.5 concentration. In addition, the significant positive coefficients for temperature indicate that higher temperatures promote the formation of PM2.5, which may be related to the acceleration of certain chemical reactions by high temperatures. The increase in SO_2 and NO_2 may promote the formation of secondary particulate matter, which in turn increases the PM2.5 concentration. Wind speed increases particulate dispersion, and rainfall may also promote the formation of secondary particulate matter from some soluble substances. The prediction formula of the PM2.5 in this State is as follows:

$$\begin{aligned}\text{PM2.5}_3 = &- 0.303\text{PM10} + 1.08\text{SO}_2 + 1.462\text{NO}_2 - 0.279\text{O3} + 18.55\text{TEMP} \\ &- 0.123\text{PRES} - 0.127\text{DEWP} + 9.057\text{RAIN} + 19.217\text{WSPM}.\end{aligned}$$

In summary, the regression coefficients for the three states reflect the effects of different environmental factors on PM2.5 concentrations. The positive and negative signs and magnitudes of these coefficients can provide scenarios on how to manage and predict PM2.5 concentrations by controlling these environmental factors under different environmental conditions. In particular, the fact that temperature, rainfall and wind speed have different effects on PM2.5 concentrations in different states suggests that PM2.5 management needs to take into account complex meteorological conditions and interactions between air pollutants.

6. Conclusions

In this paper, the variable selection for high-dimensional HMMs is studied based on the variational inference. We develop a Variational Bayesian method with the double-exponential shrinkage prior for variable selection. The proposed method can quickly and accurately estimate the posterior distributions and the unknown parameters. In the simulation studies and real data analysis, the proposed method outperforms the common methods in term of variable selection and prediction. In the Beijing Multi-Site Air-Quality analysis, we select the optimal number of the hidden stats based on the AIC and BIC methods, and fit the HMMs of the response variable PM2.5. In the current research work, we investigate variational inference for linear HMMs with high dimensional covariates; that is, the mean of the response variable is linear with respect to the high dimensional covariates. Many of the relationships between variables in practical applications may be not linear, so variational inference for nonlinear HMMs is worth studying. In addition, it is assumed that the variances in observations are the same in different hidden states in this study, but in practical applications, heteroskedasticity may be more in line with real-world data characteristics. For that reason, the heteroskedasticity issue for HMMs is also worth exploring deeply. Moreover, Ivan Gorynin's work [36] verifies that the Pairwise Markov Model (PMM) outperforms the traditional HMM in terms of accuracy when the observed variable y is highly autocorrelated or when the hidden chain is not Markovian. Unlike the HMM, which assumes that the hidden chain z is Markovian, the PMM assumes that (z, y) is Markovian. Since hidden chains are not necessarily Markovian in the PMM, it is more general than the HMM. Parameter estimation of PMM models is done using Variational Bayesian methods in the work of Katherine Morales [37]. However, the effect of including the covariate x on the target variable y was not considered in their work. Therefore, as an extension of the proposed method, which replaces the HMM with the PMM, the inclusion of high-dimensional covariates in the PMM may yield more accurate predictions.

Author Contributions: Conceptualization, Y.Z. (Yao Zhai), W.L. and Y.J.; methodology, Y.Z. (Yao Zhai), W.L. and Y.Z. (Yanqing Zhang); software, W.L., Y.Z. (Yao Zhai) and Y.J.; validation, Y.Z. (Yao Zhai), W.L., Y.J. and Y.Z. (Yanqing Zhang); formal analysis, Y.Z. (Yao Zhai) and Y.J.; investigation, Y.Z. (Yao Zhai), W.L. and Y.J.; resources, Y.Z. (Yao Zhai) and Y.J.; data curation, Y.Z. (Yao Zhai); writing—original draft preparation, Y.Z.; writing—review and editing, Y.J. and Y.Z. (Yanqing Zhang); visualization, Y.J.; supervision, Y.J.; project administration, Y.J.; funding acquisition, Y.J. and Y.Z. (Yanqing Zhang). All authors have read and agreed to the published version of the manuscript.

Funding: This research was funded by the National Key R&D Program of China (No. 102022YFA1003701), the National Natural Science Foundation of China (No. 12271472, 12231017, 12001479, 11871420), the Natural Science Foundation of Yunnan Province of China (No. 202101AU070073 and 202201AT070101), and Yunnan University Graduate Student Research and Innovation Fund Project Grant (No. KC-22221108).

Data Availability Statement: The research data are available on the website https://archive.ics.uci.edu/dataset/501/beijing+multi+site+air+quality+data (accessed on 23 March 2024).

Acknowledgments: We would like to thank the action editors and referees for insightful comments and suggestions which improve the article significantly.

Conflicts of Interest: The authors declare no conflicts of interest.

Appendix A

Appendix A.1. Variational Posterior of Parameters

We derive the optimal variational densities based on Formula (8). The complete likelihood function of the model is:

$$\begin{aligned}
&P\left(Y, z \mid X, \pi, A, \beta, \sigma^2\right) \\
&= P(z_1|\pi) \prod_{t=2}^{T} P(z_t \mid z_{t-1}, A) \prod_{t=1}^{T} P\left(y_t \mid x_t, z_t, \beta, \sigma^2\right) \\
&= \left(\prod_{k=1}^{K} \pi_k^{v_{1k}}\right) \left(\prod_{t=2}^{T} \prod_{k=1}^{K} \prod_{j=1}^{K} A_{jk}^{v_{t-1,j} v_{tk}}\right) \left(\prod_{t=1}^{T} \prod_{k=1}^{K} P\left(y_t \mid x_t, \beta_k, \sigma^2\right)^{v_{tk}}\right).
\end{aligned}$$

We derive the conditional posterior distribution of A as:

$$\begin{aligned}
P(A \mid \cdot) &\propto P(Y, z \mid X, \pi, A, \beta, \sigma^2) P(A) \\
&\propto P(z_1|\pi) \left[\prod_{t=2}^{T} P(z_t \mid z_{t-1}, A)\right] \left[\prod_{t=1}^{T} P(y_t \mid x_t, z_t, \beta, \tau)\right] P(A) \\
&\propto \left(\prod_{t=2}^{T} \prod_{k=1}^{K} \prod_{j=1}^{K} A_{jk}^{v_{t-1,j} \cdot v_{tk}}\right) \left(\prod_{j=1}^{K} \prod_{k=1}^{K} A_{jk}^{\alpha_{jk}^{(A)} - 1}\right) \\
&\propto \prod_{j=1}^{K} \left(\prod_{k=1}^{K} A_{jk}^{\sum_{t=2}^{T} v_{t-1,j} \cdot v_{tk} + \alpha_{jk}^{(A)} - 1}\right).
\end{aligned}$$

So $P(A_j \mid \cdot) \sim D_{ir}\left(\sum_{t=2}^{T} v_{t-1,j} \cdot v_{tk} + \alpha_{jk}^{(A)}\right)$ for $j = 1, \ldots K$.
According to Equation (10), The variational posterior distribution of A_j is given by:

$$\begin{aligned}
q^*(A_j) &\propto \exp\{E[\log P(A_j \mid \cdot)]\} \\
&\propto \exp\left\{E\left[\sum_{k=1}^{K} \left(\sum_{t=2}^{T} v_{t-1,j} \cdot v_{tk} + \alpha_{jk}^{(A)} - 1\right) \log A_{jk}\right]\right\} \\
&\propto \prod_{k=1}^{K} A_{jk}^{\sum_{t=2}^{T} E\left(v_{t-1,j} v_{tk}\right) + \alpha_{jk}^{(A)} - 1}.
\end{aligned}$$

So $q^*(A_j) \sim \mathrm{Dir}\left(\sum_{t=2}^{T} E\left(v_{t-1,j} v_{tk}\right) + \alpha_{jk}^{(A)}\right)$.

Similarly, we derive the conditional posterior distribution of σ^2 as:

$$\begin{aligned}
P\left(\sigma^2 \mid \cdot\right) &\propto \prod_{t=1}^{T} P\left(y_t \mid \sigma^2, x_t, z_t, \beta\right) P\left(\sigma^2\right) \\
&\propto \prod_{t=1}^{T}\prod_{k=1}^{K}\left(\frac{1}{\sqrt{2\pi\sigma}}\right)^{v_{tk}} \exp\left\{-\frac{1}{2\sigma^2}\left(y_t - x_t^\top \beta_k\right)^2\right\}^{v_{tk}} \sigma^{-2} \\
&\propto \left(\sigma^2\right)^{-\left(\frac{T}{2}+1\right)} \exp\left\{-\frac{\sum_{t=1}^{T}\sum_{k=1}^{K} v_{tk}\left(y_t - x_t^\top \beta_k\right)^2}{2\sigma^2}\right\} \\
&\sim \text{IGamma}\left(\frac{T}{2}, \frac{1}{2}\sum_{t=1}^{T}\sum_{k=1}^{K} v_{tk}\left(y_t - x_t^\top \beta_k\right)^2\right).
\end{aligned}$$

The variational posterior distribution of σ^2 is given by:

$$\begin{aligned}
&q^*\left(\sigma^2 \mid \cdot\right) \propto \exp\left\{E\left[-\left(\frac{T}{2}+1\right)\cdot \log\left(\sigma^2\right) + \left(-\frac{\sum_{t=1}^{T}\sum_{k=1}^{K} v_{tk}\left(y_t - x_t^T \beta_k\right)^2}{2\sigma^2}\right)\right]\right\} \\
&\propto \left(\sigma^2\right)^{-\left(\frac{T}{2}+1\right)} \exp\left\{-\frac{\sum_{t=1}^{T}\sum_{k=1}^{K} E(v_{tk})\cdot E\left(y_t - x_t^\top \beta_k\right)^2}{2\sigma^2}\right\} \\
&\sim \text{IGamma}\left(\frac{T}{2}, \frac{1}{2}\sum_{t=1}^{T}\sum_{k=1}^{K} E(v_{tk}) E\left(y_t - x_t^\top \beta_k\right)^2\right).
\end{aligned}$$

We derive the conditional posterior distribution of λ^2 as:

$$\begin{aligned}
P\left(\lambda^2 \mid \cdot\right) &\propto P\left(\tau_1^2, \ldots, \tau_p^2 \mid \lambda^2\right) P\left(\lambda^2\right) \\
&\propto \prod_{m=1}^{p} \frac{\lambda^2}{2} \exp\left\{-\frac{\lambda^2 \tau_m^2}{2}\right\}\left(\lambda^2\right)^{r-1} \exp\left(-\delta\lambda^2\right) \\
&\propto \left(\lambda^2\right)^{p+r-1} \cdot \exp\left\{-\left(\delta + \frac{1}{2}\sum_{m=1}^{p} \tau_m^2\right)\lambda^2\right\} \\
&\sim \Gamma\left(p+r, \delta + \frac{1}{2}\sum_{m=1}^{p} \tau_m^2\right).
\end{aligned}$$

The variational posterior distribution of λ^2 is given by:

$$\begin{aligned}
&q^*\left(\lambda^2 \mid \cdot\right) \propto \exp\left\{E\left[\log P\left(\lambda^2 \mid \cdot\right)\right]\right\} \\
&\propto \exp\left\{E\left[(p+r-1)\cdot \log \lambda^2 + \left\{-\left(\delta + \frac{1}{2}\sum_{m=1}^{p} \tau_m^2\right)\lambda^2\right\}\right]\right\} \\
&\propto \left(\lambda^2\right)^{p+r-1} \cdot \exp\left\{-\left(\delta + \frac{1}{2}\sum_{m=1}^{p} E\left(\tau_m^2\right)\right)\lambda^2\right\} \\
&\sim \Gamma\left(p+r, \delta + \frac{1}{2}\sum_{m=1}^{p} E\left(\tau_m^2\right)\right).
\end{aligned}$$

We derive the conditional posterior distribution of τ_m^2. Note that since the variational posterior of τ_m^2 is difficult to obtain, we derive the variational posterior of $\frac{1}{\tau_m^2}$ as:

$$\begin{aligned}
P\left(\tau_m^2 \mid \cdot\right) &\propto P\left(\beta_{1m}, \beta_{2m}, \ldots \beta_{Km} \mid \tau_m^2\right) P\left(\tau_m^2\right) \\
&\propto \prod_{k=1}^{K} \frac{1}{\sqrt{2\pi}\sigma\tau_m} \exp\left\{-\frac{1}{2\tau_m^2\sigma^2}\beta_{km}^2\right\} \exp\left\{-\frac{\lambda^2\tau_m^2}{2}\right\} \\
&\propto \frac{1}{\tau_m^K} \exp\left\{-\frac{1}{2}\left(\frac{\sum_{k=1}^{K}\beta_{km}^2\sigma^{-2}}{\tau_m^2} + \lambda^2\tau_m^2\right)\right\} \\
&\propto (\tau_m^2)^{-K/2} \exp\left(-\frac{1}{2}\left(\frac{\sum_{k=1}^{K}\beta_{km}^2}{\sigma^2}(\tau_m^2)^{-1} + \lambda^2\tau_m^2\right)\right).
\end{aligned}$$

The variational posterior distribution of τ_m^2 is given by:

$$\begin{aligned}
q^*\left(\tau_m^2\right) &\propto \exp\left\{E\left[\log P\left(\tau_m^2 \mid \cdot\right)\right]\right\} \\
&\propto \exp\left\{E\left[-\frac{K}{2}\log\tau_m^2 - \frac{1}{2}\left(\frac{\sum_{k=1}^{K}\beta_{km}^2}{\sigma^2}(\tau_m^2)^{-1} + \lambda^2\tau_m^2\right)\right]\right\} \\
&\propto (\tau_m^2)^{-\frac{K}{2}} \cdot \exp\left\{-\frac{1}{2}\left(E(\frac{1}{\sigma^2})\sum_{k=1}^{K}E(\beta_{km}^2)(\tau_m^2)^{-1} + E(\lambda^2)\tau_m^2\right)\right\} \\
&\sim \text{Generalized-Inverse–Gaussian}(C_{\tau_m}, a_{\tau_m}, b_{\tau_m}),
\end{aligned}$$

where $a_{\tau_m} = E(\lambda^2)$, $b_{\tau_m} = E(1/\sigma_2)\sum_{k=1}^{K}E(\beta_{km}^2)$, and $C_{\tau_m} = 1 - K/2$.

We derive the conditional posterior distribution of β as:

$$\begin{aligned}
P(\beta \mid \cdot) &\propto P(Y, z \mid X, \pi, A, \beta, \sigma^2)P(\beta) \\
&\propto \prod_{t=1}^{T} P\left(y_t \mid x_t, z_t, \beta, \sigma^2\right)P(\beta) \\
&\propto \left(\prod_{t=1}^{T}\prod_{k=1}^{K} P\left(y_t \mid x_t, \beta_k, \sigma^2\right)^{v_{tk}}\right)\left(\prod_{k=1}^{K} P\left(\beta_k \mid \sigma^2, \tau_1^2, \ldots \tau_p^2\right)\right) \\
&\propto \prod_{k=1}^{K}\left[\exp\left\{-\frac{1}{2\sigma^2}\sum_{t=1}^{T} v_{tk}\left(y_t - x_t^\top\beta_k\right)^2\right\}\exp\left\{-\frac{1}{2\sigma^2}\beta_k^\top D_\tau^{-1}\beta_k\right\}\right] \\
&\propto \prod_{k=1}^{K}\left[\exp\left\{-\frac{1}{2}\left(\beta_k^\top\left(\frac{1}{\sigma^2}\sum_{t=1}^{T} v_{tk}x_t x_t^\top + \frac{1}{\sigma^2}D_\tau^{-1}\right)\beta_k - 2\frac{1}{\sigma^2}\sum_{t=1}^{T} v_{tk}y_t\beta_k^\top x_t\right)\right\}\right] \\
&\sim \prod_{k=1}^{K} N_p(\beta_k, \mu_k, \Sigma_k),
\end{aligned}$$

$$\text{where } \mu_k = \left(\frac{1}{\sigma^2}\sum_{t=1}^{T} v_{tk}x_t x_t^\top + \frac{1}{\sigma^2}D_\tau^{-1}\right)^{-1}\left(\frac{1}{\sigma^2}\sum_{t=1}^{T} y_t v_{tk} x_t\right),$$

$$\Sigma_k = \left(\frac{1}{\sigma^2}\sum_{t=1}^{T} v_{tk}x_t x_t^\top + \frac{1}{\sigma^2}D_\tau^{-1}\right)^{-1}.$$

The variational posterior distribution of β is given by:

$$\begin{aligned}
q^*(\beta) &\propto \exp\{E[\log P(\beta \mid \cdot)]\} \\
&\propto \exp\left\{E\left[\sum_{k=1}^{K}\left\{-\frac{1}{2}\left(\beta_k^\top\left(\frac{1}{\sigma^2}\sum_{t=1}^{T} v_{tk}x_t x_t^\top + \frac{1}{\sigma^2}D_\tau^{-1}\right)\beta_k - 2\frac{1}{\sigma^2}\sum_{t=1}^{T} v_{tk}y_t\beta_k^T x_t\right\}\right]\right\} \\
&\propto \prod_{k=1}^{K}\left[\exp\left\{-\frac{1}{2}\left(\beta_k^\top\left(E\left(\frac{1}{\sigma^2}\right)\sum_{t=1}^{T} E(v_{tk})x_t x_t^\top + E\left(\frac{1}{\sigma^2}\right)\cdot E\left(D_\iota^{-1}\right)\right)\beta_k \right.\right.\right. \\
&\quad \left.\left.\left. - E\left(\frac{2}{\sigma^2}\right)\sum_{t=1}^{T} E(v_{tk})y_t\beta^\top x_t\right)\right\}\right] \\
&\sim \prod_{k=1}^{K} N_p(\beta_k;\mu_k,\Sigma_k),
\end{aligned}$$

where

$$\begin{aligned}
\mu_k &= \left(E\left(\frac{1}{\sigma^2}\right)\sum_{t=1}^{T} E(v_{tk})x_t x_t^\top + E\left(\frac{1}{\sigma^2}\right)E\left(D_\tau^{-1}\right)\right)^{-1}\left(E\left(\frac{1}{\sigma^2}\right)\sum_{t=1}^{T} y_t\cdot E(v_{tk})\cdot x_t\right), \\
\Sigma_k &= \left(E\left(\frac{1}{\sigma_2}\right)\sum_{t=1}^{T} E(v_{tk})x_t x_t^\top + E\left(\frac{1}{\sigma^2}\right)E\left(D_\tau^{-1}\right)\right)^{-1}.
\end{aligned}$$

We derive the conditional posterior distribution of π:

$$\begin{aligned}
P(\pi \mid \cdot) &\propto P\left(Y, z \mid X, \pi, A, \beta, \sigma^2\right)P(\pi) \\
&\propto P(z_1|\pi)P(\pi) \\
&\propto \prod_{k=1}^{K}\pi_k^{v_{1k}}\cdot\prod_{k=1}^{K}\pi_k^{\alpha_k^{(\pi)}-1} \\
&\sim \text{Dir}\left(z_1 + \alpha^{(\pi)}\right).
\end{aligned}$$

The variational posterior distribution of π is given by:

$$\begin{aligned}
q^*(\pi) &\propto \exp\{E[\log P(\pi|\cdot)]\} \\
&\propto \exp\left\{E\left[\sum_{k=1}^{K}\left(v_{1k} + \alpha_k^{(\pi)} - 1\right)\log\pi_k\right]\right\} \\
&\sim \text{Dir}\left(E(z_1) + \alpha^{(\pi)}\right).
\end{aligned}$$

Finally, we derive the variational posterior of z. Based on the dependencies of hidden states, we divide the variational posterior of z into the following three parts.

We derive the conditional posterior distribution of z_1 as:

$$\begin{aligned}
P(z_1 \mid \cdot) &\propto P\left(Y, z \mid X, \pi, A, \beta, \sigma^2\right) \\
&\propto P(z_1 \mid \pi)\left[\prod_{t=2}^{T} P(z_t \mid z_{t-1}, A)\right]\left[\prod_{t=1}^{T} P\left(y_t \mid x_t, z_t, \beta, \sigma^2\right)\right] \\
&\propto \left(\prod_{k=1}^{K}\pi_k^{v_{1k}}\right)\left(\prod_{k=1}^{K}\prod_{j=1}^{K} A_{jk}^{v_{1j}v_{2k}}\right)\left(\prod_{k=1}^{K} P\left(y_1 \mid x_1, \beta_k, \sigma^2\right)^{v_{1k}}\right) \\
&\propto \prod_{k=1}^{K}\left[\pi_k P\left(y_1 \mid x_1, \beta_k, \sigma^2\right)\prod_{j=1}^{K} A_{kj}^{v_{2j}}\right]^{v_{1k}} \\
&\sim \text{Mult}\left(\pi_k\cdot P\left(y_1 \mid x_1, \beta_k, \sigma^2\right)\prod_{j=1}^{K} A_{kj}^{v_{2j}}\right).
\end{aligned}$$

The variational posterior distribution of z_1 is given by:

$$\begin{aligned}
q^*(z_1) &\propto \exp\{E[\log P(z_1 \mid \cdot)]\} \\
&\propto \exp\left\{\sum_{k=1}^{K} v_{1k} E\left[\log \pi_k + \log P\left(y_1 \mid x_1, \beta_k, \sigma^2\right) + \sum_{j=1}^{K} v_{2j} \log A_{kj}\right]\right\} \\
&\propto \prod_{k=1}^{K}\left[\exp\{E[\log \pi_k]\} \exp\left\{E\left[\log P\left(y_1 \mid x_1, \beta_k, \sigma^2\right)\right]\right\} \cdot \prod_{j=1}^{K} \exp\left\{E\left[\log A_{kj}\right] \cdot E(v_{2j})\right\}\right]^{v_{1k}} \\
&\sim \text{Mult}\left(\exp\{E[\log \pi_k]\} \exp\left\{E\left[\log P\left(y_1 \mid x_1, \beta_k, \sigma^2\right)\right]\right\} \prod_{j=1}^{K} \exp\left\{E[v_{2j}] E\left[\log A_{kj}\right]\right\}\right).
\end{aligned}$$

We derive the conditional posterior distribution of z_t for $t = 2, \ldots, T-1$ as:

$$\begin{aligned}
P(z_t \mid \cdot) &\propto P\left(Y, z \mid X, \pi, A, \beta, \sigma^2\right) \\
&\propto P(z_1 \mid \pi)\left[\prod_{t=1}^{T}\prod_{k=1}^{K}\prod_{j=1}^{K} A_{jk}^{v_{t-1,j}v_{t,k}}\right]\left[\prod_{t=1}^{T}\prod_{k=1}^{K} P\left(y_t \mid x_t, \beta_k, \sigma^2\right)^{v_{tk}}\right] \\
&\propto \prod_{k=1}^{K}\left[\prod_{j=1}^{K} A_{jk}^{v_{t-1,j}} A_{kj}^{v_{t+1,j}} P\left(y_t \mid x_t, \beta_k, \sigma^2\right)\right]^{v_{tk}} \\
&\sim \text{Mult}\left(\prod_{j=1}^{k} A_{jk}^{v_{t-1,j}} A_{kj}^{v_{t+1,j}} P\left(y_t \mid x_t, \beta_k, \sigma^2\right)\right).
\end{aligned}$$

The variational posterior distribution of z_t for $t = 2, \ldots, T-1$ is given by:

$$\begin{aligned}
q^*(z_t) &\propto \exp\{E[\log P(z_t \mid \cdot)]\} \\
&\propto \exp\left\{\sum_{k=1}^{K} v_{tk} \cdot E\left[\log P\left(y_t \mid x_t, \beta_k, \sigma^2\right) + \sum_{j=1}^{K} v_{t-1,j} \log A_{jk} + \sum_{j=1}^{K} v_{t+1,j} \log A_{kj}\right]\right\} \\
&\propto \prod_{k=1}^{K}\left[\exp\left\{E\left[\log P\left(y_t \mid x_t, \beta_k, \sigma^2\right)\right]\right\} \prod_{j=1}^{K} \exp\left\{E\left[\log A_{jk}\right] E\left(v_{t-1,j}\right)\right\}\right. \\
&\left.\cdot \prod_{j=1}^{K} \exp\left\{E\left[\log A_{kj}\right] E\left(v_{t+1,j}\right)\right\}\right]^{v_{tk}} \\
&\sim \text{Mult}\left(\exp\left(E\left[\log P\left(y_t \mid x_t, \beta_k, \sigma^2\right)\right]\right) \cdot \prod_{j=1}^{K} \exp\left\{E\left[\log A_{jk}\right] E\left(v_{t-1,j}\right)\right\}\right. \\
&\left.\cdot \prod_{j=1}^{K} \exp\left\{E\left[\log A_{kj}\right] E\left(v_{t+1,j}\right)\right\}\right).
\end{aligned}$$

We derive the conditional posterior distribution of z_T as:

$$\begin{aligned}
P(z_T \mid \cdot) &\propto \prod_{k=1}^{K}\left(\prod_{j=1}^{K} A_{jk}^{v_{T-1,j}} P\left(y_T \mid x_T, \beta_k, \sigma^2\right)\right)^{v_{Tk}} \\
&\sim \text{Mult}\left(\prod_{j=1}^{K} A_{jk}^{v_{T-1,j}} P\left(y_T \mid x_T, \beta_k, \sigma^2\right)\right).
\end{aligned}$$

The variational posterior distribution of z_T is given by:

$$
\begin{aligned}
q^*(z_T) &\propto \exp\{E[\log P(z_T \mid \cdot)]\} \\
&\propto \exp\left\{ \sum_{k=1}^{K} v_{Tk} E\left[\log P\left(y_T \mid x_T, \beta_k, \sigma^2\right) + \sum_{j=1}^{K} v_{T-1,j} \log A_{jk} \right] \right\} \\
&\propto \prod_{k=1}^{K} \left[\exp\left\{ E\left[\log\left(y_T \mid x_T, \beta_k, \sigma^2\right)\right] \right\} \cdot \prod_{j=1}^{K} \exp\left\{ E\left[\log A_{jk}\right] E\left(v_{T-1,j}\right) \right\} \right]^{v_{Tk}} \\
&\sim \text{Mult}\left(\exp\left\{ E\left[\log P\left(y_T \mid x_T, \beta_k, \sigma^2\right)\right] \right\} \cdot \prod_{j=1}^{K} \exp\left\{ E\left[\log A_{jk}\right] E\left(v_{T-1,j}\right) \right\} \right).
\end{aligned}
$$

Appendix A.2. Box Plots of the Estimator Values Based on 10 Experiments under $p = 20, 40$ *and* $T = 200$

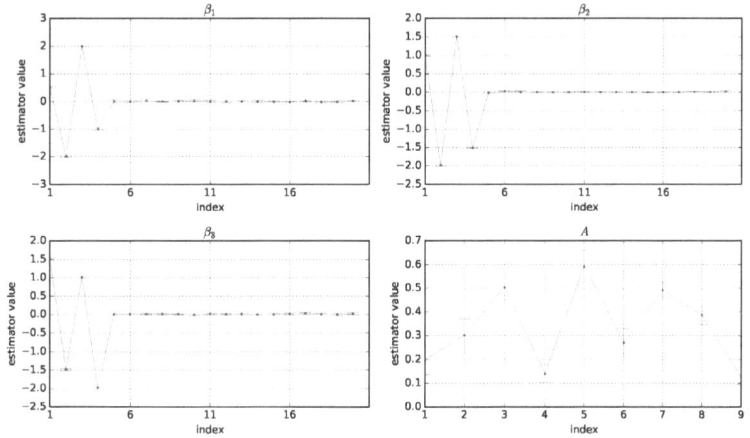

Figure A1. Box plots of the estimator values of A, β_1, β_2, and β_3 based on 50 experiments under $p = 20$ and $T = 200$. The horizontal coordinate is the index of the variables and the vertical coordinate is the values of the estimators.

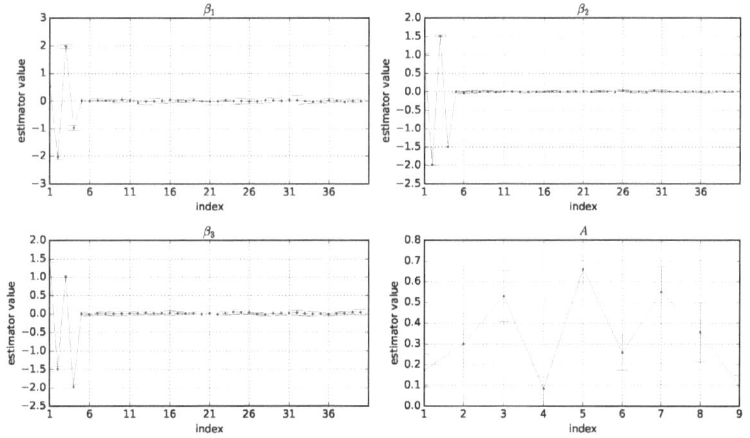

Figure A2. Box plots of the estimator values of A, β_1, β_2, and β_3 based on 50 experiments under $p = 40$ and $T = 200$. The horizontal coordinate is the index of the variables and the vertical coordinate is the values of the estimators.

Appendix A.3. Box Plots of the Estimator Values Based on 50 Experiments under $p = 60, 120$ and $T = 600$

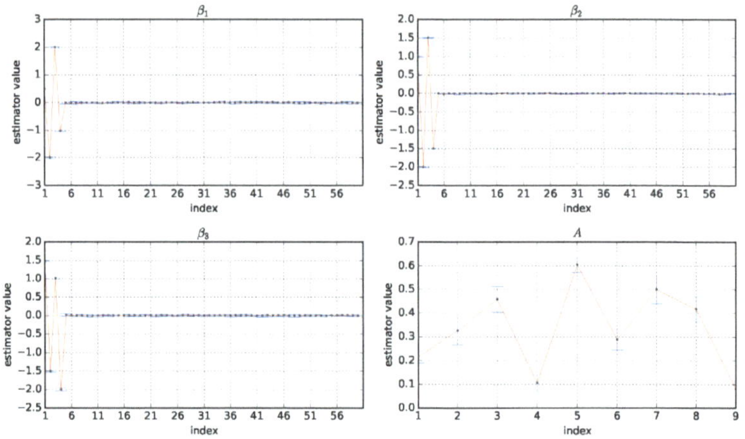

Figure A3. Box plots of the estimator values of A, β_1, β_2, and β_3 based on 10 experiments under $p = 60$ and $T = 600$. The horizontal coordinate is the index of the variables and the vertical coordinate is the values of the estimators.

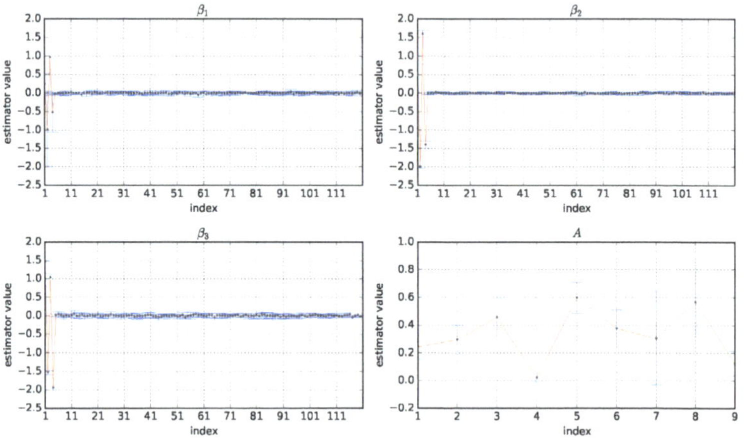

Figure A4. Box plots of the estimator values of A, β_1, β_2, and β_3 based on 10 experiments under $p = 120$ and $T = 600$. The horizontal coordinate is the index of the variables and the vertical coordinate is the values of the estimators.

Appendix A.4. Sensitivity Analysis Results for Different Hyperparameter Settings

To further understand whether the algorithm is sensitive to the choice of hyperparameters r, δ, we conduct experiments on simulated data with a sample size of 300 and a covariate dimension of 30 similar to Section 4. Consider the following experiments with three different hyperparameter settings, the first $r = 0.5, \delta = 0.5$; the second $r = 1.0, \delta = 1.0$; and the third $r = 1.5$, $\delta = 1.5$.

Table A1. The hyperparameters were set to $r = 0.5, \delta = 0.5$; $r = 1.0, \delta = 1.0$; and $r = 1.5$, $\delta = 1.5$ to compute the mean of the four metrics, with standard deviations in parentheses, based on 10 simulations under the conditions of $T = 300$, $p = 30$.

p	Method	Estimate Performance			
		MAPE	RMSE	MAE	R^2
$p = 30$	$r = 0.5, \delta = 0.5$	0.8766 (0.9195)	1.1861 (0.2441)	1.4605 (0.6291)	0.8617 (0.0739)
	$r = 1.0, \delta = 1.0$	0.8766 (0.9195)	1.1861 (0.2441)	1.4606 (0.6291)	0.8616 (0.0739)
	$r = 1.5, \delta = 1.5$	0.8766 (0.9195)	1.1861 (0.2440)	1.4606 (0.6290)	0.8616 (0.0739)

The experimental results show that the estimation results are not sensitive to the choice of the two hyperparameters r and δ. The images of the Gamma distributions for the three different hyperparameter settings are very similar in shape. It implies that the performances of the proposed method are not sensitive to these hyperparameters for a certain range of variations in r and δ values.

References

1. Baum, L.E.; Petrie, T.; Soules, G.; Weiss, N. A maximization technique occurring in the statistical analysis of probabilistic functions of Markov chains. *Ann. Math. Stat.* **1970**, *41*, 164–171. [CrossRef]
2. Forney, G.D. The viterbi algorithm. *Proc. IEEE* **1973**, *61*, 268–278. [CrossRef]
3. LeGland, F.; Mével, L. Recursive Estimation in Hidden Markov Models. In Proceedings of the 36th IEEE Conference on Decision and Control, San Diego, CA, USA, 12 December 1997; Volume 4, pp. 3468–3473.
4. Ford, J.J.; Moore, J.B. Adaptive estimation of HMM transition probabilities. *IEEE Trans. Signal Process.* **1998**, *46*, 1374–1385. [CrossRef]
5. Djuric, P.M.; Chun, J.H. An MCMC sampling approach to estimation of nonstationary hidden Markov models. *IEEE Trans. Signal Process.* **2002**, *50*, 1113–1123. [CrossRef]
6. Ma, Y.A.; Foti, N.J.; Fox, E.B. Stochastic gradient MCMC methods for Hidden Markov Models. In Proceedings of the International Conference on Machine Learning Research, Sydney, Australia, 6–11 August 2017; pp. 2265–2274.
7. Dellaportas, P.; Roberts, G.O. An introduction to MCMC. In *Spatial Statistics and Computational Methods*; Springer: Berlin/Heidelberg, Germany, 2003; pp. 1–41.
8. Neal, R.M. MCMC using Hamiltonian dynamics. *Handb. Markov Chain. Monte Carlo* **2011**, *2*, 2.
9. Box, G.E.; Tiao, G.C. *Bayesian Inference in Statistical Analysis*; John Wiley & Sons: Hoboken, NJ, USA, 2011.
10. Scott, S.L. Bayesian methods for Hidden Markov Models: Recursive computing in the 21st century. *J. Am. Stat. Assoc.* **2002**, *97*, 337–351. [CrossRef]
11. Rydén, T. EM versus Markov chain Monte Carlo for estimation of hidden Markov models: A computational perspective. *Bayesian Anal.* **2008**, *3*, 659–688. [CrossRef]
12. Brooks, S.P.; Roberts, G.O. Convergence assessment techniques for Markov chain Monte Carlo. *Stat. Comput.* **1998**, *8*, 319–335. [CrossRef]
13. Jordan, M.I.; Ghahramani, Z.; Jaakkola, T.S.; Saul, L.K. An introduction to variational methods for graphical models. *Mach. Learn.* **1999**, *37*, 183–233. [CrossRef]
14. Tzikas, D.G.; Likas, A.C.; Galatsanos, N.P. The variational approximation for Bayesian inference. *IEEE Signal Process. Mag.* **2008**, *25*, 131–146. [CrossRef]
15. Hoffman, M.D.; Blei, D.M.; Wang, C.; Paisley, J. Stochastic variational inference. *J. Mach. Learn. Res.* **2013**.
16. Blei, D.M.; Kucukelbir, A.; McAuliffe, J.D. Variational inference: A review for statisticians. *J. Am. Stat. Assoc.* **2017**, *112*, 859–877. [CrossRef]
17. Wang, Y.; Blei, D.M. Frequentist Consistency of Variational Bayes. *J. Am. Stat. Assoc.* **2019**, *114*, 1147–1161. [CrossRef]
18. Han, W.; Yang, Y. Statistical inference in mean-field Variational Bayes. *arXiv* **2019**, arXiv:1911.01525.
19. Ranganath, R.; Gerrish, S.; Blei, D. Black box variational inference. In Proceedings of the Artificial Intelligence and Statistics, Reykjavik, Iceland, 22–25 April 2014; pp. 814–822.
20. MacKay, D.J. *Ensemble Learning for Hidden Markov Models*; Technical Report; Cavendish Laboratory, University of Cambridge: Cambridge, UK, 1997.
21. McGrory, C.A.; Titterington, D. Variational Bayesian analysis for Hidden Markov Models. *Aust. N. Z. J. Stat.* **2009**, *51*, 227–244. [CrossRef]

22. Foti, N.; Xu, J.; Laird, D.; Fox, E. Stochastic variational inference for Hidden Markov Models. In Proceedings of the Advances in Neural Information Processing Systems, Vancouver, BC, Canada, 9–15 December 2014; pp. 1–9.
23. Gruhl, C.; Sick, B. Variational Bayesian inference for Hidden Markov Models with multivariate Gaussian output distributions. *arXiv* **2016**, arXiv:1605.08618.
24. Ding, N.; Ou, Z. Variational nonparametric Bayesian Hidden Markov Model. In Proceedings of the 2010 IEEE International Conference on Acoustics, Speech and Signal Processing, Dallas, TX, USA, 14–19 March 2010; pp. 2098–2101.
25. Park, T.; Casella, G. The bayesian lasso. *J. Am. Stat. Assoc.* **2008**, *103*, 681–686. [CrossRef]
26. Meinshausen, N. Relaxed lasso. *Comput. Stat. Data Anal.* **2007**, *52*, 374–393. [CrossRef]
27. Hans, C. Bayesian lasso regression. *Biometrika* **2009**, *96*, 835–845. [CrossRef]
28. Ranstam, J.; Cook, J. LASSO regression. *J. Br. Surg.* **2018**, *105*, 1348. [CrossRef]
29. Ye, L.; Beskos, A.; De Iorio, M.; Hao, J. Monte Carlo co-ordinate ascent variational inference. *Stat. Comput.* **2020**, *30*, 887–905. [CrossRef]
30. Jaakkola, T.S. Tutorial on variational approximation methods. In *Advanced Mean Field Methods: Theory and Practice*; The MIT Press: Cambridge, MA, USA, 2000; pp. 129–159.
31. Jaakkola T.S., Jordan, M. Bayesian parameter estimation via variational methods. *Stat. Comput.* **2000**, *10*, 25–37. [CrossRef]
32. Tran, M.-N.; Nott, D.J.; Kuk, A.Y.C.; Kohn, R. Parallel Variational Bayes for Large Datasets with an Application to Generalized Linear Mixed Models. *J. Comput. Graph. Stat.* **2016**, *25*, 626–646. [CrossRef]
33. Winn, J.; Bishop, C.M.; Jaakkola, T. Variational message passing. *J. Mach. Learn. Res.* **2005**, *6*, 661–694.
34. Dofadar, D.F.; Khan, R.H.; Alam, M.G.R. COVID-19 Confirmed Cases and Deaths Prediction in Bangladesh Using Hidden Markov Model. In Proceedings of the 2021 4th International Conference on Bio-Engineering for Smart Technologies (BioSMART), Paris, France, 21 January 2022; pp. 1–4.
35. Shoko, C.; Sigauke, C. Short-term forecasting of COVID-19 using support vector regression: An application using Zimbabwean data. *Am. J. Infect. Control.* **2023**, *51*, 1095–1107. [CrossRef]
36. Gorynin, I.; Gangloff, H.; Monfrini, E.; Pieczynski, W. Assessing the segmentation performance of pairwise and triplet Markov models. *Signal Process.* **2018**, *145*, 183–192. [CrossRef]
37. Morales, K.; Petetin, Y. Variational Bayesian inference for pairwise Markov models. In Proceedings of the 2021 IEEE Statistical Signal Processing Workshop (SSP), Rio de Janeiro, Brazil, 11–14 July 2021; pp. 251–255.

mathematics

Article

Evaluating the Discrete Generalized Rayleigh Distribution: Statistical Inferences and Applications to Real Data Analysis

Hanan Haj Ahmad [1,*], Dina A. Ramadan [2] and Ehab M. Almetwally [3,4,*]

1 Department of Basic Science, The General Administration of Preparatory Year, King Faisal University, Hofuf 31982, Al Ahsa, Saudi Arabia
2 Department of Mathematics, Faculty of Science, Mansoura University, Mansoura 33516, Egypt; dinaramadan21@mans.edu.eg
3 Department of Mathematics and Statistics, Faculty of Science, Imam Mohammad Ibn Saud Islamic University (IMSIU), Riyadh 11432, Riyadh, Saudi Arabia
4 Faculty of Business Administration, Delta University of Science and Technology, Gamasa 11152, Egypt
* Correspondence: hhajahmed@kfu.edu.sa (H.H.A.); emalmetwally@imamu.edu.sa (E.M.A.)

Abstract: Various discrete lifetime distributions have been observed in real data analysis. Numerous discrete models have been derived from a continuous distribution using the survival discretization method, owing to its simplicity and appealing formulation. This study focuses on the discrete analog of the newly generalized Rayleigh distribution. Both classical and Bayesian statistical inferences are performed to evaluate the efficacy of the new discrete model, particularly in terms of relative bias, mean square error, and coverage probability. Additionally, the study explores different important submodels and limiting behavior for the new discrete distribution. Various statistical functions have been examined, including moments, stress–strength, mean residual lifetime, mean past time, and order statistics. Finally, two real data examples are employed to evaluate the new discrete model. Simulations and numerical analyses play a pivotal role in facilitating statistical estimation and data modeling. The study concludes that the discrete generalized Rayleigh distribution presents a notably appealing alternative to other competing discrete distributions.

Keywords: generalized Rayleigh; maximum likelihood estimation; Bayes estimation; reliability; simulation analysis; Monte Carlo Markov chain; goodness-of-fit measures

MSC: 62E10; 62F15; 62N05; 60E05; 62P30

Citation: Haj Ahmad, H.; Ramadan, D.A.; Almetwally, E.M. Evaluating the Discrete Generalized Rayleigh Distribution: Statistical Inferences and Applications to Real Data Analysis. *Mathematics* **2024**, *12*, 183. https://doi.org/10.3390/math12020183

Academic Editor: Diana Mindrila

Received: 10 December 2023
Revised: 30 December 2023
Accepted: 3 January 2024
Published: 5 January 2024

1. Introduction

As each day passes, the volume of data in our world increases exponentially, necessitating the development of new statistical distributions to better characterize the features of many phenomena and experiments. While a great deal of lifetime data appear to be continuous, they are originally discrete. This discrepancy ensures the need for more appropriate methods to generate discrete distributions that more accurately represent the data in the experiment. Discrete distributions are frequently employed in statistical modeling for several reasons.

Discrete distributions are used to model data that can only take on a finite or countably infinite number of values, such as counts, proportions, and binary outcomes, for example, the number of customers in a store, the number of heads in a coin flip, or the number of defective items in a production line. Discrete distributions are often easy to understand and interpret as they model data that take on a limited number of values. The probability mass function (*pmf*) or probability generating function (*pgf*) of a discrete distribution is a simple function that provides the probability of each possible outcome. Also, many discrete distributions have closed-form expressions for their *pmf* or *pgf*, which makes it easy to work with them mathematically. This allows for efficient computation of probabilities and moments without the need for integration. Furthermore, discrete distributions can be used to model a wide variety of real-world phenomena, such as the distribution of species in an ecosystem, the distribution of genetic variations in a population, or the distribution of traffic on a road network.

Recently, many discrete distributions have been considered, particularly in medicine, engineering, reliability, survival analysis, and more. For more descriptions and applications of discrete distributions, refer to [1–9]. Hence, many authors have conducted much work to originate and develop discrete models from different aspects.

The characterization of continuous random variables can be performed either by their probability density function, cumulative distribution function, moments, moment-generating function, hazard rate functions, or others. Different discretization methods appeared in the literature to create an appropriate discrete distribution based on the underlying continuous model.

By deriving discrete analogs or counterparts of well-known continuous distributions, statisticians can better tailor their models to the specific nature of the data. Usually, creating a discrete analog from a continuous distribution is based on the principle of preserving one or more characteristic properties of the continuous one. Consequently, different ways to discretize a continuous distribution appear in the literature depending on the property the researcher intends to preserve; for example, Lai [10] used the survival and the hazard rate preservation methods to create discrete distributions from different continuous ones. Haj Ahmad and Almetwally [11] used the survival, hazard rate, and probability distribution function preservation methods to discretize the generalized Pareto distribution.

The benefit of using the survival discretization method is that it can maintain the statistical properties of the original distribution, including median and percentiles, in addition to the overall shape of the distribution. A drawback of this method is that it can be computationally intensive and may require numerical methods for complex distributions.

For the hazard preservation method, the main benefit is that it preserves the hazard function of the continuous distribution. This is important in applications like reliability analysis where the failure rate is a key parameter. On the other hand, mathematical complexity can be viewed using this method, especially for continuous distributions with nonlinear hazard functions. This complexity can increase computational time and resource requirements. Another drawback of the hazard preservation method is that it only preserves the hazard function, but other characteristics of the distribution (like mean, variance, or skewness) may not be as accurately retained. For more details about other discretization methods and their properties, one may refer to [12,13] who provided a review of several discretization methods.

From the previous research work, it is evident that the results look appealing and motivational to continue creating new discrete distributions to model new data.

In the present study, we obtain a discrete analog of the continuous new generalized Rayleigh distribution (NGRD) using the survival discretization method that depends on the survival function. See for example [6,7], in which the survival discretization approach was used to obtain the discrete normal and discrete Rayleigh distributions, respectively. Using the same approach, more discrete distributions have been introduced and studied; see for example [14–23].

Still, there is an enduring need to create and develop more discrete models and to generate new ones because of modeling and fitting real data, which appear and spread constantly in human life. The efficiency in discretization methods refers to the ability of a method to produce accurate and useful discretized versions of continuous data with minimal loss of information. Also, discrete distributions derived from continuous ones can inherit their flexibility and adaptability. This allows statisticians to model a wider range of data characteristics, such as skewness or kurtosis, which might be difficult with standard discrete distributions. In statistical methodology, continuous distributions may have characteristics that are missing in the discrete space; hence, creating discrete analogs can fill these gaps, providing new tools for data analysis.

The suggested discrete model with three parameters offers an immense degree of fitness to skewed, symmetric, monotone, and inverse J-shape types of data. Therefore, some statistical functions and properties are achieved, in addition to observing the submodels and limiting behavior of the proposed discrete distribution. Examining statistical inference is crucial; therefore, point and interval estimation for the unknown parameters using the maximum likelihood estimator and the Bayesian method is performed.

Simulation analysis via numerical techniques such as Monte Carlo simulation is employed to evaluate the estimators using the maximum likelihood and the Bayesian estimation methods to compare the performance of these two methods. The efficiency is assessed using the relative bias, the mean squared error, and the coverage probability of the confidence intervals. Two real datasets are analyzed to emphasize the empirical validation of the new model, where several goodness-of-fit measures are employed. The first example is related to the industrial field, where several strikes that occurred in coal mining in the UK were recorded over four weeks. Modeling and predicting the number of strikes will save human lives and money. The second example is related to the number of fires that occurred in Greece's forests in the year 1998 during the summer months. The main purposes of this study are, first, to introduce new discrete analogs of the continuous NGRD and evaluate some of its important statistical functions, second, to perform the inferential statistics related to the new distributions' parameters and compare the results, and, third, to assess the efficiency of the new discrete distribution by modeling real data examples and comparing the goodness-of-fit measures with other discrete distributions that were studied earlier in the literature.

The originality of this work emanates from the basis of exploring the creation of a new discrete analog from less commonly used continuous distributions and investigating their properties, potential applications, and how they compare to existing distributions. Also, it focuses on specific application areas, such as the industrial, engineering, and reliability fields. To our knowledge, no previous work has studied the discrete new generalized Rayleigh distribution and employed it to model real-life data examples from different scientific fields.

The authors' contributions to this study can be summarized as follows:

- Development of a New Discrete Model: Creation of a discrete analog of the continuous new generalized Rayleigh distribution (NGRD) using the survival discretization method.
- Statistical Functions and Properties: Achievement of various statistical functions and properties of the proposed discrete distribution, including observing its submodels and limiting behavior.
- Statistical Inference Examination: Conducting point and interval estimation for the unknown parameters using both the maximum likelihood estimator (MLE) and the Bayesian method.
- Simulation Analysis for Estimator Evaluation: Implementation of numerical techniques such as Monte Carlo simulation to evaluate the estimators derived from MLE and Bayesian estimation methods through relative bias, mean squared error, and coverage probability of confidence intervals
- Empirical Validation via Real Data Analysis: Analyzing two real datasets to validate the new model empirically, including modeling industrial and environmental phenomena.
- Comparison with other Distributions: Comparing the goodness of fit of the new model with other discrete distributions previously studied in the literature.

The remaining parts of this work are organized as follows: Section 2 describes the new generalized Rayleigh distribution. The discretization methods are presented in Section 3, along with some statistical functions. In Section 4, the maximum likelihood and the Bayesian inference are presented. In Section 5, simulation analysis and the tabulated results are carried out. Some real data examples are provided in Section 6. Finally, conclusions are remarked on in Section 7.

2. Model Description

The Rayleigh distribution (RD) is a continuous distribution that has much practical importance; hence, many of its statistical characteristics, inference, and reliability analysis have been studied by several authors, and numerous extended forms of the Rayleigh distribution have been proposed. For example, Ref. [24] applied the inverse Rayleigh to the failure times data. Ref. [25] introduced the transmuted Rayleigh and used it to model the amount of nicotine in the blood. In [26], the authors studied the beta-generalized Rayleigh distribution and its application. Ref. [27] obtained the transmuted inverse Rayleigh distribution to lifetime data. Ref. [28] obtained a new modified Rayleigh distribution named the Kumaraswamy generalized Rayleigh distribution with application to real data. For more information, refer to [29–31]. In this work, we are interested in studying a new form of the Rayleigh distribution called a new generalized Rayleigh distribution (NGR), which was first introduced by Shen et al. [32]. It has three parameters and it was shown that the NGR is suitable for modeling large data values rather than small data values. However, as a continuous distribution, it is restricted from describing discrete data forms. Discretizing the NGR distribution is our goal; therefore, it yields a subsequent distribution that accommodates the countable data while retaining the influential tail modeling characteristics of the NGR. In this study, we carry out a discrete version of the NGR and use it to model real data.

The probability density function (*pdf*) and the survival function (S) of the continuous NGR are provided respectively as:

$$f(x;\alpha,\beta,\theta)=\frac{2\alpha\beta\theta(\alpha-1)xe^{-\theta x^2}\left(1-e^{-\theta x^2}\right)^{\beta-1}}{\left[\alpha-\left(1-e^{-\theta x^2}\right)^{\beta}\right]^2},\quad x>0, \tag{1}$$

and

$$S(x;\alpha,\beta,\theta)=\frac{\alpha[1-\left(1-e^{-\theta x^2}\right)^{\beta}]}{\alpha-\left(1-e^{-\theta x^2}\right)^{\beta}}, \tag{2}$$

in which the parameters $\alpha>1,\ \beta>0,\ \theta>0$.

The hazard rate function is

$$h(x;\alpha,\beta,\theta)=\frac{2\beta\theta(\alpha-1)xe^{-\theta x^2}\left(1-e^{-\theta x^2}\right)^{\beta-1}}{\left[1-\left(1-e^{-\theta x^2}\right)^{\beta}\right]\left[\alpha-\left(1-e^{-\theta x^2}\right)^{\beta}\right]^2}. \tag{3}$$

To identify submodels or special distributions that arise from this general form, we can consider different values or limits of parameters α, β, and θ. Here are some special cases:

1. Standard Rayleigh Distribution: When $\theta=1$ and $\beta=1$, the term $(1-e^{-x^2})$ simplifies to the CDF of the standard Rayleigh distribution. This is observed if the parameter α also approaches infinity, which simplifies the formula to $1-e^{-x^2}$, the CDF of the standard Rayleigh distribution.
2. Exponential Distribution: If β approaches infinity, the term $\Lambda=(1-e^{-\theta x^2})^{\beta}$ can approach an exponential-like behavior for small values of x, depending on how θ is defined.
3. Modified Rayleigh Distribution: For specific fixed values of α and β, you can obtain various forms of modified Rayleigh distributions, where the behavior is measured by the degree of skewness and kurtosis determined by these parameters.
4. Weibull-like Distribution: By interchanging between θ and β, especially when β is not equal to 1, the distribution can possess Weibull-like properties.

By discretizing the continuous range of x, discrete versions of this distribution can be derived, which may be useful for certain types of count data or integer-valued measurements.

Since our goal in this work is to define a new discrete NGR distribution, we generate a discrete analog based on the survival discretization method, which is denoted by DNGR. The *pmf* and cumulative distribution function (*CDF*) are obtained. Furthermore, the moments, stress–strength function, the mean residual, and mean past lifetimes, order statistics, and L-moments are obtained. All these statistical functions are used for studying the features of the DNGR.

3. The Discrete New Generalized Rayleigh Distribution

Roy and Gupta [3,4] defined the probability mass function (*pmf*) for a discrete distribution using the survival function and it was expressed as follows:

$$P(X=k) = S(k) - S(k+1), \qquad k = 0,1,2,\ldots \tag{4}$$

where $S(x)$ is the survival function provided by Equation (2); hence, the *pmf* of the discrete analog of NGR distribution, namely DNGR, is written as

$$P(X=k) = \frac{\alpha[1-\Lambda(k;\theta,\beta)]}{\alpha-\Lambda(k;\theta,\beta)} - \frac{\alpha[1-\Lambda(k+1;\theta,\beta)]}{\alpha-\Lambda(k+1;\theta,\beta)}, \tag{5}$$

where $\Lambda(k;\theta,\beta) = \left(1-e^{-\theta k^2}\right)^{\beta}$.

The CDF of the DNGR distribution can be written as

$$P(X<k) = F(k+1) = 1 - \frac{\alpha[1-\Lambda(k+1;\theta,\beta)]}{\alpha-\Lambda(k+1;\theta,\beta)}. \tag{6}$$

The quantile function with given values of parameters as α, β, and θ of the DNGR distribution is provided by

$$x_i = \sqrt{\frac{-1}{\theta}\ln\left[1-\left(\frac{\alpha q}{\alpha+q-1}\right)^{1/\beta}\right]} - 1; \quad q=[0,1],\ i=1,\ldots,n. \tag{7}$$

The hazard rate function (*HRF*) of the DNGR distribution is provided by

$$h_{\mathrm{DNGR1}}(k) = \frac{[1-\Lambda(k;\theta,\beta)][\alpha-\Lambda(k+1;\theta,\beta)]}{[\alpha-\Lambda(k;\theta,\beta)][1-\Lambda(k+1;\theta,\beta)]} - 1. \tag{8}$$

We also observe that the reversed hazard rate function for the DNGR of this distribution is provided by

$$r_{DNGR}(k;\alpha,\beta,\theta) = \frac{f(k;\alpha,\beta,\theta)}{F(k+1;\alpha,\beta,\theta)}, \tag{9}$$

$$\begin{aligned} r_{DNGR}(k;\alpha,\beta,\theta) = & \left[\frac{\alpha[1-\Lambda(k;\theta,\beta)]}{\alpha-\Lambda(k;\theta,\beta)} - \frac{\alpha[1-\Lambda(k+1;\theta,\beta)]}{\alpha-\Lambda(k+1;\theta,\beta)}\right] \\ & \times\left[\frac{\alpha-\Lambda(k+1;\theta,\beta)}{(\alpha-\Lambda(k+1;\theta,\beta))-\alpha[1-\Lambda(k+1;\theta,\beta)]}\right] \end{aligned} \tag{10}$$

In Figure 1, the bar charts represent each parameter $\alpha, \theta,$ and β that has a specific role in the behavior of the *pmf*, and their effects are observable when we fix one and vary the others. An explanation of the effect of each parameter based on the plots is as follows:

1. Effect of α when θ and β are changeable:

When α is fixed, the variations in θ and β create different trends in the probability values. Higher values of θ tend to stretch the curve horizontally, meaning, for a given α, as θ increases, the decrease in probability values with increasing k is less steep. Higher values of β tend to amplify the curve vertically, making the probability have fewer values for higher k values. The reaction between θ and β at a fixed α demonstrates that θ affects the spread of the distribution, while β affects the sharpness of the probability decrease.

2. Effect of θ when α and β are changeable:
   With θ fixed, the changes in α and β show distinct patterns. An increase in α generally results in higher probability values across all k. This is because a higher α relative to $\Lambda(k; \theta, \beta)$ increases the numerator and decreases the denominator of the function $P(X = k)$, resulting in a larger overall value. The effect of β at a fixed θ is similar to its effect when α is fixed; it controls the sharpness of the decrease in probability values. Higher β values cause a quicker decline in probability as k increases.
3. Fixing β and varying α and θ, we can see that
   As α increases, for a fixed β, the overall probability values increase, similar to when θ is fixed. The role of θ here is nuanced; for lower values of k, the impact of changing θ is minimal, but, as k increases, higher θ values preserve higher probabilities, indicating a wider spread in the distribution. From the above explanations, it is clear that α primarily scales the probability values, β determines the rate at which the probability values decline as k increases (sharpness of the distribution), and θ controls the spread or dispersion of the distribution across k values. The combination of these three parameters can thus shape the function's distribution in various ways, and each has a distinctive role in the form of the probability curve.

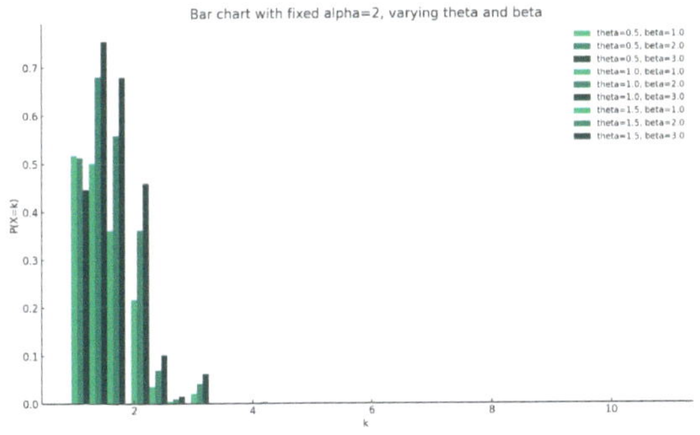

(**a**)

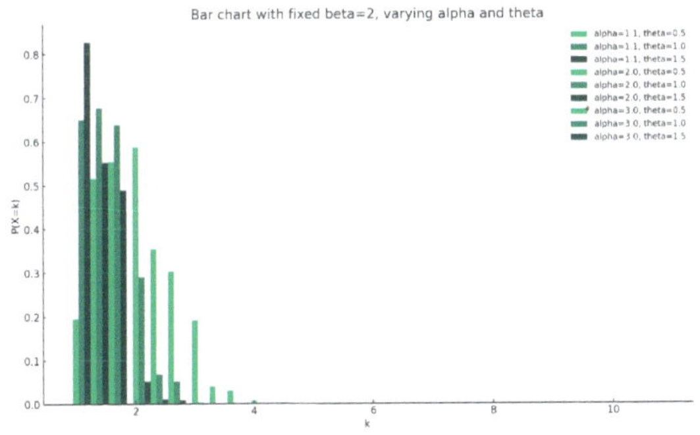

(**b**)

Figure 1. *Cont.*

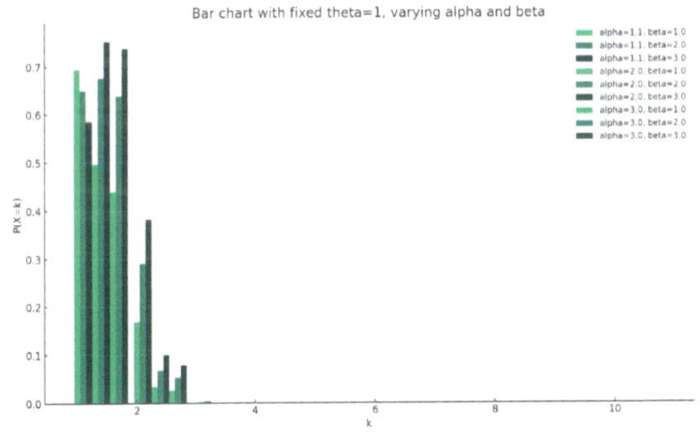

(**c**)

Figure 1. The *pmf* bar charts for the DNGR, (**a**) when $\alpha = 2$, (**b**) when $\beta = 2$, and (**c**) when $\theta = 1$.

In Figure 2, the plots represent the *HRF* of the DNGR distribution for various combinations of parameters $\alpha, \theta,$ and β. Each subplot corresponds to a different set of these parameters. The values of k range from 1 to 10. The curves are increasing for different values of the parameters; we can realize the effect of increasing the parameter θ while keeping other parameters fixed by going steeply to the left. For the effect of β, assuming other parameters are fixed, it can be figured by increasing in a slower mode when k takes small values. Finally, the effect of increasing the values of α while fixing the remaining parameters is going to the left more steeply.

The limiting behavior of DNGR for different choices in parameter values at the boundary points includes

$\lim_{k\to\infty} p(k) = 0,\ \lim_{k\to 0} p(k) = 0,$
$\lim_{\alpha\to 1} p(k) = 0,\ \lim_{\alpha\to\infty} p(k) = \Lambda(k+1;\theta,\beta) - \Lambda(k;\theta,\beta),$
$\lim_{\theta\to 0} p(k) = 0,\ \lim_{\theta\to\infty} p(k) = 0,$
$\lim_{\beta\to 0} p(k) = 0,$ and $\lim_{\beta\to\infty} p(k) = 0.$

From the above limiting behavior of the DNGR, some submodels and special cases can be derived, such as

1. Discrete standard Rayleigh Distribution: When $\theta = 1$ and $\beta = 1$, and α approaches infinity, the pmf simplifies to $(1 - e^{-(x+1)^2}) - (1 - e^{-x^2})$, which represents the pmf of the discrete Rayleigh distribution created from applying the survival discretization method.
2. Discrete Exponential-like Distribution: For large values of β and specific values of θ, the DNGR distribution might possess characteristics similar to an exponential distribution for smaller values of k, where the exponential decay behavior is more evident, since the term $\Lambda = (1 - e^{-\theta k^2})^\beta$ has a decaying form and can be considered as exponential-like function.
3. Discrete Uniform Distribution: If the parameters α, β, and θ are chosen such that the pmf becomes constant for all k within a certain range, the DNGR distribution could approximate a discrete uniform distribution.
4. Geometric-like Distribution: By adjusting θ and β, you might be able to create a distribution that behaves like the geometric distribution, especially if the probability of larger k values decays like the geometric series.

These possible submodels and special cases demonstrate the versatility and adaptability of the DNGR distribution. The ability to derive such a variety of distributions from a single distribution highlights the potential utility of the DNGR distribution in modeling a wide range of discrete data scenarios. Each submodel or special case would be suited to different types of data and could provide unique insights depending on the context of the analysis.

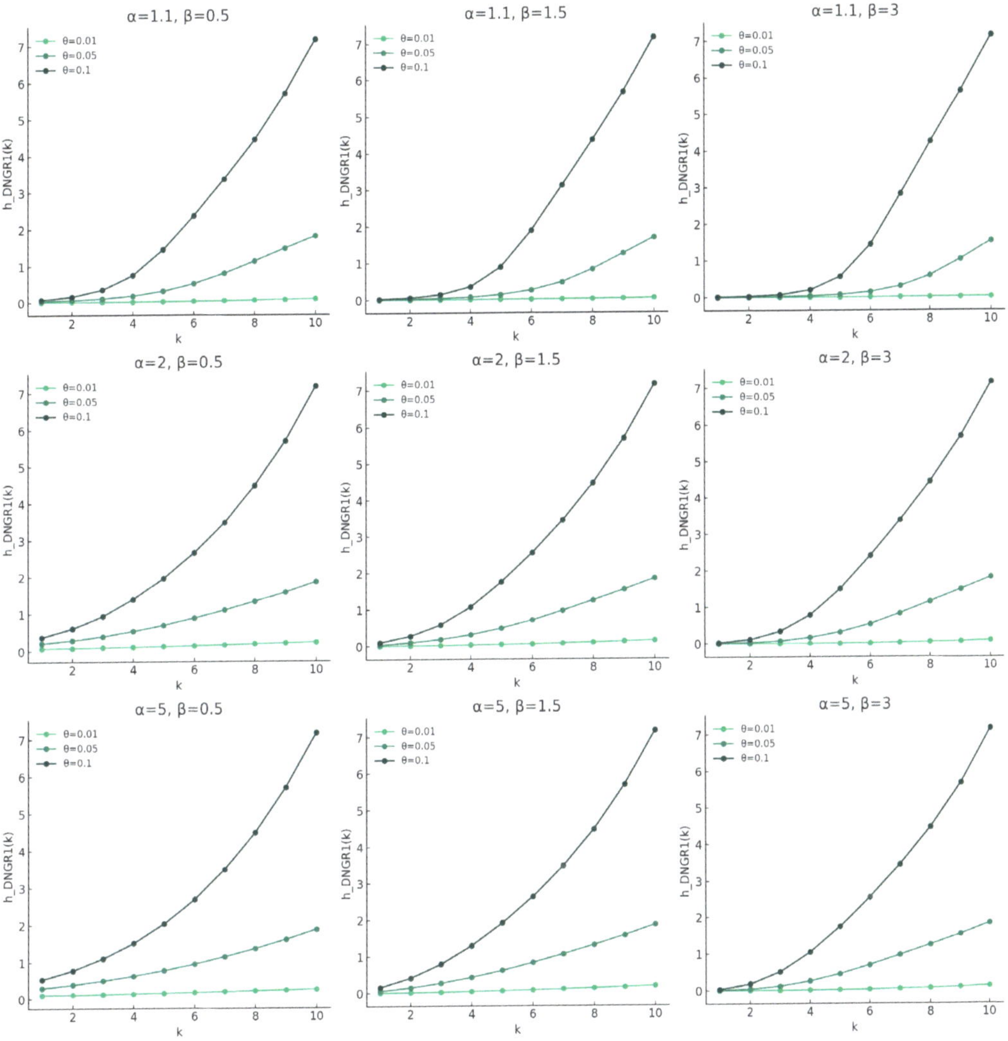

Figure 2. The HRF of the DNGR distribution.

3.1. Moments

Assume a non-negative random variable $k \sim DNGR(\alpha, \beta, \theta)$. The sth moment, say $\psi_s^{'}$, can be expressed as follows

$$\psi_s' = \sum_{k=0}^{\infty} k^s \, f(k; \alpha, \beta, \theta),$$

and then

$$\psi_s' = \sum_{k=0}^{\infty} k^s \left[\frac{\alpha[1 - \Lambda(k; \theta, \beta)]}{\alpha - \Lambda(k; \theta, \beta)} - \frac{\alpha[1 - \Lambda(k+1; \theta, \beta)]}{\alpha - \Lambda(k+1; \theta, \beta)} \right]. \quad (11)$$

It is impossible to write an exact form of the sth moment; hence, R programming with version (4.3.0) is helpful, and the moment is evaluated numerically. Equation (11) is convergent for $\alpha > 1$, $\beta > 0$, and $\theta > 0$.

Table 1 explores some functions like the minimum, mean, variance, maximum, skewness (SK), and kurtosis (Kt) for different values of α, β, and θ. In addition, the DNGR distribution is appropriate for modeling both over- and under-dispersed data since, in this model, the variance can be smaller than the mean, which is the case with some standard classical discrete distributions, in addition to the positive and negative skewness values, which show that this distribution can be skewed to the right or left. Also, a very small skew value that tends to zero indicates a symmetry possible curve for the *pmf*. A higher kurtosis means more of the variation is due to infrequent extreme deviations as opposed to frequent modestly sized deviations. By varying θ, α, and β, one can realize the distribution changes. For instance, with $\theta = 0.8$ and $\alpha = 0.5$, β changing from 0.84 to 2.73 drastically increases the kurtosis, indicating a heavier tail.

Table 1. Summary of descriptive statistics for the DNGR distribution.

θ	α	β	Minimum	Mean	Variance	Maximum	SK	Kt
0.8	1.05	0.3	0	1.5140	0.4643	3	−0.0700	1.2956
	1.05	1.2	0	1.9760	0.3338	4	−0.0563	1.3066
	1.05	2.5	0	2.1890	0.3176	4	−0.0479	1.3039
	1.05	3	1	2.2420	0.3137	4	−0.0461	1.3030
	2.2	0.3	0	0.6480	0.4305	3	0.1245	1.1141
	2.2	1.2	0	1.2470	0.3664	3	0.0229	1.2125
	2.2	2.5	0	1.5330	0.3252	3	0.0203	1.2296
	2.2	3	0	1.6020	0.3179	3	0.0212	1.2320
	3	0.3	0	0.5810	0.4098	3	0.1629	1.1205
	3	1.2	0	1.1880	0.3570	3	0.0373	1.2097
	3	2.5	0	1.4770	0.3178	3	0.0320	1.2281
	3	3	0	1.5450	0.3163	3	0.0325	1.2307
1.5	1.05	0.3	0	1.0800	0.2799	3	−0.0699	1.2957
	1.05	1.2	0	1.4600	0.2827	3	−0.0564	1.3066
	1.05	2.5	0	1.6570	0.2396	3	−0.0481	1.3039
	1.05	3	0	1.7020	0.2214	3	−0.0461	1.3031
	2.2	0.3	0	0.4520	0.2860	2	0.1244	1.1142
	2.2	1.2	0	0.9210	0.2190	2	0.0229	1.2125
	2.2	2.5	0	1.1120	0.1616	2	0.0203	1.2295
	2.2	3	0	1.1510	0.1624	2	0.0213	1.2321
	3	0.3	0	0.4010	0.2725	2	0.1629	1.1205
	3	1.2	0	0.8800	0.2258	2	0.0374	1.2096
	3	2.5	0	1.0830	0.1503	2	0.0321	1.2280
	3	3	0	1.1210	0.1485	2	0.0325	1.2306

Table 1. *Cont.*

θ	α	β	Minimum	Mean	Variance	Maximum	SK	Kt
3	1.05	0.3	0	0.8230	0.1598	2	−0.0700	1.2956
	1.05	1.2	0	0.9890	0.0689	2	−0.0563	1.3067
	1.05	2.5	0	1.0460	0.0680	2	−0.0480	1.3040
	1.05	3	0	1.0610	0.0734	2	−0.0459	1.3031
	2.2	0.3	0	0.2800	0.2038	2	0.1245	1.1143
	2.2	1.2	0	0.6820	0.2231	2	0.0229	1.2123
	2.2	2.5	0	0.8840	0.1127	2	0.0203	1.2296
	2.2	3	0	0.9200	0.0857	2	0.0210	1.2320
	3	0.3	0	0.2420	0.1856	2	0.1628	1.1204
	3	1.2	0	0.6360	0.2357	2	0.0374	1.2097
	3	2.5	0	0.8590	0.1292	2	0.0321	1.2280
	3	3	0	0.9020	0.0985	2	0.0325	1.2307

3.2. Stress–Strength Analysis

The stress–strength (reliability) analysis is an important tool in mechanical design. The idea relies on the probability of failure that is obtained from the probability of r exceeding r^*. Assume that both r and r^* are in the positive domain. The expected reliability (R^*) can be calculated by

$$R^* = P[K_r \le K_{r^*}] = \sum_{k=0}^{\infty} f_{K_r}(k) R_{K_{r^*}}(k), \tag{12}$$

in which $K_r \sim DNGR(\alpha_1, \beta_1, \theta_1)$ and $K_{r^*} \sim DNGR(\alpha_2, \beta_2, \theta_2)$, and then R^* can be expressed as follows

$$R^* = \sum_{k=0}^{\infty} \left[\frac{\alpha_1[1-\Lambda_1(k;\theta_1,\beta_1)]}{\alpha_1 - \Lambda_1(k;\theta_1,\beta_1)} - \frac{\alpha_1[1-\Lambda_1(k+1;\theta_1,\beta_1)]}{\alpha_1 - \Lambda_1(k+1;\theta_1,\beta_1)} \right] \left[\frac{\alpha_2[1-\Lambda_2(k;\theta_2,\beta_2)]}{\alpha_2 - \Lambda_2(k;\theta_2,\beta_2)} \right],$$

where $\Lambda_1(k;\theta_1,\beta_1) = \left(1 - e^{-\theta_1 k^2}\right)^{\beta_1}$ and $\Lambda_2(k;\theta_2,\beta_2) = \left(1 - e^{-\theta_2 k^2}\right)^{\beta_2}$.

We cannot obtain a closed form for the above equation; consequently, simulation analysis is utilized to obtain a numerical solution. In Section 6, numerical analysis is performed to obtain the value of the stress–strength function under two real data applications.

3.3. The Mean Residual and the Mean Past Lifetimes

In reliability and survival analysis, many lifetime measures have been discussed in the literature. They were defined to study the aging behavior of the experimental units. One of these measures is the mean residual lifetime (*MRL*), which is a helpful tool in determining burn-in and maintenance policies. For discrete distributions, the *MRL* is defined as follows:

$$\zeta(i) = E(k - i \mid k \ge i) = \frac{1}{S(i)} \sum_{j=i+1}^{l} S(j); \; i \in \mathbb{N}, \tag{13}$$

where $0 < l < \infty$.

If the random variable k follows the DNGR distribution with parameters α, β, and θ, which is denoted by $k \sim DNGR(\alpha, \beta, \theta)$, then the *MRL* is expressed as

$$\zeta(i) = \frac{\alpha - \Lambda(i;\theta,\beta)}{\alpha[1-\Lambda(i;\theta,\beta)]} \sum_{j=i+1}^{l} \frac{\alpha[1-\Lambda(j;\theta,\beta)]}{\alpha - \Lambda(j;\theta,\beta)}. \tag{14}$$

The mean past lifetime (*MPL*) is another important measure in reliability analysis. The *MPL* measures the time elapsed after the failure of K units given that the system has failed sometime earlier to i. In the discrete case, the *MPL* is defined as follows

$$\zeta^*(i) = E(i-k|k<i) = \frac{1}{F(i-1)} \sum_{m=1}^{i} F(m-1); i \in \mathbb{N} - \{0\}, \tag{15}$$

where $\zeta^*(0) = 0$; see [33].

Then,

$$\zeta^*(i) = \left[1 - \frac{\alpha - \Lambda(i;\theta,\beta)}{\alpha[1-\Lambda(i;\theta,\beta)]}\right]^{-1} \sum_{m=1}^{i} \left[1 - \frac{\alpha[1-\Lambda(m-1;\theta,\beta)]}{\alpha - \Lambda(m-1;\theta,\beta)}\right]. \tag{16}$$

3.4. Order Statistics

Let $K_1, K_2, \ldots, K_n$ be a random sample with the DNGR distribution and $K_{1:n}, K_{2:n}, \ldots, K_{n:n}$ denote the corresponding order statistics. Then, the *CDF* of ith order statistics at the value k can be written as follows

$$F_{i:n}(k;\alpha,\beta,\theta) = \sum_{i=1}^{n} \binom{n}{m} [F_i(k;\alpha,\beta,\theta)]^m [1 - F_i(k;\alpha,\beta,\theta)]^{n-m}. \tag{17}$$

By using the negative binomial theorem, then

$$F_{i:n}(k;\alpha,\beta,\theta) = \sum_{i=1}^{n} \sum_{j=1}^{n-m} \binom{n}{m} \binom{n-m}{j} (-1)^j [F_i(k;\alpha,\beta,\theta)]^{m+j}. \tag{18}$$

Therefore,

$$F_{i:n}(k;\alpha,\beta,\theta) = \sum_{i=1}^{n} \sum_{j=1}^{n-m} \binom{n}{m} \binom{n-m}{j} (-1)^j \left[1 - \frac{\alpha[1-\Lambda(k;\theta,\beta)]}{\alpha - \Lambda(k;\theta,\beta)}\right]^{m+j}. \tag{19}$$

Consequently, the *pmf* of the ith order statistic under the DNGR can be derived and expressed as follows

$$f_{i:n}(k;\alpha,\beta,\theta) = \sum_{i=1}^{n} \sum_{j=1}^{n-m} \binom{n}{m} \binom{n-m}{j} (-1)^j \left[\frac{\alpha[1-\Lambda(k;\theta,\beta)]}{\alpha - \Lambda(k;\theta,\beta)} - \frac{\alpha[1-\Lambda(k+1;\theta,\beta)]}{\alpha - \Lambda(k+1;\theta,\beta)}\right]^{m+j}.$$

So, the vth moments of $k_{i:n}$ can be written as follows

$$E(K_{i:n}^{v}) = \sum_{k=0}^{\infty} \sum_{i=1}^{n} \sum_{j=1}^{n-m} \binom{n}{m} \binom{n-m}{j} (-1)^j k^v \left[\frac{\alpha[1-\Lambda(k;\theta,\beta)]}{\alpha - \Lambda(k;\theta,\beta)} - \frac{\alpha[1-\Lambda(k+1;\theta,\beta)]}{\alpha - \Lambda(k+1;\theta,\beta)}\right]^{m+j}.$$

4. Estimation

Two estimation methods are considered in this work: frequentist maximum likelihood estimation *(MLE)* and the Bayesian estimation method. Simulation analysis and numerical techniques are performed in Section 5 to assess the performance of these estimation methods.

4.1. Maximum Likelihood Estimation

In this section, we use the maximum likelihood estimation *MLE* method to estimate the unknown parameters of the DNGR distributions. To evaluate the required estimators, numerical techniques are used, such as the well-known Newton–Raphson technique.

Let $x_1, \ldots, x_n$ be a random sample following the DNGR, and then, from *pmf* in Equation (5), the log-likelihood function is written as

$$\begin{aligned}\ell(\alpha, \beta, \theta) = \textstyle\sum_{k=1}^{n} \log(\alpha(\alpha-1)) + \log(\Lambda(x_{k+1};\theta,\beta) - \Lambda(x_k;\theta,\beta)) \\ -\log(\alpha - \Lambda(x_k;\theta,\beta)) - \log(\alpha - \Lambda(x_{k+1};\theta,\beta))\end{aligned} \tag{20}$$

The MLEs for α, β, and θ are obtained by finding the partial derivatives of $\ell(\alpha, \beta, \theta)$ for α, β, and θ, then equating the three equations to zero and solving numerically.

$$\frac{\partial \ell(\alpha,\beta,\theta)}{\partial \alpha} = \sum_{k=1}^{n} \frac{2\alpha-1}{\alpha(\alpha-1)} - \frac{1}{\alpha - \Lambda(x_k;\theta,\beta)} - \frac{1}{\alpha - \Lambda(x_{k+1};\theta,\beta)} = 0, \tag{21}$$

$$\frac{\partial \ell(\alpha,\beta,\theta)}{\partial \beta} = \sum_{k=1}^{n} \frac{\Lambda_\beta(x_{k+1};\theta,\beta) - \Lambda_\beta(x_k;\theta,\beta)}{\Lambda(x_{k+1};\theta,\beta) - \Lambda(x_k;\theta,\beta)} + \frac{\Lambda_\beta(x_k;\theta,\beta)}{\alpha - \Lambda(x_k;\theta,\beta)} + \frac{\Lambda_\beta(x_{k+1};\theta,\beta)}{\alpha - \Lambda(x_{k+1};\theta,\beta)} = 0 \tag{22}$$

and

$$\frac{\partial \ell(\alpha,\beta,\theta)}{\partial \theta} = \sum_{k=1}^{n} \frac{\Lambda_\theta(x_{k+1};\theta,\beta) - \Lambda_\theta(x_k;\theta,\beta)}{\Lambda(x_{k+1};\theta,\beta) - \Lambda(x_k;\theta,\beta)} + \frac{\Lambda_\theta(x_k;\theta,\beta)}{\alpha - \Lambda(x_k;\theta,\beta)} + \frac{\Lambda_\theta(x_{k+1};\theta,\beta)}{\alpha - \Lambda(x_{k+1};\theta,\beta)} = 0. \tag{23}$$

Such that $\Lambda_\beta(x_k;\theta,\beta) = \frac{\partial \Lambda(x_k;\theta,\beta)}{\partial \beta} = \Lambda(x_k;\theta,\beta)\log(1 - e^{-\theta {x_k}^2})$ and $\Lambda_\theta(x_k;\theta,\beta) = \frac{\partial \Lambda(x_k;\theta,\beta)}{\partial \theta} = \beta {x_k}^2 e^{-\theta {x_k}^2} \Lambda(x_k;\theta,\beta-1)$. To solve the system of nonlinear Equations (21)–(23), only numerical methods are helpful. Many numerical techniques were used in the literature; here, we use the Newton–Raphson method, and all results are illustrated in Section 5.

4.2. Bayesian Inference

The Bayesian estimation method is used in this section to estimate the unknown parameters of the DNGR. The basic assumption of the Bayesian method is that the model parameters are considered random variables that follow a distribution known as the prior distribution. Since prior information is usually only available, we must specify a suitable prior option. We choose the gamma conjugate prior distribution for the parameters α, β, and θ. It is defined by assuming gamma distributions for α, β, and θ.

Therefore, the prior distributions for α, β, and θ can be written as

$$\pi_1(\alpha) = \frac{{b_1}^{a_1}}{\Gamma(a_1)} \alpha^{a_1-1} e^{-b_1\alpha},$$

$$\pi_2(\beta) = \frac{{b_2}^{a_2}}{\Gamma(a_2)} \beta^{a_2-1} e^{-b_2\beta}$$

and

$$\pi_3(\theta) = \frac{{b_3}^{a_3}}{\Gamma(a_3)} \theta^{a_3-1} e^{-b_3\theta}$$

where a_1, a_2, a_3, b_1, b_2, and b_3 are nonnegative hyper parameters of the assumed distributions.

Hence, the joint prior for α, β, and θ is

$$\pi(\alpha,\beta,\theta) \propto \alpha^{a_1-1}\beta^{a_2-1}\theta^{a_3-1}e^{-(b_1\alpha+b_2\beta+b_3\theta)}. \tag{24}$$

The joint posterior of α, β, and θ given the data is defined as

$$\pi^*(\alpha,\ \beta,\ \theta \mid x) = \frac{1}{k} L(\underline{x} \mid \alpha,\ \beta,\ \theta)\pi(\alpha,\ \beta,\ \theta), \tag{25}$$

where $L(\underline{x} \mid \alpha,\ \beta,\ \theta)$ is the likelihood function of the DNGRD and $K = \int\int\int L(\underline{x} \mid \alpha,\ \beta,\ \theta)\pi(\alpha,\ \beta,\ \theta) d\alpha d\beta d\theta$.

The DNGRD parameters are estimated using a symmetric squared error (SE) loss function. A simulation study is used to investigate the performance of the estimators using the aforementioned loss function. As criteria for the superiority of the estimation methods, the bias, the mean square error (MSE), the average length (AL) of the confidence intervals, and the coverage probability (CP) are computed.

Under the SE loss function, Bayes estimation for the parameters α, β, and θ is defined as the mean or expected value regarding the joint posterior, provided as

$$\widehat{\alpha}_{SE} = \frac{1}{k} \iiint \alpha L(\underline{x} \mid \alpha,\ \beta,\ \theta)\pi(\alpha,\beta,\theta) d\alpha d\theta \tag{26}$$

$$\widehat{\beta}_{SE} = \frac{1}{k} \iiint \beta L(\underline{x} \mid \alpha,\ \beta,\ \theta)\pi(\alpha,\beta,\theta) d\alpha d\theta \tag{27}$$

$$\widehat{\theta}_{SE} = \frac{1}{k} \iiint \theta L(\underline{x} \mid \alpha,\ \beta,\ \theta)\pi(\alpha,\beta,\theta) d\alpha d\theta \tag{28}$$

To evaluate the expected values and triple integration in Equations (26)–(28) , numerical methods are required. We choose to use the Markov chain Monte Carlo (MCMC) technique via the Gibbs sampling method and by developing appropriate R code. The joint posterior density is

$$\pi^*(\alpha,\beta,\ \theta|\underline{x}) = \frac{1}{K}\prod_{i=1}^{n}\left[\frac{[1-\Lambda(i;\theta,\beta)]}{\alpha-\Lambda(i;\theta,\beta)} - \frac{[1-\Lambda(i+1;\theta,\beta)]}{\alpha-\Lambda(i+1;\theta,\beta)}\right]\alpha^{a_1}\beta^{a_2-1}\theta^{a_3-1}e^{-(b_1\alpha+b_2\beta+b_3\theta)} \tag{29}$$

Bayes estimation for parameters α, β, and θ under SE loss function is performed respectively using Equations (26)–(28) and the posterior density Equation (29).

The estimators are numerically evaluated simulations using R codes under the SE loss function, and their results are summarized and presented in Tables 2 and 3.

Table 2. MLE and Bayes for parameters of DNGR distribution: $\alpha = 1.5$.

$\alpha = 1.5$				MLE					Bayesian			
θ	β	n		RB	MSE	Lower	Upper	CP	RB	MSE	Lower	Upper
0.8	0.5	30	α	0.3587	1.5243	0.2760	4.8362	96.4%	−0.0637	0.1796	1.1698	1.6971
			θ	0.1163	0.3028	0.3278	1.4583	94.8%	−0.0900	0.1096	0.5657	0.8743
			β	2.8414	1.6217	0.3864	3.4549	97.4%	0.2762	0.1846	0.4248	0.8864
		70	α	0.1490	1.4933	0.3716	4.8186	96.2%	−0.1036	0.2005	1.1424	1.5800
			θ	0.0150	0.1898	0.4403	1.1837	94.0%	−0.0572	0.0651	0.6586	0.8455
			β	1.4425	1.5862	0.8267	3.6157	96.0%	0.3645	0.2151	0.4609	0.9003
		100	α	0.2670	1.2867	0.4987	4.2996	96.6%	−0.0749	0.1473	1.2081	1.5686
			θ	0.0141	0.1323	0.5527	1.0698	94.8%	−0.0573	0.0556	0.6889	0.8096
			β	0.9287	1.0745	0.9939	3.2935	96.2%	0.2963	0.1636	0.5165	0.7786
		200	α	0.2572	1.0381	0.5976	4.0743	93.4%	−0.0433	0.0803	1.3303	1.5184
			θ	−0.0401	0.1044	0.5730	0.9629	95.0%	−0.0446	0.0400	0.7257	0.7991
			β	0.8044	0.9101	1.4005	3.6436	93.2%	0.1773	0.0955	0.5222	0.6549

Table 2. *Cont.*

$\alpha = 1.5$				MLE					Bayesian			
θ	β	n		RB	MSE	Lower	Upper	CP	RB	MSE	Lower	Upper
0.8	2	30	α	−0.2630	0.4948	0.5194	1.6917	96.4%	−0.0723	0.1817	1.1206	1.6439
			θ	0.1964	0.2533	0.5674	1.3469	95.0%	−0.1627	0.1504	0.5360	0.8156
			β	0.1821	0.6169	1.3875	3.3410	95.4%	0.0193	0.1575	1.7350	2.3384
		70	α	−0.2876	0.4697	0.7042	1.4329	98.2%	−0.0650	0.1782	1.1407	1.6102
			θ	0.1781	0.2057	0.6514	1.2336	95.4%	−0.1452	0.1284	0.5898	0.8055
			β	0.1783	0.6075	1.3907	3.2658	97.6%	0.0268	0.1401	1.8049	2.2871
		100	α	−0.2344	0.4531	0.8432	1.4810	99.8%	−0.0867	0.1618	1.1862	1.5461
			θ	0.1899	0.1968	0.7066	1.1973	95.4%	−0.1317	0.1115	0.6174	0.7637
			β	0.1585	0.5414	1.4561	3.1780	96.6%	0.0243	0.0945	1.8941	2.2078
		200	α	−0.2144	0.3822	0.8523	1.4703	99.6%	−0.0418	0.0796	1.3468	1.5366
			θ	0.1631	0.1540	0.7699	1.0910	94.8%	−0.1128	0.0931	0.6638	0.7529
			β	0.1403	0.5402	1.3709	3.0190	98.4%	0.0133	0.0482	1.9454	2.0992
1.3	0.5	30	α	−0.0161	1.8375	0.1289	5.0806	95.2%	−0.0893	0.2022	1.1191	1.6715
			θ	0.2417	0.5574	0.7108	2.5176	97.0%	−0.0770	0.1671	0.9649	1.5162
			β	2.9986	1.9656	0.1495	4.4932	95.4%	0.2996	0.1973	0.4097	0.8957
		70	α	−0.0127	1.8093	0.2549	4.9173	96.0%	−0.1216	0.2207	1.1361	1.5891
			θ	0.2073	0.4650	0.8261	2.3130	96.8%	−0.0716	0.1243	1.0573	1.3653
			β	1.0203	1.8424	0.2329	4.2970	96.0%	0.4009	0.2295	0.4769	0.9082
		100	α	−0.0157	1.2760	0.2963	3.5147	95.8%	−0.0973	0.1743	1.1848	1.5360
			θ	0.1764	0.4309	0.8514	2.2450	97.2%	−0.0625	0.1035	1.0952	1.3359
			β	0.9256	0.9678	0.5612	3.8173	96.2%	0.3340	0.1821	0.5292	0.8175
		200	α	−0.0142	0.9294	0.3171	3.0604	96.4%	−0.0507	0.0912	1.3314	1.5228
			θ	0.0813	0.3153	0.8228	1.9886	98.0%	−0.0378	0.0594	1.1915	1.3146
			β	0.5671	0.6315	0.6249	3.0816	99.2%	0.1931	0.1029	0.5286	0.6659
	2	30	α	−0.3113	0.6551	0.1315	1.9347	99.4%	−0.1012	0.2185	1.0986	1.6487
			θ	0.1534	0.3004	1.0585	1.9403	93.6%	−0.1306	0.2070	0.9252	1.3786
			β	0.1334	0.7466	0.8986	3.6349	98.4%	0.0235	0.1570	1.7458	2.3302
		70	α	−0.3117	0.6555	0.1310	1.9338	99.4%	−0.1027	0.2185	1.0986	1.6487
			θ	0.1526	0.2934	1.0744	1.9225	93.8%	−0.1263	0.2067	0.9265	1.3857
			β	0.1323	0.7438	0.9007	3.6028	98.4%	0.0216	0.1529	1.7548	2.3035
		100	α	−0.3207	0.4811	1.0057	1.0321	93.8%	−0.0914	0.2108	1.1902	1.4334
			θ	0.1929	0.2756	1.3192	1.7823	100.0%	−0.1151	0.1656	1.0148	1.2705
			β	0.0887	0.1776	1.9162	2.1927	100.0%	0.0154	0.0757	1.9316	2.1534
		200	α	−0.3188	0.4782	1.0096	1.0341	93.3%	−0.0538	0.0946	1.3235	1.5159
			θ	0.1594	0.2403	1.3682	1.7462	94.7%	−0.0763	0.1043	1.1427	1.2673
			β	0.0809	0.1683	1.9258	2.1821	95.9%	0.0140	0.0496	1.9582	2.1127

Table 3. MLE and Bayes for parameters of DNGR distribution: $\alpha = 3$.

$\alpha = 3$				MLE					Bayesian			
θ	β			RB	MSE	Lower	Upper	CP	RB	MSE	Lower	Upper
1.3	0.5	30	α	0.1917	2.0185	0.2210	7.3712	96.2%	−0.0038	0.1558	2.6689	3.2541
			θ	0.5325	1.0676	0.3977	3.5868	93.2%	−0.0696	0.1621	0.9486	1.4541
			β	9.3388	5.0389	1.4536	8.8852	96.4%	0.4023	0.2277	0.5010	0.8988
		70	α	0.3736	1.8015	1.3537	6.8878	92.6%	−0.0071	0.1313	2.7384	3.2252
			θ	0.3598	0.6399	0.9110	2.6245	95.2%	−0.0784	0.1300	1.0446	1.3579
			β	9.7520	5.4079	0.7873	9.9648	96.4%	0.5596	0.2929	0.6081	0.9368
		100	α	0.3083	2.0838	0.2612	7.5884	95.4%	−0.0040	0.0897	2.8073	3.1475
			θ	0.3510	0.5847	1.0390	2.4735	96.2%	−0.0627	0.1003	1.1033	1.3265
			β	9.7169	5.4798	0.3857	10.3312	93.6%	0.4114	0.2145	0.5799	0.8143
		200	α	0.8118	3.5689	0.3173	10.5536	95.8%	−0.0021	0.0452	2.9092	3.0800
			θ	0.3223	0.4785	1.2655	2.1725	95.8%	−0.0367	0.0585	1.1795	1.3146
			β	9.6584	5.0771	2.2548	8.4036	96.0%	0.2381	0.1241	0.5548	0.6959
	0.9	30	α	−0.1503	1.6360	0.5364	5.6349	95.6%	−0.0038	0.1618	2.6992	3.3379
			θ	0.3752	0.8357	0.4565	3.1191	97.8%	−0.1104	0.1934	0.9110	1.3977
			β	6.1647	6.2502	0.8019	5.0945	97.0%	0.1488	0.1888	0.7805	1.3184
		70	α	−0.1584	1.0898	0.6006	4.4492	97.6%	−0.0134	0.1395	2.7034	3.2184
			θ	0.1830	0.4443	0.8017	2.2740	93.0%	−0.1119	0.1694	0.9717	1.2952
			β	5.3507	5.0649	2.6369	4.7944	97.8%	0.1222	0.1722	0.8992	1.2938
		100	α	0.1317	2.0370	0.6525	4.3156	94.2%	−0.0031	0.0883	2.8246	3.1589
			θ	0.2120	0.4331	0.9200	2.2311	92.0%	−0.0901	0.1313	1.0629	1.2857
			β	6.4228	6.0394	3.2486	4.1124	96.4%	0.1633	0.1653	0.8964	1.1761
		200	α	0.1701	1.6929	0.3436	3.6772	94.2%	−0.0031	0.0461	2.8961	3.0742
			θ	0.1568	0.2695	1.1578	1.8498	95.0%	−0.0485	0.0704	1.1800	1.3009
			β	6.7695	6.3441	3.5224	3.8463	97.2%	0.0836	0.0848	0.9010	1.0514
2	0.5	30	α	−0.1881	0.6957	1.6375	3.2338	95.6%	−0.0047	0.1551	2.6721	3.2679
			θ	−0.0411	0.5058	0.9386	2.8969	90.0%	−0.0553	0.1959	1.5765	2.2175
			β	4.5180	2.5199	0.5683	4.9497	96.8%	0.2992	0.1983	0.3997	0.9025
		70	α	−0.2116	0.8253	1.3305	3.3996	100.0%	−0.0067	0.1355	2.6983	3.2384
			θ	−0.0649	0.6639	0.5929	3.1475	97.2%	−0.0726	0.1869	1.6464	2.0764
			β	4.1615	2.3945	0.2558	4.9057	100.0%	0.4675	0.2563	0.5129	0.9282
		100	α	−0.2105	0.8825	1.1588	3.5784	100.0%	−0.0039	0.0929	2.8218	3.1785
			θ	−0.0837	0.6576	0.5849	3.0803	100.0%	−0.0565	0.1382	1.7356	2.0397
			β	3.8575	2.2378	0.2023	4.6552	100.0%	0.3167	0.1723	0.5349	0.7985
		200	α	−0.0747	0.3607	2.2214	3.3303	92.6%	−0.0021	0.0450	2.9162	3.0799
			θ	−0.2174	0.5774	0.8199	2.3105	95.2%	−0.0275	0.0695	1.8580	2.0248
			β	3.0823	1.7603	0.3725	3.7098	95.4%	0.1803	0.0975	0.5084	0.6579
	1.1	30	α	−0.4095	1.4678	0.1957	3.3470	99.6%	−0.0073	0.1620	2.6815	3.2895
			θ	−0.0105	0.6870	0.6319	3.3262	100.0%	−0.0911	0.2397	1.5140	2.1195
			β	2.5931	3.5294	0.1257	8.0306	99.6%	0.1034	0.1881	0.9228	1.5108
		70	α	−0.3289	1.1819	0.7368	3.2901	100.0%	−0.0127	0.1424	2.6825	3.1929
			θ	−0.1004	0.6782	0.5283	3.0700	100.0%	−0.0941	0.2211	1.5755	1.9901
			β	2.3766	2.7979	1.7584	5.6701	97.8%	0.1644	0.2140	1.0552	1.4943
		100	α	−0.2776	0.9788	1.1578	3.1769	92.8%	−0.0075	0.0892	2.8212	3.1590
			θ	−0.1742	0.6667	0.5362	2.7671	93.0%	−0.0655	0.1549	1.7024	2.0002
			β	2.3582	2.7976	1.6383	5.7497	98.2%	0.1111	0.1451	1.0470	1.3534
		200	α	−0.2445	0.8234	1.5320	3.0013	93.8%	−0.0041	0.0453	2.8966	3.0706
			θ	−0.2436	0.5653	0.9504	2.0750	97.4%	−0.0367	0.0844	1.8520	2.0008
			β	2.2435	2.5427	2.3663	4.7694	96.6%	0.0569	0.0753	1.0850	1.2456

5. Simulation Analysis

In this section, Monte Carlo simulations are performed to assess the effectiveness of the suggested estimators for the parameters α, θ, and β that were established in Section 4.1. We will sum up by providing the simulation scenario. The findings of the simulation are then offered for debate.

5.1. Simulation Scenario

In this subsection, several Monte Carlo simulation studies are carried out to assess the effectiveness of the acquired maximum likelihood estimates and Bayesian estimation of α, θ, and β. Now, we suggest the following steps to gather a sample from the DNGR model:

- Set α, θ, and β to their actual values as shown:
  In Table 2: $\alpha = 1.5, \theta = 0.8, \beta = 0.5$, $\alpha = 1.5, \theta = 0.8, \beta = 2$, $\alpha = 1.5, \theta = 1.3, \beta = 0.5$, $\alpha = 1.5, \theta = 1.3, \beta = 2$.
  In Table 3: $\alpha = 3, \theta = 1.3, \beta = 0.5$, $\alpha = 3, \theta = 1.3, \beta = 0.9$, $\alpha = 3, \theta = 2, \beta = 0.5$, $\alpha = 3, \theta = 2, \beta = 1.1$.
- Specific values for n (total test units) should be determined as 30, 70, 100, 200.
- Generate a uniform random variable within the range of 0 to 1. Utilize the quantile function described in Equation (7) to produce a random sample from the DNGR(α, θ, β) distribution. Afterwards, round the quantity of samples to the nearest whole number.
- Compute the MLEs and $100(1-\gamma)\%$ via 'maxLik' package in R program with version number (4.3.0), with Fisher information matrix (Hessian matrix).
- Use 'coda' package in R program with version number (4.3.0), to obtain the Bayes' inferences by running the MCMC sampler 12,000 times and 2000 is burn-in.
- Repeat the above steps 5000 times.
- The relative bias (RB), mean squared error (MSE), average lower, average upper, and coverage probability (CP) of the parameter are specifically determined for each group (n, or actual value of the parameter). For more details about comparing interval estimates, we discuss using the CP requirement in our evaluations. R 4.2.2 programming language is used to carry out all numerical analyses. In Tables 2 and 3, respectively, all numerical findings for α, β, and θ are obtained and presented.

5.2. Simulation Conclusion

The performance of the suggested point and interval estimate algorithms is the main topic of this subsection. We can infer the following facts from Tables 2 and 3:

- The acquired estimates of the unknown parameters α, θ, and β generally perform well in terms of lowest MSE, RB, and difference between upper and lower values with CP.
- The MSE, RB, and CI of α, θ, and β tend to decline as n rises. This result supports the associated estimates' consistency property of DNGR distribution when the necessary sample size is raised.
- As the true value of β increases, for each setting, the MSE, RB, and CL measures of unknown parameters α and β decrease, while they increase regarding unknown parameter θ.
- The MSE, RB, and CL measures of all unknown parameters α, θ, and β increase for each set as the true value of θ grows.
- For CI of Bayesian, the credible interval decreases when the sample size increases.
- Almost always, and regardless of sample size, Bayesian estimation based on the SE loss function yields the minimal RB and MSE values.

6. Real Data Examples

This section presents the analysis of two applications using different real datasets. The main goals of this section are

- Examine the usefulness and applicability of the proposed model to real phenomena;
- Show the applicability of the inferential results to a real practical situation;

- Evaluate whether the proposed model is a better choice than the other seven models.

Data I: The first dataset includes the number of strikes that occurred in the UK coal mining industry over four consecutive week periods between 1948 and 1959. It was derived from Kendall [34]. An empirical model was used to analyze this example by Ridout and Besbeas [35] and is presented in Table 4.

Table 4. Data I: The number of strikes and their frequency that occurred in the UK coal mining industry.

data	0	1	2	3	4 or more
Freq	46	76	24	9	1

Data II: The number of fires that occurred in Greece between 1 July and 31 August, 1998. We only take into account fires in forest districts. These data have a sample size of 124. The minimum value is 0, the first quartile is 2, the median value is 4, the mean value is 5.065, the third quartile's maximum value is 8, and the variance value is 18.256. The data are as follows: 2, 4, 4, 3, 3, 1, 2, 4, 3, 1, 1, 0, 5, 5, 0, 3, 1, 1, 0, 1, 0, 2, 0, 1, 2, 0, 0, 0, 0, 0, 0, 0, 0, 1, 4, 2, 2, 1, 2, 1, 2, 0, 2, 2, 1, 0, 3, 2, 1, 2, 2, 7, 3, 5, 2, 5, 4, 5, 6, 5, 4, 3, 8, 4, 3, 8, 4, 4, 3, 10, 5, 4, 5, 12, 3, 8, 12, 10, 11, 6, 1, 8, 9, 12, 9, 4, 8, 12, 11, 8, 6, 4, 7, 9, 15, 12, 15, 15, 12, 9, 16, 7, 11, 9, 11, 6, 5, 20, 9, 8, 8, 5, 7, 10, 6, 6, 5, 5, 15, 6, 8, 5, 6. These data were discussed by [36].

Based on the first and second datasets, the DNGRD probability model is contrasted and compared with the other seven competing models to show the reliability and superiority of the proposed model, including Poisson, binomial, geometric, discrete Burr (DB) by [37], discrete Marshall–Olkin Lomax (DMOL) by [38], new discrete Lindley (NDL) by [39], and discrete odd perks exponential (DOPE) by [40] distributions. To specify the best model, several criteria are used, namely: Akaike ($AIC = 2p - 2\hat{l}$), where p is the length of the model parameter and $\hat{l}$ is log-likelihood value, consistent Akaike ($CAIC = -2\hat{l} + \frac{2np}{n-p-1}$), Bayesian ($BIC = -2\hat{l} + p\ln(n)$), and Hannan–Quinn ($HQIC = -2\hat{l} + p\ln(\ln(n))$ information criteria. Along with these, the $X^2 - square$ statistic and its p-value are taken into account. If a probability model distribution has the highest p-value and the lowest values for all other metrics, it is obvious that it will provide the best fit for a particular collection of data. The maximum likelihood estimates (with their standard errors (St.Es)), as well as the fitted model selection criteria, are shown in Tables 5 and 6 using the R programming language and the 'bbmle' package in R program with version number (4.3.0), that was recommended.

To compare the performance and efficiency of the DNGR distribution with other distributions listed in Table 5, using measures of goodness of fit and p-values, we can proceed as follows:

1-DNGR versus DOPE:

DNGR shows a better fit with a lower AIC, $CAIC$, BIC, and $HQIC$. The chi-squared value is lower for DNGR, indicating a better fit. DNGR has a higher p-value, suggesting a better fit to the data than DOPE.

2-DNGR versus Binomial/Poisson:

DNGR has a higher p-value than both binomial and Poisson distributions, indicating a more suitable model for the data. The information criteria (AIC, CAIC, HQIC) for DNGR are lower compared to binomial and Poisson, suggesting a better fit than binomial and Poisson.

3-DNGR versus DMOL:

DNGR and DMOL have comparable p-values, but DNGR shows better performance in terms of information criteria.

4-DNGR versus DB:

DNGR has a higher p-value than DB, indicating that DNGR shows slightly better performance in terms of information criteria.

5-DNGR versus Geometric/NDL:

DNGR outperforms both geometric and NDL distributions in terms of p-value, indicating a significantly better fit. DNGR has lower information criteria values, further suggesting its superiority in model fitting. Overall, DNGR appears to offer a more efficient and suitable fit for Data I compared to the other listed distributions.

Table 5. MLE estimates and different measures of fit for Data I.

		Estimators	SE	AIC	CAIC	BIC	HQIC	X^2	p-Value
DNGR	α	33.0996	13.6357	382.7571	382.9150	391.9067	386.4733	3.6721	0.4522
	θ	0.3343	0.0434						
	β	0.9276	0.1425						
DOPE	α	39.3584	3.5957	387.5354	387.6933	396.6850	391.2516	5.2373	0.2638
	θ	0.2757	0.0264						
	β	3.1307	0.5099						
Binomial	θ	0.9937	0.1315	386.1302	386.1562	389.1801	387.3689	10.1078	0.0387
Poision	θ	0.9936	0.0798	386.1302	386.1562	389.1801	387.3689	9.8986	0.0422
DB	α	4.6524	0.6986	388.4190	388.4974	394.5187	390.8964	5.4076	0.2480
	β	0.5940	0.0448						
DMOL	α	21.1960	7.5845	386.4288	386.5867	395.5784	390.1450	3.9771	0.4091
	θ	1.8892	0.3088						
	β	0.0028	0.0009						
Geometric	θ	0.5017	0.0284	433.1343	433.1603	436.1842	434.3731	50.7984	0.0000
NDL	θ	0.5017	0.0284	433.1343	433.1603	436.1842	434.3731	50.7984	0.0000

Table 6. MLE estimates and different measures for Data II.

		Estimators	SE	AIC	CAIC	BIC	HQIC	X^2	p-Value
DNGRD	α	12.5091	2.5308	668.4150	668.6150	676.8759	671.8520	22.8025	0.2986
	θ	0.0122	0.0027						
	β	0.4495	0.1299						
DOPE	α	44.2111	9.3605	685.1374	685.3374	693.5983	688.5744	32.6330	0.0370
	θ	0.0134	0.0017						
	β	0.7291	0.0844						
Binomial	θ	0.9608	0.2622	821.7835	821.8163	824.6038	822.9292	12776.3387	0.0000
Poision	θ	5.0645	0.2021	821.7835	821.8163	824.6038	822.9292	26469.5465	0.0000
DB	α	2.5385	0.4910	748.2257	748.3249	753.8663	750.5170	91.3536	0.0000
	β	0.7611	0.0425						
DMOL	α	4.6349	1.8610	674.2602	674.4602	682.7210	677.6972	25.0095	0.2011
	θ	13.0180	3.2683						
	β	0.0031	0.0018						
Geometric	θ	0.1649	0.0135	675.3352	675.3679	678.1554	676.4808	27.5152	0.1214
NDL	θ	0.1649	0.0135	675.3352	675.3679	678.1554	676.4808	27.5152	0.1214

Figures 3–5 confirm these results for Data I (the black point refer to data; the pink point refer to DNGR distribution). Additionally, it is evident from Data II in Table 6 that the DNGR distribution is the best distribution among all the examined models in terms of the P-value, whereas Figures 6–8 confirm these results for Data II. Figure 3 confirms the results of MLE fitting and demonstrates the existence, uniqueness, and maximum point of likelihood value of the likelihood estimates for Data I. Figure 4 regarding associated empirical CDF and estimated CDF plot illustrates the connection between observed cumulative probability

and observation through a visual plot and also Q–Q plot for Data I. Figure 5 highlights estimated frequency by using PMF for each comparative model for Data I. Table 7 indicates survival and hazard rate functions for DNGR distribution with different values of Data I, noting that the survival value decreased when the values of Data I increased, while the hazard rate value increased when the values x of Data I increased.

Figure 6 confirms the results of MLE fitting and demonstrates the existence, uniqueness, and maximum point of likelihood value of the likelihood estimates for Data II. Figure 7 regarding associated empirical CDF and estimated CDF plot illustrates the connection between observed cumulative probability and observation through a visual plot and Q–Q plot for Data II. Figure 8 highlights estimated frequency by using PMF for each comparative model for Data II. Table 8 indicates survival and hazard rate functions for DNGR distribution with different values of Data II, noting that the survival values decreased when the values of Data II increased, while the hazard rate value increased when the values x of Data II increased.

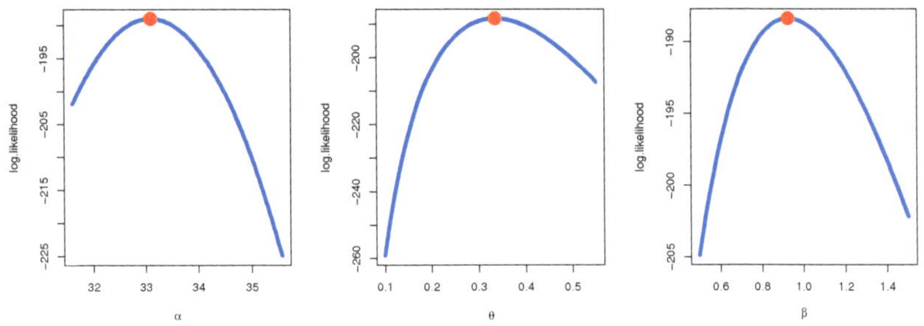

Figure 3. Likelihood profile (blue line) with the maximum likelihood estimation (red dot): Data I.

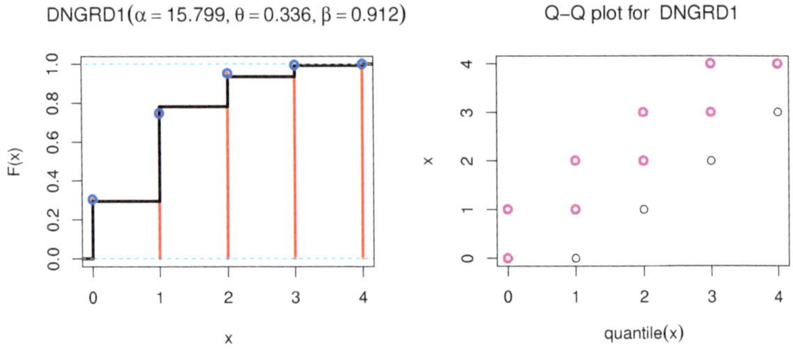

Figure 4. Estimated CDF and Q–Q plot of DNGR by using MLE: Data I.

Table 7. Survival and hazard rate functions for DNGR distribution with different values of Data I.

x	$S\ (x; \mathbf{33.0994, 0.3343, 0.9275})$	$h\ (x; \mathbf{33.0994, 0.3343, 0.9275})$
0	0.6952	0.4383
1	0.2518	1.7607
2	0.0472	4.3343
3	0.0045	9.3889
4	0.0002	19.2691

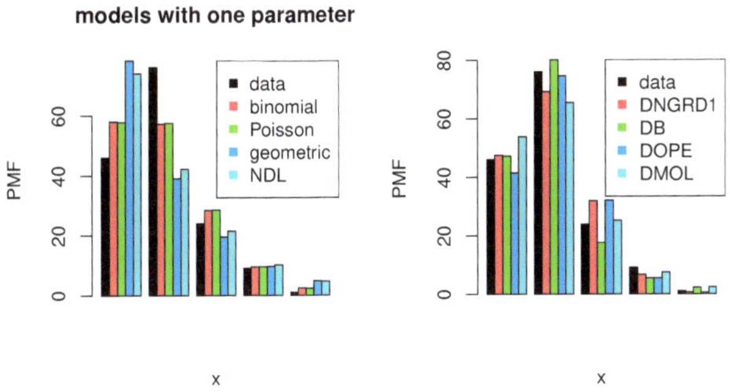

Figure 5. Estimated PMF of each comparative model by using MLE: Data I.

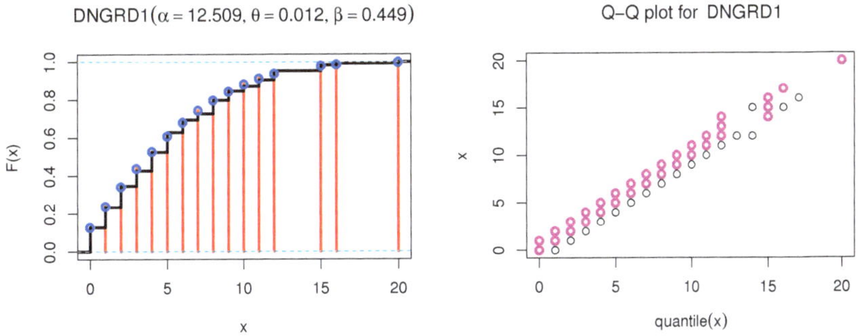

Figure 6. Estimated CDF and Q–Q plot of DNGR by using MLE: Data II.

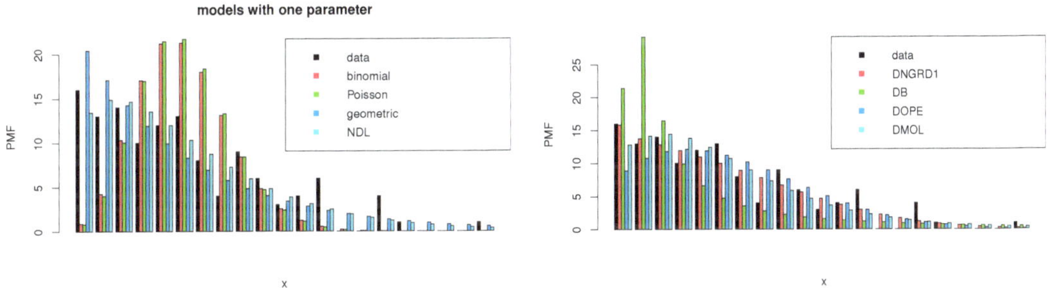

Figure 7. Estimated PMF of each comparative model by using MLE: Data II.

Table 8. Survival and hazard rate functions for DNGR distribution with different values of Data II.

x	$S\,(x; \mathbf{12.5091, 0.0122, 0.4495})$	$h\,(x; \mathbf{12.5091, 0.0122, 0.4495})$
0	0.8721	0.1466
2	0.6577	0.1572
4	0.4725	0.1879
5	0.3919	0.2059
8	0.2023	0.2686
20	0.0023	0.6490

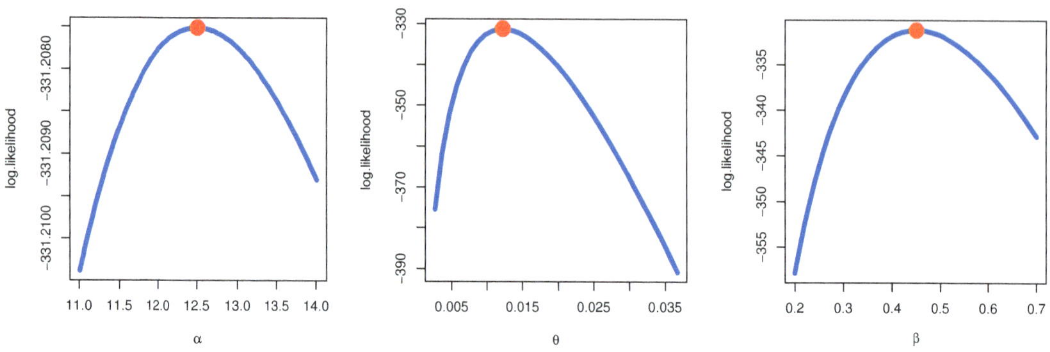

Figure 8. Likelihood profile (blue line) with the maximum likelihood estimation (red dot): Data II.

7. Conclusions

In this study, the authors successfully developed a novel discrete analog from the continuous generalized Rayleigh distribution denoted by DNGR through the application of the survival discretization method. Several key attributes of the DNGR model were studied, such as its unimodal probability mass function, which exhibits varying degrees of symmetry and skewness based on parameter selection. Comprehensive statistical measures for DNGR were derived, including moments, stress–strength function, moment-generating function, and mean residual and mean past lifetimes. The potential submodels and special cases derived from the DNGR demonstrate the versatility and adaptability of the DNGR distribution, which can be suitable for modeling different types of data and could provide unique insights depending on the context of the analysis. Furthermore, the practical applicability of the work was enhanced by conducting detailed simulation analyses and presenting the results in tabular form. Point and interval estimation using both the maximum likelihood and the Bayesian methods were obtained, supplementing these with simulation analyses executed using R code. This was complemented by a numerical analysis aimed at evaluating the estimation methods for DNGR's unknown parameters and assessing the efficiency of these methods. A significant aspect of their contribution is the application of the DNGR model to real-world data. Two real data examples were selected, one from the industrial sector concerning UK coal mining strikes and another focusing on environmental issues related to fires in Greece. Their analysis revealed that the DNGR model outperformed seven competitive discrete distributions in various goodness-of-fit measures, demonstrating its superior ability to model the given datasets effectively. This finding was further illustrated through detailed tables and figures showcasing the properties and efficiency of the new model. As a pathway for future work, the authors suggest exploring alternative discretization methods to assess their performance and applicability to a broader range of real-life data scenarios.

Author Contributions: Conceptualization, H.H.A.; Methodology, H.H.A. and D.A.R.; Software, E.M.A.; Validation, D.A.R.; Formal analysis, H.H.A., E.M.A. and D.A.R.; Investigation, H.H.A. and E.M.A.; Resources, E.M.A.; Data curation, E.M.A.; Writing—original draft, H.H.A., D.A.R. and E.M.A.; Writing—review & editing, D.A.R. and H.H.A.; Funding acquisition, H.H.A. All authors have read and agreed to the published version of the manuscript.

Funding: This work was supported by the Deanship of Scientific Research, Vice Presidency for Graduate Studies and Scientific Research, King Faisal University, Saudi Arabia [GRANT No. 5476].

Data Availability Statement: Data are contained within the article.

Acknowledgments: This work was supported by the Deanship of Scientific Research, Vice Presidency for Graduate Studies and Scientific Research, King Faisal University, Saudi Arabia [GRANT No. 5476].

Conflicts of Interest: The authors declare no conflicts of interest.

References

1. Xekalaki, E. Hazard function and life distributions in discrete time. *Commun. Stat. Theory Methods* **1983**, *12*, 2503–2509. [CrossRef]
2. Hitha, N.; Nair, U.N. Characterization of some discrete models by properties of residual life function. *Calcutta Stat. Assoc. Bull.* **1989**, *38*, 219–223. [CrossRef]
3. Roy, D.; Gupta, R.P. Classifications of discrete lives. *Microelectron. Reliab.* **1992**, *32*, 459–1473. [CrossRef]
4. Roy, D.; Gupta, R.P. Stochastic modeling through reliability measures in the discrete case. *Stat. Probab. Lett.* **1999**, *43*, 197–206. [CrossRef]
5. Roy, D. On classifications of multivariate life distributions in the discrete set-up. *Microelectron. Reliab.* **1997**, *37*, 361–366. [CrossRef]
6. Roy, D. The discrete normal distribution, Commun. *Stat. Theor. Methods* **2003**, *32*, 1871–1883. [CrossRef]
7. Roy, D. Discrete Rayleigh distribution. *IEEE. Trans. Reliab.* **2004**, *53*, 255–260. [CrossRef]
8. Roy, D.; Ghosh, T. A new discretization approach with application in reliability estimation. *IEEE. Trans. Reliab.* **2009**, *58*, 456–461. [CrossRef]
9. Lai, C.-D. Constructions and applications of lifetime distributions. *Appl. Stoch. Model. Bus. Ind.* **2013**, *29*, 127–140. . [CrossRef]
10. Lai, C.D. Issues concerning constructions of discrete lifetime models. *Qual. Technol. Quant. Manag.* **2013**, *10*, 251–262. [CrossRef]
11. Ahmad, H.H.; Almetwally, E.M. Generating optimal discrete analogue of the generalized Pareto distribution under Bayesian inference with application. *Symmetry* **2022**, *14*, 1457. . [CrossRef]
12. Bracquemond, C.; Gaudoin, O. A survey on discrete life time distributions. *Int. J. Reliabil. Qual. Saf. Eng.* **2003**, *10*, 69–98. [CrossRef]
13. Chakraborty, S. Generating discrete analogues of continuous probability distributions-A survey of methods and constructions. *J. Stat. Distrib. Appl.* **2015**, *2*, 6. [CrossRef]
14. Al-Huniti, A.A.; Al-Dayjan, G.R. Discrete Burr type III distribution. *Am. J. Math. Stat.* **2012**, *2*, 145–152. [CrossRef]
15. Bebbington, M.; Lai, C.D.; Wellington, M.; Zitikis, R. The discrete additive Weibull distribution: A bathtub-shaped hazard for discontinuous failure data. *Reliab. Eng. Syst. Saf.* **2012**, *106*, 37–44. [CrossRef]
16. Yari, G.; Tondpour, Z. Discrete Burr XII-Gamma Distributions: Properties and Parameter Estimations. *Iran J. Sci. Technol. Trans. Sci.* **2018**, *42*, 2237–2249. [CrossRef]
17. Eliwa, M.S.; Altun, E.; El-Dawoody, M.; El-Morshedy, M. A new three-parameter discrete distribution with associated INAR(1) process and applications. *IEEE Access* **2020**, *8*, 91150–91162. [CrossRef]
18. Eldeeb, A.; Haq, M.A.U.; Babar, A. A Discrete Analog of Inverted Topp-Leone Distribution: Properties, Estimation and Applications. *Int. J. Anal. Appl.* **2021**, *19*, 695–708.
19. Ahmad, H.H.; Almetwally, E. On discrete generalization of the inverse exponential distribution: Properties, characterizations and applications. *AIP Conf. Proc.* **2023**, *2738*, 020001. . [CrossRef]
20. El-Dawoody, M.; Eliwa, M. S. Bivariate Discrete Burr Lifetime Distribution: A Mathematical and Statistical Framework for Modeling Medical and Engineering Data. *Inf. Sci. Lett.* **2023**, *12*, 3199–3214
21. El-Dawoody, M.; Eliwa, M.S.; El-Morshedy, M. An Extension of the Poisson Distribution: Features and Application for Medical Data Modeling. *Processes* **2023**, *11*, 1195. [CrossRef]
22. Al-Bossly, A.; Eliwa, M.S.; Ahsan-Ul-Haq, M.; El-morshedy, M. Discrete Logistic Exponential Distribution with Application. *Stat. Optim. Inf. Comput.* **2023**, *11*, 629–639. [CrossRef]
23. Abd EL-Hady, A.E.; Hegazy, M.A.; EL-Helbawy, A.A. Discrete Exponentiated Generalized Family of Distributions. *Comput. J. Math. Stat. Sci.* **2023**, *2*, 303–327. [CrossRef]
24. Rosaiah, K.; Kantam, R.R.L. Acceptance sampling based on the inverse Rayleigh distribution. *Econ. Qual. Control* **2005**, *20*, 277–286. [CrossRef]
25. Merovci, F. Transmuted rayleigh distribution. *Aust. J. Stat.* **2013**, *42*, 21–31. [CrossRef]
26. Cordeiro, G.M.; Cristino, C.T.; Hashimoto, E.M.; Ortega, E.M. The beta generalized Rayleigh distribution with applications to lifetime data. *Stat. Pap.* **2013**, *54*, 133–161. [CrossRef]
27. Ahmad, A.; Ahmad, S.P.; Ahmed, A. Transmuted inverse Rayleigh distribution: A generalization of the inverse Rayleigh distribution. *Math. Theory Model* **2014**, *4*, 90–98.
28. Gomes, A.E.; da-Silva, C.Q.; Cordeiro, G.M.; Ortega, E.M. A new lifetime model: The Kumaraswamy generalized Rayleigh distribution. *J. Stat. Comput. Simul.* **2014**, *84*, 290–309. [CrossRef]
29. Nofal, Z.M.; Abd El Hadi, N.E. Exponentiated transmuted generalized Raleigh distribution: A new four-parameter Rayleigh distribution. *Pak. J. Stat. Oper. Res.* **2015**, *11*, 115–134.
30. Iriarte, Y.A.; Vilca, F.; Varela, H.; Gomez, H.W. Slashed generalized Rayleigh distribution. *Commun. Stat.-Theory Methods* **2017**, *46*, 4686–4699. [CrossRef]

31. Haj Ahmad, H.; Bdair, O.M.; Naser, M.F.M.; Asgharzadeh, A. The rayleigh lindley distribution: A new generalization of rayleigh distribution with physical applications. *Rev. Investig. Oper.* **2023**, *44*, 1–18.
32. Shen, Z.; Alrumayh, A.; Ahmad, Z.; Abu-Shanab, R.; Al-Mutairi, M.; Aldallal, R. A new generalized Rayleigh distribution with analysis to big data of an online community. *Alex. Eng. J.* **2022**, *61*, 11523–11535. [CrossRef]
33. Goliforushani, S.; Asadi, M. On the discrete mean past lifetime. *Metrika* **2008**, *68*, 209–217. [CrossRef]
34. Kendall, M.G. Natural law in social sciences. *J. R. Stat. Soc. Ser.* **1961**, *124*, 1–19. . [CrossRef]
35. Ridout, M.S.; Besbeas, P. An empirical model for under dispersed count data. *Stat. Model.* **2004**, *4*, 77–89. [CrossRef]
36. Karlis, D.; Xekalaki, E.; Lipitakis, E.A. On some discrete valued time series models based on mixtures and thinning. In Proceedings of the Fifth Hellenic-European Conference on Computer Mathematics and Its Applications, Athens, Greece, 20–22 September 2001; pp. 872–877.
37. Krishna, H.; Pundir, P.S. Discrete Burr and discrete Pareto distributions. *Stat. Methodol.* **2009**, *6*, 177–188. [CrossRef]
38. Ibrahim, G.M.; Almetwally, E.M. Discrete marshall–olkin lomax distribution application of COVID-19. *Biomed. J. Sci. Tech. Res.* **2021**, *32*, 25381–25390.
39. Al-Babtain, A.A.; Ahmed, A.H.N.; Afify, A.Z. A new discrete analog of the continuous Lindley distribution, with reliability applications. *Entropy* **2020**, *22*, 603. [CrossRef]
40. Elbatal, I.; Alotaibi, N.; Almetwally, E.M.; Alyami, S.A.; Elgarhy, M. On odd perks-G class of distributions: Properties, regression model, discretization, Bayesian and non-Bayesian estimation, and applications. *Symmetry* **2022**, *14*, 883. [CrossRef]

mathematics

MDPI

Article

Assessing the Risk of *APOE*-ϵ4 on Alzheimer's Disease Using Bayesian Additive Regression Trees

Yifan Xia [1] and Baosheng Liang [2,*]

[1] Institute of Medical Technology, Peking University, Beijing 100191, China; xiayifan@hsc.pku.edu.cn
[2] Department of Biostatistics, School of Public Health, Peking University, Beijing 100191, China
* Correspondence: liangbs@hsc.pku.edu.cn; Tel.: +86-010-8280-5541

Abstract: Alzheimer's disease (AD) affects about a tenth of the population aged over 65 and nearly half of those over 85, and the number of AD patients continues to grow. Several studies have shown that the ϵ4 variant of the apolipoprotein E (*APOE*) gene is potentially associated with an increased risk of AD. In this study, we aimed to investigate the causal effect of *APOE*-ϵ4 on Alzheimer's disease under the potential outcome framework and evaluate the individualized risk of disease onset for *APOE*-ϵ4 carriers. A total of 1705 Hispanic individuals from the Washington Heights-Inwood Columbia Aging Project (WHICAP) were included in this study, comprising 453 *APOE*-ϵ4 carriers and 1252 non-carriers. Among them, 265 subjects had developed AD (23.2%). The non-parametric Bayesian additive regression trees (BART) approach was applied to model the individualized causal effects of *APOE*-ϵ4 on disease onset in the presence of right-censored outcomes. The heterogeneous risk of *APOE*-ϵ4 on AD was examined through the individualized posterior survival probability and posterior causal effects. The results showed that, on average, patients carrying *APOE*-ϵ4 were 0.968 years younger at onset than those with non-carrying status, and the disease risk associated with *APOE*-ϵ4 carrying status was 3.9% higher than that for non-carrying status; however, it should be noted that neither result was statistically significant. The posterior causal effects of *APOE*-ϵ4 for individualized subjects indicate that 14.41% of carriers presented strong evidence of AD risk and approximately 38.65% presented mild evidence, while around 13.71% of non-carriers presented strong evidence of AD risk and 40.89% presented mild evidence. Furthermore, 79.26% of carriers exhibited a posterior probability of disease risk greater than 0.5. In conclusion, no significant causal effect of the *APOE*-ϵ4 gene on AD was observed at the population level, but strong evidence of AD risk was identified in a sub-group of *APOE*-ϵ4 carriers.

Keywords: Bayesian model; individualized disease risk; right-censored data; Alzheimer's disease

MSC: 62P10

Citation: Xia, Y.; Liang, B. Assessing the Risk of *APOE*-ϵ4 on Alzheimer's Disease Using Bayesian Additive Regression Trees. *Mathematics* **2023**, *11*, 3019. https://doi.org/10.3390/math11133019

Academic Editor: Diana Mindrila

Received: 20 May 2023
Revised: 26 June 2023
Accepted: 27 June 2023
Published: 7 July 2023

1. Introduction

Alzheimer's disease (AD) is a devastating neurological disease that affects millions of people around the world. About one in ten people over 65 and almost half of people over 85 suffer from AD [1], and the number of afflicted individuals continues to grow annually. It has been revealed that the apolipoprotein E locus (*APOE*) gene is associated with an increased risk of AD onset, in both sporadic and familial forms [2,3]. Particularly, among three alleles, the epsilon 4 (E4 or ϵ4) variant of *APOE* has been found to be an important factor in the etiology of more than half of all AD [2,4]. Thus, determining how to quantify the risk of *APOE*-ϵ4 on AD is critical. In previous studies, a research team from Duke University concluded that *APOE*-ϵ4 was associated with AD as a major risk factor using the Mantel–Haenszel correlation statistic and Cox proportional hazard model [2]. Another study using logistic regression has also revealed that *APOE*-ϵ4 was associated with a higher AD risk [4]. A meta-analysis showed that *APOE*-ϵ4 was a major risk factor across ethnic groups, ages, and gender [5]. In addition, a twin study suggested that multiple susceptibility genes along with *APOE*-ϵ4 contributed to around 80% of AD cases [6]. However, the above studies on the effects of the gene on AD were all based on statistical association analysis. So far, to the best of our knowledge, there have been very limited studies evaluating the risk of *APOE*-ϵ4 on AD in terms of causal effect at individualized level [2,5,6]. Assessing the AD risk using the causal effect of *APOE*-ϵ4 at the individual level could help to target patients who may be susceptible to *APOE*-ϵ4 [7,8].

Treatment effects (or risk) of specific treatments or interventions are usually evaluated at population level in randomized controlled trial (RCT) studies. However, in practice, clinical decisions are often made at the individual level. Real-world observations include large amounts of clinical information about patients, hence offering us an opportunity to infer the treatment effects for heterogeneous patients, even from a causal perspective. It is well known that the causal effect of treatment can be inferred under the potential outcome framework by Rubin [9], which usually requires strong assumptions before performing causal inference. An advantage of the potential outcome framework is that it can be employed to infer the individualized treatment effect [10], for which the causal effect of a specific treatment can be identified under the assumption that treatment is independent of potential outcome of treatment and control, given the pre-treatment covariates. The individualized treatment effect (ITE) is an important measure that has been widely investigated in the field of personalized medicine [11], which helps to quantify individualized responses to specific treatments for heterogeneous individuals by calculating the difference of outcomes between treatment and control for any patient. A major challenge in the models of the ITE method is to handle the non-linear relationship between the covariates and survival outcomes, especially in the presence of complex censoring.

Bayesian additive regression trees (BART) is an ensemble learning method by which the value of any unknown function can be approximated through the summation of a series of Bayesian regression trees. In particular, BART is flexible, powerful, and can handle the complex non-linear relationships and interactions among covariates [12,13]. More importantly, the Bayesian framework allows for the construction of 95% credible intervals for statistical inference. In practice, BART applies to both continuous and binary outcomes; hence, it has a wide range of applications. It has also recently been generalized to survival analysis [14,15] and can handle right-censored data [16], even interval-censored data [17]. Furthermore, BART is suitable for observational studies [12]. Therefore, the BART method can also be used to estimate the ITE and conduct causal inference. Finally, BART can easily be extended to various settings, and a generalized BART model that unifies extensions is called general BART [18]. Generalized BART is commonly used for non-parametric or semi-parametric problems, correlated outcomes, survey matching problems, and models with weaker distributional assumption. The flexible extensibility of BART is a particular advantage in practical applications.

In this article, we aim to assess the causal risk of *APOE*-ϵ4 on AD in the presence of right-censored observations under the potential outcome framework and examine the individualized risk of disease onset for *APOE*-ϵ4 carriers. To the best of our knowledge, the data analysis in existing studies focused on AD has concluded merely in terms of the correlation, instead of the causal association between AD and *APOE*-ϵ4. The novelty of this article lies in the investigation of causal associations between *APOE*-ϵ4 and AD using BART, a hybrid Bayesian and machine learning method, which enables us to estimate and infer the causal effect of interest at both the population and individual level. In particular, the key contributions of this paper are as follows:

- We apply the BART method to a non-parametric AFT model for right-censored data;
- We infer the causal effect of *APOE*-ϵ4 on AD at both population and individual levels under the potential outcome framework;
- We explore heterogeneous evidence of the causal effect and identify important variables associated with the causal effect.

The remainder of this article is organized as follows. The data, notation, and statistical models are described in Section 2. In Section 3, we present the results regarding the estimated gene effect of *APOE-ϵ4* on AD with respect to age at onset and onset risk for each patient. We conclude with a brief discussion in Section 4.

2. Model and Methods

2.1. Notation

Suppose there are n patients in the study. For the ith patient, let $\tilde{Y}_i$ denote the true AD onset time and C_i denote the censoring time. Denote the observed AD onset time as $Y_i = \min(\tilde{Y}_i,\ C_i)$ and the censoring indicator as $\Delta_i = I(\tilde{Y}_i \leq C_i)$. Let W_i be an indicator of carrying the *APOE-ϵ4* gene, such that $W_i = 1$ indicates assignment to the treatment group and $W_i = 0$ indicates assignment to the control group. Let X_i denote a $p \times 1$ vector of baseline covariates. Therefore, the observed data can be denoted as $O = \{O_i = (Y_i, \Delta_i, W_i, X_i) : i = 1, \cdots, n\}$. We make some regular assumptions for identifying the causal effect. First, the treatment assignment is strongly ignorable. Denote $Y_i(1)$ and $Y_i(0)$ as the potential outcomes under the treatment $W_i = 1$ and the control $W_i = 0$, respectively. We assume that the treatment W_i is independent of the potential outcome $Y_i(1)$ and $Y_i(0)$, given X_i. Furthermore, the treatment probabilities for the patients are bounded away from 0 and 1; that is, $\Pr(W_i = 1|X_i) \in (0, 1)$.

2.2. Non-parametric Accelerated Failure Time BART Model

To explore the causal effect of *APOE-ϵ4* on Alzheimer's disease using a general and flexible model, we consider a non-parametric AFT model, defined as follows

$$\log \tilde{Y} = f(W, X) + \epsilon, \tag{1}$$

where $\tilde{Y}$ is modeled using a non-linear function and the residual term ϵ satisfies $E(\epsilon|W, X) = 0$. In the following, we name (1) as the AFT-BART model.

For the model regression, we use Bayesian additive regression trees to approximate the unknown non-linear function $f(W, X)$. Let T denote a binary tree that consists of the tree structure and the interior node decision rules leading to subsequent nodes; in particular, all of the interior nodes of T have decision rules. Rules decide a (W, X) pair to either the left or right node. Let $M = \{\mu_1, \mu_2, ..., \mu_b\}$ be the parameter values (mean response of the subgroup of observations) associated with the b leaf nodes of the tree T. Given the tree model (T, M) and a pair (W, X), we can define the value obtained at the leaf node and report the value μ associated with that leaf node. BART consists of two parts: A sum-of-trees model and a regularization prior. We denote the single tree model function as $g(W, X; T, M)$. The regression function m is represented in BART as a sum of the individual tree contributions

$$f(W, X) = \sum_{j=1}^{m} g(W, X; T_j, M_j), \tag{2}$$

where each (T_j, M_j) denotes a single tree model. Let $T_{(-j)}$ be the set of all trees except for T_j, and define $M_{(-j)}$ similarly. The sum-of-tree model begins taking the fit from the first weak-learning tree, $g(W, X; T_1, M_1)$. After the fitting process, the model subtracts the first fit from the observed response and forms residuals. Then, the model fits the next tree to the residuals. The above procedure is performed m times in total. In the spirit of boosting, the number of trees in the model can be large, allowing each tree to contribute only a small part to the total fit. Over-fitting can be avoided through the use of a regularization prior, which limits the fit of each (T_j, M_j) tree. The second piece of BART is the prior. In our analysis, we used the prior settings recommended for the AFT-BART model [14]. When using BART, the AFT model is fully non-parametric, and both the regression function and error distribution are modeled non-parametrically. The random error term ϵ follows a flexible location mixture of normal densities.

In essence, Algorithm 1 is an algorithm for the non-parametric AFT model in the presence of right-censored data, which is an extension of the BART model. In particular, it assumed to be a DP mixture model for the residual distribution. Under the non-parametric AFT framework, it deals with right censoring using a data augmentation technique with truncated normal distribution.

Algorithm 1 Bayesian algorithm for the AFT-BART model.

Input: Data $D_i = (Y_i, X_i, \delta_i)$, $i = 1, 2, ..., n$, initial values for T_b, M_b, $b = 1, ..., m$, the (τ_i, σ) on the residual, $i = 1, 2, ..., n$, and other parameters variables $\theta = (m, k, \alpha, \beta)$.

1: To update $T_b^*, M_b^* \mid T_{(-b)}, M_{(-b)}, \theta, D$, transform original Y_i to $Y_i - \hat{\mu}_{AFT}$ as the responses.
2: Update $T_1, ..., T_m$ and $M_1, ..., M_m$ as in Algorithm 2.
3: Update $f(X_i) \mid T_1, ..., T_m, M_1, ..., M_m$.
4: To update the parameters related to the residual distribution:
5: Update cluster labels $S_1, ...S_n$ with probability $P(S_i = h) \propto \pi_h \phi\left(\frac{\log Y_i - f(X_i) - \tau_h}{\sigma}\right)$, let $n_h = \sum_{i=1}^n \mathbf{1}\{S_i = h\}$.
6: Sample stick-breaking weights $V_h \sim \text{Beta}(\alpha_h, \beta_h)$, $\alpha_h = 1 + n_h$, $\beta_h = M \sum_{k=h+1}^H n_k$, $h = 1, ..., H-1$,
let $V_H = 1$.
7: Set $\pi_h = V_h \prod_{k<h}(1 - V_k)$, $h = 1, ..., H$, set update mixture proportions.
8: Sample unconstrained cluster locations

$$\tau_h^* \sim N\left(\frac{\sigma_\tau^2}{n_h\sigma_\tau^2 + \sigma^2} \sum_{i=1}^n \{\log Y_i - f(X_i)\}\mathbf{1}\{S_i = h\}, \frac{\sigma_\tau^2\sigma^2}{n_h\sigma_\tau^2 + \sigma^2}\right).$$

9: Update constrained cluster locations $\tau_h = \tau_h^* - \mu_{G^*}$, where $\mu_{G^*} = \sum_{h=1}^H \pi_h \tau_h^*$.
10: Update mass parameter $M \sim \text{Gamma}\left(\psi_1 + H - 1, \psi_2 - \sum_{h=1}^{H-1} \log(1 - V_h)\right)$.
11: Update $\sigma^2 \sim \text{Inv-Gamma}(\frac{v+n}{2}, \frac{\hat{s}^2 + kv}{2})$, where $\hat{s}^2 = \sum_{h=1}^H \sum_{i=1}^n \{\log Y_i - f(X_i) - \tau_h\}^2 \mathbf{1}\{S_i = h\}$.
12: **for** $i \in \{\delta_i = 0\}$ **do**
13: Sample $\log z_i \sim \text{Truncated-Norm}(f(X_i) + \tau_{S_i}, \sigma^2; \log Y_i)$, set $Y_i = z_i$.
14: **end for**
15: Compute the final $\log \hat{Y}_i = f(X_i) + \hat{\mu}_{AFT}$.

Output: New values of T_b, M_b, $b = 1, ..., m$, and (τ_i, σ), $i = 1, 2, ..., n$.

In AFT-BART, (α, β, k, m) on f and (G, σ) on ϵ are treated as parameters in a formal statistical model. We used the prior settings recommended for AFT-BART [1]. After setting the prior on the parameters, the posterior can be computed using a Markov chain Monte Carlo (MCMC) technique; in particular, a Gibbs sampler was extended for computation of the posterior. After updating the trees and the terminal leaf node parameters, the parameters of the residual distribution can then be updated. The part of the residual distribution J can be expressed as

$$J_i \mid \tau_i, \sigma^2 \sim N(\tau_i, \sigma^2), \text{ for } i = 1, .., n, \quad \sigma^2 \sim kv/\chi^2$$
$$\tau_i \sim G, \; G \mid M \sim \text{CDP}(M, G_0), \; M \sim \text{Gamma}(\psi_1, \psi_2). \tag{3}$$

Here, the mixing distribution G is truncated to have a large, finite number of components H. $V_h \sim \text{Beta}(1, M)$ for $h = 1, ..., H-1$. We summarize the algorithm for this model as Algorithm 1. In the analysis, we set 5000 as the number of MCMC iterations to be treated as burn-in and 1000 as the number of iterations for posterior drawing. Furthermore, we set the number of trees as 200.

Base on the above models, we can estimate the individualized treatment effect (ITE), which can be expressed as the difference in expected log disease-onset time in the treatment group versus that in the control group. The ITE $\tau(x)$ for a subject with covariate x can be calculated as

$$\begin{aligned}\tau(x) &= E(\log(Y)|W=1, X=x) - E(\log(Y)|W=0, X=x) \\ &= f(1,x) - f(0,x).\end{aligned} \tag{4}$$

In this scenario, the ITE represents the difference in age at onset of AD for patients.

2.3. Onset Probability Analysis

Let the binary outcome of AD be Y, where $Y = 1$ denotes the onset endpoint of the participant and $Y = 0$ denotes the unobserved endpoint of the participant. It is straightforward to adapt or extend BART to the probit model. Define

$$p(X) = P(Y=1|X=X) = \Phi[f(X)], \tag{5}$$

where

$$f(X) = \sum_{j=1}^{m} g(X; T_j, M_j) \tag{6}$$

and $\Phi[\cdot]$ is the cumulative distribution function of standard normal distribution, T_j denotes the jth binary regression tree, and M_j denotes the associated terminal node parameters of tree j. Each probability $p(x)$ is obtained as a function of $f(x)$. This idea differs from traditional aggregate classifier approaches, which often use a majority or average vote based on an ensemble of weak learners. For posterior calculation, the latent variables $Z_1, \cdots, Z_n \overset{\text{i.i.d}}{\sim} N(G(x), 1)$ are introduced into the model [19], with $Y_i = 1$ if $Z_i > 0$ and $Y_i = 0$ if $Z_i \leq 0$. Here, $\overset{\text{i.i.d}}{\sim}$ means independent and identically distributed. Finally, we obtain $Z_i|Y_i = 1 \sim \max\{N[g(x), 1], 0\}$ and $Z_i|y_i = 0 \sim \min\{N[g(x), 1], 0\}$. We summarize the BART method [12,20] in Algorithm 2.

Algorithm 2 Bayesian back-fitting algorithm for updating BART

Input: Data $D_i = (Y_i, X_i), i = 1, 2, ..., n$, initial values for T_b, M_b, $b = 1, \cdots, m$, and other parameters/variables $\theta = (m, k, \alpha, \beta)$.

```
To update T_b^*, M_b^* | T_(b), M_(b), θ, D:
for b in 1:m do
    Compute partial residuals R_b = Y_i − Σ_{j≠b}^m g(X_i; T_b, M_b).
    Compute L(T_b; T_(b), M_(b), θ) = ∫ ( Π_{i=1}^n p(R_b | T_b, M_b, T_(b), M_(b), θ) ) p(M_b | T_b, θ) dM_b.
    Propose T_b^* = q(T_b^*; T_b).
    Set a ← [L(T_b^*; T_(b), M_(b), θ) p(T_b^*) / L(T_b; T_(b), M_(b), θ) p(T_b)] · [q(T_b; T_b^*) / q(T_b^*; T_b)].
    Sample u ~ U(0, 1)
    if u < min(a, 1) then
        T_b ← T_b^*.
    end if
    Sample M_b ~ p(M_b | T_b, T_(b), M_(b), θ, D), μ_bi ~ N(0, σ_μ^2).
end for
Draw σ | T_1, ···, T_m, M_1, ···, M_m, y, σ ~ vλ/χ_v^2.
```

Output: New values of T_b, M_b, $b = 1, \cdots, m$.

In the binary case, the ITE $\tau(x)$ for a patient with covariate vector x can be defined as

$$\tau(x) = P(Y=1|W-1, X=x) - P(Y=1|W=0, X=x). \tag{7}$$

In this scenario, the ITE represents the risk of onset of AD for a patient.

2.4. Posterior Inference Statistics

To predict the outcome Y for a particular x, we take the empirical average of the after burn-in sample $f_1^*, \cdots, f_K^*$, as follows:

$$\frac{1}{K}\sum_{k=1}^{K} f_k^*(x). \tag{8}$$

The individual-level causal effects can be estimated as

$$\frac{1}{K}\sum_{k=1}^{K} f_k^*(1,x) - f_k^*(0,x). \tag{9}$$

Given the conditions on the X values in the sample, the conditional average treatment effect can be estimated as follows

$$\frac{1}{N}\sum_{i=1}^{N} E[Y_i(1)|X_i] - E[Y_i(0)|X_i] = \frac{1}{N}\sum_{i=1}^{N} f(1,x_i) - f(0,x_i). \tag{10}$$

We utilize the posterior probabilities of the differential treatment effect to detect the presence of heterogeneous treatment effects

$$D_i = P\{\theta(\mathbf{x}_i) > 0|\mathbf{y}, \delta\}, \tag{11}$$

along with the closely related quantity

$$D_i^* = \max\{1 - 2D_i, 2D_i - 1\}. \tag{12}$$

Here, D_i denotes the posterior probability that measures whether $\theta(\mathbf{x}_i)$ is greater than or equal to 0. For patient i, there exists a strong evidence of a differential treatment effect if $D_i^* > 0.95$; that is, $D_i \geq 0.975$ or $D_i \leq 0.025$. Mild evidence of a differential treatment effect exists if $D_i^* > 0.80$; that is, $D_i \geq 0.9$ or $D_i \leq 0.1$.

Another research line involves quantifying the heterogeneous treatment effects using the proportion of individuals who benefit from treatment. The proportion of benefit measure provides an interpretation and a useful quantity for determining the presence of cross-over or qualitative interactions among variables. The treatment effect in some cases may have the opposite sign, in comparison to the overall average treatment effect. A low proportion of patients benefiting in a situation where an overall treatment benefit has been determined may indicate the existence of cross-over interactions. With the treatment differences $\theta(\mathbf{x})$, we define the benefit proportion as

$$Q = \frac{1}{n}\sum_{i=1}^{n} I\{\theta(\mathbf{x}_i) > 0\}. \tag{13}$$

Here, Q is the posterior mean, which is the average of the posterior probabilities of treatment benefit $\hat{p}_i = P\{\theta(\mathbf{x}_i) > 0|y, \delta\}$. Treatment assignment for a patient can be decided according to the posterior probabilities of treatment benefit with $\hat{p}_i > 0.5$ or $\hat{p}_i < 0.5$.

Based on the above, we summarize the methods for determining the continuous survival outcome and binary outcome in Algorithm 3. The corresponding R codes and a brief intrduction of the implementation are presented in Appendix A.

Algorithm 3 Effect Estimation of $APOE$-$\epsilon4$ on AD

Input: Two data sets in total, n training samples in each. $D_i = (Y_i, W_i, X_i), i = 1, 2, \cdots, n$; $\tilde{D}_i = (Y_i, W_i, X_i, \delta_i), i = 1, 2, \cdots, n$.

1: For continuous outcome,
predict $\log Y_{(i)} \mid T_b, M_b, b = 1, \cdots, m, (\tau_i, \sigma), \tilde{D}_i, i = 1, 2, \cdots, n$ from Algorithm 1.
2: Compute (4)
3: For classification of binary outcome,
predict $\log Z_i \mid T_b, M_b, b = 1, \cdots, m, (\tau_i, \sigma), D_i, i = 1, 2, \cdots, n$ from Algorithm 2.
4: Compute $P(Y = 1|X) = \Phi(Z_i)$.
5: Compute (7)
6: Extract information from the posterior,
7: Compute $\tau^*(x) = f_k^*(1, x) - f_k^*(0, x)$.
8: Construct credible interval $(\tau_{0.025}, \tau_{0.975}) \mid \tau^*(x)$, where $P\{\tau^*(x) < \tau_{0.025}\} = 0.025, P\{\tau^*(x) < \tau_{0.975}\} = 0.975$.
9: Compute (11) and (12).

Output: $\tau(x)$ and 95% CI of age at onset, $\tau(x)$ and 95% CI of onset risk, evidence for heterogeneity of treatment effect D_i^*.

3. Application

WHICAP is an ongoing community-based study of aging and dementia among elderly subjects residing in Northern Manhattan [21]. Proband participants were identified from Medicare records aged 65 years or older and recruited in 1992 and 1999. The prevalence of AD and dementia in proband participants was carefully monitored during the study. Dense genome-wide genotypes were collected in probands with more than two million SNPs. We focused on Hispanics, as they are one of the largest and fastest-growing ethnic groups in the United States [22]. They are generally under-studied, and the incidence of AD has been shown to increase by twofold in Hispanic elderly individuals, compared to white individuals [23]. Although WHICAP provides pedigree information and familial observations of probands, parents, and siblings, we only considered the probands in this study, as the genotypes in relatives of the proband were unobservable.

For this study, we enrolled 1705 probands of Alzheimer's disease with observed AD onset time, where 453 (27%) were $APOE$-$\epsilon4$ carriers while 1252 (73%) were non-carriers. The characteristics of probands with AD onset time are summarized in Table 1. Furthermore, there were 1720 probands whose disease status (i.e., AD or not) was observable, where 458 participants were $APOE$-$\epsilon4$ carriers and 1262 were non-carriers. We also included three baseline covariates in the model: sex, educational attainment level, and race. The survival endpoint that we examined was the age at onset of patients (reported in years). We divided educational attainment into three levels ("<-0.9", "$-0.9\sim0.5$", and "$0.5 \sim 2.0$"). For the binary response model, we only included sex and educational attainment.

3.1. Overall Causal Effect of Patients at Onset

We estimated the causal effect of $APOE$-$\epsilon4$ on Alzheimer's disease using BART [24] and a BART-based accelerated failure time model. We also compared the AFT-BART method with other existing methods under the potential outcome framework. The first method involved the application of the AFT interaction model [17]. For our application, the ITE was calculated by subtracting the estimate under control assignment from the estimate under treatment assignment. Another related method used two separate AFT models: one for the treatment group and another one for the control group. The other method was based on a survival Causal Tree and Causal Forests. We built each survival Causal Tree using the function `CausalTree` in the R package `SurvivalCausalTree` [25].

Table 1. Characteristics of probands with Alzheimer's disease for continuous age at onset.

Characteristic	*APOE-ϵ4* Carriers		Non-Carriers		Total	
Total.	453		1252		1705	
Onset age—no. (%)						
60∼70	22	(5)	60	(5)	82	(5)
70∼80	192	(42)	429	(34)	621	(36)
80∼90	203	(45)	591	(47)	794	(47)
90 ∼ 100	36	(8)	172	(14)	208	(12)
Sex—no.(%)						
male	155	(34)	432	(35)	587	(34)
female	298	(66)	820	(65)	1118	(66)
Educational—no. (%)						
<−0.9	94	(21)	266	(21)	360	(21)
−0.9∼0.5	242	(53)	620	(50)	862	(51)
0.5∼2.0	117	(26)	366	(29)	483	(28)
Race—no. (%)						
Race-1	113	(25)	425	(34)	538	(32)
Race-2	174	(38)	363	(29)	537	(31)
Race-3	161	(36)	441	(35)	602	(35)
Race-4	5	(1)	23	(2)	28	(2)

The causal effects of *APOE-ϵ4* on Alzheimer's disease, according to the models, are presented in Table 2. The analysis causal effect using AFT-BART indicated that the conditional average effect of the *APOE-ϵ4* gene on Alzheimer's disease was −0.032 in log years difference; that is, patients with the *APOE-ϵ4* gene presented 0.032 log years earlier age at onset than patients without *APOE-ϵ4*, on average. From the results using AFT-BART and BART to analyze the non-censored data, the age at onset was 0.001 and 0.003 log years earlier than those without *APOE-ϵ4*, respectively.

Table 2. The causal effects of *APOE-ϵ4* on Alzheimer's disease according to BART and BART-based accelerated failure time models (unit: log years).

Methods	Mean	2.5%	97.5%
AFT-BART	−0.032	−0.059	0.024
AFT	0.079	0.056	0.102
Two-AFT	0.044	−0.015	0.103
SCT	−0.013	−−	−−

Note: AFT-BART denotes non-parametric Bayesian accelerated failure time model, AFT denotes the method based on one AFT model, Two-AFT denotes the method based on two separate AFT models, SCT denotes the method based on survival Causal Tree.

The survival time posteriors for patients with and without *APOE-ϵ4* are presented in Figure 1. The red line is the posterior survival time of patients with *APOE-ϵ4*, while the black line is the posterior survival time of patients without *APOE-ϵ4*. It can be seen that the two lines do not overlap completely, which directly indicates that patients with *APOE-ϵ4* tend to present an earlier onset of Alzheimer's disease, compared to those without *APOE-ϵ4*.

Table 3 presents the difference in AD onset risk associated to *APOE-ϵ4*. The results show that patients with the *APOE-ϵ4* gene have an onset risk of AD of 0.166, while those without *APOE-ϵ4* gene have an onset risk of AD of 0.127. Thus, the *APOE-ϵ4* gene increases the mean onset risk by 0.039 for patients with *APOE-ϵ4*, compared with those without it.

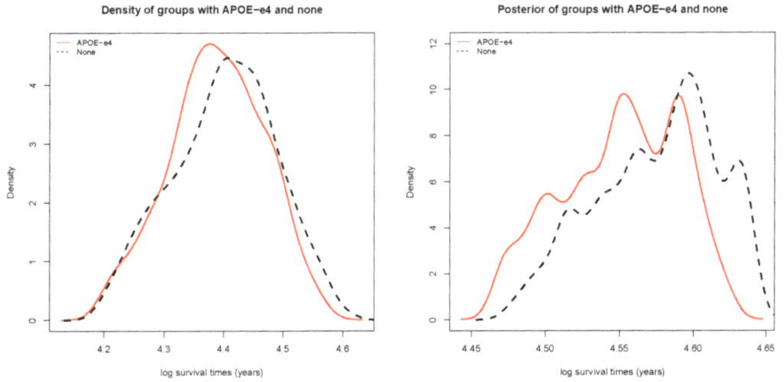

Figure 1. (**left**) Density of survival time for groups with and without *APOE*-ϵ4; and (**right**) posterior of survival time for groups with and without *APOE*-ϵ4.

Table 3. The estimated treatment effects of *APOE*-ϵ4 on AD by BART for onset risk with 95% credible interval.

Value	Mean	2.5%	97.5%
Risk diff	0.039	−0.002	0.075
Gene prob	0.166	0.058	0.361
None prob	0.127	0.052	0.292

3.2. Distribution of Causal Effect for Patients

To characterize the variation in the causal effect of *APOE*-ϵ4 on AD, we plotted the histogram and distribution of causal effect for patients, as presented in Figure 2. Smooth posterior estimates provide the causal effect distribution of *APOE*-ϵ4 on Alzheimer's disease for all patients. The histogram was constructed using all point estimates from both patients with and without the *APOE*-ϵ4 gene. The blue part indicates the total treatment effect for patients with *APOE*-ϵ4, while the red part indicates the treatment effect for patients without *APOE*-ϵ4. Three peaks can be observed in the histogram, both for all patients and for the individual groups. The major patients with *APOE*-ϵ4 presented an earlier age at onset than those without *APOE*-ϵ4: about 0.06 and 0.01 log years earlier at onset. However, a minority of patients presented opposite results. Among these patients, the patients with *APOE*-ϵ4 had about 0.03 log years earlier time of AD onset than patients without *APOE*-ϵ4. It seems that these patients presented Alzheimer's disease onset at a later age, or were affected by the existence of cross-over interactions. Overall, the majority of patients showed an earlier age of onset associated to *APOE*-ϵ4.

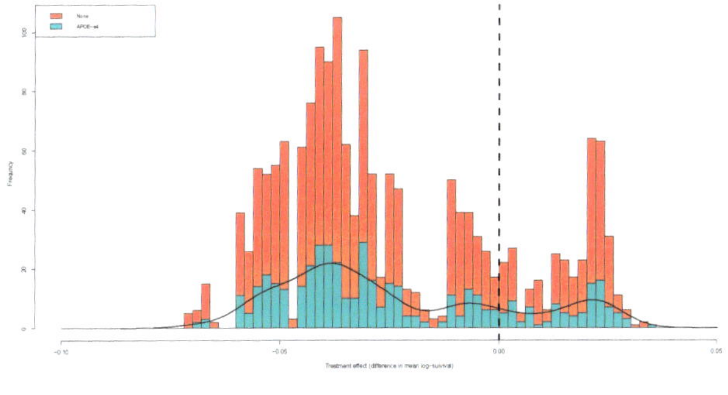

Figure 2. Distribution of causal effect on *APOE*-ϵ4.

3.3. Individualized Treatment Effect

Figure 3 presents the individualized treatment effect estimates for the 1705 patients, clearly indicating an overall earlier age at onset associated to *APOE-ϵ4* for patients. The estimates consist of posterior means of treatment effect with corresponding 95% credible intervals for all patients. There are two obvious groups of patients, according to the difference in onset time. The patients whose treatment effect was less than 0 had an earlier age at onset due to the *APOE-ϵ4* gene. It is clear that some patients had the treatment effect and 95% credible intervals below zero. The causal effect of *APOE-ϵ4* on Alzheimer's disease in these patients presented significant statistical significance. The variation in the treatment effects suggests substantial heterogeneity in response to *APOE-ϵ4*, which may be due to some individualized characteristics.

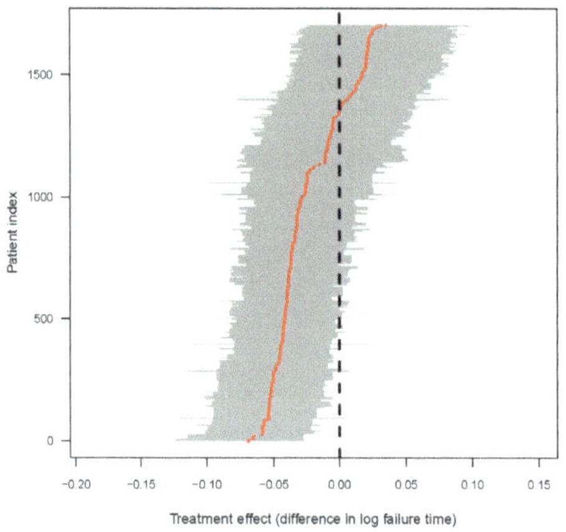

Figure 3. Posterior of causal effect for individual patients, where the red line shows the posterior mean treatment effect for all of the patients, and the gray area show the 95% credible interval of each individualized *APOE-ϵ4* gene effect on AD.

The patients which presented a significant causal effect caused by *APOE-ϵ4* were extracted, for 515 patients in total. Table 4 presents the patients with and without significant ITE, grouped by sex, race, and education level. In particular, 171 patients were male and 344 were female; in terms of the education level of patients, 116 patients received education of low level, 259 patients received education of middle level, and 140 patients received high level education; as for race, the number of patients characterized by the four races were 138, 176, 191, and 10, respectively.

3.4. Covariate-Specific Treatment Effects

We constructed partial dependence plots for survival time (in years) of patients, along with the posterior distributions of treatment effect in male and female groups, each of the four races, and sub-groups defined according to educational attainment level. For the male and female groups, the posterior of survival time for male and female patients and difference in survival time between male and female patients are presented in Figure 4. The posterior of onset distribution and treatment effect in the male group were not distinct from those in the female group.

Next, we examined the four race groups of patients, and the posterior survival time and difference in survival time for the four groups are presented in Figure 4. The onset distributions and treatment effects in the first three race groups were highly similar, but distinct from those for the fourth race group. The possible explanation is that the sample size of fourth race group was very small (28 patients), and only accounted for 28%.

Figure 4 presents the posterior of the survival time and difference in survival time for patients grouped by educational attainment level. The partial dependence plots clearly show differences between patients with and without *APOE*-ϵ4 in the posterior distribution, except for a crossover point, where the sample size may have not been large enough. In the posterior of treatment effect, the median curves for both patients with and without *APOE*-ϵ4 were below zero, clearly indicating the earlier age at onset caused by the *APOE*-ϵ4 gene.

Table 4. Patients with and without significant ITE, grouped by sex, race, and education level.

	Significant		Not Significant	
Characteristic	**Count**	**Percentage (%)**	**Count**	**Percentage (%)**
Total	515	30	1190	70
Sex—no. (%)				
male	171	33	415	35
female	344	67	774	65
Education—no. (%)				
<−0.9	116	23	244	21
−0.9∼0.5	259	50	603	51
0.5∼2.0	140	27	343	29
Race—no. (%)				
Race-1	138	27	400	34
Race-2	176	34	361	30
Race-3	191	37	411	35
Race-4	10	2	18	2

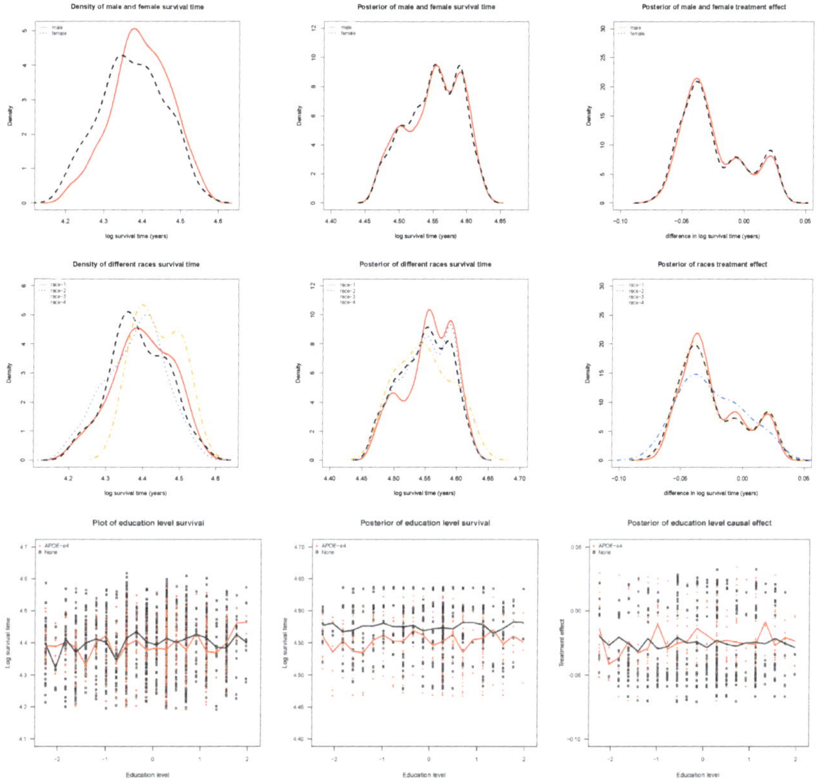

Figure 4. (**top-left**) Density of survival time by sex; (**top-middle**) Posterior of survival time by sex; (**top-right**) Difference in survival time by sex; (**mid-left**) Density of survival time by race;

(**mid-middle**) Posterior of survival time by race; (**mid-right**) Difference in survival time by race; (**bottom-left**) Density of survival time by education level; (**bottom-middle**) Posterior of survival time by education level; and (**bottom-right**) Difference in survival time by education level.

3.5. Individual Survival Curves

Figure 5 displays the individual posterior survival curves; in particular, there were 1705 individual survival curves associated to patients. The gray and black lines indicate the survival curves for patients with and without $APOE$-$\epsilon 4$, respectively. Although the survival curves of the two groups overlap to some extent, the patients without $APOE$-$\epsilon 4$ had a higher survival proportion than those with $APOE$-$\epsilon 4$, overall. At the same age, the patients with $APOE$-$\epsilon 4$ presented higher onset probability than those without $APOE$-$\epsilon 4$. The red and green lines are the posterior mean survival curves for patients with and without $APOE$-$\epsilon 4$, respectively; it can be seen that the red line lies above the green line. This indicates that patients with $APOE$-$\epsilon 4$ are more likely to have an earlier onset of AD than those without $APOE$-$\epsilon 4$.

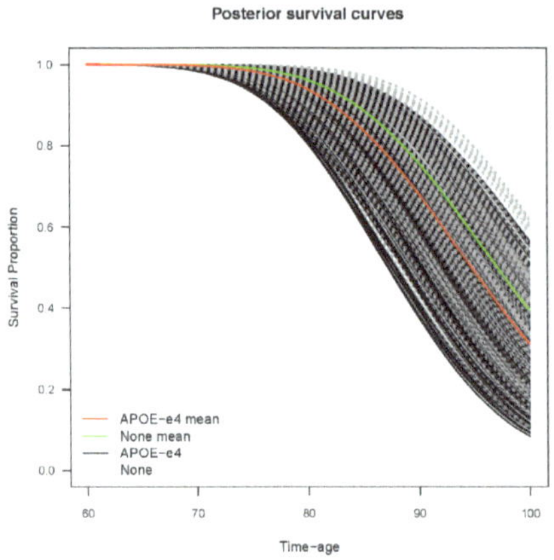

Figure 5. Individual posterior survival curves for patients.

3.6. Evidence for Heterogeneous Treatment Effects

The posterior probabilities of treatment benefit are provided in Table 5. Table 5 shows that, among patients with $APOE$-$\epsilon 4$, 29.80% of patients presented strong evidence of a differential treatment effect, while approximately 54.97% of patients presented mild evidence. Among patients without $APOE$-$\epsilon 4$, approximately 30.35% of patients presented strong evidence of a differential treatment effect, while 56.39% presented mild evidence. For the proportion of patients who benefited from treatment, 77.93% of patients with $APOE$-$\epsilon 4$ and 78.83% of patients exhibited a posterior probability of benefit greater than 0.5. These patients are more likely to have an earlier age at onset caused by the $APOE$-$\epsilon 4$ gene.

Table 5 also shows that the probability of difference of onset among patients with and without $APOE$-$\epsilon 4$. Among patients with $APOE$-$\epsilon 4$, 14.41% of patients presented strong evidence of Alzheimer's disease onset risk, while approximately 38.65% presented mild evidence. Among patients without $APOE$-$\epsilon 4$, approximately 13.71% of patients presented strong evidence of Alzheimer's disease onset risk, while 40.89% presented mild evidence. Furthermore, posterior probabilities of treatment benefit can be used for treatment assignment for patients with $\hat{p}_i > 1/2$ or $\hat{p}_i < 1/2$ when estimating onset risk. It was found that 79.26% of patients with $APOE$-$\epsilon 4$ and 82.57% of patients exhibited a posterior probability of benefit greater than 0.5. These patients are more likely to have higher onset risk caused by the $APOE$-$\epsilon 4$ gene.

Table 5. Posterior probabilities of *APOE-ϵ4* carrier benefit and differential treatment effect among subjects with and without *APOE-ϵ4*.

Measurement	Posterior Probabilities	*APOE-ϵ4*	None
Onset age	$P\{\theta(\mathbf{x}_i) < 0 \mid \mathbf{y}, \delta\} \in (0.99, 1]$	15.45	14.86
	$P\{\theta(\mathbf{x}_i) < 0 \mid \mathbf{y}, \delta\} \in (0.95, 0.99]$	23.18	24.92
	$P\{\theta(\mathbf{x}_i) < 0 \mid \mathbf{y}, \delta\} \in (0.75, 0.95]$	27.37	25.96
	$P\{\theta(\mathbf{x}_i) < 0 \mid \mathbf{y}, \delta\} \in (0.50, 0.75]$	11.92	13.10
	$P\{\theta(\mathbf{x}_i) < 0 \mid \mathbf{y}, \delta\} \in (0.25, 0.50]$	11.92	10.78
	$P\{\theta(\mathbf{x}_i) < 0 \mid \mathbf{y}, \delta\} \in (0, 0.25]$	10.15	10.38
	$D_i^* > 0.95$	29.80	30.35
	$D_i^* > 0.80$	54.97	56.39
Onset probability	$P\{\theta(\mathbf{x}_i) > 0 \mid \mathbf{y}, \delta\} \in (0.99, 1]$	9.83	8.64
	$P\{\theta(\mathbf{x}_i) > 0 \mid \mathbf{y}, \delta\} \in (0.95, 0.99]$	9.83	12.76
	$P\{\theta(\mathbf{x}_i) > 0 \mid \mathbf{y}, \delta\} \in (0.75, 0.95]$	41.49	43.34
	$P\{\theta(\mathbf{x}_i) > 0 \mid \mathbf{y}, \delta\} \in (0.50, 0.75]$	18.12	17.83
	$P\{\theta(\mathbf{x}_i) > 0 \mid \mathbf{y}, \delta\} \in (0.25, 0.50]$	20.74	17.43
	$P\{\theta(\mathbf{x}_i) > 0 \mid \mathbf{y}, \delta\} \in (0, 0.25]$	0	0
	$D_i^* > 0.95$	14.41	13.71
	$D_i^* > 0.80$	38.65	40.89

3.7. Important Factors

To explore important factors or features driving the differences in treatment effect, we proposed the use of BART to select important variables through identifying the most frequently used variables in the model. In this way, we may identify those predictors which have the most significant influence on the response. The number of trees was set as 50, and the frequencies of variables used are presented in Figure 6. The median used frequency of the sex variable was 20 and the 95% interval was [13,26]. The median used frequency of the education level variable was 24 and the 95% interval was [16,34]. Therefore, the education level variable is a more important predictor than the sex variable.

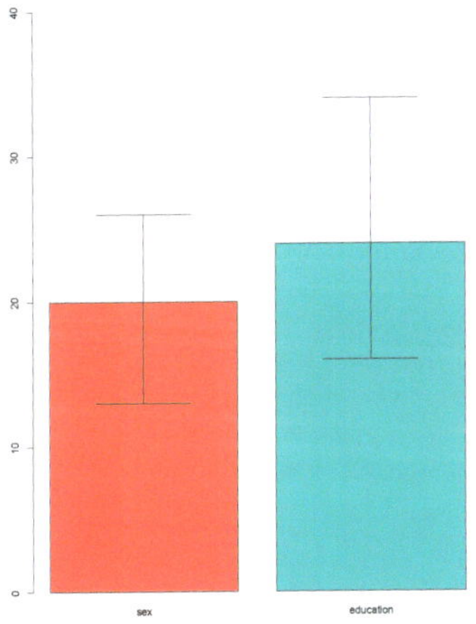

Figure 6. The importance of variables using BART.

4. Discussion

In this study, we estimated the effect of the *APOE-ϵ4* gene on onset risk of AD at the individual level. The individualized effects were qualified by constructing a credible interval for every patient. In particular, in this way, the individualized effects for any patient and their credible interval can be inferred, instead of those at the population level. This may help to better target those patients who are more significantly affected by $APOE$-$\epsilon4$. Furthermore, we can estimate the effects of $APOE$-$\epsilon4$ at the population level, based on the individualized effects. We inferred the effect of $APOE$-$\epsilon4$ on AD using causal inference. As such, assumptions for observational data were necessary, such as strong ignorability, which may induce treatment selection bias in the observational data. Further, in order to perform causal inference on observational data, the assumptions of overlap and no hidden confounders had to be made.

According to the causal effects for all patients, the causal effect of $APOE$-$\epsilon4$ on AD was not statistically significant at the population level. However, we observed a sub-population of patients presenting significant causal effects. Compared with the patients without significant causal effects, this sub-population had a higher proportion of female patients. Patients with low educational attainment level tended to present significant causal effects. In terms of the race of patients, patients of race 2 and race 3 in the sub-population accounted for higher proportions than in those without significant causal effects.

In the data analysis, we used BART to estimate the causal effects of $APOE$-$\epsilon4$ on AD for patients at the individual level. BART has been shown to be efficient and flexible, and has better or comparable performance to non-Bayesian competitors such as Boosting, LASSO, neural networks, and random forests [13]. BART has been shown to have good prediction performance and performs well for causal inference in various scenarios. Furthermore, it is necessary to quantify the outcome, especially in clinical research. In this context, Bayesian methods can provide natural credible intervals for outcomes. Although it is based on the potential outcome framework, our method may contribute to the identification of potential factors associated to the outcome at the causal level, which may help to determine the front node and directed path in the construction of the Bayesian network.

There are several metrics used for evaluation in this work. First, the prediction accuracy and the quantified uncertainty of prediction results are the most important metrics in clinical applications. In this line, we provided the estimate bias of the causal effect of $APOE$-$\epsilon4$ on AD and the 95% credible interval. As we handled right-censored data in this work, the effect of the censoring rate on the accuracy and efficiency of inference can be evaluated using Monte Carlo simulation techniques.

There were some limitations to our study; for example, there were no more than three baseline variables. We only included three variables and two variables for time-to-event data and binary outcome data, respectively. The inference for causal effects was limited by the few variables, as they only provided limited information. When analyzing data employing BART, as an MCMC technique, it can be computationally demanding; as such, the method was computationally expensive and required a significant amount of time for execution.

Author Contributions: Conceptualization, B.L.; methodology, Y.X. and B.L.; software, Y.X.; validation, Y.X. and B.L.; formal analysis, Y.X. and B.L.; investigation, B.L.; resources, B.L.; data curation, B.L.; writing—original draft preparation, Y.X.; writing—review and editing, B.L.; visualization, B.L.; supervision, B.L.; project administration, B.L.; funding acquisition, B.L. All authors have read and agreed to the published version of the manuscript.

Funding: This research was supported by the National Natural Science Foundation of China [11901013], Beijing Natural Science Foundation [1204031], Scientific Research and Development Funds of Peking University People's Hospital [RDX2021-05], and the project 2020BD029 supported by PKU-Baidu Fund.

Informed Consent Statement: Not applicable, as this research uses a publicly available data set.

Data Availability Statement: The data presented in this study are available upon request from the corresponding author.

Conflicts of Interest: The authors declare no conflict of interest.

Appendix A. Implementation

The AFT-BART Model is a non-parametric Bayesian AFT model which combines a sum-of-trees model for the regression function and a DP mixture model for the residual distribution. This method was implemented based on the `AFTrees` package of the R software (version R-4.3.1).

To install and use the `AFTrees` package in R software, the development version of the package can be obtained from the GitHub website.The package can be installed directly from github, or downloaded and installed from the local files. For remote installation, the following commands should be run:

```
install.packages("devtools")
library(devtools)
install_github("nchenderson/AFTrees")
```

First, we processed the data set and constructed the data frame for the model. The data consisted of n independent measurements $D = \{Y_i, \delta_i, W_i, X_i\}$. We split the data set into three folds and analyzed the data three times. Each time, two folds were used as the training set and the remaining fold was used as the testing set.

```
library(caret)
library(AFTrees)
source("SurvivalProb-AD.R")
# loading data ...
set.seed(1)
data <- read.csv('AD_Data.csv')
censor_data <- data
n <- nrow(censor_data)
d <- 3
X <- cbind(censor_data$X.1, censor_data$X.2, censor_data$X.3)
# treatment indicators
W <- censor_data$G_i
Y <- censor_data$Y
status <- censor_data$delta
# prepare data
colnames(X) <- colnames(X, do.NULL = FALSE, prefix = "x")
AD_data <- data.frame(X, W = W, Y = Y, status = status)
n <- nrow(AD_data)
# data split
set.seed(10)
fold_idx <- createFolds(y = AD_data$W, k=3)
```

We split the data into training and testing sets, and used the Bayesian non-parametric AFT Model to estimate the conditional average treatment effect by employing BART. In BART, the number of trees was set as 200. In the MCMC iterations, we set 5000 iterations to be treated as burn-in and 1000 as the number for posterior drawing. The implementation details are as follows:

```
for(i in 1:3){
  cat("\n NO.", i, "fold analysis ...\n")
  train_data <- AD_data[-fold_idx[[i]], ]
  est_data <- AD_data[fold_idx[[i]], ]
  # IndivAFT ...
  bart.tot <- IndivAFT(x.train = as.matrix(xtrain),
                       y.train = train_data$Y,
                       status = train_data$status,
                       Trt = xtrain$W,
```

```
                            x.test = as.matrix(xtest),
                            ntree = 200,
                            ndpost = 1000,
                            nskip = 5000)
  ite <- colMeans(bart.tot$Theta.test)
}
```

The posterior of individual treatment effects could then be obtained. The result was a matrix with posterior drawn times rows and test case size columns. In order to obtain the ITE posterior means, we averaged the output values in a column-wise manner.

References

1. Evans, D.A.; Funkenstein, H.H.; Albert, M.S.; Scherr, P.A.; Cook, N.R.; Chown, M.J.; Hebert, L.E.; Hennekens, C.H.; Taylor, J.O. Prevalence of Alzheimer's disease in a community population of older persons: Higher than previously reported. *JAMA* **1989**, *262*, 2551–2556. [CrossRef] [PubMed]
2. Corder, E.H.; Saunders, A.M.; Strittmatter, W.J.; Schmechel, D.E.; Gaskell, P.C.; Small, G.W.; Roses, A.D.; Haines, J.L.; Pericak-Vance, M.A. Gene dose of apolipoprotein e type 4 allele and the risk of alzheimer's disease in late onset families. *Science* **1993**, *261*, 921–923. [CrossRef] [PubMed]
3. Ortega-Rojas, J.; Arboleda-Bustos, C.E.; Guerrero, E.; Neira, J.; Arboleda, H. Genetic Variants and Haplotypes of TOMM40, APOE, and APOC1 are Related to the Age of Onset of Late-onset Alzheimer Disease in a Colombian Population. *Alzheimer Dis. Assoc. Disord.* **2022**, *36*, 29–35. [CrossRef]
4. Corder, E.H.; Saunders, A.M.; Risch, N.J.; Strittmatter, W.J.; Schmechel, D.E.; Gaskell, P.C., Jr.; Rimmler, J.B.; Locke, P.A.; Conneally, P.M.; Schmader, K.E.; et al. Protective effect of apolipoprotein E type 2 allele for late onset Alzheimer' disease. *Nat. Genet.* **1994**, *7*, 180–184. [CrossRef]
5. Farrer, L.A.; Cupples, L.A.; Haines, J.L.; Hyman, B.; Kukull, W.A.; Mayeux, R.; Myers, R.H.; Pericak-Vance, M.A.; Risch, N.; van Duijn, C.M.; Effects of age, sex and ethnicity on the association between apolipoprotein E genotype and Alzheimer' disease. A meta analysis. APOE and Alzheimer' disease Meta Analysis Consortium. *JAMA* **1997**, *278*, 1349–1356. [CrossRef]
6. Gatz, M.; Reynolds, C.A.; Fratiglioni, L.; Johansson, B.; Mortimer, J.A.; Berg, S.; Fiske, A.; Pedersen, N.L. Role of genes and environments for explaining Alzheimer' disease. *Arch. Gen. Psychiatry* **2006**, *63*, 168–174. [CrossRef]
7. Robins, C.; Wingo, A.P.; Meigs, J.; Duong, D.; Cutler, D.J.; De Jager, P.L.; Lah, J.J.; Bennett, D.A.; Seyfried, N.T.; Wingo, T.S.; et al. Identifying novel causal genes and proteins in Alzheimer's disease. *Alzheimer's Dement.* **2020**, *16*, e043523. [CrossRef]
8. Zhang, W.; Jiao, B.; Xiao, T.; Liu, X.; Liao, X.; Xiao, X.; Guo, L.; Yuan, Z.; Yan, X.; Tang, B.; et al. Association of rare variants in neurodegenerative genes with familial Alzheimer's disease. *Ann. Clin. Transl. Neurol.* **2020**, *7*, 1985–1995. [CrossRef] [PubMed]
9. Rubin, D.B. Estimating causal effects of treatments in randomized and nonrandomized studies. *J. Educ. Psychol.* **1974**, *66*, 688–701. [CrossRef]
10. Nguyen, T.L.; Collins, G.S.; Landais, P.; Manach, Y.L. Counterfactual clinical prediction models could help to infer individualized treatment effects in randomized controlled trials - An illustration with the International Stroke Trial. *J. Clin. Epidemiol.* **2020**, *125*, 47–56. [CrossRef]
11. Dorresteijn, J.A.N.; Visseren, F.L.J.; Ridker, P.M.; Wassink, A.M.J.; Paynter, N.P.; Steyerberg, E.W.; van der Graaf, Y.; Cook, N.R. Estimating treatment effects for individual patients based on the results of randomised clinical trials. *BMJ* **2011**, *343*, d5888. [CrossRef]
12. Jennifer, L.H. Bayesian nonparametric modeling for causal inference. *J. Comput. Graph. Stat.* **2011**, *20*, 217–240. [CrossRef]
13. Chipman, H.A.; George, E.I.; Mcculloch, R.E. BART: Bayesian additive regression trees. *Ann. Appl. Stat.* **2010**, *4*, 266–298. [CrossRef]
14. Henderson, N.C.; Louis, T.A.; Rosner, G.L.; Varadhan, R. Individualized treatment effects with censored data via fully nonparametric Bayesian accelerated failure time models. *Biostatistics* **2018**, *21*, 5–68. [CrossRef]
15. Bonato, V.; Baladandayuthapani, V.; Broom, B.M.; Sulman, E.P.; Aldape, K.D.; Do, K.A. Bayesian ensemble methods for survival prediction in gene expression data. *Bioinformatics* **2011**, *27*, 359–367. [CrossRef]
16. Sparapani, R.A.; Logan, B.R.; Mcculloch, R.E.; Laud, P.W. Nonparametric survival analysis using bayesian additive regression trees (BART). *Stat. Med.* **2016**, *35*, 2741–2753. [CrossRef]
17. Basak, P.; Linero, A.; Sinha, D.; Lipsitz, S. Semiparametric analysis of clustered interval-censored survival data using soft Bayesian additive regression trees (SBART). *Biometrics* **2022**, *78*, 880–893. [CrossRef]
18. Tan, Y.V.; Roy, J. Bayesian additive regression trees and the General BART model. *Stat. Med.* **2019**, *38*, 5048–5069. [CrossRef]
19. Albert, J.H.; Chib, S. Bayesian analysis of binary and polychotomous response data. *Publ. Am. Stat. Assoc.* **1993**, *88*, 669–679. [CrossRef]
20. Hill, J.; Linero, A.; Murray, J. Bayesian Additive Regression Trees: A Review and Look Forward. *Annu. Rev. Stat. Its Appl.* **2021**, *7*, 251–278. [CrossRef]

21. Mayeux, R.; Reitz, C.; Brickman, A.M.; Haan, M.N.; Manly, J.J.; Glymour, M.M.; Weiss, C.C.; Yaffe, K.; Middleton, L.; Hendrie, H.C.; et al. Operationalizing diagnostic criteria for Alzheimer's disease and other age-related cognitive impairment—Part 1. *Alzheimers Dement.* **2011**, *7*, 15–34. [CrossRef]
22. González Burchard, E.; Borrell, L.N.; Choudhry, S.; Naqvi, M.; Tsai, H.J.; Rodriguez-Santana, J.R.; Chapela, R.; Rogers, S.D.; Mei, R.; Rodriguez-Cintron, W.; et al. Latino populations: A unique opportunity for the study of race, genetics, and social environment in epidemiological research. *Am. J. Public Health* **2005**, *95*, 2161–2168.
23. Tang, M.X.; Stern, Y.; Marder, K.; Bell, K.; Gurl, B.; Lantigua, R.; Andrews, H.; Feng, L.; Tycko, B.; Mayeux, R. The *APOE*-e4 allele and the risk of Alzheimer disease among African Americans, whites, and Hispanics. *JAMA* **1998**, *279*, 751–755. [CrossRef]
24. Sparapani, R.; Spanbauer, C.; McCulloch, R. Nonparametric machine learning and efficient computation with Bayesian additive regression trees: The BART R Package. *J. Stat. Softw.* **2021**, *97*, 1–66. [CrossRef]
25. Zhang, W.; Le, T.D.; Liu, L.; Zhou, Z.; Li, J. Mining heterogeneous causal effects for personalized cancer treatment. *Bioinformatics* **2017**, *33*, 2372–2378. [CrossRef]

mathematics

MDPI

Article

Bayesian Latent Class Analysis: Sample Size, Model Size, and Classification Precision

Diana Mindrila

Department of Leadership, Research, and School Improvement, University of West Georgia, Carrollton, GA 30112, USA; dmindril@westga.edu

Abstract: The current literature includes limited information on the classification precision of Bayes estimation for latent class analysis (BLCA). (1) Objectives: The present study compared BLCA with the robust maximum likelihood (MLR) procedure, which is the default procedure with the M*plus* 8.0 software. (2) Method: Markov chain Monte Carlo simulations were used to estimate two-, three-, and four-class models measured by four binary observed indicators with samples of 1000, 750, 500, 250, 100, and 75 observations, respectively. With each sample, the number of replications was 500, and entropy and average latent class probabilities for most likely latent class membership were recorded. (3) Results: Bayes entropy values were more stable and ranged between 0.644 and 1. Bayes' average latent class probabilities ranged between 0.528 and 1. MLR entropy values ranged between 0.552 and 0.958. and MLR average latent class probabilities ranged between 0.539 and 0.993. With the two-class model, BLCA outperformed MLR with all sample sizes. With the three-class model, BLCA had higher classification precision with the 75-sample size, whereas MLR performed slightly better with the 750- and 1000-sample sizes. With the 4-class model, BLCA underperformed MLR and had an increased number of unsuccessful computations, particularly with smaller samples.

Keywords: Bayes estimation; BLCA; latent class analysis; structural equation modeling; latent variable modeling; person-oriented analyses

MSC: 60E05; 62H05; 62E10; 62F10; 62F15; 62P05

Citation: Mindrila, D. Bayesian Latent Class Analysis: Sample Size, Model Size, and Classification Precision. *Mathematics* **2023**, *11*, 2753. https://doi.org/10.3390/math11122753

Academic Editor: Manuel Alberto M. Ferreira

Received: 26 April 2023
Revised: 7 June 2023
Accepted: 15 June 2023
Published: 17 June 2023

1. Introduction

Bayesian analysis is a statistical approach that incorporates prior knowledge or beliefs with observed data to make probabilistic inferences and update our knowledge. It is named after the Reverend Thomas Bayes, an 18th-century British statistician, and theologian who developed the foundational principles of this method [1].

In Bayesian analysis, the main focus is on estimating and updating the posterior probability distribution of parameters of interest, given the observed data and any prior information. This is done using Bayes' theorem, which mathematically expresses the relationship between the prior probability, likelihood, and posterior probability. The prior probability represents our initial beliefs about the parameters, and the likelihood quantifies the compatibility between the observed data and the parameter values. By combining these elements, Bayesian analysis provides a coherent framework for inference [1,2].

One of the key advantages of Bayesian analysis is its ability to incorporate prior knowledge. This is particularly useful when there is limited data available or when expert opinions and historical information are valuable in making predictions or decisions. The use of prior information allows for a more nuanced and flexible analysis, accommodating subjective judgments and external evidence [3,4].

Bayesian analysis finds applications in a wide range of fields, including but not limited to:

(1) Medicine and Healthcare: Bayesian methods are employed in clinical trials, diagnostic tests, epidemiology, and personalized medicine to quantify uncertainty and make informed decisions.
(2) Finance and Economics: Bayesian analysis is used in risk assessment, portfolio optimization, forecasting, and economic modeling to account for uncertainty and update beliefs.
(3) Engineering: Bayesian techniques are applied in reliability analysis, optimization, and decision-making under uncertainty in various engineering domains.
(4) Machine Learning and Artificial Intelligence: Bayesian inference is used in probabilistic modeling, Bayesian networks, and Bayesian optimization to reason under uncertainty and provide robust predictions.
(5) Environmental Science: Bayesian analysis is utilized in environmental modeling, ecological studies, and climate change research to integrate diverse data sources and quantify uncertainty in predictions [5].

In social and behavioral sciences, Bayesian data analysis has been more frequently used since software packages popular among social scientists supported model fitting for Bayesian models and Markov chain Monte Carlo simulations (MCMC). These developments are facilitated by the availability of tutorials, software programs, and introductory textbooks on practical analytic skills [6,7].

1.1. Bayesian Latent Variable Modeling

Bayesian latent variable modeling refers to a class of statistical modeling techniques that involve unobserved or latent variables. Latent variables are variables that are not directly measured or observed but are inferred based on observed data. Bayesian methods are particularly well-suited for latent variable modeling because they allow for the incorporation of prior beliefs and uncertainty in estimating the latent variables and their relationships with the observed variables [1,8,9].

In Bayesian latent variable modeling, the goal is to estimate the values of the latent variables and their associated parameters, given the observed data and any prior knowledge. This is typically done by specifying a probabilistic model that describes the relationships between the latent variables and the observed variables. The model parameters are then estimated using Bayesian inference, which involves updating the prior beliefs to obtain the posterior distribution of the parameters given the observed data [9].

Bayesian latent variable modeling has wide-ranging applications in various fields, including psychology, social sciences, econometrics, and machine learning. It allows researchers to capture and analyze complex relationships, account for measurement errors, handle missing data, and make predictions or inferences about the latent variables [10,11].

1.2. Bayesian Factor Analysis

In factor analysis, the Bayesian method is used to uncover latent variables or factors that underlie a set of observed variables. It combines the principles of factor analysis, which aims to identify common patterns or underlying dimensions in observed data, with Bayesian inference, which allows for the incorporation of prior beliefs and uncertainty in parameter estimation [8,10,12].

In Bayesian factor analysis, the goal is to estimate the factor loadings, which represent the relationships between the latent factors and the observed variables, and the factor scores, which indicate the values of the latent factors for each individual. The method assumes that the observed variables are linearly related to the latent factors and that the observed variables are influenced by both specific (unique) factors and common factors shared across variables [12,13].

The Bayesian approach to factor analysis allows for the incorporation of prior information about the factor loadings and the factor scores. It also provides posterior distributions for the estimated parameters, which reflect both the observed data and the prior beliefs.

This posterior distribution can be used to make inferences about the latent factors and their relationships with the observed variables [12–14]

With factor models, Bayes estimation outperformed the mean- and variance-adjusted weighted least squares procedure with ordinal data [15,16]. This method incorporates prior information, thus increasing the accuracy of parameter estimates and reducing the number of Heywood solutions [17–19].

1.3. Bayesian Latent Class Analysis

Bayesian latent class analysis (BLCA) is a statistical method used to identify unobserved subgroups or latent classes within a population based on observed categorical variables [20]. It combines the principles of latent class analysis (LCA), which seeks to identify homogeneous subgroups within a population, with Bayesian inference techniques, which allow for the incorporation of prior beliefs and uncertainty in parameter estimation [12,20,21].

In BLCA, the goal is to estimate the latent class membership probabilities and the conditional response probabilities for each observed categorical variable given the latent class membership. The latent class membership probabilities indicate the likelihood of each individual belonging to each latent class, while the conditional response probabilities describe the probability of observing each response category for each variable within each latent class [20].

The Bayesian approach to latent class analysis allows for the integration of prior information about the latent class membership probabilities and the conditional response probabilities. It also provides posterior distributions for the estimated parameters, which reflect both the observed data and the prior beliefs. This posterior distribution can be used to make inferences about the latent classes and their relationships with the observed categorical variables [20–23]. While several studies investigated the effectiveness of the Bayesian method in factor analysis [17–19], few studies examined this estimation procedure's performance with latent class models.

Specifically, the classification precision of BLCA is an area that has received limited research attention. Despite the growing popularity of Bayesian methods in other areas of statistics, there has been a dearth of studies specifically examining the classification precision of BLCA.

Compared to traditional frequentist approaches, BLCA offers several advantages, such as the ability to incorporate prior information, handle missing data more effectively, and provide uncertainty estimates through posterior distributions. However, the specific performance of BLCA in terms of classification precision, as measured by metrics, such as entropy and average latent class probabilities, remains relatively unexplored.

The lack of research in this area can be attributed to various factors. First, BLCA involves complex modeling and estimation procedures, which require specialized knowledge and computational resources. This complexity may have deterred researchers from exploring the classification precision of BLCA in depth. Second, the focus of previous studies on LCA has predominantly been on model selection, identifying the appropriate number of latent classes, and examining the substantive interpretation of latent classes rather than evaluating classification precision. As a result, the evaluation of classification precision has often taken a backseat. Third, the availability of user-friendly software and computational tools for BLCA has been relatively limited compared to frequentist counterparts. This may have hindered researchers from conducting comprehensive studies on classification precision using Bayesian approaches.

Given the potential advantages of BLCA and the importance of classification precision in understanding latent class membership, there is a need for more research in this area. Future studies could explore the performance of BLCA under various conditions, compare it with frequentist approaches, and investigate the impact of different prior specifications on classification precision. By addressing these gaps in the literature, researchers can gain a deeper understanding of the strengths and limitations of BLCA in accurately classifying individuals into latent classes, ultimately enhancing the quality and applicability of latent class analysis in various fields.

2. Theoretical Framework

Latent Class Analysis (LCA) is a statistical method used to identify unobserved subgroups or latent classes within a population based on observed categorical variables [24]. It is a form of finite mixture modeling where the population is assumed to be composed of distinct latent classes, and individuals are probabilistically assigned to these classes based on their responses to the observed variables [21]. LCA is sometimes referred to as "mixture modeling based clustering" [25], "mixture-likelihood approach to clustering" [26], or "finite mixture modeling" [27,28]. In fact, "finite mixture modeling" is a more general term for latent variable modeling where latent variables are categorical. The latent categories represent a set of sub-populations of individuals, and individuals' memberships to these sub-populations are inferred based on patterns of variation in the data [26–29].

In LCA, the goal is to estimate the latent class membership probabilities and the conditional response probabilities for each observed categorical variable given the latent class membership. The latent class membership probabilities indicate the likelihood of each individual belonging to each latent class, while the conditional response probabilities describe the probability of observing each response category for each variable within each latent class [30,31].

The estimation of LCA parameters can be done using maximum likelihood estimation (MLE) or Bayesian methods. MLE involves finding the parameter values that maximize the likelihood of the observed data, while Bayesian methods incorporate prior information and uncertainty in the estimation process, typically using iterative techniques, such as the Expectation-Maximization (EM) algorithm [31].

LCA has applications in various fields, including psychology, sociology, marketing, and public health. It allows researchers to identify meaningful subgroups within a population, understand the relationships between variables, and examine the predictors or consequences of latent class membership [21,27,31].

2.1. The LCA Model

A mixture model includes a measurement model and a structural model. LCA is the measurement model, which consists of a set of observed variables, also referred to as observed indicators, regressed on a latent categorical variable [21]. LCA explains the relationships between a set of r observed indicators i and an underlying categorical variable C [31–33].

Observed variables can be continuous, counts, ordered categorical, binary, or unordered categorical variables [31–33]. When estimating a latent variable C with k latent classes (C = k; k = 1, 2, ... k), the "marginal item probability" for item ij = 1 can be expressed as:

$$P(i,j=1) = \sum_{k=1}^{K} P(C=k)P(ij=1|C=k) \quad (1)$$

Assuming that the assumption of local independence is met, the joint probability for all observed variables can be expressed as:

$$P(i1, i2, \ldots.., ir) = \sum_{k=1}^{K} P(C=k)P(C=k)P(C=k)\ldots P(ir|C=k) \quad (2)$$

The computation procedures used for estimating model parameters are based on the type of variables used as observed indicators (Table 1).

Table 1. Computation Procedures by Variable Type.

Variable Type	Computation Procedure
Continuous	Linear regression equations
Censored	Censored-inflated normal regression
Count	Poisson or zero-inflated Poison regression equations
Ordered categorical	Logistic regression
Binary	Logistic regression
Nominal	Multinomial logistic regression

2.2. Estimation Procedures

LCA assigns individuals to latent classes using an iterative procedure. Researchers can specify starting values or use automatic, random starts. This process is similar to selecting seed values for the *k*-means clustering algorithm. Estimation iterates until the exact solution results from multiple sets of starting values, at which point parameters are considered most likely representative of a latent class [34].

Estimated model parameters include item means and variances by latent class. Results also include, for each case, the probability of membership to each class. These probabilities add up to one across latent classes and are referred to as "posterior probabilities" [31]. Latent class memberships result from a modal assignment, consisting of placing each person in the latent class for which the probability of membership is the highest [35].

The robust maximum likelihood (MLR) estimation procedure uses "log-likelihood functions derived from the probability density function underlying the latent class model" [29]. The statistical software employed in the current study was M*plus*. This software allows users to use other estimation procedures, such as the Bayesian estimation, which can be specified using the ESTIMATOR = BAYES option of the ANALYSIS command. Although MLR corrects standard errors and test statistics, it would be reasonable to hypothesize that other estimators, such as BAYES, may provide more accurate results with small sample sizes, ordinal data, and non-normal continuous variables [36,37].

2.3. The Bayesian Approach

Traditionally, LCA models were estimated using the maximum likelihood procedure using the expectation-maximization (EM) algorithm [38]. The new developments in statistical software now allow researchers to employ estimation procedures that are computationally more complex and used to take an extended amount of time [15,39]. For instance, Asparouhov and Muthen [40] developed an algorithm that permits the computation of a correlation matrix using Bayesian estimation. Using this correlation matrix, the LCA model can be estimated with more flexibility because the estimation procedure no longer requires within-class indicators to be independent [40] and allows researchers to increase estimation precision by taking into account prior information [15].

The Bayes estimation allows the use of both informative and non-informative priors. Informative priors are used when researchers have prior information about model parameters based on theory, expert opinion, or previous research [6]. The Bayes theorem for continuous parameters specifies that "the posterior is proportional to the prior times the likelihood" [41]. This statement very clearly explains how the Bayes approach inverts the likelihood function to estimate the probability p of a parameter θ given and observed distribution of a variable y, as indicated in the following formula:

$$p(\theta \mid y) \propto p(y \mid \theta) \times p(\theta). \quad (3)$$

Bayes estimation also allows non-informative or diffuse priors when researchers do not have sufficient information about the parameters of interest [6]. Nevertheless, as the amount of information about parameters increases through repeated applications of the data generation process, the precision of the posterior distributions continues to grow. Eventually, it overwhelms the effect of the non-informative priors [41].

Frequentist procedures such as ML estimate model parameters by deriving point estimates that have asymptotic properties. ML estimation assumes that point estimates have an asymptotic normal distribution and are consistent and efficient [36,42]. In contrast, Bayesian inference focuses on estimating the model parameter's posterior distribution features, such as point estimates and posterior probability intervals. Summarizing posterior distributions requires the calculation of expectations. Such computations become very complex with high-dimensional problems which require multiple integrals. For this reason, researchers rely on Monte Carlo integration to draw samples from the posterior distributions and summarize the distribution formed by the extracted samples [6].

2.4. Bayesian LCA

One of the advantages of employing Bayesian estimation is using information from prior distributions. This allows researchers to use prior knowledge to inform current analyses. In the context of Bayesian LCA (BLCA), researchers could use prior information regarding individuals' response patterns to help increase estimation accuracy [43].

In the case of BLCA, two parameters are of special interest. The first one refers to the proportion of observations in the C latent classes. The proportion of observations in the C latent classes (π_C) has a Dirichlet distribution, which can be notated as:

$$\pi_C \sim D[d_1,..,d_C], \tag{4}$$

where parameters $d_1 \ldots d_C$ determine the uniformity of the D distribution. When $d_1 \ldots d_C$ have relatively equal values, the identified latent classes are similar in size and have similar probabilities [43].

The second parameter of interest is the response probability ($\rho_{V,rv\,|\,C}$). The Bayesian estimation calculates this parameter in two ways. The response probability can be calculated as a probability as follows:

$$\rho_{V,rv\,|\,C} \sim D[d_1,..,d_C]. \tag{5}$$

where D is the Dirichlet distribution with its parameters $d_1 \ldots d_C$.

Furthermore, response probabilities can be calculated using a probit link function as indicated below:

$$[probit]\rho_{V,rv\,|\,C} \sim N[\mu_\rho, \sigma^2{}_\rho], \tag{6}$$

where N is the Normal distribution with its mean μ_ρ and variance $\sigma^2{}_\rho$ parameters. Depending on the software used for estimation, the variance parameter may be referred to as precision [43].

The Bayesian approach can be used to increase estimation accuracy and allows for more flexibility in the construction of LCA models [43]. The frequentist approach relies on the assumption of independent observed indicators within each class and specifies non-correlating indicators in the within-class correlation matrix. Nevertheless, this assumption is rarely met with real data, particularly in social sciences, which may lead to biased parameter estimates, increased classification errors, and poor model fits [43]. In contrast, the Bayesian estimation relaxes this restriction and only assumes approximate independence [40,43]. Asparouhov and Muthen describe near-zero correlations as hybrid parameters, which are not quite fixed nor free parameters [20]. This flexibility of BLCA may limit the degree of model misspecification which may occur when within-class correlations are fixed to zero [40].

2.5. Label Switching

Label switching is a potential issue that may pose problems with models relying on Markov Chain Monte Carlo (MCMC) procedures. Label switching occurs when the order of classes arbitrarily changes across the MCMC chains [44,45]. Reordering may occur because LCA models do not specify the order of classes. This change may affect the estimated posterior and may lead to convergence issues. Label switching often occurs with mixture models; therefore, it is critical to be aware of its causes and proposed solutions such as reparameterization techniques, relabeling algorithms, and label invariant loss functions [46,47].

2.6. Classification Precision

With exploratory LCA, the researcher does not know a priori the number of classes of the latent categorical variable. The selection of the optimal model often relies on criteria, such as (a) the interpretability of the latent class solutions [35]; (b) measures of model fit (e.g., Bayesian Information Criteria [BIC], the sample-size adjusted BIC, the Akaike Information Criteria, the Lo-Mendell-Rubin (LMR) likelihood ratio test, etc.); and (c) measures of classification precision (e.g., entropy, average latent class probabilities, etc.).

Measures of classification precision help address the issue of class separation. The interpretability of item loadings is a critical criterion in selecting the optimal latent class model. This criterion is essential to ensure a strong theoretical and practical support for the latent class solution. For instance, in the context of an educational psychology study, one group of participants may have very low loadings on extrinsic motivation items and very high loadings on intrinsic motivation items, whereas another group may have the opposite characteristics. In such situations, latent class separation is clear. Nevertheless, as the number of latent classes increases, the separation between groups may not be as clear. For instance, a three-class model may yield another group with slightly above average intrinsic motivation and slightly below average extrinsic motivation. In such situations, the separation between groups is not as clear and using measures of fit and classification precision is essential.

For every observation, LCA calculates the probability of membership to each one of the classes specified in the LCA model. When membership probabilities are close to one for one class and close to zero for all other groups, the model has a high level of classification precision. Membership probabilities for the entire sample are summarized in a $k \times k$ table, where k is the number of latent classes specified in the LCA model. The diagonal elements of these tables represent the average probabilities of membership to the assigned class or the proportions of correctly classified cases.

The average probability of membership in Latent Class Analysis (LCA) represents the average likelihood of an individual belonging to each latent class based on their observed categorical responses. It provides information about the strength of membership in each latent class for each individual. The average probability of membership is computed by taking the average of the individual posterior probabilities across all individuals and classes. Hagenaars and McCutcheon [44] specified the formula for calculating the average probability of membership in LCA is as follows:

$$P(k) = (1/N) \times \Sigma\, P(k \mid i), \tag{7}$$

where N represents the total number of individuals in the sample, $P(k \mid i)$ represents the posterior probability of belonging to class k given the observed responses for individual i, and the summation is taken over all individuals in the sample. This formula computes the average across all individuals for each latent class, providing a measure of the overall probability of membership in each class. The specific formula for calculating the average probability of membership may vary slightly depending on the software or algorithm used for LCA estimation; therefore, it is always recommended to consult the software documentation or specific references provided by the software developers for accurate formulas and implementation details. Average probabilities of membership are considered indices of classification certainty and should be close to one [35]. The off-diagonal elements of the $k \times k$ table represent the proportions of misclassified cases and should be close to zero [35]. For instance, in a well-fitting model with four latent classes may have the $k \times k$ table represented in Table 2.

Table 2. Average Latent Class Probabilities and Misclassification Probabilities for a Hypothetical 4 × 4 Latent Class Model.

	Class 1	Class 2	Class 3	Class 4
Class 1	**0.980**	0.010	0.000	0.010
Class 2	0.030	**0.961**	0.000	0.009
Class 3	0.020	0.040	**0.890**	0.050
Class 4	0.020	0.049	0.010	**0.921**

Note: The diagonal elements are the average latent class probabilities and are marked in bold. The off-diagonal elements represent the misclassification probabilities.

Another indicator of classification certainty is entropy. In LCA, entropy is a commonly used measure to assess the quality of classification or the uncertainty in assigning individuals to latent classes. Entropy provides an indication of how well the latent classes differentiate individuals based on their observed responses. It is an omnibus index of classification certainty, which relies on the class posterior probabilities reported in the $k \times k$ table. This index shows the degree to which the entire LCA model accurately predicts individual class memberships [48], or the extent to which latent classes are distinct [49]. Higher entropy values indicate a better separation between classes, whereas lower entropy values suggest a more ambiguous or overlapping classification. The formula for calculating entropy in LCA is as follows:

$$\text{Entropy} = -\Sigma\ (P(k \mid i) \times \log(P(k \mid i))), \quad (8)$$

where $P(k \mid i)$ represents the posterior probability of belonging to class k given the observed responses for individual i, and the summation is taken over all individuals in the sample [34]. This formula computes the entropy for each individual and class and sums the contributions across all individuals. The negative sign is used to ensure that entropy values are positive. Entropy values can range from zero to one, where values closer to one indicate superior classification precision [29].

3. Objectives

Although MLR corrects standard errors and test statistics, based on related research, the researcher hypothesized that Bayes estimation might provide more accurate results with small sample sizes, ordinal data, and non-normal continuous variables [32,36]. The proposed study aimed to examine and compare the classification precision of the MLR and Bayes estimation methods, as measured by entropy and average latent class probabilities for most likely latent class membership, with binary observed indicators and samples of varying sizes, and models with different numbers of latent classes.

4. Simulation Study

Using a population with a known structure allows researchers to investigate the performance of an estimation method under different conditions. In other words, researchers can determine whether an estimation procedure can identify the underlying latent class memberships.

The Monte Carlo technique is a mathematical procedure that uses multiple probability simulation to estimate the outcome of uncertain events. This computational algorithm predicts a set of outcomes using an estimated range of values instead of a given series of fixed values. Therefore, this technique yields a model of plausible results by using a specified probability distribution (e.g., Normal distribution, Uniform distribution, etc.) of a variable with an uncertain outcome. Numerous sets of randomly generated values that follow the specified distribution are used to repeatedly estimate likely outcomes. This procedure consists of three steps:

1. Specify the predictive model including the independent and dependent variables.

2. Specify the distribution of the independent variables (based on historical information and theory.
3. Use multiple sets of randomly generated values following the specified distribution to calculate a representative sample of results [50].

A Markov chain is a model that describes a series of likely events, where the probability of one event depends on the probability of the antecedent event [51]. Markov chain Monte Carlo (MCMC) procedures rely on computer simulations of Markov chains. Markov chains are specified so that the posterior distribution of the inferred parameters is the asymptotic distribution.

In applied statistics, MCMC simulations can be used for several purposes, including (1) comparing statistics across samples given a set of realistic conditions, and (2) provide random samples for posterior Bayesian distributions [52]. The present study used MCMC simulations to compare Bayes and MLR classification precision under the same conditions. Specifically, the researcher compared three LCA models (with 2, 3, and 4 latent classes) measured by four binary observed indicators. The three LCA models were estimated using the Bayes with non-informative priors and the MLR procedures using samples of 1000, 750, 500, 250, 100, and 75 observations ($3 \times 2 \times 6$) with 500 replications. Entropy and average latent class probabilities were recorded and compared for each condition. The researcher used the M*plus* 8.0 statistical package to conduct all analyses. The code for Monte Carlo simulations followed example 7.3 from the M*plus* User's Guide [37] for generating a categorical latent variable with binary indicators. The example was modified to vary the sample sizes, the estimation method, and the number of classes. A sample code for the two-class model with Bayes estimation and a sample of 500 observations is included below:

```
Title:
    Example of LCA model with binary;
    latent class indicators using automatic;
    starting values with random starts;
  Montecarlo:
    NAMES = u1-u4;
    generate = u1-u4(1);
    categorical = u1-u4;
    genclasses = c(2);
    classes = c(2);
    nobs = 500;
    seed = 3454367;
    nrep = 500;
    save = resultsfile.dat;
  Analysis:
    type = mixture;
    estimator bayes;
  Model population:
    %overall%
    [c#1*1];
    %c#1%
    [u1$1*2 u2$1*2 u3$1*-2 u4$1*-2];
    %c#2%
    [u1$1*-2 u2$1*-2 u3$1*2 u4$1*2];
  Model:
    %overall%
    [c#1*1];
    %c#1%
    [u1$1*2 u2$1*2 u3$1*-2 u4$1*-2];
    %c#2%
```

[u1$1*-2 u2$1*-2 u3$1*2 u4$1*2];
Output:
tech8 tech9;

5. Results

With the Bayes estimation method, entropy values relatively ranged between (a) 0.997 and 1 for the 2-class model, (b) 0.802 and 0.848 for the 3-class model, and (c) 0.644 and 0.818 for the 4-class model. The Bayes and MLR entropy values for the two-, three-, and four-class models are represented in Figure 1, Figure 2, and Figure 3, respectively. Figure 4 illustrates all entropy values in relation to sample size, estimation method, and model size.

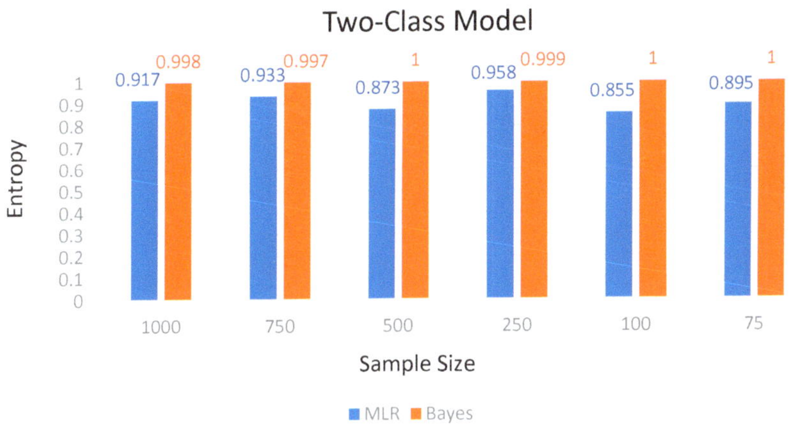

Figure 1. Bayes and MLR entropy values for the two-class model.

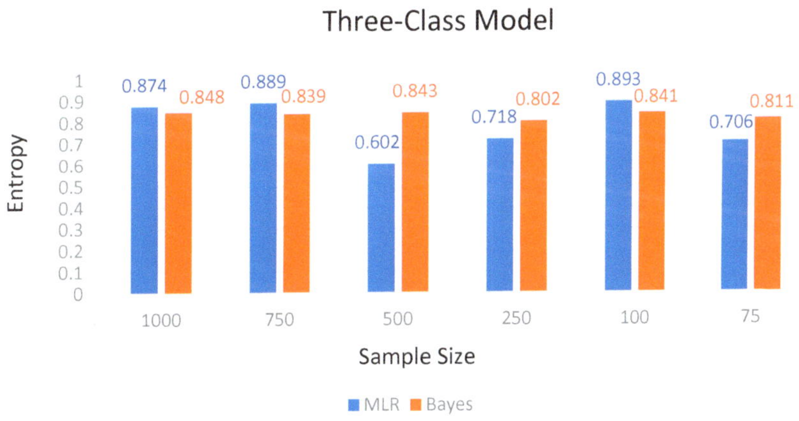

Figure 2. Bayes and MLR entropy values for the three-class model.

Figure 3. Bayes and MLR entropy values for the four-class model.

Figure 4. Bayes and MLR entropy values in relation to sample size and model size.

Overall, average latent class probabilities for most likely latent class membership ranged between 0.540 (4-class model) and 1 (2-class model) (Table 3). Figure 5 illustrates all recorded average latent class probabilities for most likely latent class membership in reference to sample size and the number of classes specified in the latent class model.

Table 3. Indices of Classification Precision by Model and Sample Size.

LCA Model	Estimator	Sample Size	Average Latent Class Probabilities for Most Likely Latent Class Membership			
			Class 1	Class 2	Class 3	Class 4
2 Class Model	Bayes	1000	0.999	0.999		
		750	0.999	0.999		
		500	0.999	0.999		
		250	1.000	0.999		
		100	0.999	0.999		
		75	1.000	1.000		
	MLR	1000	0.974	0.982		
		750	0.974	0.981		
		500	0.975	0.978		
		250	0.993	0.987		
		100	0.984	0.967		
		75	0.987	0.968		
3 Class Model	Bayes	1000	0.941	0.938	0.987	
		750	0.939	0.939	0.989	
		500	0.940	0.939	0.993	
		250	0.935	0.943	0.995	
		100	0.916	0.948	0.993	
		75	0.910	0.948	0.993	
	MLR	1000	0.867	0.848	0.67	
		750	0.874	0.855	0.695	
		500	0.882	0.868	0.735	
		250	0.889	0.884	0.807	
		100	0.915	0.914	0.872	
		75	0.921	0.922	0.905	
4 Class Model	Bayes	1000	0.548	0.874	0.768	0.742
		750	0.560	0.882	0.788	0.770
		500	0.535	0.889	0.801	0.741
		250	0.540	0.887	0.834	0.731
		100	0.528	0.913	0.756	0.780
		75	0.574	0.925	0.808	0.815
	MLR	1000	0.821	0.756	0.599	0.539
		750	0.832	0.77	0.621	0.570
		500	0.845	0.793	0.664	0.616
		250	0.866	0.823	0.752	0.707
		100	0.891	0.881	0.855	0.835
		75	0.911	0.901	0.887	0.868

With the smallest sample size (N = 75), Bayes estimation showed greater classification precision for the 2-class and the 3-class models, but MLR outperformed Bayes with the 4-class model. With the largest sample size (N = 1000), Bayes estimation had greater precision with the 2-class model and was comparable to MLR for the 3-class and the 4-class models (Table 2).

As the complexity of the model increased, the number of successful computations decreased for Bayes estimation, particularly for the 4-class model (Figures 6–8). Additionally, the time required to estimate the 4-class model was significantly longer, particularly for larger sample sizes.

Figure 5. Bayes and MLR average latent class probabilities for the most likely latent class membership in relation to sample size and model size.

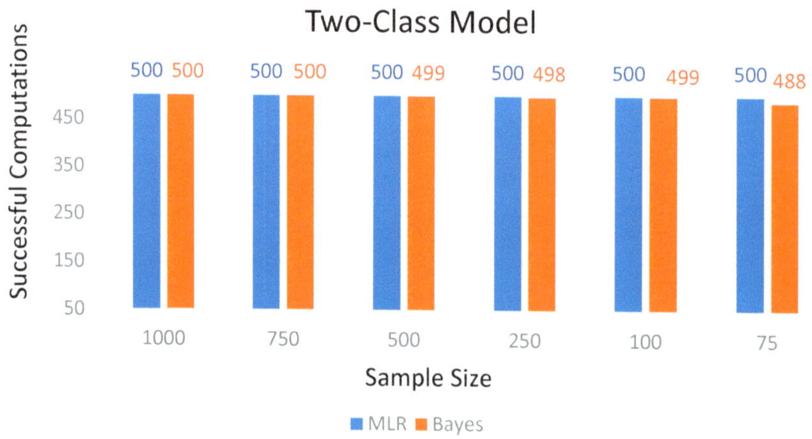

Figure 6. Number of successful computations by sample size for the two-class model.

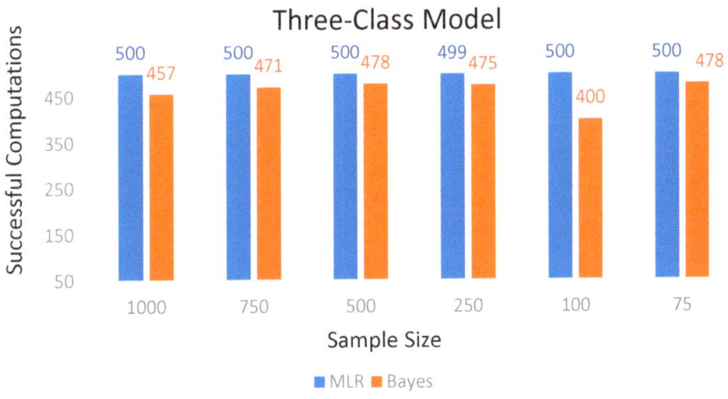

Figure 7. Number of successful computations by sample size for the three-class model.

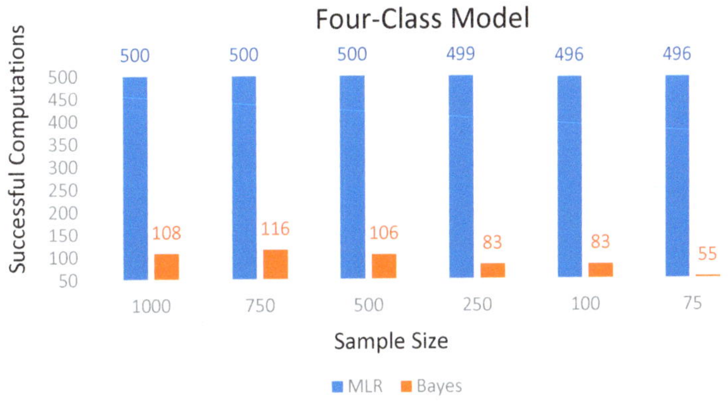

Figure 8. Number of successful computations by sample size for the four-class model.

6. Discussion and Conclusions

There is a noticeable gap in the existing research literature when it comes to studying the classification precision of BLCA. Despite the growing popularity of Bayesian methods in various fields, such as psychology, sociology, and marketing, there has been relatively limited attention given to the evaluation and comparison of classification accuracy specifically within the context of BLCA.

While LCA itself has been extensively studied and applied, much of the existing research has focused on traditional frequentist estimation methods, such as maximum likelihood estimation. BLCA offers unique advantages, such as the ability to incorporate prior knowledge, handle missing data, and provide probabilistic inferences. However, there is a lack of comprehensive empirical studies that directly investigate the classification precision of BLCA and compare it to other estimation approaches.

The limited research in this area may be attributed to several factors. First, Bayesian methods, including B LCA, often require advanced statistical knowledge and specialized software, which may deter some researchers from exploring these techniques. Second, there may be a perception that the computational complexity and longer execution times associated with Bayesian estimation hinder the feasibility of large-scale studies. Additionally, the absence of standardized guidelines or benchmarks for assessing the classification precision of BLCA further contributes to the scarcity of research in this domain.

As a result, more empirical studies are needed to address this gap in the literature. Such studies could compare the classification accuracy of BLCA with other popular estimation methods, evaluate its performance across different sample sizes and data characteristics, and provide insights into the factors that may influence the precision of BLCA classifications. These investigations would not only enhance our understanding of the strengths and limitations of BLCA but also provide researchers and practitioners with valuable guidance for selecting appropriate estimation methods in latent class analysis.

The primary objective of this study was to address this gap in the literature by investigating and comparing the accuracy of classification between two existing estimation methods: MLR and Bayes. MLR is the default M*plus* estimation procedure for categorical variables. Despite its assumed benefits, the Bayes option, which is also available, is less frequently used and needs to be specified in the M*plus* code. The current study aimed to determine whether using the default estimation settings, as most users do, may impact LCA classification precision.

Evaluating the classification precision of Bayes and MLR was based on the measurement of entropy and the average latent class probabilities for the most likely latent class membership. The study used binary observed indicators and included samples of different sizes and models with two–four latent classes.

Results suggest that for models with two latent classes, regardless of sample size, the Bayes method consistently outperforms the MLR procedure. Specifically, Bayesian entropy values ranged between 0.997 and 1, whereas MLR entropy values ranged between 0.855 and 0.958. Similarly, Bayesian average latent class probabilities for latent class memberships ranged between 0.999 and 1, whereas and MLR average latent class probabilities ranged between 0.974 and 0.993.

With three-class models, the Bayes method showed higher overall levels of classification precision with the sample of 75 (Bayesian entropy = 0.811, Bayes average latent class probabilities between 0.910 and 0.993; MLR entropy = 0.706, MLR average latent class probabilities between 0.905 and 0.922) and 500 samples (Bayesian entropy = 0.843, Bayesian average latent class probabilities between 0.940 and 0.993; MLR entropy = 0.602, MLR average latent class probabilities between 0.735 and 0.882). Nevertheless, the MLR procedure had slightly higher overall levels of classification precision with the larger samples (n = 750 and n = 1000). With the 750-sample size, the MLR entropy value was 0.889, whereas the Bayes entropy was 0.839; similarly, with the 1000-sample size, the MLR entropy was 0.874, whereas the Bayes entropy was 0.848.

When the model included four classes, MLR outperformed Bayes estimation with smaller samples (n = 100 and n = 75). With the 75-sample size, MLR entropy was 0.866, and MLR average latent class probabilities ranged between 0.868 and 0.911, whereas the Bayes entropy was only 0.664 and Bayes average latent class probabilities ranged between 0.574 and 0.925. Similarly, with the 100-sample size, MLR entropy was 0.860, and average latent class probabilities ranged between 0.835 and 0.891, whereas Bayes entropy was 0.727, and average latent class probabilities ranged between 0.528 and 0.913.

Although some researchers suggest that the Bayes method may be more effective with smaller sample sizes [43], results from the current study showed that this was only true for the smaller models, and classification precision varied mostly by model size than sample size. Overall, Bayes estimation provided more stable results, whereas MLR showed greater variations in average latent class probabilities for most likely latent class membership and entropy estimates. Nevertheless, the Bayes estimation had a much smaller number of successful computations than the four-class model. Furthermore, the Bayes estimation took extended time (days) with the four-class model. These computational difficulties may pose practical issues in using the Bayes procedure for applied research projects.

Based on these results, when working with binary observed indicators, researchers are advised to avoid deferring to the default M*plus* settings and select an appropriate estimation procedure based on both sample size and model size. Specifically, with smaller models, users are advised to use the Bayes estimation, which seems to have greater classification precision even with very small samples. In contrast, as the number of classes specified in the LCA model increases, users can defer to MLR, particularly with smaller sample sizes. In these conditions, the Bayes method does not seem to yield the same level of classification precision as MLR and yields an increased number of unsuccessful computations.

In conclusion, the Bayesian procedure can benefit the classification precision of mixture models when models have fewer classes. Additionally, non-reliance on the assumption of independence may reduce estimation bias. Furthermore, the option to specify informative prior may increase estimation accuracy [43]. Nevertheless, Bayesian estimation may encounter issues related to label switching [4], lead to unsuccessful computations, and take extended time.

The essential contribution of this study is providing information on the classification precision LCA models with binary indicators using the Bayes and MLR estimation methods. Although some research in exploratory factor analysis indicates that this estimation method is effective with small sample sizes and ordinal data [36], no research has assessed the precision of Bayes estimation for latent class models. Furthermore, the current study considered the complexity of the model by comparing models with different numbers of latent classes.

Although this information is helpful to applied researchers, this study is only a first step in comparing the effectiveness of the Bayes and MLR estimation procedures in latent class modeling. Additional simulation studies are needed to investigate the effectiveness of Bayes estimation compared to other estimators, such as maximum likelihood, and under other conditions, such as different types of observed indicators (ordered categorical, continuous, etc.), correctly specified versus miss-specified models, classes with varying prevalence, and with informative versus non-informative priors. Furthermore, we also encourage researchers to use BLCA with real data, particularly when estimating smaller LCA models.

Funding: This research received no external funding.

Data Availability Statement: Data for this study can be generated using the simulation code provided. For further assistance please contact the author.

Conflicts of Interest: The author declares no conflict of interest.

References

1. Gelman, A.; Carlin, J.B.; Stern, H.S.; Dunson, D.B.; Vehtari, A.; Rubin, D.B. *Bayesian Data Analysis*, 3rd ed.; CRC Press: Boca Raton, FL, USA, 2014.
2. Kruschke, J.K. *Doing Bayesian Data Analysis: A Tutorial with R, JAGS, and Stan*; Academic Press: Cambridge, MA, USA, 2014.
3. McElreath, R. *Statistical Rethinking: A Bayesian Course with Examples in R and Stan*; Chapman and Hall/CRC: Boca Raton, FL, USA, 2016.
4. Carlin, B.P.; Louis, T.A. *Bayesian Methods for Data Analysis*, 3rd ed.; CRC Press: Boca Raton, FL, USA, 2009.
5. Barber, D. *Bayesian Reasoning and Machine Learning*; Cambridge University Press: Cambridge, UK, 2012.
6. Kaplan, D. *Bayesian Statistics for the Social Sciences*; Guilford Publications: New York, NY, USA, 2014.
7. Gill, J. *Bayesian Methods: A Social and Behavioral Sciences Approach*; Chapman and Hall/CRC: Boca Raton, FL, USA, 2014.
8. Ghahramani, Z. Probabilistic machine learning and artificial intelligence. *Nature* **2015**, *521*, 452–459. [CrossRef] [PubMed]
9. Bishop, C.M. *Pattern Recognition and Machine Learning*; Springer: New York, NY, USA, 2006.
10. Lee, M.D.; Wagenmakers, E.J. *Bayesian Cognitive Modeling: A Practical Course*; Cambridge University Press: Cambridge, UK, 2014.
11. van de Schoot, R.; Kaplan, D.; Denissen, J.; Asendorpf, J.B.; Neyer, F.J.; van Aken, M.A. A gentle introduction to Bayesian analysis: Applications to developmental research. *Child Dev.* **2014**, *85*, 842–860. [CrossRef] [PubMed]
12. Muthén, B.; Asparouhov, T. Bayesian structural equation modeling: A more flexible representation of substantive theory. *Psychol. Methods* **2012**, *17*, 313–335. [CrossRef] [PubMed]
13. Wang, W.; Hancock, G.R. Bayesian factor analysis for structural equation modeling. *J. Educ. Behav. Stat.* **2010**, *35*, 22–50.
14. DeCarlo, L.T. On the analysis of factorial surveys by Bayesian confirmatory factor analysis. *Sociol. Methods Res.* **2012**, *41*, 313–337.

15. Asparouhov, T.; Muthén, B. *Bayesian Analysis of Latent Variable Models Using Mplus*; Technical Report; Version 4; Muthén & Muthén: Los Angeles, CA, USA, 2010. Available online: http://www.statmodel.com/download/BayesAdvantages18.pdf (accessed on 5 May 2023).
16. Asparouhov, T.; Muthén, B. *Bayesian Analysis Using Mplus: Technical Implementation (Technical Appendix)*; Muthén & Muthén: Los Angeles, CA, USA, 2010. Available online: http://www.statmodel.com/download/BayesAdvantages18.pdf (accessed on 5 May 2023).
17. Lee, S.Y. A Bayesian approach to confirmatory factor analysis. *Psychometrika* **1981**, *46*, 153–160. [CrossRef]
18. Martin, J.K.; McDonald, R.P. Bayes estimates in restricted factor analysis: A treatment of Heywood cases. *Psychometrika* **1975**, *40*, 505–517. [CrossRef]
19. Mayekawa, S. *Bayesian Factor Analysis (ONR Technical Report No. 85-3)*; CadaResearch Group, University of Iowa: Iowa City, IA, USA, 1985.
20. Albert, J.H.; Chib, S. Bayesian analysis of binary and polychotomous response data. *J. Am. Stat. Assoc.* **1993**, *88*, 669–679. [CrossRef]
21. Vermunt, J.K.; Magidson, J. Latent class cluster analysis. In *The Handbook of Advanced Multilevel Analysis*; HoX, J.J., Roberts, J.K., Eds.; Routledge: Oxfordshire, UK, 2016; pp. 141–160.
22. Friel, N.; Wyse, J. Estimating the number of classes in a finite mixture model. *J. R. Stat. Soc. Ser. B* **2012**, *74*, 411–438.
23. Celeux, G.; Soromenho, G. An entropy criterion for assessing the number of clusters in a mixture model. *J. Classif.* **1996**, *13*, 195–212. [CrossRef]
24. Hagenaars, J.A.; McCutcheon, A.L. (Eds.) *Applied Latent Class Analysis*; Cambridge University Press: Cambridge, UK, 2002.
25. Banfield, J.D.; Raftery, A.E. Model-based Gaussian and non-Gaussian clustering. *Biometrics* **1993**, *49*, 803–821. [CrossRef]
26. Everitt, B.S. *Cluster Analysis*; Edward Arnold: London, UK, 1993.
27. McLachlan, G.; Peel, D. *Finite Mixture Models*; John Wiley & Sons: New York, NY, USA, 2000.
28. Everitt, B.S.; Hand, D.J. Finite mixture models. In *Handbook of Markov Chain Monte Carlo*; Gelman, A., Rubin, D.B., Eds.; CRC Press: Boca Raton, FL, USA, 2011; pp. 79–110.
29. Vermunt, J.K.; Magidson, J. Latent class cluster analysis. In *Applied Latent Class Analysis*; Hagenaars, J.A., McCutcheon, A.L., Eds.; Cambridge University Press: Cambridge, UK, 2002; pp. 89–106.
30. Nylund-Gibson, K.; Choi, A.Y. Ten frequently asked questions about latent class analysis. *Transl. Issues Psychol. Sci.* **2018**, *4*, 440–461. [CrossRef]
31. Muthén, B. Beyond SEM: General latent variable modeling. *Behaviormetrika* **2002**, *29*, 81–117. [CrossRef]
32. Muthén, B.; Bayesian analysis in Mplus: A brief introduction. *Mathematics* 2010, *Unpublished manuscript*. Available online: www.statmodel.com/download/IntroBayesVersion,203 (accessed on 5 May 2023).
33. Geiser, C. *Data Analysis with Mplus (Methodology in the Social Sciences)*; Guilford Press: New York, NY, USA, 2013.
34. Collins, L.M.; Lanza, S.T. *Latent Class and Latent Transition Analysis for the Social, Behavioral, and Health Sciences*; Wiley: New York, NY, USA, 2010.
35. DiStefano, C. Cluster analysis and latent class clustering techniques. In *Handbook of Developmental Research Methods*; The Guilford Press: New York, NY, USA, 2012; pp. 645–666.
36. Finney, S.J.; DiStefano, C. Non-normal and categorical data in structural equation modeling. *Struct. Equ. Model. Second Course* **2006**, *10*, 269–314.
37. Muthén, L.K.; Muthén, B.O. *Mplus User's Guide*; Muthén and Muthén: Los Angeles, CA, USA, 2017.
38. Goodman, L.A. The analysis of systems of qualitative variables when some of the variables are unobservable. Part IA modified latent structure approach. *Am. J. Sociol.* **1974**, *79*, 1179–1259. [CrossRef]
39. Elliott, M.R.; Gallo, J.J.; Ten Have, T.R.; Bogner, H.R.; Katz, I.R. Using a Bayesian latent growth curve model to identify trajectories of positive affect and negative events following myocardial infarction. *Biostatistics* **2005**, *6*, 119–143. [CrossRef] [PubMed]
40. Asparouhov, T.; Muthén, B. Using Bayesian priors for more flexible latent class analysis. In Proceedings of the 2011 Joint Statistical Meeting, Miami Beach, FL, USA, 30 July–4 August 2011; American Statistical Association: Alexandria, VA, USA, 2011.
41. Jackman, S. *Bayesian Analysis for the Social Sciences*; John Wiley & Sons: New York, NY, USA, 2009; Volume 846.
42. Silvey, S.D. *Statistical Inference*; CRC Press: Boca Raton, FL, USA, 1975; Volume 7.
43. Depaoli, S. The Latent Class Model. In *Bayesian Structural Equation Modeling*; The Guilford Press: New York, NY, USA, 2021.
44. Redner, R.A.; Walker, H.F. Mixture densities, maximum likelihood and the EM algorithm. *SIAM Rev.* **1984**, *26*, 195–239. [CrossRef]
45. Stephens, M. Dealing with label Switching in mixture models. *J. R. Stat. Soc.* **2000**, *62*, 795–809. [CrossRef]
46. Jasra, A.; Holmes, C.C.; Stephens, D.A. *Markov Chain Monte Carlo Methods and the Label Switching Problem in Bayesian Mixture Modeling*; Mathematical Statistics: Shaker Heights, OH, USA, 2005.
47. Farrar, D. *Approaches to the Label-Switching Problem of Classification Based on Partition-Space Label Invariant Visualization (Technical Report)*; Virginia Polytechnic Institute and State University: Blacksburg, VA, USA, 2006.
48. Akaike, H. *On Entropy Maximization Principle*; Krishnaiah, P.R., Ed.; Applications of Statistics; North Holland Publishing Company: Amsterdam, The Netherlands, 1977; pp. 27–47.
49. Ramaswamy, V.; Desarbo, W.S.; Reibstein, D.J. An empirical pooling approach for estimating marketing mix elasticities with PIMS data. *Mark. Sci.* **1993**, *12*, 103–124. [CrossRef]

50. Kroese, D.P.; Brereton, T.; Taimre, T.; Botev, Z.I. Why the Monte Carlo method is so important today. *WIREs Comput. Stat.* **2014**, *6*, 386–392. [CrossRef]
51. Gagniuc, P.A. *Markov Chains: From Theory to Implementation and Experimentation*; John Wiley & Sons: Hoboken, NJ, USA, 2017; pp. 1–235, ISBN 978-1-119-38755-8.
52. Sawilowsky, S.; Fahoome, G.C. *Statistics via Monte Carlo Simulation with Fortran*; JMASM: Rochester Hills, MI, USA, 2003; ISBN 978-0-9740236-0-1.

mathematics

Article

Bayesian Spatial Split-Population Survival Model with Applications to Democratic Regime Failure and Civil War Recurrence

Minnie M. Joo [1,*], Brandon Bolte [2], Nguyen Huynh [2] and Bumba Mukherjee [2]

[1] Department of Political Science, University of Massachusetts Lowell, Lowell, MA 01854, USA
[2] Department of Political Science, Pennsylvania State University, University Park, PA 16802, USA; blb72@psu.edu (B.B.); nkh8@psu.edu (N.H.); bumba.mukherjee@psu.edu (B.M.)
* Correspondence: minhyung_joo@uml.edu

Abstract: The underlying risk factors associated with the duration and termination of biological, sociological, economic, or political processes often exhibit spatial clustering. However, existing nonspatial survival models, including those that account for "immune" and "at-risk" subpopulations, assume that these baseline risks are spatially independent, leading to inaccurate inferences in split-population survival settings. In this paper, we develop a Bayesian spatial split-population survival model that addresses these methodological challenges by accounting for spatial autocorrelation among units in terms of their probability of becoming immune and their survival rates. Monte Carlo experiments demonstrate that, unlike nonspatial models, this spatial model provides accurate parameter estimates in the presence of spatial autocorrelation. Applying our spatial model to data from published studies on authoritarian reversals and civil war recurrence reveals that accounting for spatial autocorrelation in split-population models leads to new empirical insights, reflecting the need to theoretically and statistically account for space and non-failure inflation in applied research.

Keywords: Bayesian inference; estimation in survival analysis and censored data; spatial autocorrelation; split-population models; Monte Carlo; democratic survival; civil wars

MSC: 62F15; 62N02; 62H11; 91D25

Citation: Joo, M.M.; Bolte, B.; Huynh, N.; Mukherjee, B. Bayesian Spatial Split-Population Survival Model with Applications to Democratic Regime Failure and Civil War Recurrence. *Mathematics* **2023**, *11*, 1886. https://doi.org/10.3390/math11081886

Academic Editor: Diana Mindrila

Received: 16 March 2023
Revised: 11 April 2023
Accepted: 13 April 2023
Published: 16 April 2023

1. Introduction

Originally used to study human survival rates following the onset of a disease or the administration of medical treatment, parametric and semi-parametric tools for modeling time-to-event data or "survival times" have been used to study innumerable biological, industrial, psychological, social, and political phenomena. However, two common types of heterogeneity in the data generation process (d.g.p.) of many time-to-event applications violate the core assumptions of conventional survival models. The first is the presence of non-failure cases resulting from "immunity" to a failure event or being "cured" from that event due to some treatment. Cases that will never experience the event of interest violate the assumption that all right-censored observations eventually experience the failure event even if the failure is not observed. The frequent need to relax this assumption in applied settings has given rise to a class of split-population (SP) survival models that first estimate the probability of being immune or at risk of experiencing the event and subsequently estimate the time until that event occurs, conditional upon not being immune to the event. In other words, SP survival models do not assume that every observation will eventually experience the event. "Instead, the model splits the population into two groups—one that will experience the event and one that will not" [1] (p. 148). The probability of a case being immune to the event is estimated in the first (split) stage as a binary process modeled with a specified set of covariates, then the survival stage is modeled with a specified baseline function representing the time until those cases at risk of experiencing the event actually do so, again conditional upon covariates.

These tools have been useful for modeling a wide range of phenomena, including oncological studies of the survival of breast cancer patients [2] and melanoma relapse [3], the occurrence of interstate war [4], susceptibility to and mortality from parasitic infection among river salmon [5], and criminal recidivism [6]. SP survival models themselves have been extended to incorporate independent and identically distributed (i.i.d.) frailties [7], to account for random right-censoring [8], to address misclassified failures [9], and to account for "triadic duration" independence [10]; however, existing formulations often ignore the effects of spatial clustering among units, at least in the first stage [11].

The spatial clustering of common unobserved characteristics among units that may affect their baseline risk of experiencing a failure event violates a second core assumption of conventional survival models, namely, that units are conditionally independent. Such spatial autocorrelation differs from spatial dependence in that it cannot be accounted for with i.i.d. frailty terms or spatial lags [12,13]. Spatially weighted frailties have been appropriately incorporated into conventional survival models via Bayesian estimation [12,14–16]. However, existing spatial survival models cannot accommodate any heterogenous mixture of immune and at-risk populations, nor have they included "covariates and spatial random effects as regressors in the cure rate portion of the model, instead of just the log-relative risk portion" [11] (p. 274).

In this article, we develop a parametric Bayesian spatial split-population (SP) survival model that can incorporate time-varying covariates. Rather than adopting the frequentist maximum likelihood estimation approach to finding parameter values, Bayesian estimation more flexibly utilizes Bayes' Rule to estimate parameters based on iterated updates to pre-specified priors that define baseline expectations about the probability distribution of the phenomenon of interest. This method is particularly useful for estimating parameters in a split-population model, which can place high demands on the observed data.

Similar to a conventional split-population survival model, our approach consists of a split-stage equation that estimates the probability of a unit being immune from a failure event and a second-stage equation that estimates the survival probability conditioned upon the subject being at risk of failure. In our spatial frailty model, however, each of the two equations may include spatially autocorrelated frailties with a joint distribution that is interpretable in a spatial context. This allows analysts to eschew the assumption that the frailties themselves are i.i.d. The hierarchical model leverages the flexibility of Bayesian estimation using a Markov Chain Monte Carlo (MCMC) sampling algorithm whereby the frailties of "neighboring" units and any spatial autocorrelations among them are incorporated into each equation via a conditionally autoregressive (CAR) prior. MCMC algorithms are tools for estimating analytically complex probability densities by randomly and repeatedly drawing samples from a distribution (Monte Carlo) such that each sample depends on the prior one but not those before it (Markov Chain). Thus, unlike regular SP survival models, the Bayesian spatial SP survival model can account for spatial autocorrelation in a unit's propensity for being at risk of experiencing an event as well as the time it takes for that event to occur.

After presenting the Bayesian spatial SP survival model and describing the slice-sampling algorithm used for estimation in a Bayesian framework, we illustrate its properties through a series of Monte Carlo experiments. The results reveal that (i) the Bayesian spatial SP model reliably retrieves its true parameter values regardless of the size of the immune fraction or degree of spatial autocorrelation and that (ii) nonspatial models produce biased parameter estimates if the true d.g.p. includes spatial clustering. We then apply the spatial SP survival model to replication data from two prominent studies in political science. The first is a previous application of a nonspatial split-population survival model used to study whether democratic countries consolidate, survive, or revert to a dictatorship [17]. The second is a conventional survival analysis of whether civil wars are more or less likely to recur after terminating [18]. We discuss theoretical reasons to expect spatial clustering in each of these contexts, then demonstrate empirically that spatial clustering does indeed exist in the replication data. Then, we show that applying our Bayesian spatial SP survival model to these data significantly alters the previously reported results. In light of the evidence of spatial autocorrelation in each application, the new results from our spatial model indicate that faulty inferences can result from ignoring spatial heterogeneity when modeling survival processes.

2. (Spatial) Split-Population Survival Model

2.1. Model Development

Supposing that $i = \{1, 2, \ldots, N\}$ are the units in the survival data, we define $f(t)$ as the probability density function and $F(t)$ as the cumulative distribution function. Thus, $S(t) = 1 - F(t)$ is the survival distribution and $h(t) = \frac{f(t)}{S(t)}$ is the hazard rate. The general likelihood of the conventional survival model is proportional to

$$L = \prod_{i=1}^{N} [f(t_i)]^{\widetilde{C}_i} [S(t_i)]^{1-\widetilde{C}_i}, \tag{1}$$

where $\widetilde{C}_i = 1$ are the units that fail and $\widetilde{C}_i = 0)$ are the units that do not fail, and as such are "right-censored." Two subpopulations can potentially exist in the survival data that researchers use for empirical analysis: an "at-risk" fraction of cases that can fail, and an "immune" fraction of cases that will not fail, implying that units in this fraction do not experience the event of interest [19–21]. These two subpopulations are accounted for in split-population survival models (with or without unit-specific frailties) by estimating the probability of a given unit being in the immune fraction and the influence of covariates on the at-risk fraction's hazard rate [19,22,23].

The split-population survival model for a duration t that splits the sample in the manner described above is constructed as follows. First, we define $\alpha_i = Pr(Y_i = 1)$ as the probability of units entering the immune fraction, which can be estimated via logit:

$$\alpha_i = \frac{\exp(\mathbf{Z}_i \gamma + V_i)}{1 + \exp(\mathbf{Z}_i \gamma + V_i)} \tag{2}$$

where $\mathbf{Z}_i$ are p2-dimensional covariates, γ is the corresponding parameter vector in $\mathbb{R}^{p_2}$, and $V_i \sim N(0, \sigma^2)$ are the *nonspatial* unit-specific frailties (random effects), which are assumed to be independent and identically distributed (i.i.d.). The *nonspatial* i.i.d. unit-specific frailties V_i in the model's split-stage accounts for unobserved heterogeneity on α_i while being independent of other random effects. In the split-population model's survival stage, however, $W_i \sim N(0, \sigma^2)$ denotes the *nonspatial* i.i.d. unit-specific frailties. As such, these nonspatial frailties capture unobserved factors that potentially influence the units' distinct risks of experiencing the event of interest. Hence, the proportional hazards of the split-population survival model with nonspatial unit-specific i.i.d. frailties are

$$h(t_i | \mathbf{X}_i \boldsymbol{\beta}, W_i) = h_0(t_i) \omega_i \exp(\mathbf{X}_i \boldsymbol{\beta}) = h_0(t_i) \exp(\mathbf{X}_i \boldsymbol{\beta} + W_i), \tag{3}$$

where $h_0(t_i)$ is the baseline hazard (which can be Weibull, log-logistic, or log-normal distributed), $\log \omega_i = W_i$, $\mathbf{X}_i$ represents the p1-dimensional covariates, and $\boldsymbol{\beta}$ is the corresponding parameter vector in $\mathbb{R}^{p_1}$. As discussed below, while we use the Weibull distribution in the presented Monte Carlo experiments and applications, the experiments perform similarly when using either of the other two parametric distributions.

Suppose that we need to incorporate time-varying covariates in our split population. Let $t0$ be the unique "entry time" and let t be the "exit time" for each period. Suppose that j is the beginning of the time period. Then, each unit i's elapsed time from inception until (i) j can be denoted as $t0_{ij}$ and (ii) the end of period j is t_{ij}. In this case, $\tilde{C}_{ij} = 0$ implies that the observation can be censored, while $\tilde{C}_{ij} = 1$ indicates that the observation has failed at t_{ij}. The probability of survival until period j is now $S_i(t0) = 1 - F(t0)$, where

$F(t0) = \int_0^{t0} f(t0)$. In this case, both subpopulations contribute to the log-likelihood of the split-population survival model with nonspatial i.i.d. frailties as follows: $S(t0) = 1 - F(t0)$ (where $F(t0) = \int_0^{t0} f(u)du$) is the probability of survival until j. Thus, the log-likelihood of the split-population survival model with nonspatial i.i.d. frailties is proportional to

$$\ln L = \sum_{i=1}^{N} \left\{ \widetilde{C}_{ij} \ln \left[(1 - \alpha_{ij}) \frac{f(t_{ij}|\mathbf{X}_{ij}\boldsymbol{\beta}, W_i)}{S(t0_{ij}|\mathbf{X}_{ij}\boldsymbol{\beta}, W_i)} \right] + (1 - \widetilde{C}_{ij}) \ln \left[\alpha_i + (1 - \alpha_i) \frac{S(t_{ij}|\mathbf{X}_{ij}\boldsymbol{\beta}, W_i)}{S(t0_{ij}|\mathbf{X}_{ij}\boldsymbol{\beta}, W_i)} \right] \right\} \tag{4}$$

where $\alpha_{ij} = \frac{\exp(\mathbf{Z}_{ij}\gamma + V_i)}{1 + \exp(\mathbf{Z}_{ij}\gamma + V_i)}$ is the split-stage equation; the model's survival stage estimates the effect of covariates $\mathbf{X}_{ij}$ on the probability of survival conditional on each unit being at-risk along with the baseline hazard. Here, V_i and W_i are the nonspatial i.i.d. unit-specific frailties in the cure model's split and survival stages, respectively. If $V_i = W_i = 0$, then Equation (4) reduces to the log-likelihood of the nonspatial "pooled" split-population survival model (*without* unit-specific frailties) with time-varying covariates [22,24]. Suppose, however, that unobserved unit-specific heterogeneity influences the units' survival time or probability of entering the immune fraction (or both). Unobserved heterogeneity of this sort is addressed by incorporating the i.i.d. split-stage and survival-stage frailty terms (V_i and W_i) into the split-population survival model. In a Bayesian framework, the following exchangeable normal prior is employed to assess these frailties in each stage of the model:

$$W_i \sim N(0, 1/\tau) \text{ and } V_i \sim N(0, 1/\tau) \tag{5}$$

where τ is the precision parameter and each unit is specified as exchangeable to generate the prior [11,12,25].

If researchers believe that the effect of each unit-specific frailty on the unit's risk-propensity or probability entering the immune fraction is independent of neighboring units' frailty effects, then the nonspatial split-population survival model should be estimated with i.i.d. unit-specific frailties. It is possible, however, for the frailties to exhibit spatial heterogeneity, meaning that each unit's propensity to be in the immune fraction as well as its survival time is influenced by unobserved factors among its neighboring units. Spatial weights can be assigned to the unit-specific frailties in the split-population survival's split and survival stage to model this spatial autocorrelation. These spatially weighted frailties can then be incorporated via the conditionally autoregressive (CAR) approach previously developed in [26].

In the Bayesian split-population survival model, the CAR prior accounts for spatial autocorrelation in the frailties by allowing these frailities to be autocorrelated across, e.g., geographically adjacent units, where "adjacency" can be defined by the researcher based on the context. More specifically, spatial data are often represented by a lattice, in which the spatial surface is divided into a grid of units that, depending on the empirical context, can be counties, districts, countries, or other areal units. The spatially weighted frailties are then incorporated via the CAR prior by defining the relevant spatial relationship among all geographically adjacent units in an adjacency matrix $\mathbf{A}$, where each element is denoted as $a_{ii'}$.

Note that $a_{ii'} = 1$ in $\mathbf{A}$ if units i and i' are "neighbors". If i and i' are not neighbors, then $a_{ii'} = 0$. The assignment of spatial weights is incorporated into the CAR prior in order to model spatially autocorrelated frailties between adjacent units. After doing this, the frailties V_i and W_i are collected into the vectors $\mathbf{V} = \{V_1, \ldots, V_N\}$ and $\mathbf{W} = \{W_1, \ldots, W_N\}$, respectively. This facilitates the use of separate CAR priors for $\mathbf{V}$ and $\mathbf{W}$, which in turn produces the following CAR model structure:

$$\mathbf{V}|\lambda \sim \text{CAR}(\lambda) \text{ and } \mathbf{W}|\lambda \sim \text{CAR}(\lambda) \tag{6}$$

where λ is the precision parameter [11,26]. The CAR(λ) prior for $\mathbf{V}$ and $\mathbf{W}$ has a joint distribution, which is formally characterized in [14] and is described in our paper's Supplementary Materials. The resulting conditional distributions of the spatial frailties for **V** and **W** are

$$V_i|V_{i'\neq i} \sim N(\overline{V_i}, 1/(\lambda m_i)), \quad W_i|W_{i'\neq i} \sim N(\overline{W_i}, 1/(\lambda m_i)), \tag{7}$$

where: $\overline{W_i} = m_i^{-1}\sum_{i'\ adj\ i} W_{i'}$; $\overline{V_i} = m_i^{-1}\sum_{i'\ adj\ i} V_{i'}$; $\overline{W_i}$ and $\overline{V_i}$ are the averages of the neighboring $W_{i'\neq i}$ and $V_{i'\neq i'}$, respectively, where i' adj i denotes that i' is adjacent to i given the matrix $\mathbf{A}$; and m_i is the number of these adjacencies [25,27]. Using this CAR prior approach, we can then define the log-likelihood of the spatial split-population survival model by substituting $\mathbf{V} = \{V_i\}$ and $\mathbf{W} = \{W_i\}$ in Equation (4), where $\alpha_{ij} = \frac{\exp(\mathbf{Z}_{ij}\gamma, \mathbf{V})}{1+\exp(\mathbf{Z}_{ij}\gamma, \mathbf{V})}$ is the split-stage equation.

The log-likelihood of the pooled ("nonfrailty"), nonspatial i.i.d. frailty, and spatial split-population survival models are compatible with any commonly employed parametric survival distribution. For our empirical applications, we assume a Weibull distribution for the baseline hazard, in which ρ denotes the shape parameter. The density, survival function, and the hazard rate for the Weibull distribution are defined in our Supplementary Material. We use the Geweke [28] convergence test and Heidelberger and Welch [29] stationarity test in our empirical applications below to assess whether the obtained Markov chains converge to their respective stationary distributions.

2.2. *Markov Chain Monte Carlo Estimation*

Following standard practice for Bayesian inference [30], we assign the Multivariate Normal (MVN) prior to $\boldsymbol{\beta} = \{\beta_1, \ldots, \beta_{p_1}\}$ and $\boldsymbol{\gamma} = \{\gamma_1, \ldots, \gamma_{p_2}\}$, and the Gamma prior for ρ with shape and scale parameters a_ρ and b_ρ for the Bayesian pooled (nonfrailty), nonspatial (i.i.d.) frailty, and spatial split-population parametric (Weibull) survival models:

$$\begin{aligned} \rho \sim \text{Gamma}(a_\rho, b_\rho), \quad \boldsymbol{\beta} \sim \text{MVN}_{p_1}(\mu_\beta, \Sigma_\beta), \quad \boldsymbol{\gamma} \sim \text{MVN}_{p_2}(\mu_\gamma, \Sigma_\gamma) \\ \Sigma_\beta \sim \text{IW}(S_\beta, \nu_\beta),\ \Sigma_\gamma \sim \text{IW}(S_\gamma, \nu_\gamma), \end{aligned} \tag{8}$$

where a_ρ, b_ρ, S_β, ν_β, S_γ, and ν_γ are the hyperparameters in (8) and μ_β and μ_γ are random variables. Here, Σ_β and Σ_γ are estimated in a Bayesian hierarchical framework using the Inverse Wishart (IW) distribution when employing the MVN (weakly informative) prior. For Bayesian MCMC estimation of the spatial split-population parametric (e.g., Weibull) model, we additionally assign the hyperprior $p(\lambda)$ to λ in light of the CAR prior approach. Specifically, we assign the Gamma hyperprior $\lambda \sim \text{Gamma}(a_\lambda, b_\lambda)$ for λ [11,12]. We specify the vague prior $(a_\lambda, b_\lambda) = (0.001, 1/0.001) = (0.001, 1000)$, as for the case of for ρ. To estimate the nonspatial frailty split-population survival model in this case, we assign the normal prior for the model's split and survival-stage frailties (V_i, W_i), and use the prior described above for the model's $\boldsymbol{\beta}$, $\boldsymbol{\gamma}$, and ρ parameters. To identify the nonspatial frailty and spatial split-population model intercepts, we impose the constraint that the frailties sum to zero, i.e., $\sum_i V_i = 0$ and $\sum_i W_i = 0$.

The joint posterior distribution of the Bayesian spatial split-population Weibull model with time-varying covariates is

$$\begin{aligned} \pi(\boldsymbol{\beta}, \boldsymbol{\gamma}, \rho, \mathbf{W}, \mathbf{V}, \lambda, \Sigma_\beta, \Sigma_\gamma | \mathbf{C}, \mathbf{X}, \mathbf{Z}, \mathbf{t}, \mathbf{t0}, \gamma) \quad \propto \quad & L(\boldsymbol{\beta}, \boldsymbol{\gamma}, \rho, \mathbf{W}, \mathbf{V} | \mathbf{C}, \mathbf{X}, \mathbf{Z}, \mathbf{t}, \mathbf{t0}) \\ & \pi(\mathbf{W}|\lambda)\pi(\mathbf{V}|\lambda)\pi(\beta|\mu_\beta, \Sigma_\beta)\pi(\gamma|\mu_\gamma, \Sigma_\gamma)\pi(\rho)\pi(\lambda)\pi(\Sigma_\beta)\pi(\Sigma_\gamma), \end{aligned} \tag{9}$$

where the likelihood $L(\boldsymbol{\beta}, \boldsymbol{\gamma}, \rho, \mathbf{W}, \mathbf{V}|\mathbf{C}, \mathbf{X}, \mathbf{Z}, \mathbf{t}, \mathbf{t0})$ is from Equation (4) with frailties V_i collected into $\mathbf{V} = \{V_1, \ldots, V_N\}$ and W_i into $\mathbf{W} = \{W_1, \ldots, W_N\}$; here, $\mathbf{C}$ represents the vector of censored observations. The density, survival function, and hazard rate for this likelihood are defined in the Supplementary Materials for the Weibull case; $\pi(\mathbf{W}|\lambda)$ and $\pi(\mathbf{V}|\lambda)$ are defined via their respective conditional distributions in Equation (9), $\pi(\beta|\mu_\beta, \Sigma_\beta)$, $\pi(\gamma|\mu_\gamma, \Sigma_\gamma)$, $\pi(\rho)$, $\pi(\Sigma_\beta)$, and $\pi(\Sigma_\gamma)$ are defined in Equation (8), and $\pi(\lambda)$ is the Gamma hyperprior for the spatial split-population parametric survival models. From (9), we can formally state the joint posterior distribution of the time-varying nonspatial frailty split-population parametric (Weibull) model by incorporating the frailties V_i and W_i defined in Equation (6) instead of $\mathbf{W}$, $\mathbf{V}$, and their respective CAR priors. The conditional posterior distributions for $\boldsymbol{\beta}, \gamma$, and ρ in the pooled (nonfrailty) parametric model with time-varying covariates are

$$\begin{aligned} P(\boldsymbol{\beta} \,|\mathbf{C}, \mathbf{X}, \mathbf{Z}, \mathbf{t}, \mathbf{t0}, \gamma, \rho) &\propto P(\mathbf{C}, \mathbf{X}, \mathbf{Z}, \mathbf{t}, \mathbf{t0}, \boldsymbol{\beta}, \gamma, \rho) \times P(\boldsymbol{\beta} \,|\Sigma_\beta), \\ P(\gamma \,|\mathbf{C}, \mathbf{X}, \mathbf{Z}, \mathbf{t}, \mathbf{t0}, \boldsymbol{\beta}, \rho) &\propto P(\mathbf{C}, \mathbf{X}, \mathbf{Z}, \mathbf{t}, \mathbf{t0}, \boldsymbol{\beta}, \gamma, \rho) \times P(\gamma \,|\Sigma_\gamma), \\ P(\rho \,|\mathbf{C}, \mathbf{X}, \mathbf{Z}, \mathbf{t}, \mathbf{t0}, \boldsymbol{\beta}, \gamma) &\propto P(\mathbf{C}, \mathbf{X}, \mathbf{Z}, \mathbf{t}, \mathbf{t0}, \boldsymbol{\beta}, \gamma, \rho) \times P(\rho \,|a_\rho, b_\rho), \end{aligned} \tag{10}$$

where $P(\mathbf{C}, \mathbf{X}, \mathbf{Z}, \mathbf{t}, \mathbf{t0}, \boldsymbol{\beta}, \gamma, \rho)$ is the likelihood obtained from Equation (4) after excluding the frailty terms and $P(\boldsymbol{\beta} \,|\Sigma_\beta)$, $P(\gamma \,|\Sigma_\gamma)$, and $P(\rho \,|a_\rho, b_\rho)$ are the priors defined in Equation (8).

The pooled (nonfrailty), nonspatial (i.i.d.) frailty, and spatial split-population survival model using the Weibull distribution can be estimated using an MCMC algorithm for Bayesian inference. To begin with, because closed form distributions for the posterior distributions of $\boldsymbol{\beta}, \gamma, \rho, \lambda, \mathbf{W}$ and $\mathbf{V}$ are not available for the spatial split-population survival model, our MCMC method's update scheme in this case incorporates Gibbs Sampling (for estimating λ), the Metropolis–Hastings algorithm (for $\mathbf{W}$ and $\mathbf{V}$ given λ), and slice-sampling [31] for updating $\boldsymbol{\beta}, \gamma, \rho$. We use Gibbs sampling for λ, as it is easier to sample from the conditional distribution (which is known) in this case and because the joint distribution is not known explicitly. We employ the Metropolis–Hastings algorithm for $\mathbf{W}$ and $\mathbf{V}$ given λ because it is difficult to sample from the conditional distribution for $\mathbf{W}$ and $\mathbf{V}$. Finally, we use slice-sampling for updating $\boldsymbol{\beta}, \gamma$ and ρ, as it requires little tuning because the slice width adapts quickly to the distribution and sampler performance. Furthermore, considering that slice-sampling draws from the posterior samples from any prior distribution as long as these distributions have a reasonable value range of parameters, this sampling algorithm provides researchers with flexibility in the choice of prior distribution. This permits the use of informative, weakly informative, or non-informative priors.

The MCMC algorithm described above proceeds as follows:

1. Choose a starting point $\beta_0, \gamma_0, \rho_0, \lambda_0$ and corresponding $\mathbf{W}_0 = \{W_1, \ldots, W_N\}$ and $\mathbf{V}_0 = \{V_1, \ldots, V_N\}$, then set $k = 0$.
2. Update $\Sigma_\beta \sim \pi(\Sigma_\beta|\beta)$, $\Sigma_\gamma \sim \pi(\Sigma_\gamma|\gamma)$, $\lambda \sim \pi(\lambda|\mathbf{W}, \mathbf{V})$ using Gibbs sampling. The closed form of the full conditional distributions for $\pi(\Sigma_\beta|\beta)$, $\pi(\Sigma_\gamma|\gamma)$, $\pi(\lambda|\mathbf{W}, \mathbf{V})$ are derived and defined in the Supplementary Materials.
3. Update $\beta \sim \pi(\beta|\mathbf{C}, \mathbf{X}, \mathbf{Z}, \mathbf{t}, \mathbf{W}, \mathbf{V}, \gamma, \rho, {}^{-}{}_{\beta}, \Sigma_\beta)$, $\gamma \sim \pi(\gamma|\mathbf{C}, \mathbf{X}, \mathbf{Z}, \mathbf{t}, \mathbf{W}, \mathbf{V}, \beta, \rho, {}^{-}{}_{\gamma}, \Sigma_\gamma)$, and $\rho \sim \pi(\rho|\mathbf{C}, \mathbf{X}, \mathbf{Z}, \mathbf{t}, \mathbf{W}, \mathbf{V}, \beta, \gamma, a_\rho, b_\rho)$ using the slice sampler with stepout and shrinkage (Neal, 2003); see the Supplementary Materials for details on performing the slice sampling operation in this step.
4. Update $\mathbf{W} \sim \pi(\mathbf{W}|\mathbf{C}, \mathbf{X}, \mathbf{Z}, \mathbf{t}, \mathbf{V}, \beta, \gamma, \rho, \lambda)$ and $\mathbf{V} \sim \pi(\mathbf{V}|\mathbf{C}, \mathbf{X}, \mathbf{Z}, \mathbf{t}, \mathbf{W}, \beta, \gamma, \rho, \lambda)$ via Metropolis–Hastings.
5. Set $k = k + 1$, then return to Step 2 and repeat for K iterations.

The MCMC algorithm for estimation of the nonspatial frailty model is similar to the steps delineated above, with the exception of the nonspatial i.i.d. frailties V_i and W_i in this model being updated via Metropolis–Hastings with the proposal variance as the conditional prior variance for these frailties. To estimate the pooled (nonfrailty) model, we use the following MCMC algorithm, as closed forms for the posterior distributions of ρ, $\boldsymbol{\beta}$, and γ are not available:

1. Choose the initial values of $\boldsymbol{\beta}, \boldsymbol{\gamma}$, and ρ, then set $m = 0$.
2. Update Σ_β and Σ_γ via Metropolis–Hastings; see the Supplementary Material for the closed form of the full conditional distributions for Σ_β and Σ_γ.
3. Update $\boldsymbol{\beta}$, $\boldsymbol{\gamma}$, and ρ using the slice sampler with stepout and shrinkage, as described in the Supplementary Materials.
4. Repeat Steps 2 and 3 until the chain converges.
5. After M iterations, summarize the parameter estimates using posterior samples.

3. Monte Carlo Simulations

We conducted three Monte Carlo (MC) experiments to compare the performance of the nonspatial SP Weibull models with and without i.i.d. frailties to our spatial split-population Weibull model. The design of our MC experiments and the results from these experiments are presented in greater detail in the Supplementary Materials. We focus on the Weibull case here, as the empirical applications below use the Weibull survival distribution; however, our Monte Carlo simulation results hold for other parametric (e.g., log-logistic) distributions as well.

More specifically, our MC experiments simulate a split-population Weibull distributed outcome variable that exhibits spatial autocorrelation across neighboring units in each stage. For all experiments, we consider sample sizes of $N = 100, 400, 1000, 1500$, and 2000. Note that $N = 100$ and $N = 400$ respectively correspond to a small and moderate sample size, while $N = 1000$, 1500, and 2000 represent a relatively larger sample. For each model in our MC experiments, we include one survival-stage covariate $\mathbf{x}_1$and two split-stage covariates $\mathbf{z}_1$ and $\mathbf{z}_2 = \mathbf{x}_1$, as the same covariate may be included in both stages. We incorporate information about the spatial relationship between units in our simulated data via an adjacency matrix $\mathbf{A}$. To generate $\mathbf{A}$, we consider a hypothetical space with five areal units (e.g., countries), with each unit having at least one adjacent "neighbor." This spatial relational information is then incorporated into the simulated data generation process, which follows an SP Weibull distribution (see Supplementary Materials for details).

Next, recalling that the split-stage equation in the spatial split-population survival model is provided by a binary response function that captures the effect of covariates $\mathbf{Z}_i$ and the associated parameter vector γ on the probability of units entering the immune fraction (α), we have a case in which the more likely a greater share of units is to enter the immune fraction, the higher the immune fraction level. Hence, for the MC experiments, we set the true γ values that affect the immune fraction (via α)—calculated as the mean value of the binary response function $\alpha_i = \frac{\exp(\mathbf{Z}_i\gamma + V_i)}{1+\exp(\mathbf{Z}_i\gamma + V_i)}$) for all i in our N-sample data—using the pre-set true γ value and the randomly generated variables $\mathbf{Z}_i$ (as well as the $\mathbf{V}$ spatial frailties for the spatial d.g.p. This permits us to adjust the immune fraction level in a way that is consistent with the model's split-stage. Finally, for each experiment, we use 500 iterations in the MCMC, 100 burn-ins, and a thinning of 1, and assess the convergence of the Markov chain via trace-plots and the Geweke convergence test.

Using these experimental conditions, we now turn to assessing our nonspatial and spatial models of interest for three experiments. In the first MC experiment, we compare the performance of our nonspatial and spatial split-population Weibull models when the fraction of the immune subpopulation is fixed at 25% and the proportion of units that share spatial frailties is held at 40%. The results from this MC experiment reveal that our spatial split-population Weibull model outperforms both the nonspatial split-population Weibull models in retrieving the true theoretical values of the split-stage (γ) and survival-stage (β) covariates (Figure 1) along with the spatial frailties in both stages (Figure S1, Supplementary Material) for small and moderate sample sizes as well as for the relatively larger sample sizes listed above. Thus, the spatial split-population Weibull model should be favored over the nonspatial split-population models even when a low number of observations have spatially-dependent frailties in a split-population survival framework. The trace plots show stability, and the models pass the Geweke convergence test.

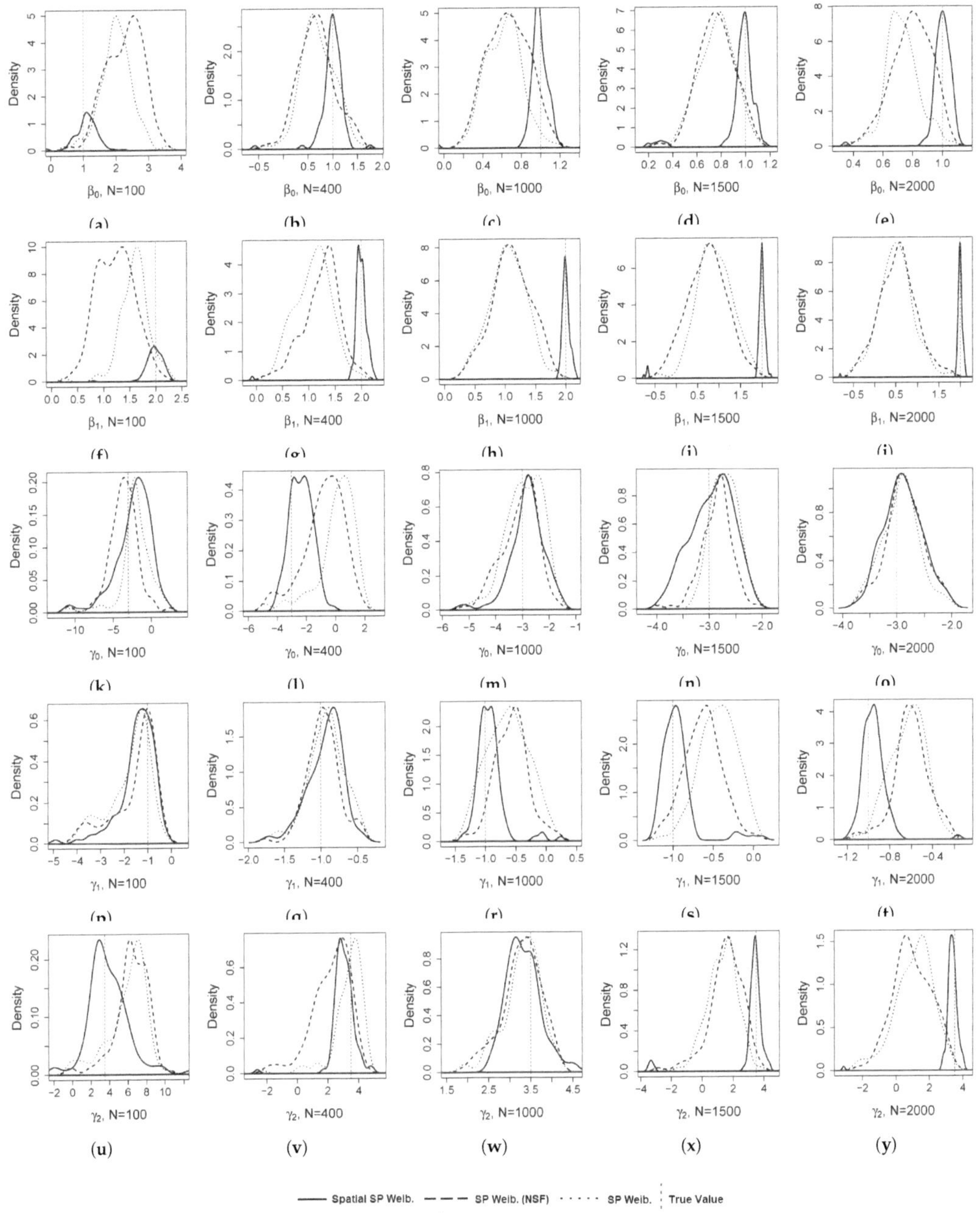

Figure 1. MC Experiment 1 β,γ densities for SP Weibull, NS Frailty Weibull, and Spatial SP Weibull models for: **(a)** $\hat{\beta_0}, N = 100$, **(b)** $\hat{\beta_0}, N = 400$, **(c)** $\hat{\beta_0}, N = 1000$, **(d)** $\hat{\beta_0}, N = 1500$, **(e)** $\hat{\beta_0}, N = 2000$, **(f)** $\hat{\beta_1}, N = 100$, **(g)** $\hat{\beta_1}, N = 400$, **(h)** $\hat{\beta_1}, N = 1000$, **(i)** $\hat{\beta_1}, N = 1500$, **(j)** $\hat{\beta_1}, N = 2000$, **(k)** $\hat{\gamma_0}, N = 100$, **(l)** $\hat{\gamma_0}, N = 400$, **(m)** $\hat{\gamma_0}, N = 1000$, **(n)** $\hat{\gamma_0}, N = 1500$, **(o)** $\hat{\gamma_0}, N = 2000$, **(p)** $\hat{\gamma_1}, N = 100$, **(q)** $\hat{\gamma_1}, N = 400$, **(r)** $\hat{\gamma_1}, N = 1000$, **(s)** $\hat{\gamma_1}, N = 1500$, **(t)** $\hat{\gamma_1}, N = 2000$, **(u)** $\hat{\gamma_2}, N = 100$, **(v)** $\hat{\gamma_2}, N = 400$, **(w)** $\hat{\gamma_2}, N = 1000$, **(x)** $\hat{\gamma_2}, N = 1500$, **(y)** $\hat{\gamma_2}, N = 2000$.

In the second MC experiment, we compare the nonspatial and spatial split-population Weibull models' performance when the immune fraction of the simulated split-population Weibull-distributed outcome variable remains at 25% and the proportion of units that share spatial frailties is 30%, 40%, 60%, and 80%. Results from Figure 2, which displays the mean RMSEs for $\hat{\beta}$ and $\hat{\gamma}$ for $N = 100$, 400, 1000, 1500, and 2000, demonstrate that the spatial split-population Weibull model substantially outperforms the nonspatial split-population Weibull models at all levels of spatial autocorrelation; the mean RMSEs for the spatial SP Weibull models are always negligible (close to 0) for all the sample sizes that are examined in the experiment, while those of the two nonspatial SP Weibull models are critically high, indicating a considerable level of bias. Moreover, the results show that the nonspatial split-population Weibull models' split and survival-stage covariates and nonspatial frailties exhibit deteriorating coverage and lower efficiency at all levels of spatial autocorrelation. By contrast, the spatial split-population Weibull model recovers the true theoretical values of the split and survival-stage covariates and the spatial frailties in both stages of the model with more accuracy, coverage, and efficiency. Therefore, for the various sample sizes considered here, the spatial split-population Weibull model outperforms the nonspatial models at all levels of spatial autocorrelation. Again, the trace plots show stability, and the models pass the Geweke convergence test.

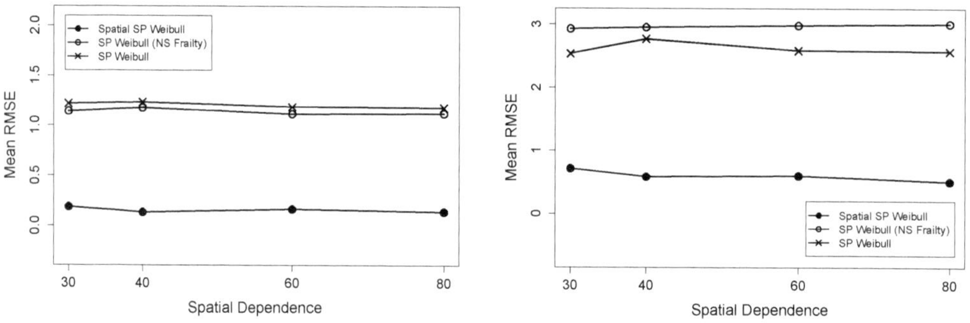

(a) (b)

Figure 2. MC Experiment 2 mean RMSE comparison between SP Weibull, NS Frailty SP Weibull, and spatial SP Weibull models for (**a**) $\hat{\beta}$ coefficients and (**b**) $\hat{\gamma}$ coefficients with spatial dependence changing from 30% to 80% of the data.

The third MC experiment applies the d.g.p. from Experiment 1, where the share of units with spatially dependent frailties is held at 40% and the size of the immune fraction is varied from 25% to 33%, 40%, 48%, and finally 60%. Figure 3a,b, which additionally shows the mean RMSEs for $\hat{\beta}$ and $\hat{\gamma}$ with $N = 100$, 400, 1000, 1500, and 2000, reveals that the split and survival stage results strongly favor the spatial split-population Weibull model over the other two nonspatial split-population Weibull models for all the sample sizes examine here and at all immune fraction levels; the mean RMSEs are close to 0, and are negligible in the spatial SP Weibull models, while those for the two nonspatial SP Weibull models are considerably higher. This indicates that the $\hat{\beta}$ and $\hat{\gamma}$ parameters are biased in the latter models. Furthermore, the retrieved values of the spatial frailties from the spatial split-population Weibull model rapidly converge to their true values with high coverage probabilities. Thus, if the true d.g.p. is split-population Weibull in which the unit-specific frailties exhibit spatial autocorrelation, then the spatial split-population Weibull should be estimated when the size of the immune fraction is 25% or above for small, moderate, and even relatively large sample sizes.

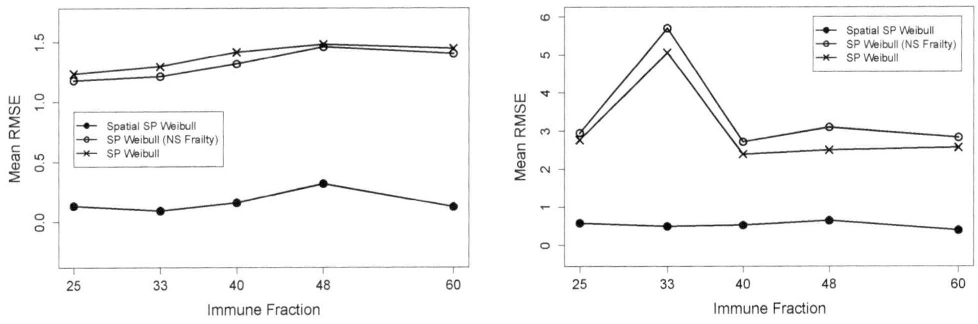

(**a**) (**b**)

Figure 3. MC Experiment 3 mean RMSE comparison between SP Weibull, NS Frailty SP Weibull, and spatial SP Weibull models for (**a**) $\hat{\beta}$ coefficients and (**b**) $\hat{\gamma}$ coefficients with the immune fraction changing from 25% to 60% of the data.

We conducted three additional MC experiments that we do not report here in order to save space; these are briefly presented in the Supplementary Materials (see Figure S2 and Tables S3 and S4) . Briefly, in one set of MC experiments (Experiment 4) we increased both the immune fraction and the share of units with spatially dependent frailties. In another experiment (Experiment 5), we re-evaluated our primary MC results using an alternative prior. In the third set of MC experiments (Experiment 6), we compared our model's performance to a Bayesian spatial SP model that incorporates spatial frailties in just the survival stage. For each of these additional experiments, we set $sims = 100$ and evaluated model performance for a variety of different sample sizes. These additional MC experiments revealed that, unlike the nonspatial models, both the retrieved values of the split and survival-stage parameters and the spatial frailties from the spatial SP Weibull model converge to their true values with high coverage probabilities when the true d.g.p. is SP Weibull in which the unit-specific frailties exhibit spatial autocorrelation.

4. Empirical Applications

Our MC experiments suggest that if spatial autocorrelation exists in the true d.g.p., then failing to account for it leads to faulty inferences. In applied settings, this means that if there are a priori theoretical reasons to suspect that the survival times and immune fractions of interest are spatially clustered, where our spatial frailty approach is superior to nonfrailty or i.i.d. frailty split-population models. Below, we apply our Bayesian spatial split-population Weibull model to survival data from two published studies in Political Science about (1) democratic regime survival [17] and (2) the duration of post-civil war peace (i.e., before civil war recurs) [18]. In both cases, we discuss theoretical reasons and empirical evidence suggesting that these processes exhibit spatial clustering, then compare results from our Bayesian spatial frailty model to those of nonspatial specifications.

4.1. Democratic Consolidation and Survival

Comparative political scientists often conceptually distinguish between *transitional* democracies, which can revert to authoritarian rule, and *consolidated* democracies, in which the "democratic regime becomes sufficiently durable that democratic breakdown is no longer likely" [32] (p. 743). Previous empirical analyses of democratic *consolidation*, which typically employ discrete choice models, consistently find that wealth (measured by GDP per capita) has a positive and highly significant effect on the probability of democratic consolidation [32,33]. On the other hand, presidential systems (as opposed to parliamentary systems) have a negative but weakly significant or insignificant impact on democratic consolidation, while the association between past authoritarian institutions and democratic consolidation is inconsistent [32,34]. Research on the *survival* of democratic regimes (which often uses conventional parametric survival models) usually finds that, unlike presidential systems, economic growth and parliamentary systems help democracies to endure [35,36], while democracies preceded by military rule revert to dictatorships more quickly [34,37]. The most consistent finding, however, is that GDP per capita has a strong positive influence on survival of democracy, leading a number of scholars to infer that democratic survival "increases monotonically with per capita income" and then endures *indefinitely* after GDP per capita reaches approximately USD 6000 [38] (p. 165).

Although these insights are important, Ref. [17] emphasizes that by employing standard duration or discrete choice models these studies assume that *all* democracies face the same baseline risk of reversal to authoritarianism. This assumption is unjustified, as the population of democracies includes "at-risk" transitional democracies along with an "immune fraction" of fully consolidated democracies for which the risk of authoritarian reversal is negligible. Hence, the observed survival of democracy results from two separate processes: "democracies that survive because they are consolidated and those democracies that are not consolidated but survive because of some favorable circumstances" [17] (p. 153). Considering these two subpopulations, Ref. [17] then re-examines extant findings about democratic consolidation and survival by estimating via MLE parametric (Weibull) nonspatial split-population survival models (with and without i.i.d. frailties) on a dataset of democratic spells across 133 countries between 1789 and 2001. All right-censored observations in his data are either consolidated or transitional democracies that have not yet reverted to authoritarian rule. Thus, the split-stage in his SP survival model estimates the probability of democratic consolidation (62% of his cases are right-censored), while the survival stage estimates the duration of democracy among cases that eventually experience an authoritarian reversal. He incorporates seven covariates in both stages: *GDP per capita*, *GDP growth*, *Presidential* and *Parliamentary* systems, and previous *Military*, *Civilian*, and *Monarchical* dictatorships.

Briefly, [17] found that GDP per capita has a positive and statistically significant effect on the probability of democratic consolidation in the split-stage, while presidential systems and democracies preceded by military dictatorships are less likely to consolidate. Economic *growth* helps "transitional" democracies to survive longer, though there is no statistically significant relationship between GDP per capita and democratic survival among these at-risk regimes. Most other covariates are statistically insignificant, though the insignificant coefficient for presidential systems is notably positive in the survival stage.

Despite this important contribution to the extensive literature on the durability of democracies, the nonspatial split-population survival models employed in the original study assumed that *neither* the likelihood of democratic consolidation *nor* the prospects for democratic survival exhibit spatial autocorrelation. This assumption may be untenable, as democracies tend to cluster in space. Indeed, "since 1815, the probability that a randomly chosen country will be a democracy is about 0.75 if the majority of its neighbors are democracies, but only 0.14 if the majority of its neighbors are nondemocracies" [39] (p. 916). Research in political science has found that geographical proximity to (consolidated) democracies not only encourages democratic transition in authoritarian regimes, it increases the odds of consolidation and survival of nascent democracies, as stable democracies create a "regional production chain" of democratic institutions, practices, and norms that are conducive to democracy [40] (p. 25; and see [32,39,41]). Thus, democratic clustering reinforces democratic norms, making it costly for elites to engage in democratic backsliding [41]. In this way, democratic neighborhood effects may have important latent influences on both democratic reversal and democratic survival.

Although there are clear theoretical reasons to expect spatial autocorrelation in the d.g.p. of democratic regime survival and consolidation, we can additionally use common tools in spatial statistics to diagnose the extent of spatial clustering in each year of the data. Specifically, we conduct two pre-estimation tests by calculating (1) the join count and (2) the Global Moran's I statistic for each cross-section of democracies in the data. The join count is a measure of the extent to which the number of observed areal units that are *adjacent* and *of the same category* is greater than or less than what is expected if the spatial distribution of those categories were random [42]. In general, in a setting with two discrete categories A and B, the join count test statistic is

$$Z(AB) = \frac{AB - E(AB)}{\sqrt{\sigma^2_{AB}}}, \tag{11}$$

where AB and $E(AB)$ are the observed and expected counts of adjacent units in categories A and B, respectively, and $\sigma^2_{AB} = E(AB^2) - E(AB)^2$. Positive statistics indicate spatial dispersion (units of the same category are further from each other than expected by chance), while negative statistics indicate positive spatial clustering (units of the same category are more likely to be adjacent than what is expected by chance.

For this application, we construct a separate cross-sectional adjacency matrix with elements $a_{ii'}$ for each year in the data, wherein proximate pairs of democratic countries (within 800 km of each other) are assigned a weight of 1 ($a_{ii'} = 1$). Our outcome of interest for the join count analysis is whether a country is identified as being "at risk" of democratic reversal in the original dataset.

In addition, we use the Global Moran's I statistic to assess the number of years that democratic regimes survive as a group of clusters in space. Global Moran's I is an inferential statistic that measures the direction and degree of spatial clustering in continuous data [43]. Positive statistics indicate positive spatial clustering of similar values of the continuous variable of interest, while negative statistics indicate that dissimilar values are more likely to be proximate than if they were distributed randomly in space. Using the same adjacency matrix described above, we use the Global Moran's I to evaluate whether democracies that have survived for similar periods of time exhibit spatial heterogeneity (clustering or dispersion).

We report the join count and Moran's I tests in detail in the Appendix A (Figure A1) provided at the end of the paper. Briefly, the results clearly indicate significant spatial clustering in both the probability of democratic reversal and the survival rates of democracies, particularly in the post-World War 2 period. Next, we replicate the above analysis using our Bayesian spatial SP survival model in order to compare our results to the original nonspatial models with and without i.i.d. frailties. Because the original analysis of these data used maximum likelihood estimation, we used the same for the nonspatial frailty and nonfrailty models in order to exactly replicate the previous results. Our spatial SP Weibull model incorporates spatially-weighted frailties across neighboring democracies via the adjacency matrix **A**. We construct a matrix **A** with elements $a_{ii'}$ such that $a_{ii'} = 1$ for each year if the capital of country i is less than 800 km from the capital of country i' and $a_{ii'} = 0$ if countries i and i' are greater than 800 km from each other. Using geographic proximity as the spatial relationship of interest is appropriate, as it allows the frailties to be correlated with those of neighboring democracies rather than assuming spatial independence even within the same regions. Considering our Bayesian MCMC estimation approach, we incorporate the spatial information in **A** by employing separate CAR priors for the frailty terms vector **V** (split-stage) and **W** (survival-stage), which implies a CAR structure of $\mathbf{V}|\lambda \sim \text{CAR}(\lambda)$ and $\mathbf{W}|\lambda \sim \text{CAR}(\lambda)$. The spatial SP Weibull model is estimated based on the sample from [17] using the MVN prior and our MCMC algorithm described earlier and assigning the Gamma hyperprior for λ. Here, we use the hyperparameters $a = 1, b = 1$, $S_\beta = I_{p1}$, $S_\gamma = I_{p2}$, $\nu_\beta = p1$, $\nu_\gamma = p2$. Recall that Σ_β is the variance of the MVN prior of the vector β for p1-dimensional survival stage covariates and that Σ_γ is the MVN's prior of the vector γ for p2-dimensional split-stage covariates. Hence, when we employ the Inverse Wishart (IW) distribution to estimate both Σ_β, in which ν_β is the hyperparameter, and Σ_γ, in which ν_γ is the hyperparameter, we adopt the values p1 for ν_β and p2 for ν_γ. Finally, $\lambda \sim \text{Gamma}(a_\lambda, b_\lambda)$ with a vague prior $(a_\lambda, b_\lambda) = (0.001, 1/0.001)$. Our Bayesian SP Weibull model results are based on a set of 50,000 iterations after 4000 burn-in scans and thinning of 10.

We begin our analysis of the split-stage results by examining choropleth maps (Figure 4a,b) that illustrate the posterior means of the spatial frailties obtained from the spatial SP Weibull model. The split-stage map (Figure 4a) reveals that there are distinct spatial bands in the frailties, which range from −0.725 to 0.716 with a corresponding standard deviation of 0.31. The spatial patterns in the map suggest that there is strong spatial clustering in the underlying factors linked to democratic consolidation, as states with a higher baseline risk for democratic consolidation are in similar geographic neighborhoods, whereas those with lower propensities are clustered in *separate* regions.

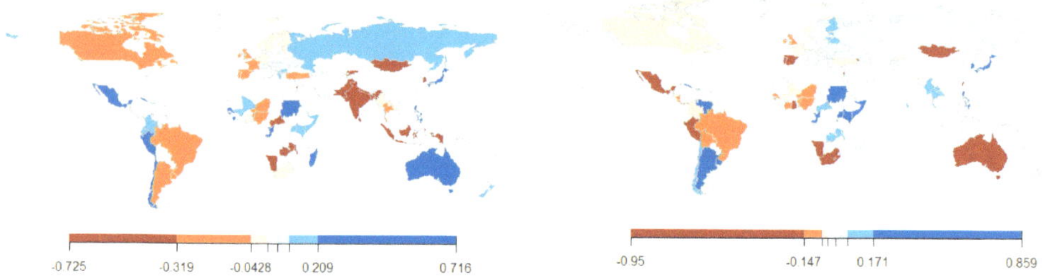

(**a**) (**b**)

Figure 4. Democratic survival application spatial frailty maps: (**a**) depicts the posterior mean estimates of **V** (split−stage spatial frailties) and (**b**) depicts the posterior mean estimates of **W** (survival−stage spatial frailties).

Figure 5 displays the results for each covariate from the replicated models and our spatial frailty model. For the nonspatial models, the points represent coefficient estimates and the bars represent 90% confidence intervals. For admittedly rough comparability purposes, the dots in the figure represent posterior means for the spatial frailty models, while the bars represent symmetric 90% credible intervals. Although these are certainly not perfect comparisons, our goal here is simply to illustrate the applicability of our model and the way in which accounting for spatial autocorrelation can affect inferences.

The dot-whisker plots in Figure 5 show that while Svolik's results for *GDP growth*, *Monarchy*, and *Civilian* are similar across the nonfrailty and nonspatial frailty SP Weibull models, the differences in the split-stage results between these nonspatial models and our spatial frailty model are more pronounced. For instance, the *Presidential* and *Military* covariates are each negative and highly significant in the nonspatial SP Weibull models with and without i.i.d. frailties. By contrast, the negative estimates for *Presidential* and *Military* are each highly unreliable in the spatial SP Weibull model's split-stage equation. Thus, after we explicitly account for spatial autocorrelation in the split-stage of the SP survival model, the relationship between each of the covariates noted above and the probability of democratic consolidation is considerably attenuated. In nonspatial SP Weibull models from [17], the *Parliamentary* dummy's split-stage estimate is positive, albeit insignificant. However, the estimate of *Parliamentary* in the spatial SP Weibull model's split-stage is *negative* (though not reliably so). Thus, the results from our spatial model raises doubts about prior claims that parliamentary systems are strongly associated with democratic consolidation.

Finally, we consider the split-stage parameter estimate for *GDP per capita*, which is positive and statistically significant in the original nonspatial SP Weibull models with and without i.i.d. frailties. In contrast, the split-stage estimate of *GDP per capita* is negative (though insignificant) in our spatial SP Weibull model. Hence, the widely accepted positive association between higher per capita income and democratic consolidation is neither consistent nor robust when accounting for spatial autocorrelation among neighboring democracies.

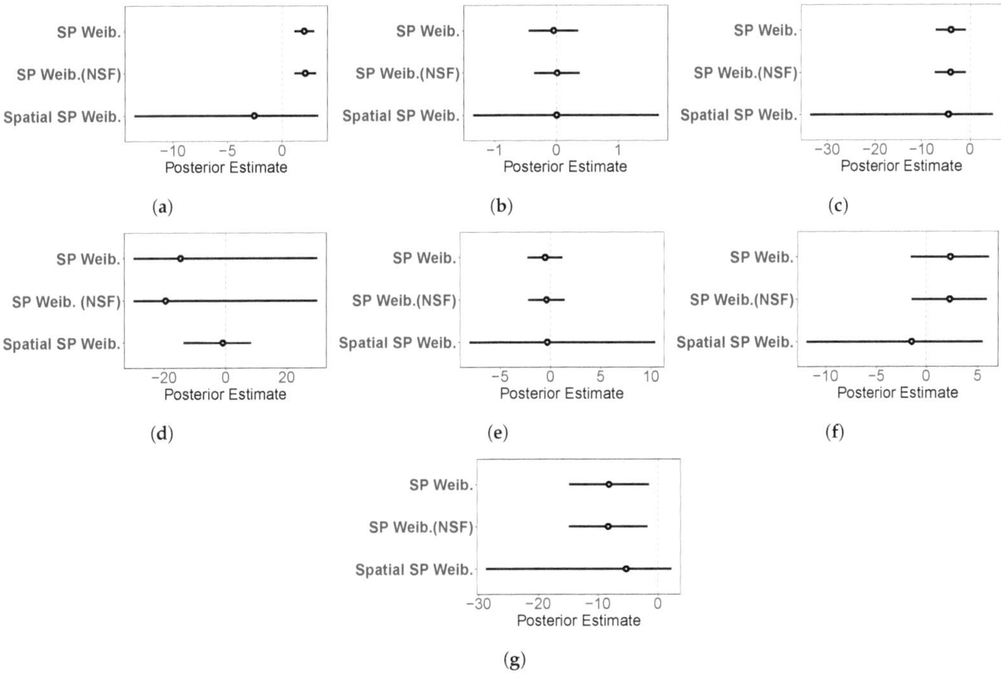

Figure 5. Democratic consolidation stage ($\hat{\gamma}$) coefficient results from SP Weibull, NS Frailty SP Weibull, and spatial SP Weibull models for the following covariates: (**a**) GDP/cap, (**b**) GDP growth, (**c**) military government, (**d**) monarchy, (**e**) civilian government, (**f**) parliamentary government, and (**g**) presidential government.

Turning to the survival stage results, we first consider the choropleth map in Figure 4b, which illustrates posterior means of the spatial frailties obtained from the Spatial SP Weibull model. The spatial frailty values vary from -0.95 to 0.859, with a corresponding standard deviation of 0.313. These maps again reveal spatial clustering associated with democratic regime survival; those democracies with greater underlying propensity for democratic survival are located near countries with similar propensities, while those with a lower propensity for democratic survival are located in disparate geographic areas.

The plots in Figure 6 reveal additional differences between the original nonspatial split-population model results and the new results from the Bayesian spatial split-population model. For instance, although the original study found that the survival stage estimate of *Monarchy* is positive and highly significant in the nonspatial models, this relationship is insignificant and *negative* in the spatial SP Weibull model. Hence, the association between democracies that were previously ruled by a monarch and democratic durability is tenuous after accounting for the influence of spatial autocorrelation on democratic survival. Next, the original survival stage estimate for *Military* in both the nonspatial SP Weibull models is negative (though insignificant). In the spatial SP Weibull model, *Military* is again negativ; however, unlike the nonspatial models, in this case it is statistically reliable. This suggests that not accounting for neighborhood democracies can lead researchers to underestimate the relationship between democratic durability and democratic states that were preceded by military rule. Finally, while [17] challenged the confidence of previous findings with respect to the relationship between *GDP per capita* and democratic survival (e.g., [35,36]), the influence of per capita income on democratic survival in our spatial SP Weibull model is positive and statistically reliable, consistent with the previous literature. Taken together, a re-examination of these data on democratic survival and consolidation using the spatial SP survival model leads to new inferences about important political phenomena.

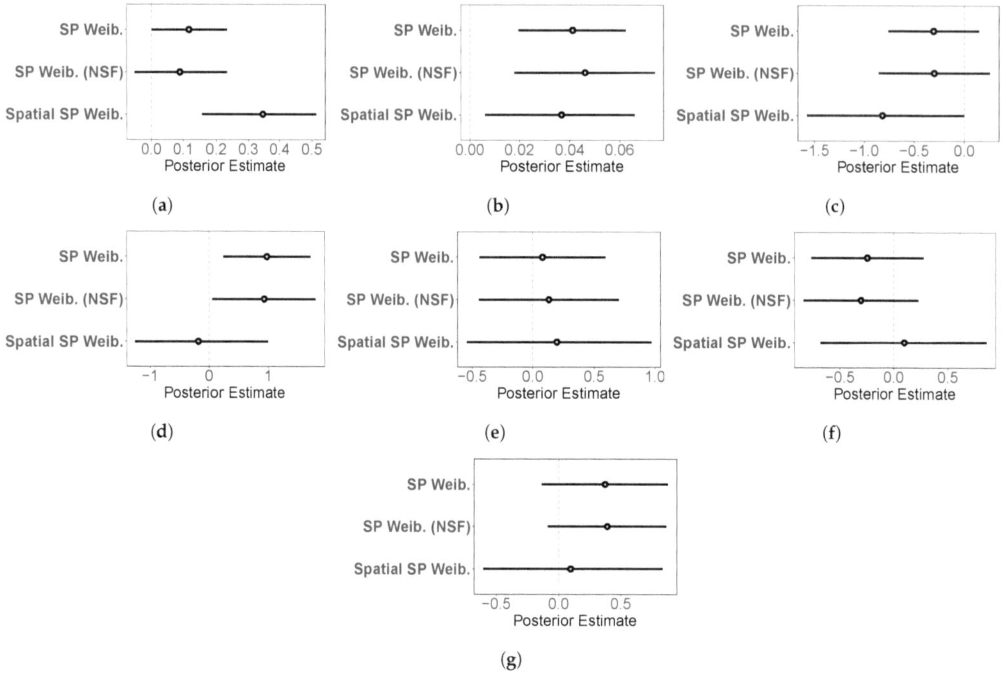

Figure 6. Democratic survival stage ($\hat{\beta}$) coefficient results from SP Weibull, NS Frailty SP Weibull, and Spatial SP Weibull models for the following covariates: (**a**) GDP/cap, (**b**) GDP growth, (**c**) military government, (**d**) monarchy, (**e**) civilian government, (**f**) parliamentary government, and (**g**) presidential government.

4.2. Post-Civil War Peace Duration

To further demonstrate the applicability of our framework, in this section we use our spatial SP survival model to re-investigate previous findings that suggest information transparency and other political freedoms can increase post-civil war peace survival [18]. The normative importance of consolidating peace after civil war has motivated a wide body of research in Political Science on why certain civil wars recur and others do not. Much of this literature has focused on the characteristics of the initial war and termination [44,45] and the characteristics of the post-war environment, including the presence of third-party or U.N. intervention [46–48]; ref. [18] contributes to this literature by arguing that increased political accountability and civil liberties in the post-war period can augment the costs of reneging on an agreement and allow governments to credibly commit to not resuming violence. Her primary testable expectation is that "civil wars that are fought against governments with limited accountability should be more likely to repeat themselves than civil wars in countries with highly accountable governments" [18] (p. 1248). Estimating conventional survival models using a sample of 77 post-civil war peace spells during 1945–2009, Ref. [18] finds support for her hypotheses in various measures of civil liberties, democratic institutions, and rule of law; however, one of her primary variables of interest, *Press Freedom*, does not reach standard frequentist levels of statistical significance.

The conventional survival models used in the aforementioned study, however, assume that all wars recur at some point. A split-population survival model may be more appropriate for studying post-war peace survival, as certain wars are at risk for recurring while others are structurally distinct cases in which peace is "consolidated" or one side is eliminated entirely, in which the same conflict cannot recur. Moreover, there are theoretical and empirical reasons to believe that spatial autocorrelation may influence both the risk of peace consolidation or of renewed war in these data. Previous research has shown that diffusion processes may lead to *conflict contagion* (e.g., [13,49,50]), while *peace stability* is regionally clustered due to the clustering of observable and unobservable political attributes (e.g., [39,51,52]). In the split stage, post-war countries might never return to violence if they are surrounded by similarly peaceful countries that have demonstrated the institutional capability and political interest to prevent the resurgence of violence in the region [49], or if latent localized interests among elites or civilian populations to contain violence—for instance, to prevent war recurrence in their own contiguous countries—are clustered in space [53]. In the survival stage, stable institutions [51] or other latent geographic factors that transcend the borders of a single state may influence the time that rebels take to remobilize or that governments take to re-engage with dormant dissident movements. If any of these unobserved factors are not *randomly* distributed in space, then failing to account for this heterogeneity will lead to faulty inferences [13].

As in the previous application, we conducted Moran's I and join count tests to evaluate whether spatial autocorrelation exists between peace survival rates and countries that experience a civil war and never experience another. The results (see Figure A2 in the Appendix A) clearly indicate spatial clustering and dispersion, especially after 1960 and prior to 1900. This pre-estimation empirical evidence suggests that spatial autocorrelation should be taken into account when modeling the survival and consolidation of post-civil war peace.

In our analysis of post-civil war peace, we use replication data from the aforementioned study and specify an SP model with four covariates in the split-stage: *Press Freedom*; whether the previous conflict ended in an outright *Victory*; percentage of *Mountainous* terrain; and *GDP/capita*. In the survival stage, we include our main variable of interest, *Press Freedom*, and control for *GDP/capita*, whether the previous conflict ended in a *Peace Agreement*, the *Intensity* of the previous conflict, *Ethnic Factionalization*, the presence of *UN Peacekeeping* forces, whether the previous conflict was over *Territory*, whether the country has some *Non-Contiguous* territory that can be used to expedite rebel mobilization, and the percentage of *Mountainous* terrain. We focus on *Press Freedom* because it closely captures information transparency, which is an important mechanism of interest in the original theory, and because the results for this variable were inconsistent with the other findings in the study.

To examine the effects of these covariates on both the probability of peace consolidation and the survival of peace, we estimate a nonfrailty SP Weibull model and a spatial SP Weibull model. Following the original analysis of these data, we estimate the nonfrailty model using maximum likelihood. For the spatial model, we again define spatial proximity as $a_{ii'} = 1$ if the distance between the capitals of state i and i' is less than 800 km and $a_{ii'} = 0$ otherwise [54], though we found similar results when increasing these distances to 2000 and 2500 km. We allow the frailties between neighboring units to be spatially correlated by employing separate CAR priors for the frailty vectors $\mathbf{V}$ and $\mathbf{W}$, which implies a CAR structure of $\mathbf{V}|\lambda \sim \text{CAR}(\lambda)$ and $\mathbf{W}|\lambda \sim \text{CAR}(\lambda)$. The spatial SP Weibull model is estimated using the multivariate normal prior. For the slice-sampling (MCMC) algorithm, we specify the hyperparameters as $a = 1, b = 1, S_\beta = I_{p1}, S_\gamma = I_{p2}, \nu_\beta = p1$, and $\nu_\gamma = p2$, and assign the Gamma hyperprior $\lambda \sim \text{Gamma}(a_\lambda, b_\lambda)$ for λ with vague prior $(a_\lambda, b_\lambda) = (0.001, 1/0.001)$. We estimated the model with 50,000 iterations and 38,000 burn-ins. Every parameter passes the Geweke [28] convergence test and Heidelberger and Welch [29] stationarity test.

Turning first to the choropleth maps of the spatial frailty values from the spatial SP Weibull model in Figure 7a,b, the split-stage frailties ($\mathbf{V}$) range from -1.19 to 1.49 with a standard deviation of 0.5503, and the survival-stage frailties ($\mathbf{W}$) range from -1.21 to 0.727 with a standard deviation of 0.4498. In both stages, there seem to be regional clusters of frailty values.

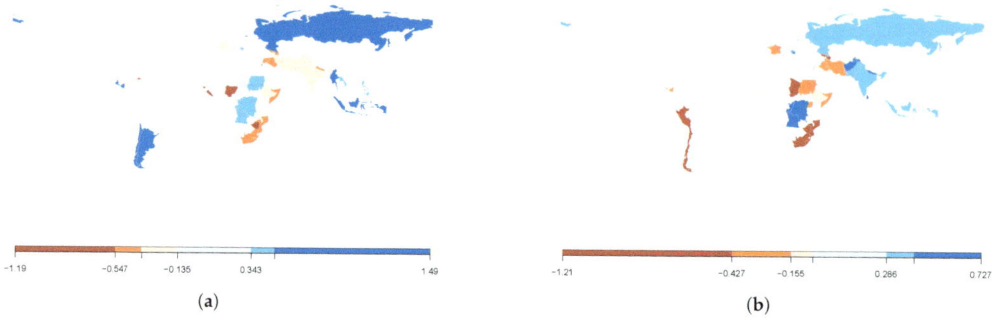

Figure 7. Post−war peace duration application spatial frailty maps: (**a**) depicts the posterior mean estimates of **V** (split−stage spatial frailties) and (**b**) depicts the posterior mean estimates of **W** (survival−stage spatial frailties).

The β and γ results for both the nonspatial and spatial SP Weibull models are reported in Figures 8 and 9. The dots and bars are interpreted in the same way as in the previous application. Much like the original analysis, the nonspatial SP model reveals no clear evidence that information transparency affects the survival of peace after conflict; the coefficient estimates for *Press Freedom* are statistically insignificant in both stages. In fact, the only (weakly) significant result in the split-stage of the nonspatial model is *GDP/capita*, indicating limited evidence that increased economic prosperity can increase the probability of a civil conflict never recurring. *Mountains* and previous *Victory* appear to have no relationship with peace consolidation, contrary to extant findings (e.g., [44]). The survival stage results of the nonspatial Weibull model reveal that conflicts over *Territory* and *Ethnic Factionalization* are associated with longer post-conflict peace periods, while *Non-Contiguous* territory and previous war *Intensity* are associated with shorter peace periods, at least among countries at risk of recurrent conflict.

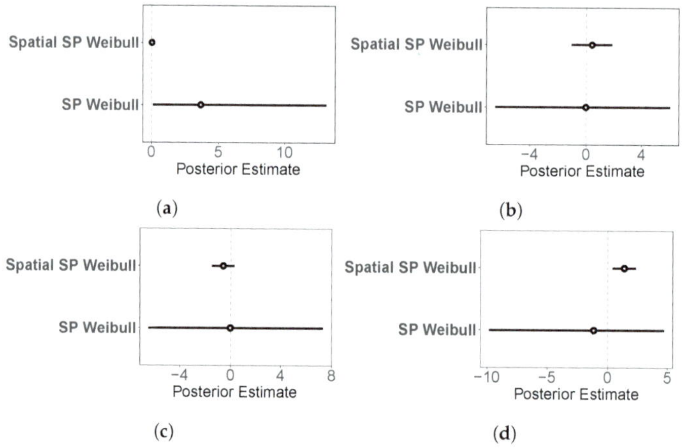

Figure 8. Peace consolidation ($\hat{\gamma}$) coefficient results from SP Weibull and Spatial SP Weibull models for the following covariates: (**a**) press freedom, (**b**) victory, (**c**) mountains, and (**d**) GDP/cap.

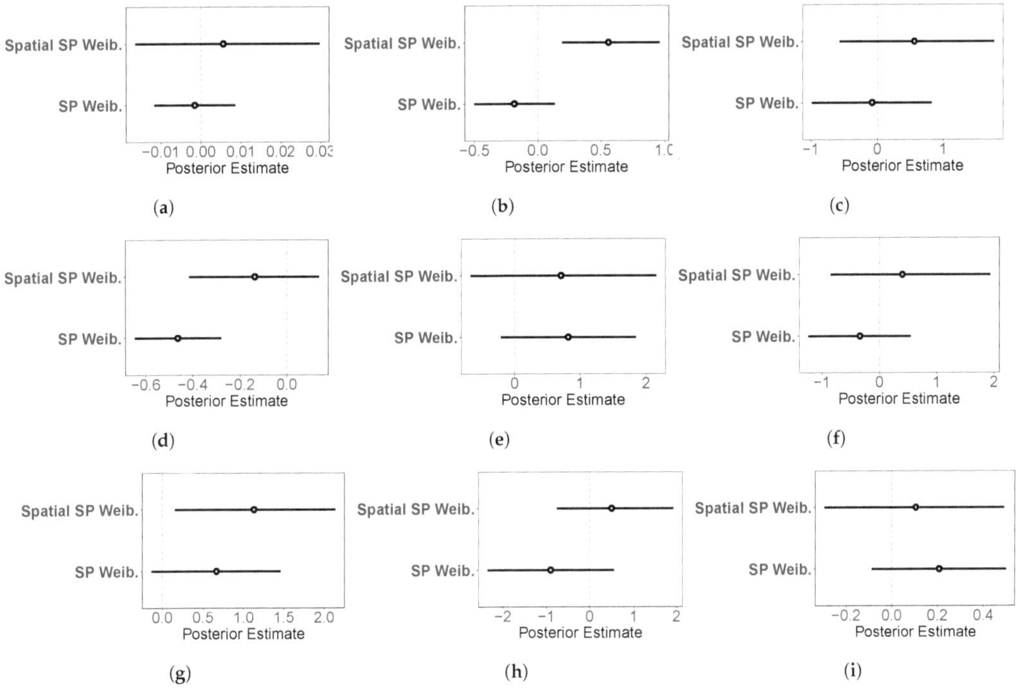

Figure 9. Peace survival stage ($\hat{\beta}$) coefficient results from SP Weibull and Spatial SP Weibull models for the following covariates: (**a**) press freedom, (**b**) GDP/cap, (**c**) peace agreement, (**d**) intensity, (**e**) ethnic factionalization, (**f**) UN peacekeeping, (**g**) territory, (**h**) non-contiguous, and (**i**) mountains.

When we account for spatially autocorrelated frailties, however, we find remarkably different results. First, although conflicts over *Territory* remain positively associated with post-war peace survival, we cannot reliably conclude that *Non-Contiguous* territory, *Intensity*, or *Ethnic Factionalization* have systematic relationships with peace in at-risk conflict locations. In addition, we find that better economic opportunities are no longer reliably associated with post-war peace consolidation, though they do have a reliable statistical relationship with the survival of peace in at-risk countries. This finding is important because it suggests that while a stronger economy can make it more difficult for rebels to re-mobilize, it is not a panacea for permanently exiting the "conflict trap," as previous research has suggested [55].

Finally, although *Press Freedom* appears to have no real relationship with post-war peace survival among *at-risk* cases, it *is* positively and reliably associated with the *consolidation* of peace after war. In other words, although a country having a higher level of press freedom may not elongate a temporary peace when conflict is likely to recur, press freedom *is* associated with a significant decrease in a country's overall *susceptibility* to renewed war. Although there remains considerable debate over whether democratization after a conflict hinders or helps peace to endure (e.g., [56]), this evidence suggests that information transparency and an independent media may lead to a far more durable, if not permanent, period of domestic stability after a conflict ends.

5. Discussion and Conclusions

This has article developed a parametric spatial Split-Population Survival model that accounts for both the probability that some observations are immune to an event of interest and the tendency of underlying risk factors associated with political processes to cluster in space. While the model builds upon previous work [11,12,57], it is unique in that it allows for spatially autocorrelated frailties in the split *and* survival stages. The model incorporates time-varying covariates in both the split and survival stages. These features allow researchers to explicitly model and statistically account for spatial heterogeneity that may influence the probability of an observation becoming immune from an event *as well as* the duration of a process among units considered "at-risk" in panel survival data. Our innovation is distinct from extant works on spatial statistics that address different types of spatial dependence in settings with continuous or binary dependent variables or in conventional survival models [12,58,59], as it relaxes the assumption that all observations eventually experience the event of interest. Our MC experiments reveal that, unlike nonspatial models, our spatial split-population survival model provides accurate estimates when SP survival data exhibit spatial autocorrelation. Finally, we apply our model to previously published data on widely studied phenomena in political science in the contexts of democratic regime survival and the durability of post-civil war peace. After accounting for the immune fraction and spatial autocorrelation in these applications, we find evidence contrary to previous studies' original findings, particularly in the first-stage "cured" fraction equation.

Future work can build upon our model to devise estimation routines for survival data with multinomial outcomes (e.g., competing risks), recurrent events, or variable time trends; a particularly useful next step would be to apply this approach to a semi-parametric setting [3,60]. Extant work on nonparametric solutions could benefit from this approach [61]. Although our contribution is significant in Bayesian survival modeling by allowing for spatial frailties in the split stage rather than only in the survival stage (e.g., [11]), future extension could increase the flexibility of our model by incorporating spatial frailties *only* in the estimation of the immune fraction, while the survival stage includes only i.i.d. frailties or no frailties. This could be useful in applied cases where a researcher believes that the processes leading to immunity are spatially clustered while the survival probabilities of those at risk of the event are not.

Of course, any Bayesian analysis can be sensitive to the choice of the prior distribution; thus, applied researchers should take care to not limit themselves to the priors implemented here depending on their topic of study. Future analytical work could continue to investigate our model's performance using alternative priors, such as the Gamma distribution [62]. Furthermore, further application of the model could benefit from the development of goodness-of-fit tests for distributional assumptions in the data [63] and additional tools for model specification [60].

Beyond the applications presented here, this Bayesian spatial split-population survival model can be a useful tool for researchers interested in anything from success rates of vaccines or other medical treatments, to customer analytics around the purchase of new products, to coups d'état around the world. All of these phenomena and more tend to involve some fraction of the population under investigation being probabilistically immune from the event for some measurable reason(s) while being influenced by the spatial clustering of unobserved yet meaningful factors.

Supplementary Materials: The following supporting information can be downloaded at: https://www.mdpi.com/article/10.3390/math11081886/s1 and minniejoo.com/research/, accessed on 15 April 2023, Figures S1–S2 and Tables S1–S4. References [11,12,14,25,26,31] are cited in Supplementary Materials.

Author Contributions: Conceptualization, M.M.J., B.B. and B.M.; methodology, M.M.J., B.B. and B.M.; software, M.M.J., B.B. and N.H.; validation, M.M.J., B.B. and N.H.; formal analysis, M.M.J. and B.M.; investigation, B.B., M.M.J. and N.H.; resources, B.M.; data curation, B.B. and N.H.; writing—B.M., B.B.; writing—review and editing, B.B., B.M., M.M.J.; visualization, M.M.J., B.B. and N.H.; supervision, B.M.; project administration, B.M.; funding acquisition, B.M. and M.M.J. All authors have read and agreed to the published version of the manuscript.

Funding: This research received no external funding.

Data Availability Statement: Data will be made available for download on Minnie M. Joo's personal website upon publication. Link: minniejoo.com/research/.

Acknowledgments: We would like to thank Bomin Kim and Benjamin E. Bagozzi for helpful insights at the early stage of this project. The authors are grateful for generous funding and support for this project from the College of the Liberal Arts at Penn State University.

Conflicts of Interest: The authors declare no conflict of interest.

Abbreviations

The following abbreviations are used in this manuscript:

CAR	Conditionally Autoregressive
CP	Convergence Probabilities
d.g.p.	Data Generation Process
i.i.d.	Independent and Identically Distributed
IW	Inverse Wishart
MC	Monte Carlo
MCMC	Markov Chain Monte Carlo
MCSE	Monte Carlo Standard Error
MVN	Multivariate Normal
NS	Non-Spatial
NSF	Non-Spatial Frailty
RMSE	Root Mean Square Error
SP	Split Population

Appendix A

To conduct the join count test and Moran's I test for assessing spatial autocorrelation among countries possibly 'at risk' of democratic reversal and the survival of democracy in the sample from [17], we constructed a separate cross−sectional adjacency matrix with elements $a_{ii'}$ for each year in the data, wherein proximate pairs of democratic countries (within 800 km of each other) are assigned a weight of 1 ($a_{ii'} = 1$). Our outcome of interest for the join count analysis is whether a country is 'at risk' of democratic reversal in the data; the outcome for the Moran's I test is the number of years each democracy has survived. Figure A1a plots the results of the join count tests, with the difference between the observed and expected join counts displayed with 95% confidence intervals. The figure shows significant spatial clustering of countries possibly at risk of democratic failure, particularly during the twentieth century. Figure A1b reports the resulting Moran's I statistics for each year. Clearly, a large proportion of the sampling period, most notably between 1925–1945 and from 1952–2001, exhibits positive spatial clustering in democratic survival.

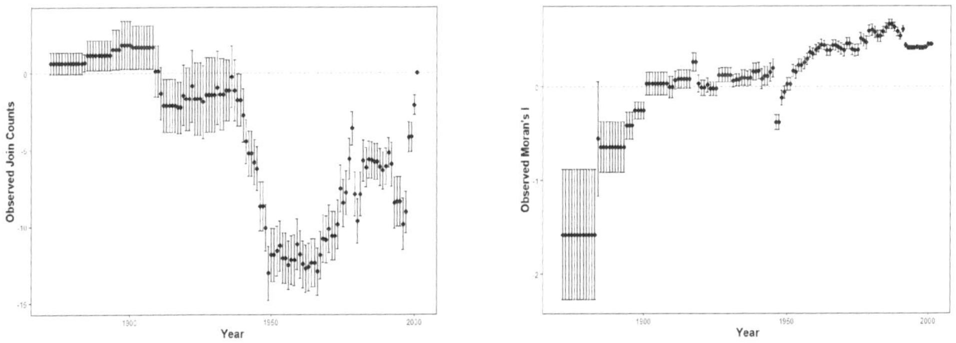

(**a**) (**b**)

Figure A1. Results from the two autocorrelation diagnostics for the democratic survival application: (**a**) join count and (**b**) Moran's I.

As with the democratic regime survival application, we employed two tests, join count analysis and Moran's I, to evaluate the possibility of spatial autocorrelation in the split and survival stages of the replication data on post−war peace survival. Figure A2a plots the observed−expected join counts of "at risk" (non-censored) conflicts in the dataset from [18] for each year along with their 90% confidence intervals. Negative values indicate clustering (positive spatial autocorrelation) and positive values indicate spatial dispersion. The figure depicts clear spatially correlated patterns of countries at risk of recurrent conflict, though the direction of the autocorrelation varies over time. Figure A2b depicts the Moran's I values for spatial autocorrelation in post-war peace duration. Again, we observe significant degrees of some variety of spatial autocorrelation in the majority of years between 1975 and 2010.

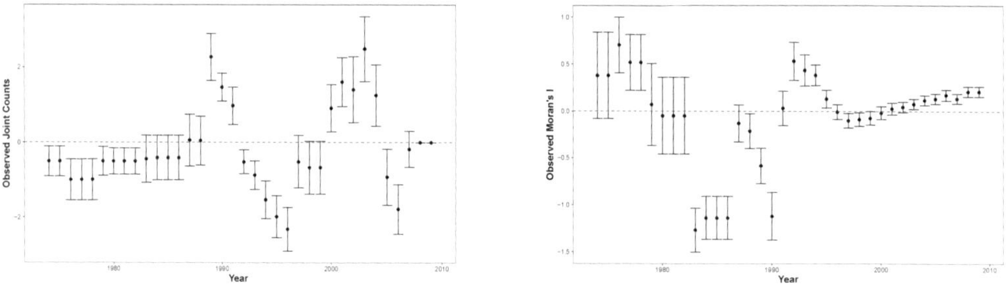

(**a**) (**b**)

Figure A2. Results from the two autocorrelation diagnostics for the post-war peace duration application: (**a**) join count and (**b**) Moran's I.

References

1. Box-Steffensmeier, J.M.; Jones, B.S. *Event History Modeling: A Guide for Social Scientists*; Cambridge University Press: New York City, NY, USA, 2004.
2. Wang, Y.; Klijn, J.G.; Zhang, Y.; Sieuwerts, A.M.; Look, M.P.; Yang, F.; Talantov, D.; Timmermans, M.; Meijer-van Gelder, M.E.; Yu, J.; et al. Gene-expression profiles to predict distant metastasis of lymph-node-negative primary breast cancer. *Lancet* **2005**, *365*, 671–679. [CrossRef] [PubMed]
3. Bremhorst, V.; Lambert, P. Flexible estimation in cure survival models using Bayesian P-splines. *Comput. Stat. Data Anal.* **2016**, *93*, 270–284. [CrossRef]
4. Clark, D.H.; Regan, P.M. Opportunities to fight: A statistical technique for modeling unobservable phenomena. *J. Confl. Resolut.* **2003**, *47*, 94–115. [CrossRef]
5. Ray, R.A.; Perry, R.W.; Som, N.A.; Bartholomew, J.L. Using cure models for analyzing the influence of pathogens on salmon survival. *Trans. Am. Fish. Soc.* **2014**, *143*, 387–398. [CrossRef]
6. Schmidt, P.; Witte, A.D. Predicting criminal recidivism using 'split population'survival time models. *J. Econom.* **1989**, *40*, 141–159. [CrossRef]
7. Peng, Y.; Taylor, J.M. Mixture cure model with random effects for the analysis of a multi-center tonsil cancer study. *Stat. Med.* **2011**, *30*, 211–223. [CrossRef]
8. Patilea, V.; Van Keilegom, I. A general approach for cure models in survival analysis. *Ann. Stat.* **2020**, *48*, 2323–2346. [CrossRef]
9. Bagozzi, B.E.; Joo, M.M.; Kim, B.; Mukherjee, B. A Bayesian Split Population Survival Model for Duration Data With Misclassified Failure Events. *Political Anal.* **2019**, *27*, 415–434. [CrossRef]
10. Chiba, D.; Metternich, N.W.; Ward, M.D. Every story has a beginning, middle, and an end (but not always in that order): Predicting duration dynamics in a unified framework. *Political Sci. Res. Methods* **2015**, *3*, 515–541. [CrossRef]
11. Banerjee, S.; Carlin, B.P. Parametric spatial cure rate models for interval-censored time-to-relapse data. *Biometrics* **2004**, *60*, 268–275. [CrossRef]
12. Darmofal, D. Bayesian Spatial Survival Models for Political Event Processes. *Am. J. Political Sci.* **2009**, *53*, 241–257. [CrossRef]
13. Darmofal, D. *Spatial Analysis for the Social Sciences*; Cambridge University Press: New York City, NY, USA, 2015.
14. Banerjee, S.; Wall, M.M.; Carlin, B.P. Frailty Modeling for Spatially Correlated Survival Data, with Application to Infant Mortality in Minnesota. *Biostatistics* **2003**, *4*, 123–142. [CrossRef] [PubMed]
15. Taylor, B.; Rowlingson, B. spatsurv: An R package for Bayesian inference with spatial survival models. *J. Stat. Softw.* **2017**, *77*, 1–32. [CrossRef]

16. Muff, S.; Signer, J.; Fieberg, J. Accounting for individual-specific variation in habitat-selection studies: Efficient estimation of mixed-effects models using Bayesian or frequentist computation. *J. Anim. Ecol.* **2020**, *89*, 80–92. [CrossRef] [PubMed]
17. Svolik, M. Authoritarian Reversals and Democratic Consolidation. *Am. Political Sci. Rev.* **2008**, *102*, 153–168. [CrossRef]
18. Walter, B.F. Why bad governance leads to repeat civil war. *J. Confl. Resolut.* **2015**, *59*, 1242–1272. [CrossRef]
19. Maller, R.A.; Zhou, X. *Survival Analysis with Long-Term Survivors*; Wiley: New York, NY, USA, 1996; Volume 525.
20. Yin, G.; Ibrahim, J.G. Cure rate models: A unified approach. *Can. J. Stat.* **2005**, *33*, 559–570. [CrossRef]
21. Beger, A.; Hill, D.W.; Metternich, N.W.; Minhas, S.; Ward, M.D. Splitting it up: The spduration split-population duration regression package for time-varying covariates. *R J.* **2017**, *9*, 474–486. [CrossRef]
22. Lu, W. Efficient estimation for an accelerated failure time model with a cure fraction. *Stat. Sin.* **2010**, *20*, 661. [PubMed]
23. Peng, Y.; Taylor, J. Cure models. In *Handbook of Survival Analysis*; Klein, J.P.; Van Houwelingen, H.C.; Ibrahim, J.G.; Scheike, T.H., Eds.; Chapman and Hall: Boca Raton, FL, USA, 2014; p. 113–34.
24. Ibrahim, J.G.; Chen, M.H.; Sinha, D. Criterion-based methods for Bayesian model assessment. *Stat. Sin.* **2001**, *11*, 419–443.
25. Bernardinelli, L.; Montomoli, C. Empirical Bayes versus fully Bayesian analysis of geographical variation in disease risk. *Stat. Med.* **1992**, *11*, 983–1007. [CrossRef] [PubMed]
26. Besag, J.; York, J.; Mollié, A. Bayesian Image Restoration, with Two Applications in Spatial Statistics. *Ann. Inst. Stat. Math.* **1991**, *43*, 1–20. [CrossRef]
27. Thomas, A.; Best, N.; Lunn, D.; Arnold, R.; Spiegelhalter, D. GeoBUGS User Manual, Version 1.2. Available online: http: www.mrc-bsu.cam.ac.uk/bugs/ (accessed on 15 January 2021) .
28. Geweke, J. Evaluating the Accuracy of Sampling-based Approaches to the Calculation of Posterior Moments. In *Bayesian Statistics*; Bernardo, J.; Berger, J.; Dawid, A.; Smith, A., Eds.; Clarendon Press: Oxford, UK, 1992.
29. Heidelberger, P.; Welch, P.D. Simulation run length control in the presence of an initial transient. *Oper. Res.* **1983**, *3*, 1109–1144. [CrossRef]
30. Carlin, B.; Louis, T. *Bayes and Empirical Bayes Methods for Data Analysis*; Chapman and Hall/CRC: New York, NY, USA, 2000.
31. Neal, R.M. Slice sampling. *Ann. Stat.* **2003**, *31*, 705–767. [CrossRef]
32. Gasiorowski, M.J.; Power, M. The Structural Determinants of Democratic Consolidation. *Comp. Political Stud.* **1998**, *31*, 740–771. [CrossRef]
33. Gassebner, M.; Lamla, M.J.; Vreeland, J.R. Extreme Bounds of Democracy. *J. Confl. Resolut.* **2013**, *57*, 171–197. [CrossRef]
34. Cheibub, J.A. *Presidentialism, Parliamentarism, and Democracy*; University Press,Cambridge, NY, USA, 2007.
35. Przeworski, A.; Alvarez, M.; Cheibub, J.A.; Limongi, F. *Democracy and Development: Political Institutions and Economic Performance, 1950–1999*; Cambridge University: New York City, NY, 2000.
36. Treisman, D. Is Democracy in Danger? A Quick Look at the Data. Prepared for the Conference on "Democratic Backsliding and Electoral Authoritarianism. Available online: https://www.danieltreisman.org/s/draft-june-7.pdf (accessed on 8 November 2018).
37. Boix, C.; Stokes, S.C. Endogenous democratization. *World Politics* **2003**, *55*, 517–549. [CrossRef]
38. Przeworski, A.; Limongi, F. Modernization: Theories and facts. *World Politics* **1997**, *49*, 155–183. [CrossRef]
39. Gleditsch, K.S.; Ward, M.D. Diffusion and the International Context of Democratization. *Int. Organ.* **2006**, *60*, 911–933. [CrossRef]
40. Kopstein, J.S.; Reilly, D.A. Geographic Diffusion and the Transformation of the Postcommunist World. *World Politics* **2000**, *53*, 1–37. [CrossRef]
41. Brinks, D.; Coppedge, M. Diffusion is no illusion: Neighbor emulation in the third wave of democracy. *Comp. Political Stud.* **2006**, *39*, 463–489. [CrossRef]
42. Cliff, A.D.; Ord, J.K. *Spatial Processes: Models & Applications*; Pion: London, UK. 1981.
43. Moran, P.A. Notes on continuous stochastic phenomena. *Biometrika* **1950**, *37*, 17–23. [CrossRef] [PubMed]
44. Toft, M.D. Ending civil wars: A case for rebel victory? *Int. Secur.* **2010**, *34*, 7–36. [CrossRef]
45. Licklider, R. The consequences of negotiated settlements in civil wars, 1945–1993. *Am. Political Sci. Rev.* **1995**, *89*, 681–690. [CrossRef]
46. Fortna, V.P. Does peacekeeping keep peace? International intervention and the duration of peace after civil war. *Int. Stud. Q.* **2004**, *48*, 269–292. [CrossRef]
47. Walter, B.F. Does conflict beget conflict? Explaining recurring civil war. *J. Peace Res.* **2004**, *41*, 371–388. [CrossRef]
48. Gates, S.; Graham, B.A.; Lupu, Y.; Strand, H.; Strøm, K.W. Power sharing, protection, and peace. *J. Politics* **2016**, *78*, 512–526. [CrossRef]
49. Braithwaite, A. Resisting infection: How state capacity conditions conflict contagion. *J. Peace Res.* **2010**, *47*, 311–319. [CrossRef]
50. Buhaug, H.; Gleditsch, K.S. Contagion or confusion? Why conflicts cluster in space. *Int. Stud. Q.* **2008**, *52*, 215–233. [CrossRef]
51. Gates, S.; Hegre, H.; Jones, M.P.; Strand, H. Institutional inconsistency and political instability: Polity duration, 1800–2000. *Am. J. Political Sci.* **2006**, *50*, 893–908. [CrossRef]
52. Elbadawi, E.; Sambanis, N. Why are there so many civil wars in Africa? Understanding and preventing violent conflict. *J. Afr. Econ.* **2000**, *9*, 244–269. [CrossRef]
53. Doyle, M.W.; Sambanis, N. International peacebuilding: A theoretical and quantitative analysis. *Am. Political Sci. Rev.* **2000**, *94*, 779–801. [CrossRef]
54. Murdoch, J.C.; Sandler, T. Civil wars and economic growth: Spatial dispersion. *Am. J. Political Sci.* **2004**, *48*, 138–151. [CrossRef]

55. Collier, P.; Elliot, V.; Hegre, H.; Hoeffler, A.; Reynal-Querol, M.; Sambanis, N. *Breaking the Conflict Trap: Civil War and Development Policy*; The World Bank: Washington, DC, USA, 2003.
56. Walter, B.F. Designing transitions from civil war: Demobilization, democratization, and commitments to peace. *Int. Secur.* **1999**, *24*, 127–155. [CrossRef]
57. Hays, J.C.; Schilling, E.U.; Boehmke, F.J. Accounting for right censoring in interdependent duration analysis. *Political Anal.* **2015**, *23*, 400–414. [CrossRef]
58. Beck, N.; Gleditsch, K.S.; Beardsley, K. Space is more than geography: Using spatial econometrics in the study of political economy. *Int. Stud. Q.* **2006**, *50*, 27–44. [CrossRef]
59. Franzese, R.J.; Hays, J.C. Spatial Econometric Models of Cross-Sectional Interdependence in Political Science Panel and Time-Series-Cross-Section Data. *Political Anal.* **2007**, *15*, 140–164. [CrossRef]
60. Wagner, H. Bayesian estimation and stochastic model specification search for dynamic survival models. *Stat. Comput.* **2011**, *21*, 231–246. [CrossRef]
61. Sparapani, R.A.; Logan, B.R.; McCulloch, R.E.; Laud, P.W. Nonparametric survival analysis using Bayesian additive regression trees (BART). *Stat. Med.* **2016**, *35*, 2741–2753. [CrossRef]
62. Rizki, S.W.; Mara, M.N.; Sulistianingsih, E. Survival bayesian estimation of exponential-gamma under LINEX loss function. *Proc. J. Phys. Conf. Ser. IOP* **2017**, *855*, 012036. [CrossRef]
63. Mahdizadeh, M.; Zamanzade, E. Goodness-of-fit testing for the Cauchy distribution with application to financial modeling. *J. King Saud-Univ.-Sci.* **2019**, *31*, 1167–1174. [CrossRef]

mathematics

MDPI

Article

Bayesian and Non-Bayesian Risk Analysis and Assessment under Left-Skewed Insurance Data and a Novel Compound Reciprocal Rayleigh Extension

Mohamed Ibrahim [1,*], Walid Emam [2], Yusra Tashkandy [2], M. Masoom Ali [3] and Haitham M. Yousof [4]

[1] Department of Applied, Mathematical and Actuarial Statistics, Faculty of Commerce, Damietta University, Damietta 34517, Egypt
[2] Department of Statistics and Operations Research, Faculty of Science, King Saud University, P.O. Box 2455, Riyadh 11451, Saudi Arabia
[3] Department of Mathematical Sciences, Ball State University, Muncie, IN 47306, USA
[4] Department of Statistics, Mathematics and Insurance, Benha University, Benha 13518, Egypt
* Correspondence: mohamed_ibrahim@du.edu.eg

Abstract: Continuous probability distributions can handle and express different data within the modeling process. Continuous probability distributions can be used in the disclosure and evaluation of risks through a set of well-known basic risk indicators. In this work, a new compound continuous probability extension of the reciprocal Rayleigh distribution is introduced for data modeling and risk analysis. Some of its properties including are derived. The estimation of the parameters is carried out via different techniques. Bayesian estimations are computed under gamma and normal prior. The performance and assessment of all techniques are studied and assessed through Monte Carlo experiments of simulations and two real-life datasets for applications. Two applications to real datasets are provided for comparing the new model with other competitive models and to illustrate the importance of the proposed model via the maximum likelihood technique. Numerical analysis for expected value, variance, skewness, and kurtosis are given. Five key risk indicators are defined and analyzed under Bayesian and non-Bayesian estimation. An extensive analytical study that investigated the capacity to reveal actuarial hazards used a wide range of well-known models to examine actuarial disclosure models. Using actuarial data, actuarial hazards were evaluated and rated.

Keywords: actuarial risks analysis; asymmetric actuarial data; insurance-claims; likelihood; value-at-risk; reciprocal rayleigh

MSC: 60E05; 62H05; 62E10; 62F10; 62F15; 62P05

Citation: Ibrahim, M.; Emam, W.; Tashkandy, Y.; Ali, M.M.; Yousof, H.M. Bayesian and Non-Bayesian Risk Analysis and Assessment under Left-Skewed Insurance Data and a Novel Compound Reciprocal Rayleigh Extension. *Mathematics* **2023**, *11*, 1593. https://doi.org/10.3390/math11071593

Academic Editor: Diana Mindrila

Received: 28 January 2023
Revised: 21 March 2023
Accepted: 23 March 2023
Published: 25 March 2023

1. Introduction

Actuarial science is a mathematical branch that deals with the financial consequences of uncertain future events. It employs statistical and mathematical methods to evaluate and manage risks in the finance and insurance industries. Actuaries use probability distributions to model and measure the likelihood of different outcomes and determine the anticipated future losses. Probability distribution is a function that describes the probability of different outcomes for a random variable. Actuaries utilize various probability distributions, such as Poisson, normal, exponential, and log-normal, to model diverse types of risks, including morbidity, mortality, and property damage, based on the nature of the risk being modeled and the data available for the modeling. Actuaries use probability distributions to compute the expected future losses, which are used to set insurance premiums, design insurance products, and assess investment strategies. Actuaries also use simulation techniques to test their models and evaluate the financial results of insurance policies and investments under various scenarios.

The adequacy of probability-based distributions in describing risk exposure is a common practice in the field of risk management. Typically, risk exposure statistics are defined by one or a small group of numbers that are functions of a specific model, commonly referred to as key risk indicators (KRIs) (Lane [1]; Klugman et al. [2]). These KRIs provide actuaries and risk managers with valuable information about a company's exposure to specific types of risk. Several KRIs, including value-at-risk (VARK), tail-value-at-risk (TVARK) or conditional tail expectation (CTE), conditional-value-at-risk (CVARK), tail variance (TV), mean excess loss (MEL), and tail mean-variance (TMV), have been developed and can be analyzed (Shrahili et al. [3]; Mohamed [4]). In particular, VARK is commonly used to estimate the quantile distribution of aggregate losses. Actuaries and risk managers focus on calculating the probability of a bad outcome, measured by the VARK indicator at a specific probability or confidence level. This indicator is used to estimate the amount of capital required to manage potential unfavorable events. The ability of an insurance company to handle such situations is highly valued by actuaries, authorities, investors, and rating agencies (Wirch [5]; Artzner [6]; Tasche [7]; Acerbi [8]; Landsman [9]; Furman and Landsman [10]). In summary, probability-based distributions and KRIs are essential tools for evaluating risk exposure in companies. The use of such indicators is widespread, and the ability to manage risk effectively is highly valued in the field of risk management.

For the left-skewed insurance-claims data, this work suggests certain KRI variables, such as VARK, TVARK, TV, MEL, and TMV, using a new model termed the exponentiated generalized reciprocal Rayleigh Poisson (EGRRP) distribution. Statistical methods frequently employed in actuarial risk analysis include:

I. Actuaries employ probability distributions to simulate the possibility of a variety of events, including claims, fatalities, and policy cancellations. The Poisson distribution, the exponential distribution, and the Weibull distribution are frequently used distributions in actuarial science.

II. Modeling the time until a specific event, such as a death or policy cancellation, is conducted using survival analysis. This method is employed to compute life expectancy and assess the likelihood of survival for a specific time period.

III. Modeling with stochastic processes: stochastic modeling is used to simulate unpredictable events such as insurance-claims and policy cancellations. This method is employed to compute the variability of these estimations and to estimate the expected value of upcoming claims.

IV. Loss distributions: Loss distributions are used to simulate how losses are distributed as a result of things such as claims and insurance cancellations. The estimated value of potential losses is calculated using this method, and the risk involved with such losses is also identified.

V. Actuaries employ statistical approaches, such as portfolio optimization and hedging strategies, to evaluate and manage financial risks.

The Reciprocal Rayleigh (RR), also known as the inverse Rayleigh distribution, is an important probability distribution that is widely used in many fields, including reliability engineering, signal processing, and wireless communications. The RR distribution is a flexible distribution that can model a wide range of phenomena. It is a continuous distribution with support on the positive real line, and it has two parameters that control the location and scale of the distribution. The RR distribution can be used to model data that are positively skewed and have long tails, which are common in many real-world applications. The RR distribution is commonly used in reliability engineering to model the lifetime of a system or component. In this context, the RR distribution is used to model the time until failure, and it has been shown to provide a good fit to many types of failure data. The RR distribution is used in signal processing to model the amplitude of a random signal. In this context, the RR distribution is used to model the probability density function of the envelope of a narrowband Gaussian noise signal, which is commonly used in wireless communications. The RR distribution is used in statistical inference to model the distribution of the inverse of a random variable. In this context, the RR distribution is used to model the distribution of the ratio of two independent Rayleigh-distributed random variables, which is commonly used in wireless communications and signal processing.

The RR distribution is used in reliability engineering to model the lifetime of a system or component. In this context, the RR distribution is used to model the time until failure, and it has been shown to provide a good fit to many types of failure data. The RR distribution has also been used in finance to model the distribution of returns on investment portfolios. In this context, the RR distribution can be used to model the distribution of the inverse of returns, which can be useful in portfolio risk management. Overall, the RR distribution has a wide range of applications in many fields, including reliability engineering, wireless communications, signal processing, statistical inference, and finance. Its flexibility, reliability modeling capabilities, and usefulness in modeling the distribution of the inverse of a random variable make it a valuable tool for researchers and practitioners in many different areas.

The RR distribution can be used to model the distribution of losses in insurance-claims. This is useful in estimating the risk of losses and setting premiums. It can also be used to model the risk of many events in insurance, such as natural disasters or accidents. This can help insurance companies to better understand and manage their risk exposure. Moreover, the RR distribution can be used to model the behavior of policyholders in insurance, such as the frequency and severity of claims. This can help insurance companies to design policies that are better suited to their customers' needs. Finally, the RR distribution can be used in actuarial modeling to estimate the probability of future events based on historical data. This is useful in predicting the likelihood of future claims and setting reserves. Overall, the RR distribution has several applications in actuarial sciences and insurance. Its flexibility and usefulness in modeling the distribution of losses and risk make it a valuable tool for insurance companies and actuaries. The RR distribution, also known as the inverse Rayleigh (IR) distribution, has several applications in actuarial sciences and insurance. The RR distribution is considered as a distribution for a lifetime random variable (r.v.). The probability density function (PDF) and cumulative distribution function (CDF) of the RR model are given by

$$g(x) = g(x;\theta) = 2\theta^2 x^{-3} exp[-(\theta/x)^2], \tag{1}$$

and

$$G(x) = G(x;\theta) = exp[-(\theta/x)^2], \tag{2}$$

respectively, where $\theta > 0$ is a scale parameter $x > 0$. The exponentiated-generalized-Poisson (EGP) family of distributions is a new flexible compound family of distributions that Aryal and Yousof [11] introduced and explored. The EGP family's CDF and PDF are provided by:

$$F(x;\alpha,\beta) = \frac{1}{c_{(\lambda)}}\left\{1 - exp\left(-\lambda\{1 - [1 - G(x;\theta)]^{\alpha}\}^{\beta}\right)\right\}, \tag{3}$$

and

$$f(x;\alpha,\beta) = \alpha\beta\lambda\frac{g(x;\theta)[1 - G(x;\theta)]^{\alpha-1}\{1 - [1 - G(x;\theta)]^{\alpha}\}^{\beta-1}}{c_{(\lambda)}exp\left(\lambda\{1 - [1 - G(x;\theta)]^{\alpha}\}^{\beta}\right)}, \tag{4}$$

respectively, where $\alpha, \beta > 0$, $\lambda \in \mathrm{R}-\{0\}$, $x > 0$ and $c_{(\lambda)} = 1 - exp(-\lambda)$. For $\beta = 1$ we have the exponentiated-G Poisson (EGP) class of distribution, and for $\alpha = 1$ we have the generalized Poisson (GP) class of distribution, both of which are embedded in EGP class. Since (2) refers to the baseline CDF of the RR model and (3) refers to the baseline CDF of the EGP family, then, substituting (2) in (3), we derive a new compound RR distribution called EGRRP with CDF which can be expressed as

$$F(x;\underline{\mathcal{H}}) = \frac{1}{c_{(\lambda)}}\Big\{1 - exp\Big(-\lambda\{1 - [1 - \Delta_\theta(x)]^\alpha\}^\beta\Big)\Big\}, \tag{5}$$

where $\underline{\mathcal{H}} = (\alpha, \beta, \lambda, \theta)$, $\alpha, \beta, \theta > 0, \lambda \in \mathrm{R}-\{0\}$, $x > 0$, and $\Delta_\theta(x) = exp[-(\theta/x)^2]$. The corresponding PDF can be written as

$$f(x;\underline{\mathcal{H}}) = 2\alpha\beta\lambda\theta^2 c_{(\lambda)}^{-1} \frac{x^{-3}\Delta_\theta(x)(1 - [1 - \Delta_\theta(x)]^\alpha)^{\beta-1}}{\{1 - \Delta_\theta(x)\}^{1-\alpha} exp\Big(\lambda\{1 - [1 - \Delta_\theta(x)]^\alpha\}^\beta\Big)}. \tag{6}$$

Figure 1 illustrates that the PDF of the EGRRP model may exhibit various shapes, such as right-skewed, left-skewed, and unimodal. On the other hand, Figure 2 shows that the hazard rate function (HRF) of the EGRRP model may be decreasing and upside down. Moreover, there are several notable extensions of the RR distribution, including Voda [12], Mukerjee and Saran [13], Nadarajah and Kotz [14], Nadarajah and Gupta [15], Barreto-Souza et al. [16], Krishna et al. [17], Mahmoud and Mandouh [18], Mead et al. [19], Chakraborty et al. [20], and Cordeiro et al. [21], among others.

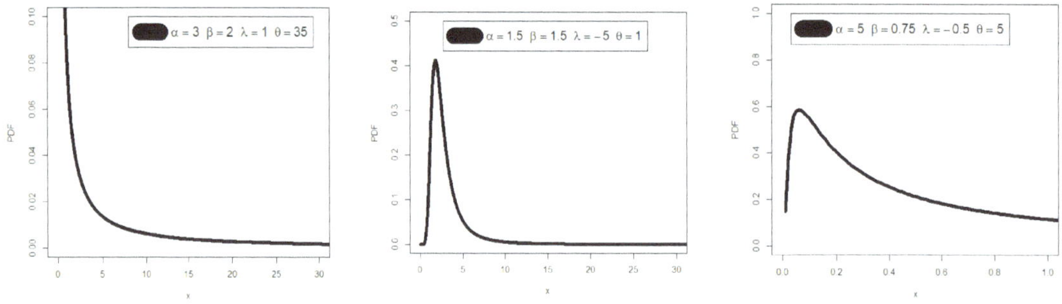

Figure 1. Graphs of the EGRRP PDF for selected parameter values.

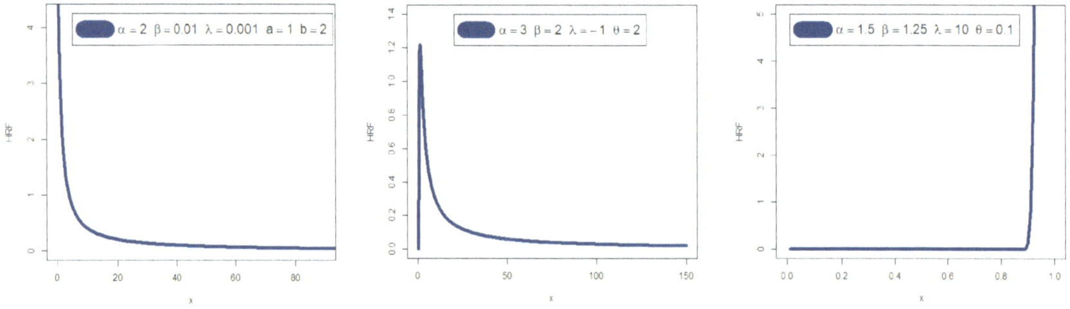

Figure 2. Graphs of the EGRRP HRF for selected parameter values.

2. Properties and Numerical Analysis

Using the power series expansion of $exp(x)$ the PDF in (6) can be expressed as

$$f(x;\underline{\mathcal{H}}) = 2\alpha\beta\theta^2 c_{(\lambda)}^{-1} x^{-3}\Delta_\theta(x)\sum_{i=0}^{\infty}(-1)^i \frac{\{1 - [1 - \Delta_\theta(x)]^\alpha\}^{\beta(i+1)-1}}{i!\lambda^{-i-1}\{1 - \Delta_\theta(x)\}^{-\alpha+1}}.$$

Using

$$\left(1-\frac{\delta_1}{\delta_2}\right)^{\delta_4}=\sum_{\delta_3=0}^{\infty}\left(-\frac{\delta_1}{\delta_2}\right)^{\delta_3}\frac{\Gamma(1+\delta_4)}{(\delta_3)!\Gamma(1+\delta_4-\delta_3)},$$

the last equation of $f(x;\underline{\mathcal{H}})$ can be expressed as

$$f(x;\underline{\mathcal{H}})=\sum\nolimits_{k=0}^{\infty}\xi_k g_{k^\cdot}(x;\theta)|_{k^\cdot=k+1}, \tag{7}$$

where

$$\xi_k=\frac{\alpha\beta(-1)^k}{c_{(\lambda)}k^\cdot}\sum_{i,j=0}^{\infty}\frac{(-1)^{i+j}}{i!\lambda^{-i-1}}\binom{\beta(i+1)-1}{j}\binom{\alpha(j+1)-1}{k},$$

and $g_{k^\cdot}(x;\theta)$ is the RR density with scale parameter $\theta\sqrt{k^\cdot}$. By integrating (7), we obtain another simple results of $F(x)$ as $F(x)=\sum_{k=0}^{\infty}\xi_k G_{k^\cdot}(x;\theta)$ where $G_{k^\cdot}(x;\theta)$ is the CDF of the RR distribution with scale parameter $\theta\sqrt{k^\cdot}$. The rth ordinary moment of X is given by $\mu'_{r,X}=E(X^r)=\int_{-\infty}^{\infty}x^r f(x)dx$. Using (7), we obtain

$$\mu'_{r,X}=\Gamma\left(1-\frac{r}{2}\right)\sum\nolimits_{k=0}^{\infty}\xi_k\left[\theta\sqrt{k^\cdot}\right]^r|_{2>r}, \tag{8}$$

where $\Gamma(1+\eta)|_{(\eta\in R^+)}=\eta!=\prod_{w=0}^{\eta-1}(\eta-w)$ and $\Gamma(\eta)=\int_0^{\infty}x^{\eta-1}exp(-x)dx$. Setting $r=1$ in (8), we have the mean of X as $E(X)=\mu'_{1,X}=\theta\Gamma\left(\frac{1}{2}\right)\sum_{k=0}^{\infty}\xi_k\sqrt{k^\cdot}$. We can find the MGF, say $M_X(t)=E(e^{tX})$ by

$$M_X(t)=\sum_{r=0}^{\infty}\frac{t^r}{r!}\mu'_r=\Gamma\left(1-\frac{r}{2}\right)\sum_{k,r=0}^{\infty}\frac{t^r}{r!}\xi_k\left[\theta\sqrt{k^\cdot}\right]^r|_{2>r}.$$

The sth incomplete moments, say $I_{s,X}(t)$, is given by $I_s(t)=\int_{-\infty}^{t}x^s f(x)dx$. Using (7), we obtain

$$I_{s,X}(t)=\sum_{k=0}^{\infty}\xi_k\left[\theta\sqrt{k^\cdot}\right]^s\gamma\left(1-\frac{s}{2},\left(\frac{\theta}{t}\right)^2k^\cdot\right)|_{2>s}, \tag{9}$$

where

$$\gamma(\eta,q)|_{(\eta\neq0,-1,-2,\ldots)}=\int_0^q t^{\eta-1}exp(-t)dt=\sum_{k=0}^{\infty}q^{\eta+k}\frac{(-1)^k}{k!(\eta+k)},$$

and $\Gamma(\eta,x)|_{(x>0)}=\int_q^{\infty}t^{\eta-1}exp(-t)dt$. The $(s,r)^{\text{th}}$ probability weighted moments (PWMs) of X following the EGRRP model, say $\rho_{s,r,X}$, is formally defined by $\rho_{s,r,X}=E\{X^sF(X)^r\}$. Using Equations (5) and (6), we can write $f(x;\underline{\mathcal{H}})F(x;\underline{\mathcal{H}})^r=\sum_{k=0}^{\infty}w_k g_{k^\cdot}(x;\theta)$ where

$$w_k=\frac{\alpha\beta}{c_{(\lambda)}^{r+1}k^\cdot}\sum_{w,i,j=0}^{\infty}\frac{(-1)^{w+i+j+k}}{i!\lambda^{-i-1}(1+w)^{-i}}\binom{r}{w}\binom{\beta(i+1)-1}{j}\binom{\alpha(1+j)-1}{k}.$$

Then, the $(s,r)^{\text{th}}$ PWM of X can be obtained and summarized as

$$\rho_{s,r,X} = \Gamma\left(1 - \frac{s}{2}\right) \sum_{k=0}^{\infty} w_k \left[\theta\sqrt{k^{\cdot}}\right]^s |_{2>s}.$$

The nth moment of the residual life, say $\tau_{n,X}(t) = E\left[(X-t)^n \,|_{n=1,2,\ldots}^{X>t}\right]$, uniquely determine $F(x)$. Therefore

$$\tau_{n,X}(t) = \frac{1}{1-F(t)} \sum\nolimits_{k=0}^{\infty} \xi_k^* \left[\theta\sqrt{k^{\cdot}}\right]^n \Gamma\left(1 - \frac{n}{2}, \left(\frac{\theta}{t}\right)^2 k^{\cdot}\right) |_{2>n}, \quad (10)$$

where $\xi_k^* = \xi_k \sum_{r=0}^{n} \binom{n}{r} (-t)^{n-r}$ and $\Gamma(\eta, x) = \Gamma(\eta) - \gamma(\eta, x)$. The nth moment of the reversed residual life, say $\omega_n(t) = E\left[(t-X)^n \,|_{n=1,2,\ldots}^{X\leq t \text{ and } t>0}\right]$. Therefore

$$\omega_{n,X}(t) = \frac{1}{F(t)} \sum_{k=0}^{\infty} \xi_k^{**} \left[\theta\sqrt{k^{\cdot}}\right]^n \gamma\left(1 - \frac{n}{2}, \left(\frac{\theta}{t}\right)^2 k^{\cdot}\right) |_{2>n}.$$

where $\xi_k^{**} = \xi_k \sum_{r=0}^{n} (-1)^r \binom{n}{r} t^{n-r}$. Table 1 lists a few sub models from the EGRRP model. A numerical investigation of the E(X), Variance (V(X)), skewness (Ske(X)), and kurtosis (Ku(X)) is shown in Table 2. According to Table 2, the proposed model's skewness can have both positive and negative values. The proposed model's kurtosis ranges from greater than three to fewer than three.

Table 1. Some sub models from the EGRRP model.

N	α	β	θ	λ	Reduced Model
1		1			EIRP
2		1		λ→0	EIR
3		1	1		EIRP
4		1	1	λ→0	EIR
5	1				GIRP
6	1			λ→0	GIR
7	1		1		GIRP
8	1		1	λ→0	GIR
9	1	1			IRP
10	1	1		λ→0	IR
11	1	1	1		IRP

Table 2. E(X), VARK(X), Ske(X), and Ku(X) of the EGRRP distribution.

α	β	λ	θ	E(X)	VARK(X)	Ske(X)	Ku(X)
5	1.5	−1	3	2.570593	0.454841	−70.64223	1102.708
3				3.163345	1.331691	−29.92022	368.7397
3	1	1	0.5	0.39404	0.021403	165,229.7	−1,780,082
	1.5			0.44028	0.023666	479,669.1	−5,491,216
	2.5			0.50186	0.026875	1,835,150	−2,247,195
	3.5			0.54450	0.029267	4,431,099	−5,641,370
	5			0.59161	0.032095	11,249,343	−1,485,963

Table 2. *Cont.*

α	β	λ	θ	E(X)	VARK(X)	Ske(X)	Ku(X)
3	1.5	−30	1.5	2.988585	0.711030	−55.15369	840.369
		−20		2.761101	0.634484	−52.04927	778.093
		−10		2.395671	0.530971	−45.4008	648.572
		−5		2.056147	0.458989	−37.0731	494.613
		1		1.32084	0.21299	17,734.68	−202,978
		2		1.214951	0.155364	258.9211	−3135.18
		5		1.020418	0.054401	−69.3133	1288.51
3	2	1	0.5	0.474498	0.025416	102,1606	−12,162,466
			1	0.948997	0.101664	127,669.8	−1,519,896
			5	4.744983	2.541611	986.3462	−11,691.72
			10	9.489967	10.16644	92.41038	−1049.147
			20	18.97993	40.66578	−19.3316	281.174
			30	28.4699	91.49799	−30.56492	414.910
			50	47.44983	254.1611	−34.2731	459.057
			100	94.89967	1016.644	−35.16704	469.699
			200	189.7993	4066.578	−35.27878	471.030
			500	474.4983	25,416.11	−35.29372	471.208
			1000	948.9967	101,664.4	−35.29461	471.218

3. Actuarial Indicators for Risk Analysis and Management

A specific insurance policy or the insurance sector as a whole may be affected by future occurrences, and risk analysis in insurance data refers to the process of assessing and estimating the chance of such events. Identification, evaluation, and development of solutions to control or mitigate potential risks are the objectives of risk analysis. To ascertain the degree of risk associated with a certain policy or portfolio of policies, this procedure entails gathering, evaluating, and interpreting data regarding a variety of elements, including demographic data, insurance-claims history, and economic indicators. Insurance firms use the findings of risk analysis to determine rates, make underwriting judgements, and create loss mitigation plans. KRIs are an essential tool for risk management as they provide a clear, quantifiable, and actionable measure of an organization's key risks, allowing organizations to take proactive steps to manage these risks and avoid negative consequences. The selection of KRIs is important and should be based on the specific risks faced by the organization and its overall risk management strategy.

3.1. VARK Indicator

The VARK, a widely utilized financial term, is a measure of the maximum expected loss that a portfolio or investment may incur over a given time period and is commonly employed as a risk management tool by financial institutions to assess market risk. This single-number indicator provides a concise summary of the potential loss of a portfolio or investment. For example, a portfolio with a VARK of USD 1 million at a 95% confidence level implies a 5% probability that the portfolio will experience a loss exceeding USD 1 million during the specified time frame. Risk exposure is an inevitable aspect of the operations of insurance organizations, and actuaries have developed Various risk indicators to statistically evaluate it. VARK is used to determine the most probable maximum amount of capital that might be lost over a specified duration. However, a loss that is unbounded or at least equal to the value of the portfolio is not necessarily informative. The risk profiles of different portfolios with the same maximum loss can vary substantially. Therefore, the VARK typically depends on the probability distribution of the loss random variable, which is influenced by the overall distribution of the risk factors that affect the portfolio. Then, for EGRRP distributions, we can simply write

$$\Pr(X > Q(X)) = \begin{cases} 0.01|_{q=99\%} \\ 0.05|_{q=95\%} \\ \vdots \end{cases}. \tag{11}$$

The $Q(X) = F^{-1}(x; \underline{\mathcal{H}})$ refers to the quantile of the EGRRP model. The VARK is a practical instrument for risk management since it gives a clear and succinct assessment of the potential loss of an investment or portfolio. It has drawbacks, too, as it simply offers a point estimate of the probable loss and ignores tail risks or extreme events. Financial organizations frequently combine VARK with other risk management tools such as stress testing and scenario analysis to address these constraints.

3.2. TVARK Risk Indicator

The $\text{TVARK}(X; q, \underline{\mathcal{H}})$ at the $100q\%$ confidence level, can be defined as the expected losses given that the losses exceed the $100q\%$ of the distribution of X. Then, the $\text{TVARK}(X; q, \underline{\mathcal{H}})$ can be then calculated as

$$\text{TVARK}(X; q, \underline{\mathcal{H}}) = \mathbf{E}(X|X > \pi(q)) = \frac{1}{1-q} \int_{\pi(q)}^{+\infty} x f_{\underline{V}}(x; \underline{\mathcal{H}}) dx. \tag{12}$$

Then, we have

$$\text{TVARK}(X; q, \underline{\mathcal{H}}) = \frac{\theta}{1-q} \sum_{k=0}^{\infty} \xi_k^* \sqrt{k^{\cdot}}\, \Gamma\left(\frac{1}{2}, \left(\frac{\theta}{t}\right)^2 k^{\cdot}\right), \tag{13}$$

So, $\text{TVARK}(X; q, \underline{\mathcal{H}})$ can be calculated as average all the $\text{VARK}(X; q, \underline{\mathcal{H}})$ values over the confidence level q. That means that the indicator of $\text{TVARK}(X; q, \underline{\mathcal{H}})$ gives us many more information about the tail of the EGRRP distribution and its properties. Generally, the $\text{TVARK}(X; q, \underline{\mathcal{H}})$ can also be written as

$$e(X; q, \underline{\mathcal{H}}) = \text{TVARK}(X; q, \underline{\mathcal{H}}) - \text{VARK}(X; q, \underline{\mathcal{H}}), \tag{14}$$

where $e(X; q, \underline{\mathcal{H}})$ is the MEL function evaluated at the $100q^{th}$ quantile.

3.3. The TV Indicator

The TV indicator (TV $(X; q, \underline{\mathcal{H}})$) can be expressed as

$$\text{TV}(X; q, \underline{\mathcal{H}}) = E\left(X^2|X > \pi(q)\right) - [\text{TVARK}(X; q, \underline{\mathcal{H}})]^2.$$

For the EGRRP model, the quantity $E(X^2|X > \pi(q))$ is not exist; however, we dealt with this amount using numerical techniques to find the closest possible value for it. It is known that numerical techniques represent the best solution in many of the problems of estimation and specialist modeling, where $\text{TVARK}(X; q, \underline{\mathcal{H}})$ is given in (13).

3.4. TMV Risk Indicator

The TMV risk indicator (TMV (X)) for the EGRRP model can then be derived as

$$\text{TMV}(X; q, \underline{\mathcal{H}}; \pi)|0 < \pi < 1 = \text{TVARK}(X; q, \underline{\mathcal{H}}) + \pi \text{TV}(X; q, \underline{\mathcal{H}}). \tag{15}$$

Then, for any loss random variable, $(\text{TMVX}; q, \underline{\mathcal{H}}; \pi) > \text{TV}(X; q, \underline{\mathcal{H}})$, and for $\pi = 1$ the $(\text{TMVX}; q, \underline{\mathcal{H}}; \pi) = \text{TVARK}(X; q, \underline{\mathcal{H}})$. Some other common examples of KRSIs can be mentioned such as:

I. Indicators of the frequency and size of losses resulting from different risks, including accidents, losses from fraud, or losses from natural catastrophes, are measured by these KRSIs.
II. Volatility indicators: These KRSIs measure the level of volatility in various financial markets, such as the stock market, currency market, or commodities market.
III. Credit risk indicators: These KRSIs measure the credit risk of various borrowers, such as individuals or organizations, based on their credit history and financial information.
IV. Operational risk indicators: These KRSIs measure the level of operational risk associated with various processes, such as supply chain disruptions, IT failures, or human errors.
V. Market risk indicators: These KRSIs measure the level of market risk associated with investments, such as stocks, bonds, or commodities.

4. Estimation

4.1. Classical Estimation

4.1.1. Maximum Likelihood Technique

For determining the maximum likelihood estimation (MLE) of $\underline{\mathcal{H}}$, we formulate the log-likelihood function as follows

$$\ell(x_i;\underline{\mathcal{H}}) = nlog2 + nlog\alpha + nlog\beta + nlog\lambda - nlog[1-exp(-\lambda)] + n2log\theta - 3\sum_{i=1}^{n} logx_i$$
$$-\sum_{i=1}^{n}(\theta/x)^2 + (\alpha-1)\sum_{i=1}^{n} log[1-\Delta_\theta(x_i)] + (\beta-1)\sum_{i=1}^{n} log\{1-[1-\Delta_\theta(x_i)]^\alpha\} - \lambda\sum_{i=1}^{n}\{1-[1-\Delta_\theta(x_i)]^\alpha\}^\beta.$$

The components of the score vector components are easily derived and then solved. To solve these equations, it is usually more convenient to use nonlinear optimization techniques such as the quasi-Newton algorithm to numerically maximize $\ell(x_i;\underline{\mathcal{H}})$. A popular numerical optimization approach for maximizing functions is the quasi-Newton algorithm. It is an iterative approach that computes the search direction at each iteration using approximations of the Hessian matrix. The fundamental goal of the quasi-Newton approach is to update a Hessian matrix approximation based on gradient data gathered from evaluating the objective function at various locations. Using the updated Hessian approximation, the technique determines a search direction at each iteration and moves in that direction to find a new location to evaluate the objective function. The Broyden–Fletcher–Goldfarb–Shanno (BFGS) algorithm is the most used quasi-Newton algorithm. The rank-two update formula used by the BFGS algorithm to update the Hessian approximation is made to keep the approximation's positive definiteness. When there are many variables in an optimization issue, the BFGS algorithm is frequently used.

4.1.2. Bootstrapping Technique

The bootstrapping technique is a potent statistical technique, particularly useful when dealing with small sample sizes. In traditional scenarios, assuming a normal or t-distribution is not feasible when working with less than 40 samples. However, bootstrap techniques are well-suited for sample sizes of less than 40 as they involve resampling and do not make any assumptions about the data distribution. With the increasing availability of computing resources, bootstrapping has gained popularity as a practical approach that requires the use of a computer. The following section will illustrate how this technique operates. This is due to the necessity of using a computer for bootstrapping to be useful. In the section that follows, we will examine how these function. There are several different types of bootstrapping techniques, including:

I. Non-parametric bootstrapping: In non-parametric bootstrapping, the statistic of interest is calculated directly from the resampled data without making any assumptions about the underlying probability distribution. This is the most commonly used type of bootstrapping and can be used for a wide range of estimators.

II. Parametric bootstrapping: In parametric bootstrapping, the resampled data are generated from a specific parametric distribution that is assumed to describe the data. This can be useful when the underlying distribution is known or can be reasonably assumed and can lead to more accurate estimates than non-parametric bootstrapping.

III. Bootstrap aggregating (bagging): In bagging, multiple copies of the original dataset are created by resampling, and then a separate model is trained on each of these new datasets. The final estimate is then obtained by averaging the estimates of the individual models. Bagging is commonly used in machine learning and can improve the accuracy and stability of models that are prone to overfitting.

IV. Cross-validation bootstrapping: In cross-validation bootstrapping, the original dataset is divided into several subsets, and then a separate model is trained on each subset while using the remaining data for validation. The final estimate is then obtained by averaging the estimates of the individual models. Cross-validation bootstrapping is commonly used in machine learning and can help to prevent overfitting by reducing the variance of the estimate.

V. Bootstrapping has become a widely used technique for statistical inference and estimation, and it has been applied to a wide range of fields, including finance, engineering, social sciences, and natural sciences. Bootstrapping can be implemented using various statistical software packages, including R, Python, MATLAB, and SAS.

4.1.3. Technique of Cramér–von Mises

The Cramér–von Mises estimation (CVME) technique of the parameters is based on the theory of minimum distance estimation. The CVME of the parameters α, β, λ, and θ are obtained by minimizing the following expression with respect to the parameters α, β, λ, and θ, respectively, where

$$\mathrm{CVM}_{(\underline{\mathcal{H}})} = \frac{1}{12n} + \sum_{i=1}^{n}\left[F(x)|_{\alpha,\beta,\theta,\lambda}^{x_{i:n}} - k(i,n)\right]^2,$$

where $k(i,n) = (2i-1)/n$ and

$$\mathrm{CVM}_{(\underline{\mathcal{H}})} = \frac{1}{12n} + \sum_{i=1}^{n}\left\{c^{-1}{}_{(\lambda)}\left[1 - exp\left(-\lambda\{1-[1-\Delta_\theta(x_i)]^\alpha\}^\beta\right)\right] - k(i,n)\right\}^2.$$

Then, CVME of the parameters are obtained by solving the following non-linear equations

$$\sum_{i=1}^{n}\xi_\alpha(x_i|\underline{\mathcal{H}})\left\{c^{-1}{}_{(\lambda)}\left[1 - exp\left(-\lambda\{1-[1-\Delta_\theta(x_i)]^\alpha\}^\beta\right)\right] - k(i,n)\right\} = 0,$$

$$\sum_{i=1}^{n}\xi_\beta(x_i|\underline{\mathcal{H}})\left\{c^{-1}{}_{(\lambda)}\left[1 - exp\left(-\lambda\{1-[1-\Delta_\theta(x_i)]^\alpha\}^\beta\right)\right] - k(i,n)\right\} = 0,$$

$$\sum_{i=1}^{n}\xi_\lambda(x_i|\underline{\mathcal{H}})\left\{c^{-1}{}_{(\lambda)}\left[1 - exp\left(-\lambda\{1-[1-\Delta_\theta(x_i)]^\alpha\}^\beta\right)\right] - k(i,n)\right\} = 0,$$

and

$$\sum_{i=1}^{n}\xi_\theta(x_i|\underline{\mathcal{H}})\left\{c^{-1}{}_{(\lambda)}\left[1 - exp\left(-\lambda\{1-[1-\Delta_\theta(x_i)]^\alpha\}^\beta\right)\right] - k(i,n)\right\} = 0,$$

where $\xi_\alpha\left(x_i|_{\underline{\mathcal{H}}}\right)$, $\xi_\beta(\underline{\mathcal{H}})$, $\xi_\lambda(\underline{\mathcal{H}})$, and $\xi_\theta(\underline{\mathcal{H}})$ are the values of the first derivatives of the CDF of EGRRP distribution with respect to $\alpha, \beta, \lambda, \theta$, respectively.

4.2. Bayesian Estimation

In this part, we build estimators for the EGRRP distribution's unknown parameters using Bayesian techniques. The maximum likelihood estimator frequently fails to converge, particularly in models with larger dimensions. In these situations, Bayesian approaches are sought after. Bayesian approaches initially appear to be highly complicated because the estimators entail unsolvable integrals. Here we assume the gamma priors of the parameters $(\alpha, \beta, \lambda, \theta)$ of the following forms

$$\pi_1(\alpha) \sim Gamma(\xi_1, d_1), \pi_2(\beta) \sim Gamma(\xi_2, d_2), \pi_3(\theta) \sim Gamma(\xi_3, d_3), \pi_4(\lambda) \sim Normal\left(\xi_4, d_4^2\right),$$

where, Gamma $(\xi_i, d_i)|_{(i=1,2,3)}$ stands for gamma distribution with shape parameter ξ_i and scale parameter d_i, and normal (ξ_4, d_4^2) stands for the normal distribution with shape parameter ξ_4 and d_4^2.

The Gamma distribution is a conjugate prior for several common likelihood functions, including the Poisson, exponential, and normal distributions. This means that if we choose a Gamma prior, the resulting posterior distribution will also be a Gamma distribution. This makes the Bayesian inference computationally efficient and enables us to obtain the posterior distribution in closed form. The Gamma distribution is a distribution over positive values only, which makes it a natural choice for modeling quantities that are inherently positive, such as rates, counts, or durations. The Gamma distribution is a flexible distribution that can take on a wide range of shapes, including skewed, unimodal, and multimodal shapes. This makes it a good choice for modeling a wide range of different data types. The parameters of the Gamma distribution have clear and intuitive interpretations, which makes it easy to incorporate prior knowledge into the model. For example, the shape parameter of the Gamma distribution can be interpreted as the number of prior observations, and the scale parameter can be interpreted as the prior sum of the observations. The Gamma distribution is relatively robust to deviations from the assumed model, which makes it a good choice when the data are noisy or when there is uncertainty about the model specification.

It is further assumed that the parameters are to be independently distributed. The joint prior distribution is given by

$$\pi(\alpha, \beta, \lambda, \theta) = \frac{d_1^{\xi_1}}{\Gamma(\xi_1)} \frac{d_2^{\xi_2}}{\Gamma(\xi_2)} \frac{d_3^{\xi_3}}{\Gamma(\xi_3)} \alpha^{\xi_1 - 1} \beta^{\xi_2 - 1} \theta^{\xi_3 - 1} \left(2\pi d_4^2\right)^{-\frac{1}{2}} e^{-(\alpha d_1 + \beta d_2 + \lambda d_3) - [(\lambda - \xi_4)^2/(2d_4^2)]}.$$

The posterior distribution of the parameters is defined by $\pi(\alpha, \beta, \lambda, \theta|\underline{x}) \propto$ likelihood $(\alpha, \beta, \lambda, \theta|\underline{x}) \times \pi(\alpha, \beta, \lambda, \theta)$. As a consequence, we recommend employing Markov chain Monte Carlo (MCMC) methods, particularly the Gibbs sampler and the Metropolis Hastings (MH) technique. We implemented a hybrid MCMC approach to draw samples from the joint posterior of the parameters because it is not possible to collect the conditional posteriors of the parameters in any basic structures. To implement the Gibbs algorithm, the full conditional posteriors of α, λ, θ, and θ are given by

$$\pi(\alpha, \beta, \lambda, \theta) = \frac{d_1^{\xi_1}}{\Gamma(\xi_1)} \frac{d_2^{\xi_2}}{\Gamma(\xi_2)} \frac{d_3^{\xi_3}}{\Gamma(\xi_3)} \alpha^{\xi_1 - 1} \beta^{\xi_2 - 1} \theta^{\xi_3 - 1} d_4^{-1} (2\pi)^{-\frac{1}{2}} e^{-(\alpha d_1 + \beta d_2 + \lambda d_3) - \frac{1}{2}\left(\frac{\lambda - \xi_4}{d_4}\right)^2}.$$

$$\pi_1(\alpha|\beta, \lambda, \theta, \underline{x}) \propto \alpha^{n + \xi_1 - 1} e^{-(\alpha d_1)} \prod_{i=1}^{n} \Lambda_i, \pi_2(\beta|\alpha, \lambda, \theta, \underline{x}) \propto \beta^{n + \xi_2 - 1} e^{-(\beta d_2)} \prod_{i=1}^{n} \Lambda_i,$$

$$\pi_3(\theta|\alpha,\beta,\lambda,\underline{x}) \propto \theta^{n+\xi_3-1}e^{-(\theta d_3)}\prod_{i=1}^{n}\Lambda_i, \pi_4(\lambda|\alpha,\beta,\lambda,\underline{x}) \propto d_4^{-n}e^{-\frac{1}{2}(\frac{\lambda-\xi_4}{d_4})^2}\prod_{i=1}^{n}\Lambda_i,$$

where

$$\Lambda_i = x_i^{-3}\frac{\Delta_\theta(x_i)}{[c_{(\lambda)}]\{1-\Delta_\theta(x_i)\}^{1-\alpha}}\frac{(1-[1-\Delta_\theta(x_i)]^\alpha)^{\beta-1}}{exp[\lambda(1-[1-\Delta_\theta(x_i)]^\alpha)^\beta]}.$$

The simulation algorithm we followed is given by

1. Provide the initial values, say $\alpha_{(0)}, \beta_{(0)}, \lambda_{(0)}$, and $\theta_{(0)}$ where $\alpha_{(0)} > 0$, $\beta_{(0)} > 0, \lambda_{(0)} \in$ R$-\{0\}$ and $\theta_{(0)} > 0$ (the initial values are randomly determined by the selection of the researcher, provided that it is within the specified range) then at i^{th} stage;
2. Using MH algorithm, generate $\alpha_{(i)} \sim \pi_1(\alpha_{(i-1)}|\beta_{(i-1)}, \lambda_{(i-1)}, \theta_{(i-1)})$;
3. Then, using the well-known algorithm MH, generate $\beta_{(i)} \sim \pi_2(\beta_{(i-1)}|\alpha_{(i)}, \lambda_{(i-1)}, \theta_{(i-1)})$;
4. Then, using the well-known algorithm MH, generate $\theta_{(i)} \sim \pi_3(\theta_{(i-1)}|\alpha_{(i)}, \beta_{(i)}, \lambda_{(i-1)})$;
5. Then, using the well-known algorithm MH, generate $\lambda_{(i)} \sim \pi_3(\lambda_{(i-1)}|\alpha_{(i)}, \beta_{(i)}, \theta_{(i)})$;
6. Repeat steps 2–5, $M = 100,000$ times to obtain the samples of size M from the corresponding posteriors of interest.

Obtain the Bayesian estimates of α, β, λ and θ using the following formulae $\hat{h}_{\text{Bayesian}} = \frac{1}{M-M_0}\sum_{j=1+M_0}^{M} h_j \,|_{h=\alpha,\ \beta,\ \lambda \text{ and } \theta}$ where $M_0(\approx 50,000)$ is the burn-in period of the generated Markov chains.

5. Simulations for Comparing Bayesian and Classical Approaches

Simulation studies are a crucial tool for assessing and contrasting various statistical techniques, including traditional estimation techniques. In simulation studies, data are generated based on a given model, and the efficacy of various estimating techniques is evaluated using the generated data. We shall talk about the value of simulation studies for contrasting traditional estimating techniques in this essay. Simulations are an important tool for comparing Bayesian and classical estimation techniques because they allow us to systematically evaluate the performance of different techniques under a wide range of conditions. Simulations allow us to examine the performance of Bayesian and classical techniques under different sample sizes, which can be particularly useful when the sample size is small. By simulating data with different sample sizes, we can assess how well the different techniques perform under conditions of low data availability. Simulations allow us to evaluate the performance of Bayesian and classical techniques under different parameter values, which can be important when the parameters of interest are not known a priori. By simulating data with different parameter values, we can assess how well the different techniques perform under conditions of parameter uncertainty. Simulations allow us to evaluate the robustness of Bayesian and classical techniques to deviations from the assumed model. By simulating data that deviate from the assumed model, we can assess how well the different techniques perform under conditions of model misspecification. Simulations allow us to compare the accuracy and precision of Bayesian and classical techniques under different conditions. By simulating data with known parameter values, we can assess how accurately and precisely the different techniques estimate the true parameters. Simulations allow us to assess the computational efficiency of Bayesian and classical techniques under different conditions. By simulating data with different sample sizes and parameter values, we can assess how well the different techniques scale to larger and more complex datasets.

The mean squared error (MSE) is a performance indicator that is frequently used in simulation studies to assess the precision of a statistical model or estimator. The average of the squared discrepancies between the estimated values and the actual values of the parameter being estimated is known as the MSE. In simulation research, MSE is chosen over measures of dispersion and biases for a number of reasons. It is a thorough evaluation. The MSE accounts for both the estimator's bias and variability. Measurements of bias simply reflect the discrepancy between the estimator and the true value, while measures of dispersion, such as variance or standard deviation, only record the estimator's variability. Because it takes into account both types of error, the MSE offers a more complete evaluation of the estimator's performance. It is simple to understand. The MSE is simple to read because it uses the same units as the parameter being estimated.

A MCMC simulation study is conducted and performed in this section to assess and compare the performance of the different estimators of the unknown parameters of the EGRRP distribution. This performance is assessed using the average values (AVs) of estimates and the MSEs. First, we generated 1000 samples of the EGRRP distribution, where $n = (20,\ 50,\ 100,\ 200)$ and choosing

	α	β	λ	θ
I	2	1.5	−1.5	1.2
II	0.6	2	1.5	0.5

The tables present the AVs and MSEs of various parameter estimators, namely MLEs, Bootstrap, CVMEs, and Bayesian estimators. To evaluate the Bayesian estimators, the MCMC technique is used with a flexible gamma prior under the SELF for all parameters, except for parameter λ, which uses a normal prior. Hyperparameters are assumed to be known and selected to have a prior mean equal to the initial value and a prior variance of one. Results from Tables 3–6 demonstrate that all estimators exhibit consistency, as evidenced by the decreasing MSEs as the sample size increases. Moreover, the Bayesian estimators have lower MSEs than the other estimators, and in some cases, the MSEs of the Bayesian and MLEs are very similar. The computations in this section were performed using the Mathcad program, version 15.0.

Table 3. The results of the AVs and their corresponding MSEs (in parentheses) for $n = 50$.

Initials	Bayesian	MLE	Bootstrap	CVM
$\alpha = 2$	1.97517	2.01461	2.04059	2.01152
	(0.04134)	(0.05516)	(0.06786)	(0.05536)
$\beta = 1.5$	1.53674	1.53098	1.50903	1.54462
	(0.06150)	(0.06659)	(0.07169)	(0.08980)
$\lambda = -1.5$	−1.26544	−1.52255	−1.45553	−1.54183
	(0.27332)	(0.27681)	(0.34521)	(0.29966)
$\theta = 1.2$	1.20801	1.20402	1.19769	1.20571
	(0.004192)	(0.00425)	(0.00475)	(0.00508)
$\alpha = 0.6$	0.58200	0.60577	0.61474	0.60646
	(0.00444)	(0.00697)	(0.00968)	(0.00595)
$\beta = 2$	1.95015	2.02998	1.97490	2.02591
	(0.05002)	(0.05832)	(0.05927)	(0.07796)
$\lambda = 1.5$	1.09199	1.51026	1.59306	1.52918
	(0.36095)	(0.38682)	(0.54474)	(0.31561)
$\theta = 0.5$	0.49998	0.50463	0.49484	0.50403
	(0.00140)	(0.00161)	(0.00166)	(0.00238)

Table 4. The results of the AVs and their corresponding MSEs (in parentheses) for $n = 100$.

Initials	Bayesian	MLE	Bootstrap	CVM
$\alpha = 2$	2.01445	2.01479	2.08685	2.00714
	(0.01989)	(0.02742)	(0.03198)	(0.02675)
$\beta = 1.5$	1.43578	1.50725	1.43233	1.51994
	(0.03121)	(0.03105)	(0.02795)	(0.04102)
$\lambda = -1.5$	−1.67548	−1.49394	−1.35285	−1.51710
	(0.08359)	(0.13440)	(0.11286)	(0.14247)

Table 4. *Cont.*

Initials	Bayesian	MLE	Bootstrap	CVM
$\theta = 1.2$	1.19235	1.19991	1.18056	1.20238
	(0.00192)	(0.00207)	(0.00201)	(0.00246)
$\alpha = 0.6$	0.62188	0.60024	0.60884	0.60335
	(0.00259)	(0.00321)	(0.00530)	(0.00298)
$\beta = 2$	1.88655	2.02283	1.99553	2.01315
	(0.02794)	(0.02865)	(0.03892)	(0.03870)
$\lambda = 1.5$	1.72490	1.48553	1.52840	1.51536
	(0.07370)	(0.18452)	(0.25946)	(0.15856)
$\theta = 0.5$	0.48597	0.50362	0.49892	0.50205
	(0.00088)	(0.00079)	(0.00106)	(0.00119)

Table 5. The results of the AVs and their corresponding MSEs (in parentheses) for $n = 200$.

Initials	Bayesian	MLE	Bootstrap	CVM
$\alpha = 2$	1.95326	2.00543	2.08507	2.00823
	(0.01153)	(0.01489)	(0.02350)	(0.01327)
$\beta = 1.5$	1.50538	1.50683	1.42706	1.50400
	(0.01469)	(0.016659)	(0.01938)	(0.01954)
$\lambda = -1.5$	−1.33473	−1.502215	−1.33426	−1.49740
	(0.04993)	(0.07365)	(0.09376)	(0.06982)
$\theta = 1.2$	1.19859	1.20070	1.17954	1.19973
	(0.00112)	(0.00113)	(0.00146)	(0.00121)
$\alpha = 0.6$	0.60456	0.60010	0.60245	0.59973
	(0.00128)	(0.00169)	(0.00165)	(0.00141)
$\beta = 2$	2.02997	2.01185	2.00300	2.01313
	(0.01397)	(0.01465)	(0.01382)	(0.01806)
$\lambda = 1.5$	1.45423	1.49191	1.51023	1.49367
	(0.05122)	(0.09747)	(0.09543)	(0.07502)
$\theta = 0.5$	0.50809	0.50188	0.50038	0.50219
	(0.00041)	(0.00041)	(0.00038)	(0.00056)

Table 6. The results of the AVs and their corresponding MSEs (in parentheses) for $n = 500$.

Initials	Bayesian	MLE	Bootstrap	CVM
$\alpha = 2$	1.95330	2.00201	2.04971	1.99759
	(0.00520)	(0.00555)	(0.00827)	(0.00487)
$\beta = 1.5$	1.48986	1.50247	1.45521	1.50826
	(0.00489)	(0.00629)	(0.00784)	(0.00750)
$\lambda = -1.5$	−1.44498	−1.50082	−1.40077	−1.51180
	(0.02654)	(0.02772)	(0.03621)	(0.02653)
$\theta = 1.2$	1.20043	1.20025	1.18770	1.20158
	(0.00042)	(0.00043)	(0.00057)	(0.00046)
$\alpha = 0.6$	0.58994	0.60066	0.57935	0.60102
	(0.00052)	(0.00062)	(0.00105)	(0.00056)
$\beta = 2$	1.99303	2.00223	2.06060	2.00119
	(0.00509)	(0.00531)	(0.01004)	(0.00716)
$\lambda = 1.5$	1.38815	1.50180	1.33988	1.50565
	(0.02737)	(0.03596)	(0.06247)	(0.02962)
$\theta = 0.5$	0.49877	0.50033	0.50989	0.50016
	(0.00015)	(0.00015)	(0.00027)	(0.00022)

6. Applications for Comparing Bayesian and Classical Estimations

Two real-life datasets for applications are introduced and analyzed to for some purposed including comparing Bayesian and classical estimations. In these applications, we recommend and consider the Cramér–von Mises (W^*), the Anderson–Darling (A^*) and the Kolmogorov–Smirnov (KS) test statistic for comparing methods. The 1st dataset consists of 100 observations of breaking stress of carbon fibers (see Nichols and Padgett [22]). Table 7 gives the values of estimators of α, β,, and θ, the KS test statistics and its p-value, and W* and A^* for all techniques using the 1st dataset. From Table 7 we conclude that the Bayesian technique is the best technique with KS = 0.067, p-value = 0.766, W* = 0.066, and A^* = 0.52. However, all other techniques performed well. For the 2nd dataset (see Smith and Naylor [23]), these data were originally obtained by workers at the UK National Physical Laboratory. Table 8 gives the values of estimators of α, β, λ, and θ, the KS test statistics and its p-value, and W^* and A^* for all techniques using the 2nd dataset. From Table 8 we conclude all other techniques performed well, and according to these results we cannot select a technique as the best one.

Table 7. The estimated parameters, KS, p-values, W*, and A^* for all estimation techniques using the 1st dataset.

Technique	$\hat{\alpha}$	$\hat{\beta}$	$\hat{\lambda}$	$\hat{\theta}$	KS	p-Value	W^*	A^*
ML	2.714	4.804	−9.936	0.749	0.072	0.683	0.073	0.55
Bayesian	2.899	5.119	−7.038	0.843	0.067	0.766	0.066	0.52
Bootstrap	2.899	5.163	−7.088	0.844	0.070	0.717	0.066	0.52
CVM	2.660	6.705	−17.703	0.607	0.067	0.756	0.087	0.623

Table 8. The estimated parameters, KS, p-values, W*, and A^* for all estimation techniques using the 2nd dataset.

Technique	$\hat{\alpha}$	$\hat{\beta}$	$\hat{\lambda}$	$\hat{\theta}$	KS	p-Value	W^*	A^*
ML	3.206	5.73	−19.87	0.726	0.069	0.926	0.059	0.47
Bayesian	4.804	3.528	0.270	1.705	0.067	0.943	0.067	0.57
Bootstrap	4.780	3.602	0.235	1.710	0.08	0.798	0.067	0.56
CVM	2.954	15.506	1.755	1.107	0.078	0.84	0.058	0.43

7. Applications for Comparing Competitive Distributions

This section presents two real-life applications of demonstrating the EGRRP distribution using real datasets. We compare the results of the new EGRRP distribution with the Weibull-inverse-Weibull (WIW), exponentiated-inverse-Weibull (EIW) (see Nadarajah and Kotz [14]), Kumaraswamy-inverse-Weibull (KumIW) (see Mead and Abd-Eltawab [19]), beta-inverse-Weibull (BIW) (Barreto-Souza et al. [16]), transmuted-inverse-Weibull (TIW) (see Mahmoud and Mandouh [18]), gamma extended-inverse-Weibull (see GEIW) (Silva et al. [24]), Marshall-Olkin-inverse-Weibull (MOIW) (see Krishna et al. [17]), and Reciprocal Weibull (RW) distributions.

The unknown parameters of these PDFs are all positive real numbers, except for the TIW distribution. To compare the distributions, we use various criteria such as the maximized Log-Likelihood, AIC (Akaike Information Criterion), CAIC (Consistent Akaike Information Criterion), BIC (Bayesian Information Criterion), and HQIC (Hannan–Quinn Information Criterion). All computations are conducted using the R PROGRAM. Additionally, the TTT graph is used to graphically verify whether the data can be fit to a specific distribution or not, which is an important graphical approach (Aarset [25]). A straight diagonal TTT graph indicates a constant HRF; a concave TTT graph indicates an increasing (or decreasing) HRF; and a U-shaped (bathtub) TTT graph indicates a unimodal HRF, while other shapes indicate otherwise. The TTT graphs for the two real datasets are presented in Figure 3, where we conclude that the empirical HRFs of the two datasets are increasing.

We contrast the EGRRP model with strong competitive distributions in Tables 9–12. Table 9 gives -2ℓ, AIC, BIC, HQIC, and CAIC for the 1st dataset. Table 10 lists MLEs and their standard errors (in parentheses) for the 1st dataset. Table 11 gives -2ℓ, AIC, BIC, HQIC, and CAIC for the 2nd dataset. Table 12 lists MLEs and their standard errors (in parentheses) for the 2nd dataset. Of all models fitted to the two real-life datasets, the EGRRP model provides the best values for the AIC, BIC, HQIC, and CAIC statistics. So, it may be picked as the best option. For the first set of data, Figure 4 shows graphs for estimated CDFs, estimated PDFs, P-P graph, and the Kaplan–Meier graph. The graphs of estimated CDFs estimated PDFs, P-P graph, and the Kaplan–Meier survival graph for the second dataset are shown in Figure 5. These graphs suggest that for both datasets, the suggested distribution provides a better match than other non-nested and nested distributions.

Table 9. -2ℓ, AIC, BIC, HQIC, and CAIC for the 1st dataset.

Model			**Criteria**		
	-2ℓ	**AIC**	**BIC**	**HQIC**	**CAIC**
EGRRP	105.26	113.26	123.68	117.48	113.68
WIW	287.53	295.50	305.92	299.71	295.90
EIW	284.77	296.74	304.51	299.92	297.00
KumIW	286.13	298.22	308.50	302.35	298.50
BIW	304.12	312.10	322.60	316.40	313.60
GEIW	305.12	313.03	333.46	318.22	314.40
IW	345.33	348.30	354.50	352.40	349.44
TIW	346.55	353.55	359.36	354.61	352.74
MOIW	344.33	352.30	358.13	352.50	353.68

Table 10. MLEs and their standard errors (in parentheses) for the 1st dataset.

Model		**Estimates**		
EGRRP (α,β,λ,θ)	2.7144	4.8045	−9.9361	0.7499
	(0.0784)	(0.537)	(0.996)	(0.018)
WIW (α,β,a,b)	2.22313	0.3552	6.97213	4.91791
	(11.409)	(0.411)	(113.83)	(3.7562)
KumIW (α,β,a,b)	2.05562	0.4654	6.28153	224.188
	(0.0716)	(0.007)	(0.0603)	(0.1646)
BIW (α,β,a,b)	1.60973	0.4046	22.0143	29.7624
	(2.4982)	(0.108)	(21.432)	(17.488)
GEIW (α,β,a,b)	1.36962	0.4776	27.6452	17.4584
	(2.0178)	(0.133)	(14.136)	(14.823)
EIW (α,β,a)	69.1489	0.5019	145.328	
	(57.349)	(0.086)	(122.924)	
TIW (α,β,a)	1.93156	1.7435	0.08195	
	(0.0975)	(0.076)	(0.1984)	
MOIW (α,β,a)	2.30666	1.5796	0.5995	
	(0.4982)	(0.1666)	(0.3093)	
IW (α,β)	1.87054	1.7779		
	(0.1126)	(0.11346)		

Table 11. -2ℓ, AIC, BIC, HQIC, and CAIC for the 2nd dataset.

Model	Criteria				
	-2ℓ	AIC	BIC	HQIC	CAIC
EGRRP	39.611	47.661	56.231	51.031	48.351
EGIWP	39.635	49.166	58.077	52.373	50.291
PBXIW	41.403	49.588	58.105	52.934	50.366
BIW	60.601	68.603	77.201	72.029	69.324
GEIW	61.605	69.603	78.101	72.991	70.333
IW	93.773	97.722	102.01	99.404	97.903
TIW	94.144	100.12	106.54	102.666	100.500
MOIW	95.722	101.78	108.28	104.299	102.132

Table 12. MLEs and their standard errors for the 2nd dataset.

Model	Estimates			
EGIWP (α,β,λ,θ)	3.206	5.73431	−19.8321	0.72639
	(0.58)	(9.22421)	(26.6332)	(0.32141)
PBXIW (λ,θ,α,β)	4.4921	19.9982	0.3833	0.5063
	(1.7783)	(9.2431)	(0.1438)	(0.1094)
BIW (α,β,a,b)	2.05181	0.6464	15.078	36.944
	(0.9861)	(0.1633)	(12.061)	(22.77)
GEIW (α,β,a,b)	1.66275	0.74213	32.1121	13.324
	(0.9521)	(0.1979)	(17.393)	(9.974)
TIW (α,β,a)	1.30629	2.7844	0.12982	
	(0.0334)	(0.165)	(0.2082)	
MOIW (α,β,a)	1.54413	2.38767	0.48163	
	(0.226)	(0.253)	(0.2525)	
IW (α,β)	1.26443	2.88873		
	(0.0592)	(0.23476)		

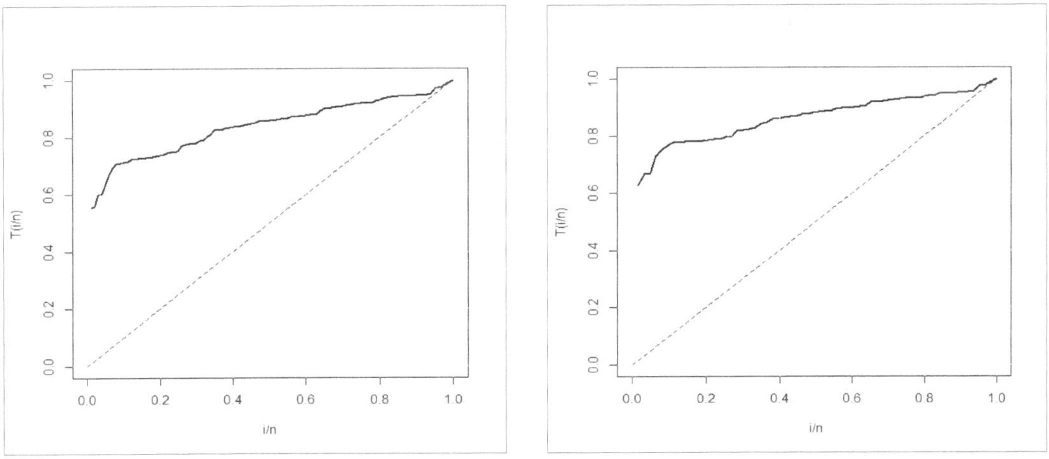

Figure 3. TTT plot for breaking stress of carbon fibers (**left**) and TTT plot for the strengths data of the glass fibers (**right**).

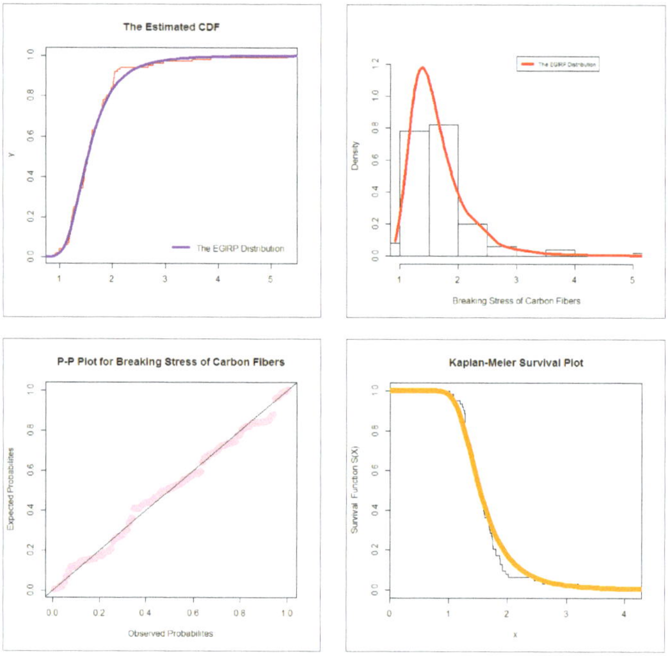

Figure 4. Estimated CDF (**top left**), estimated PDF (**top right**), P-P graph (**bottom left**), and Kaplan–Meier survival (**bottom right**) for the 1st dataset.

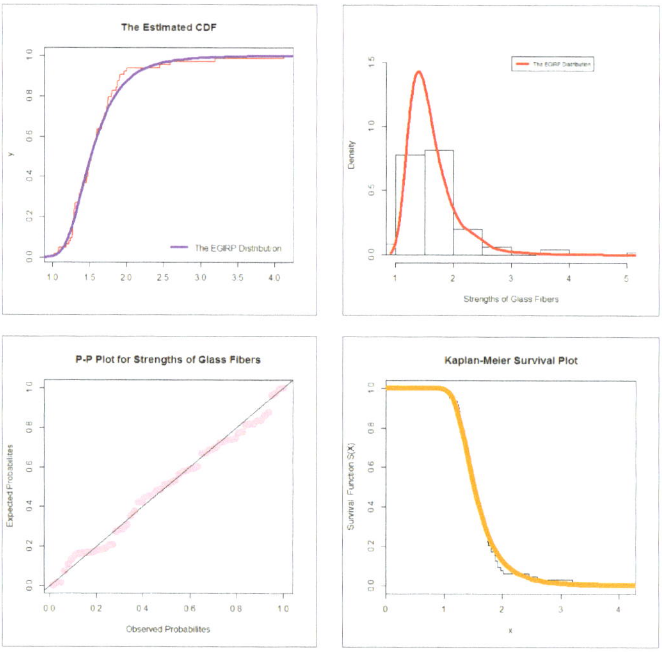

Figure 5. Estimated CDF (**top left**), estimated PDF (**top right**), P-P graph (**bottom left**), and Kaplan–Meier survival (**bottom right**) for the 2nd dataset.

8. Risk Analysis for Insurance-Claims Data

In insurance data analysis, the temporal growth of claims over time for each relevant exposure period is often presented in a triangle format. The exposure period can refer to the year the insurance policy was purchased or the time frame in which the loss occurred. It should be noted that the origin period need not be annual and can be monthly or quarterly. The claim age or claim lag is the duration between the origin period and when the claim is made. To identify consistent trends, division levels, or risks, data from various insurance policies are often combined. In this study, we utilize a U.K. Motor Non-Comprehensive account as an example of an insurance-claims payment triangle. For convenience, we set the origin period between 2007 and 2013 (see Shrahili et al. [3]; Mohamed [4]). The claims data are presented in an insurance-claims payment data frame in a standard database format, with the first column listing the development year, the incremental payments, and the origin year spanning from 2007 to 2013. It is crucial to note that a probability-based distribution was initially used to analyze this insurance-claims data. The data are analyzed using numerical and graphical techniques. The numerical approach involves fitting theoretical distributions such as the normal, uniform, exponential, logistic, beta, lognormal, and Weibull distributions. These distributions are then examined using graphical tools such as the skewness-kurtosis graph (or the Cullen and Frey graph) (see Figure 6). Figure 6 shows that our data are left-skewed and have a kurtosis of less than three.

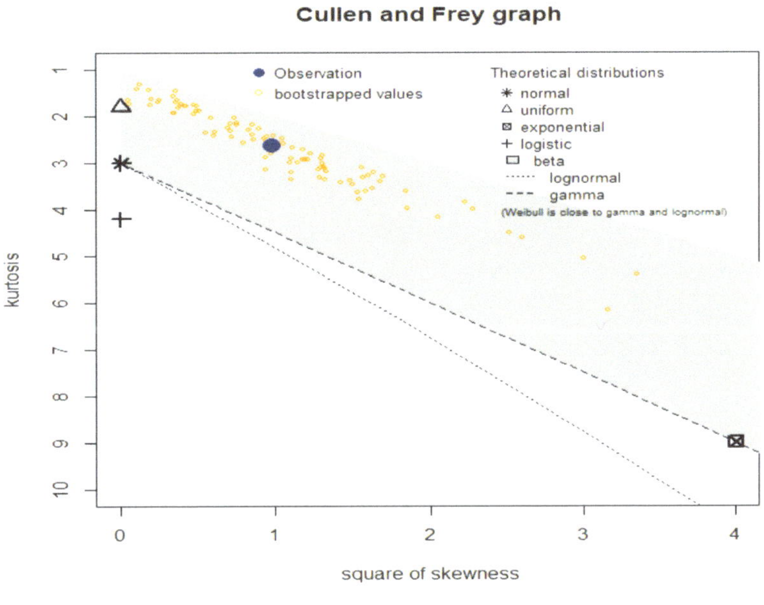

Figure 6. Cullen-Frey graph for the actuarial claims data.

Different approaches are utilized to examine multiple aspects of the insurance-claims data, which are presented in Figure 7. The NKDE method is implemented to analyze the initial shape of the insurance-claims density, while the Q-Q graph is used to evaluate the normality of the data. The TTT graph is employed to assess the initial shape of the empirical HRF, and the "box graph" is utilized to identify explanatory variables.

Figure 7 displays the results of the various graphs. The initial density is demonstrated to be an asymmetric function with a left tail in the top left graph. The bottom right graph indicates that there are no extreme claims. The bottom left graph shows that the HRF for models explaining the data should have a monotonically increasing trend. Scattergrams for the insurance-claims data are presented in Figure 8. The autocorrelation function (ACF) and partial autocorrelation function (partial ACF) for the data are depicted in Figure 9.

To assess risk for the insurance-claims data, measures such as VARK, TVARK, TV, TMV, and MEL are employed at various confidence levels, which are listed in Table 13. Table 14 gives the estimated parameters for the EGRRP under insurance-claims data. Table 15 lists the results of the KRIs for the IR under insurance-claims data. Table 16 gives the estimated parameters for the IR under insurance-claims data. The EGRRP and RR distributions are compared using five measures, while the KRIs for the EGRRP and IR are listed in Tables 13 and 15, respectively. Table 14 provides the estimators and ranks for the EGRRP model under the claims data for all estimation methods, while Table 16 lists the KRIs for the RR distribution under the claims data for all estimation methods. The RR distribution is chosen as the baseline distribution for comparison with the new EGRRP distribution. Based on these tables, the following results can be highlighted:

1. For all risk assessment Bayesian and non-Bayesian techniques $\mid q = 0.6,\ 0.7,\ 0.8,\ 0.9,\ 0.95,\ 0.99$, and 0.999:

$$\mathrm{VARK}(X;q,\underline{\mathcal{H}})|_{q=0.6} < \mathrm{VARK}(X;q,\underline{\mathcal{H}})|_{q=0.7}\ldots < \mathrm{VARK}(X;q,\underline{\mathcal{H}})|_{q=0.999}.$$

2. For all risk assessment Bayesian and non-Bayesian techniques $\mid q = 0.6,\ 0.7,\ 0.8,\ 0.9,\ 0.95,\ 0.99$, and 0.999:

$$\mathrm{TVARK}(X;q,\underline{\mathcal{H}})|_{q=0.6} < \mathrm{TVARK}(X;q,\underline{\mathcal{H}})|_{q=0.7}\ldots < \mathrm{TVARK}(X;q,\underline{\mathcal{H}})|_{q=0.999}.$$

3. For most risk assessment techniques $\mid q = 0.6,\ 0.7,\ 0.8,\ 0.9,\ 0.95,\ 0.99$, and 0.999:

$$\mathrm{TV}(X;q,\underline{\mathcal{H}})|_{q=0.6} < \mathrm{TV}(X;q,\underline{\mathcal{H}})|_{q=0.7}\ldots < \mathrm{TV}(X;q,\underline{\mathcal{H}})|_{q=0.999}.$$

4. For all risk assessment techniques $\mid q = 0.6,\ 0.7,\ 0.8,\ 0.9,\ 0.95,\ 0.99$, and 0.999:

$$\mathrm{TMV}(X;q,\underline{\mathcal{H}})|_{q=0.6} < \mathrm{TMV}(X;q,\underline{\mathcal{H}})|_{q=0.7}\ldots < \mathrm{TMV}(X;q,\underline{\mathcal{H}})|_{q=0.999}.$$

5. For all risk assessment Bayesian and non-Bayesian techniques $\mid q = 0.6,\ 0.7,\ 0.8,\ 0.9,\ 0.95,\ 0.99$, and 0.999:

$$\mathrm{MEL}(X;q,\underline{\mathcal{H}})|_{q=0.6} > \mathrm{MEL}(X;q,\underline{\mathcal{H}})|_{q=0.7}\ldots > \mathrm{MEL}(X;q,\underline{\mathcal{H}})|_{q=0.999}.$$

6. Under the EGRRP model and the MLE technique: The $\mathrm{VARK}(X;\widehat{\underline{\mathcal{H}}})$ is a consistently growing indicator which starts with $2602.272196|_{q=0.6}$ and terminates with $145{,}993.327739|_{q=0.999}$; the $\mathrm{TVARK}(X;q,\widehat{\underline{\mathcal{H}}})$ is a consistently growing indicator which starts with $7534.159674|_{q=0.6}$ and terminates with $65{,}808.922847|_{q=0.999}$. However, the $\mathrm{TV}(X;q,\widehat{\underline{\mathcal{H}}})$, the $\mathrm{TMV}(X;q,\widehat{\underline{\mathcal{H}}})$, and the $\mathrm{MEL}(X;q,\widehat{\underline{\mathcal{H}}})$ are monotonously reducing.
7. Under the RR distribution and the MLE technique: The $\mathrm{VARK}(X;q,\widehat{\underline{\mathcal{H}}})$ is a consistently growing indicator which starts with $1628.234966|_{q=0.6}$ and terminates with $36{,}791.27132|_{q=0.999}$; the $\mathrm{TVARK}(X;q,\widehat{\underline{\mathcal{H}}})$ is a consistently growing indicator which starts with $3547.122526|_{q=0.6}$ and terminates with $73{,}594.813549|_{q=0.999}$; the $\mathrm{TV}(X;q,\widehat{\underline{\mathcal{H}}})$, the $\mathrm{TMV}(X;q,\widehat{\underline{\mathcal{H}}})$, and the $\mathrm{MEL}(X;q,\widehat{\underline{\mathcal{H}}})$ are monotonously reducing indicators.
8. Under the EGRRP model and the bootstrapping technique: The $\mathrm{VARK}(X;q,\widehat{\underline{\mathcal{H}}})$ is a consistently growing indicator which starts with $2832.283001|_{q=0.6}$ and terminates with $123{,}576.386617|_{q=0.999}$; the $\mathrm{TVARK}(X;q,\widehat{\underline{\mathcal{H}}})$ is a consistently growing indicator which starts with $7560.430006|_{q=0.6}$ and terminates with $84{,}874.267755|_{q=0.999}$. However, the $\mathrm{TV}(X;q,\widehat{\underline{\mathcal{H}}})$, the $\mathrm{TMV}(X;q,\widehat{\underline{\mathcal{H}}})$, and the $\mathrm{MEL}(X;q,\widehat{\underline{\mathcal{H}}})$ are monotonously reducing.
9. Under the RR distribution and the OLSE technique: The $\mathrm{VARK}(X;q,\widehat{\underline{\mathcal{H}}})$ is a consistently growing indicator which starts with $1692.496862|_{q=0.6}$ and terminates with $38{,}243.320243|_{q=0.999}$; the $\mathrm{TVARK}(X;q,\widehat{\underline{\mathcal{H}}})$ is a consistently growing indicator which

starts with $3687.117754|_{q=0.6}$ and terminates with $76{,}499.395591|_{q=0.999}$. Additionally, the TV $(X;q,\widehat{\underline{\mathcal{H}}})$, the TMV $(X;q,\widehat{\underline{\mathcal{H}}})$, and the MEL$(X;q,\widehat{\underline{\mathcal{H}}})$ are monotonously reducing indicators.

10. Under the EGRRP model and the CVM technique: The VARK$(X;q,\widehat{\underline{\mathcal{H}}})$ is a consistently growing indicator which starts with $2770.998013|_{q=0.6}$ and terminates with $123{,}130.65901|_{q=0.999}$; the TVARK$(X;q,\widehat{\underline{\mathcal{H}}})$ is a consistently growing indicator which starts with $7453.230383|_{q=0.6}$ and terminates with $84{,}870.02498|_{q=0.999}$. However, the TV$(X;q,\widehat{\underline{\mathcal{H}}})$, the TMV$(X;q,\widehat{\underline{\mathcal{H}}})$, and the MEL$(X;q,\widehat{\underline{\mathcal{H}}})$ are monotonously reducing.
11. Under the RR distribution and the CVM technique: The VARK$(X;q,\widehat{\underline{\mathcal{H}}})$ is a consistently growing indicator which starts with $2635.229755|_{q=0.6}$ and terminates with $59545.123954|_{q=0.999}$; the TVARK$(X;q,\widehat{\underline{\mathcal{H}}})$ is a consistently growing indicator which starts with $5740.868751|_{q=0.6}$ and terminates with $119{,}110.108205|_{q=0.999}$. However, the TV$(X;q,\widehat{\underline{\mathcal{H}}})$, the TMV$(X;q,\widehat{\underline{\mathcal{H}}})$, and the MEL$(X;q,\widehat{\underline{\mathcal{H}}})$ are monotonously reducing indicators.
12. Under the EGRRP model and the Bayesian technique: The VARK$(X;q,\widehat{\underline{\mathcal{H}}})$ is a consistently growing indicator which starts with $2463.713921\ |_{q=0.6}$ and terminates with $129{,}788.635|_{q=0.999}$; the TVARK$(X;q,\widehat{\underline{\mathcal{H}}})$ is a consistently growing indicator which starts with $6896.600374|_{q=0.6}$ and terminates with $28{,}585.213639|_{q=0.999}$. However, the TV$(X;q,\widehat{\underline{\mathcal{H}}})$, the TMV$(X;q,\widehat{\underline{\mathcal{H}}})$, and the MEL$(X;q,\widehat{\underline{\mathcal{H}}})$ are monotonously reducing.
13. Under the RR distribution and the Bayesian technique: The VARK$(X;q,\widehat{\underline{\mathcal{H}}})$ is a consistently growing indicator which starts with $1628.247409|_{q=0.6}$ and terminates with $36{,}791.552475|_{q=0.999}$; the TVARK$(X;q,\widehat{\underline{\mathcal{H}}})$ is a consistently growing indicator which starts with $3547.149632|_{q=0.6}$ and terminates with $73{,}595.375934|_{q=0.999}$; the TV$(X;q,\widehat{\underline{\mathcal{H}}})$, the TMV$(X;q,\widehat{\underline{\mathcal{H}}})$, and the MEL$(X;q,\widehat{\underline{\mathcal{H}}})$ are monotonously reducing indicators.
14. For the EGRRP model and its corresponding RR base line model, the Bayesian approach is recommended since it offers the most acceptable risk exposure analysis.

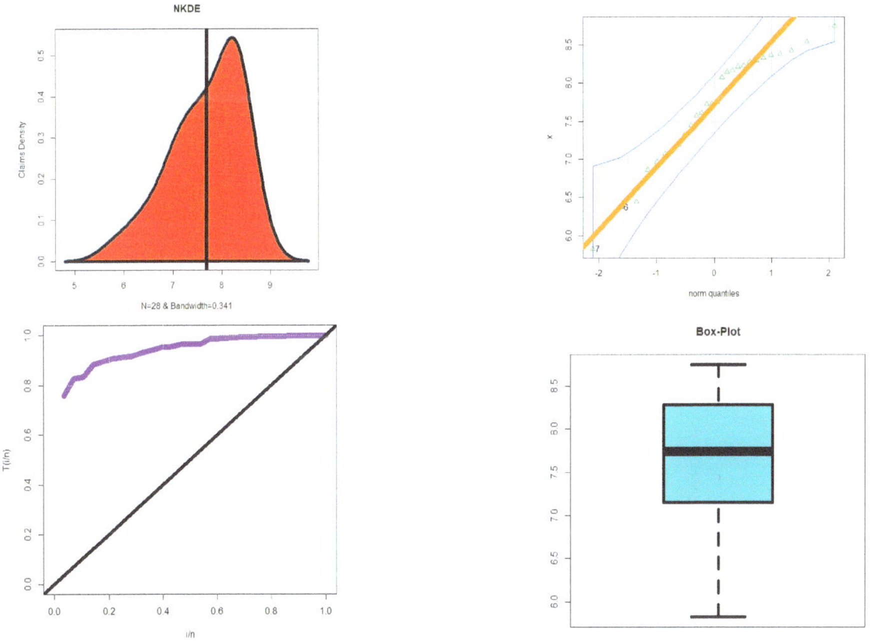

Figure 7. The NKDE graph (**top left graph**), the Q-Q graph (**top right graph**), the TTT graph (**bottom left graph**), and the box graph (**bottom right graph**) for the claims data.

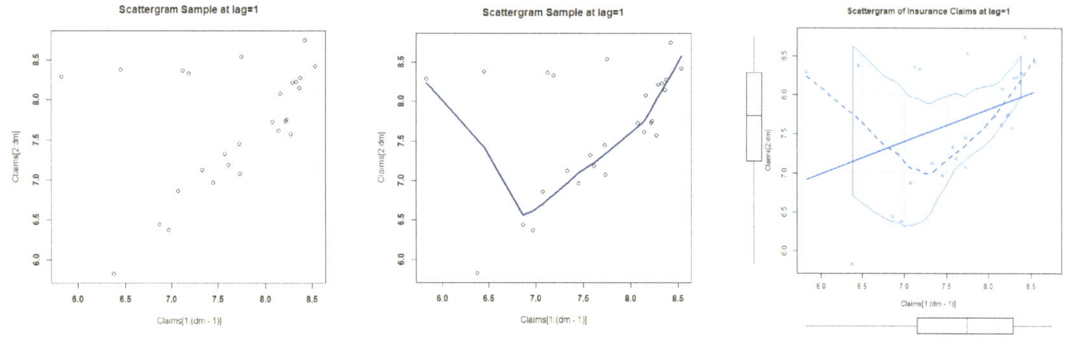

Figure 8. The initial scattergram (**left graph**), the fitted scattergram (**middle graph**), and smoothed scattergram (**right graph**).

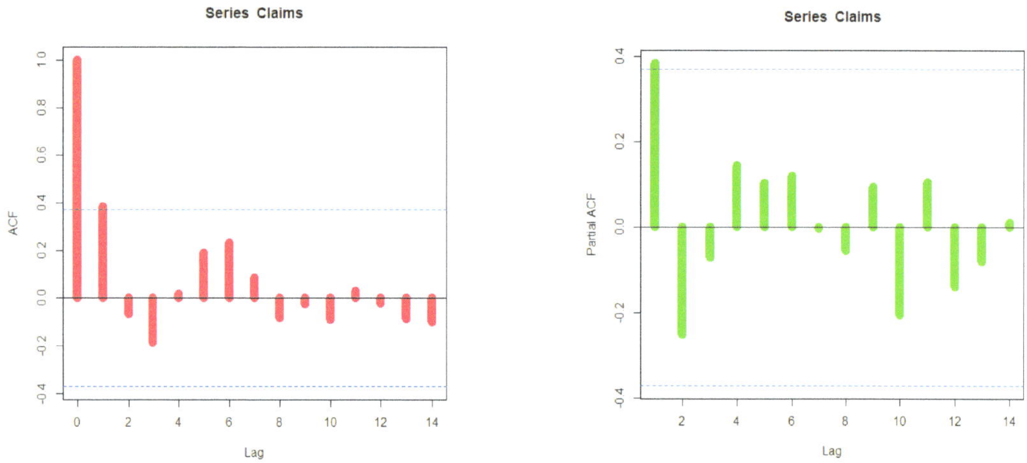

Figure 9. The ACF (**left graph**), and the partial ACF (**right graph**) for the insurance-claims data.

Table 13. The results of the EGRRP under insurance-claims data.

	q	**VARK**$(X;q,\widehat{\underline{\mathcal{H}}})$	**TVARK**$(X;q,\widehat{\underline{\mathcal{H}}})$	**TVq**$(X;q,\widehat{\underline{\mathcal{H}}})$	**TMVq**$(X;q,\widehat{\underline{\mathcal{H}}})$	$MEL(X;q,\widehat{\underline{\mathcal{H}}})$
MLE	0.6	2602.272196	7534.159674	121,153,998.83755	60,584,533.57845	4931.887478
	0.7	3314.099309	9070.251654	152,485,547.01594	76,251,843.75962	5756.152345
	0.8	4519.426502	11,679.71693	209,525,473.41096	104,774,416.4224	7160.290435
	0.9	7368.466882	17,637.56762	338,212,668.58184	169,123,971.8585	10,269.10074
	0.95	11,723.900439	26,230.44642	544,732,714.27725	272,392,587.5850	14,506.54598
	0.99	33,355.286264	58,883.98974	1,269,594,815.6707	634,856,291.8251	25,528.70348
	0.999	145,993.327739	65,808.92284	6,908,619,382.4808	3,454,375,500.163	−80,184.40489
Bayesian	0.6	2463.713921	6896.600374	86,857,109.259539	43,435,451.230144	4432.886453
	0.7	3125.83370	8271.666098	108,234,173.14921	54,125,358.240703	5145.832398
	0.8	4242.10144	10,594.482279	146,113,725.35165	73,067,457.158106	6352.380839
	0.9	6863.70244	15,873.911604	235,935,854.18517	117,983,801.00419	9010.209163
	0.95	10,841.7879	23,261.286766	361,466,587.16925	180,756,554.87139	12,419.49881
	0.99	30,345.4697	49,970.512506	816,811,981.90482	408,455,961.46491	19,625.04280
	0.999	129,788.635	28,585.213639	3,170,744,914.9386	1,585,401,042.6829	−101,203.422

Table 13. *Cont.*

	q	VARK($X;q,\widehat{\mathcal{H}}$)	TVARK($X;q,\widehat{\mathcal{H}}$)	TVq($X;q,\widehat{\mathcal{H}}$)	TMVq($X;q,\widehat{\mathcal{H}}$)	$MEL(X;q,\widehat{\mathcal{H}})$
Bootstrap	0.6	2832.283001	7560.430006	93,823,029.752448	46,919,075.30623	4728.147005
	0.7	3555.945318	9027.338287	116,455,425.56374	58,236,740.120162	5471.392969
	0.8	4759.231239	11,489.206663	156,469,913.82581	78,246,446.119569	6729.975424
	0.9	7529.018485	17,084.073298	249,678,104.66656	124,856,136.40658	9555.054813
	0.95	11,636.63127	24,950.966796	374,329,321.09034	187,189,611.51196	13,314.33552
	0.99	30,996.14226	55,050.720255	643,890,093.87870	322,000,097.65960	24,054.57799
	0.999	123,576.3866	84,874.267755	1,367,037,379.7192	683,603,564.12738	−38,702.11886
CVM	0.6	2770.998013	7453.230383	92,883,930.94896	46,449,418.704863	4682.232369
	0.7	3483.913233	8903.322882	115,420,157.1217	57,718,981.883749	5419.409649
	0.8	4670.531885	11,346.47232	155,165,515.4901	77,594,104.217413	6675.940442
	0.9	7406.183510	16,897.92006	248,097,708.6971	124,065,752.26862	9491.736551
	0.95	11,470.78194	24,724.77277	372,360,756.4232	186,205,102.98440	13,253.990839
	0.99	30,692.35944	54,850.39454	630,810,155.0580	315,459,927.92355	24,158.035102
	0.999	123,130.6590	84,870.02498	1,379,680,759.512	689,925,249.78127	−38,260.63403

Table 14. The estimated parameters for the EGRRP under insurance-claims data.

Techniques	$\hat{\alpha}$	$\hat{\beta}$	$\hat{\theta}$	$\hat{\lambda}$
MLE	0.78171	33.01598	79.08656	−3.78474
Bayesian	0.79412	33.01607	79.08662	−3.78437
Bootstrap	0.83451	60.04148	80.82928	−3.31509
CVM	0.83077	42.24841	100.49062	−3.05291

Table 15. The results of the KRIs for the IR under insurance-claims data.

	q	VARK($X;\hat{\theta}$)	TVARK($X;\hat{\theta}$)	TV($X;\hat{\theta}$)	TMV($X;\hat{\theta}$)	$MEL(X;\hat{\theta})$
MLE	0.6	1628.234966	3547.122526	76,135,623.630709	38,071,358.93788	1918.887559
	0.7	1948.574941	4136.781558	100,120,217.39896	50,064,245.48104	2188.206617
	0.8	2463.549221	5114.338228	147,302,529.07972	73,656,378.87808	2650.789007
	0.9	3585.208936	7297.644053	284,971,253.77263	142,492,924.5303	3712.435117
	0.95	5,138.343308	10,364.98915	550,930,629.98247	275,475,679.9803	5226.645843
	0.99	11,608.153078	23,255.23381	2,535,785,437.5515	1,267,915,974.009	11,647.08073
	0.999	36,791.27132	73,594.81354	22,237,318,310.631	11,118,732,750.12	36,803.54223
Bayesian	0.6	1628.247409	3547.149632	91,723,332.2753950	45,865,213.287330	1918.902223
	0.7	1948.589832	4136.813171	120,902,095.198909	60,455,184.412625	2188.223339
	0.8	2463.568047	5114.377311	178,487,896.633266	89,249,062.693944	2650.809265
	0.9	3585.236334	7297.699821	347,361,234.231741	173,687,914.81569	3712.463487
	0.95	5138.382575	10,365.068359	675,631,280.252160	337,826,005.19443	5226.685784
	0.99	11,608.241786	23,255.41153	3,159,263,800.17968	1,579,655,155.5013	11,647.16974
	0.999	36,791.552475	73,595.375934	28,472,994,153.3060	14,236,570,672.028	36,803.82345
Bootstrap	0.6	1692.496862	3687.117754	98,832,130.6276450	49,419,752.431576	1994.620892
	0.7	2025.479762	4300.048794	130,269,755.140288	65,139,177.618938	2274.569032
	0.8	2560.778641	5316.186898	192,293,925.060678	96,152,278.717237	2755.408257
	0.9	3726.707139	7585.661722	374,182,772.362154	187,098,971.84279	3858.954583
	0.95	5341.139395	10,774.066441	727,822,354.386544	363,921,951.25971	5432.927046
	0.99	12,066.294522	24,173.05307	3,402,703,365.70689	1,701,375,855.9065	12,106.758548
	0.999	38,243.320243	76,499.395591	30,654,515,544.6792	15,327,334,271.735	38,256.075348
CVM	0.6	2635.229755	5740.868751	190,889,214.50589	95,450,348.121700	3105.638997
	0.7	3153.686520	6695.206821	250,869,428.56578	125,441,409.48971	3541.520301
	0.8	3987.150714	8277.341152	368,764,811.87516	184,390,683.27873	4290.190438
	0.9	5802.509749	11,810.92971	712,295,720.90702	3,561,59671.38322	6008.419968
	0.95	8316.192353	16,775.29873	1,374,796,679.7789	687,415,115.18820	8459.106384
	0.99	18,787.30712	37,637.61690	6,300,638,164.1197	3,150,356,719.6767	18,850.309781
	0.999	59,545.12395	119,110.1082	54,832,802,342.864	27,416,520,281.540	59,564.984251

Table 16. The estimated parameters for the IR under insurance-claims data.

Techniques	$\hat{\theta}$
MLE	1163.73317
Bayesian	1163.74207
Bootstrap	1209.66248
CVM	1883.45315

9. Conclusions

Probability-based distributions are used by actuaries to determine the expected values of unexpected risk, which are then used to determine insurance premiums, create insurance products, and assess investment plans. Actuaries also employ simulation tools to test their algorithms and assess how different scenarios may affect the financial results of investments and insurance policies. Probability-based distributions can be used to explain risk exposure, often expressed as one or a few key risk indicators derived from a specific probability model. These indicators provide valuable insights for actuaries and risk managers to understand a company's exposure to various types of risks, such as value-at-risk, tail-value-at-risk, conditional value-at-risk, tail variance, mean excess loss, and tail mean-variance. Different types of data can be analyzed using probability distributions in the modeling process. A new extension of the Reciprocal Rayleigh distribution is introduced and analyzed, including properties such as moments, incomplete moments, probability-weighted moments, moment generating function, residual life, and reversed residual life functions. Parameter estimation is performed through various techniques, including Bayesian estimators under gamma and normal priors using the squared error loss function. The performance of all estimation techniques is evaluated through Monte Carlo simulations and two real data applications. These applications compare the new model with other competitive models and demonstrate the importance of the proposed model via the maximum likelihood technique. Numerical analysis for expected value, variance, skewness, and kurtosis is provided. An extensive analytical study is conducted to evaluate and rate actuarial hazards using a wide range of well-known models for actuarial disclosure. Actuarial data are used to reveal and assess these hazards.

Based on risk analysis, the following results can be highlighted:

1. For all risk assessment Bayesian and non-Bayesian techniques $\mid q = 0.6, 0.7, 0.8, 0.9, 0.95, 0.99,$ and 0.999:

$$\mathrm{VARK}(X; q, \underline{\mathcal{H}})|_{q=0.6} < \ldots < \mathrm{VARK}(X; q, \underline{\mathcal{H}})|_{q=0.999}.$$

2. For all risk assessment Bayesian and non-Bayesian techniques $\mid q = 0.6, 0.7, 0.8, 0.9, 0.95, 0.99,$ and 0.999:

$$\mathrm{TVARK}(X; q, \underline{\mathcal{H}})|_{q=0.6} < \ldots < \mathrm{TVARK}(X; q, \underline{\mathcal{H}})|_{q=0.999}.$$

3. For most risk assessment techniques $\mid q = 0.6, 0.7, 0.8, 0.9, 0.95, 0.99,$ and 0.999:

$$\mathrm{TV}(X; q, \underline{\mathcal{H}})|_{q=0.6} < \ldots < \mathrm{TV}(X; q, \underline{\mathcal{H}})|_{q=0.999}.$$

4. For all risk assessment techniques $\mid q = 0.6, 0.7, 0.8, 0.9, 0.95, 0.99,$ and 0.999:

$$\mathrm{TMV}(X; q, \underline{\mathcal{H}})|_{q=0.6} > \ldots > \mathrm{TMV}(X; q, \underline{\mathcal{H}})|_{q=0.999}.$$

5. For all risk assessment Bayesian and non-Bayesian techniques $\mid q = 0.6, 0.7, 0.8, 0.9, 0.95, 0.99,$ and 0.999:

$$\mathrm{MEL}(X; q, \underline{\mathcal{H}})|_{q=0.6} > \ldots > \mathrm{MEL}(X; q, \underline{\mathcal{H}})|_{q=0.999}.$$

6. Under the EGRRP model and the MLE technique: The VARK$(X;\widehat{\underline{\mathcal{H}}})$ is a consistently growing indicator which starts with $2602.272196|_{q=0.6}$ and terminates with $145{,}993.327739|_{q=0.999}$; the TVARK$(X;q,\widehat{\underline{\mathcal{H}}})$ is a consistently growing indicator which starts with $7534.159674|_{q=0.6}$ and terminates with $65{,}808.922847|_{q=0.999}$. However, the TV$(X;q,\widehat{\underline{\mathcal{H}}})$, the TMV$(X;q,\widehat{\underline{\mathcal{H}}})$, and the MEL$(X;q,\widehat{\underline{\mathcal{H}}})$ are monotonously reducing.
7. Under the EGRRP model and the bootstrapping technique: The VARK$(X;q,\widehat{\underline{\mathcal{H}}})$ is a consistently growing indicator which starts with $2832.283001|_{q=0.6}$ and terminates with $123{,}576.386617|_{q=0.999}$; the TVARK$(X;q,\widehat{\underline{\mathcal{H}}})$ is a consistently growing indicator which starts with $7560.430006|_{q=0.6}$ and terminates with $84{,}874.267755|_{q=0.999}$. However, the TV$(X;q,\widehat{\underline{\mathcal{H}}})$, the TMV$(X;q,\widehat{\underline{\mathcal{H}}})$, and the MEL$(X;q,\widehat{\underline{\mathcal{H}}})$ are monotonously reducing.
8. Under the EGRRP model and the CVM technique: The VARK$(X;q,\widehat{\underline{\mathcal{H}}})$ is a consistently growing indicator which starts with $2770.998013|_{q=0.6}$ and terminates with $123{,}130.65901|_{q=0.999}$; the TVARK$(X;q,\widehat{\underline{\mathcal{H}}})$ is a consistently growing indicator which starts with $7453.230383|_{q=0.6}$ and terminates with $84{,}870.02498|_{q=0.999}$. However, the TV$(X;q,\widehat{\underline{\mathcal{H}}})$, the TMV$(X;q,\widehat{\underline{\mathcal{H}}})$, and the MEL$(X;q,\widehat{\underline{\mathcal{H}}})$ are monotonously reducing.
9. Under the EGRRP model and the Bayesian technique: The VARK$(|_{q=0.6})$ is a consistently growing indicator which starts with $2463.713921\ |_{q=0.6}$ and terminates with $129{,}788.635|_{q=0.999}$; the TVARK$(X;q,\widehat{\underline{\mathcal{H}}})$ is a consistently growing indicator which starts with $6896.600374|_{q=0.6}$ and terminates with $28{,}585.213639|_{q=0.999}$. However, the TV$(X;q,\widehat{\underline{\mathcal{H}}})$, the TMV$(X;q,\widehat{\underline{\mathcal{H}}})$, and the MEL$(X;q,\widehat{\underline{\mathcal{H}}})$ are monotonously reducing.
10. For the EGRRP model and its corresponding RR base line model, the Bayesian approach is recommended since it offers the most acceptable risk exposure analysis.
11. For all q values and risk approaches, the EGRRP model outperforms the RR distribution in terms of performance. Despite the probability distributions having the same number of parameters, the new distribution performs the best when modeling insurance-claims reimbursement data and calculating actuarial risk.

Author Contributions: M.I.: review and editing, software, validation, writing the original draft preparation, conceptualization. W.E.: validation, writing the original draft preparation, conceptualization, data curation, formal analysis, software. Y.T.: conceptualization, software. M.M.A.: review and editing, conceptualization, supervision. H.M.Y.: review and editing, software, validation, writing the original draft preparation, conceptualization, supervision. All authors have read and agreed to the published version of the manuscript.

Funding: The study was funded by Researchers Supporting Project number (RSP2023R488), King Saud University, Riyadh, Saudi Arabia.

Institutional Review Board Statement: Not applicable.

Informed Consent Statement: Not applicable.

Data Availability Statement: The dataset can be provided upon requested.

Acknowledgments: The study was funded by Researchers Supporting Project number (RSP2023R488), King Saud University, Riyadh, Saudi Arabia.

Conflicts of Interest: The authors declare no conflict of interest.

References

1. Lane, M.N. Pricing risk transfer transactions1. *ASTIN Bull. J. IAA* **2000**, *30*, 259–293. [CrossRef]
2. Klugman, S.A.; Panjer, H.H.; Willmot, G.E. *Loss Models: From Data to Decisions*, 3rd ed.; John Wiley & Sons: Hoboken, NJ, USA, 2012; Volume 715.
3. Shrahili, M.; Elbatal, I.; Yousof, H.M. Asymmetric Density for Risk Claim-Size Data: Prediction and Bimodal Data Applications. *Symmetry* **2021**, *13*, 2357. [CrossRef]
4. Mohamed, H.S.; Cordeiro, G.M.; Minkah, R.; Yousof, H.M.; Ibrahim, M. A size-of-loss model for the negatively skewed insurance-claims data: Applications, risk analysis using different techniques and statistical forecasting. *J. Appl. Stat. Forthcom.* **2022**. [CrossRef]
5. Wirch, J.L. Raising value at risk. *N. Am. Actuar. J.* **1999**, *3*, 106–115. [CrossRef]

6. Artzner, P. Application of Coherent Risk Measures to Capital Requirements in Insurance. *N. Am. Actuar. J.* **1999**, *3*, 11–25. [CrossRef]
7. Tasche, D. Expected Shortfall and Beyond. *J. Bank. Financ.* **2002**, *26*, 1519–1533. [CrossRef]
8. Acerbi, C.; Tasche, D. On the coherence of expected shortfall. *J. Bank. Finance* **2002**, *26*, 1487–1503. [CrossRef]
9. Landsman, Z. On the Tail Mean–Variance optimal portfolio selection. *Insur. Math. Econ.* **2010**, *46*, 547–553. [CrossRef]
10. Furman, E.; Landsman, Z. Tail Variance premium with applications for elliptical portfolio of risks. *ASTIN Bull. J. IAA* **2006**, *36*, 433–462. [CrossRef]
11. Aryal, G.R.; Yousof, H.M. The Exponentiated Generalized-G Poisson Family of Distributions. *Stoch. Qual. Control.* **2017**, *32*, 7–23. [CrossRef]
12. Voda, V.G.H. On the Reciprocal Rayleigh Distributed Random Variable. *Rep. Statis. App. Res. JUSE.* **1972**, *19*, 13–21.
13. Mukerjee, S.P.; Saran, L.K. BiVARKiate Reciprocal Rayleigh distributions in reliability studies. *J. Ind. Statist. Assoc.* **1984**, *22*, 23–31.
14. Nadarajah, S.; Kotz, S. The exponentiated Fréchet distribution. *Interstat Electron. J.* **2003**, *14*, 1–7.
15. Nadarajah, S.; Gupta, A.K. The Beta Fréchet Distribution. *Far East J. Theor. Stat.* **2004**, *14*, 15–24.
16. Barreto-Souza, W.; Cordeiro, G.M.; Simas, A.B. Some results for beta Fréchet distribution. *Commun. Stat. Theory Methods* **2011**, *40*, 798–811. [CrossRef]
17. Krishna, E.; Jose, K.K.; Alice, T.; Ristic, M.M. The Marshall-Olkin Fréchet Distribution. *Commun. Stat. Theory Tech.* **2013**, *42*, 4091–4107. [CrossRef]
18. Mahmoud, M.R.; Mandouh, R.M. On the Transmuted Fréchet Distribution. *J. Appl. Sci-Ences Res.* **2013**, *9*, 5553–5561.
19. Mead, M.E.; Abd-Eltawab, A.R. A note on Kumaraswamy-Fréchet Distribution. *Aust. J. Basic Appl. Sci.* **2014**, *8*, 294–300.
20. Chakraborty, S.; Handique, L.; Altun, E.; Yousof, H.M. A new statistical model for extreme values: Mathematical properties and applications. *Int. J. Open Probl. Comput. Sci. Math.* **2018**, *12*, 1–18.
21. Cordeiro, G.M.; Yousof, H.M.; Ramires, T.G.; Ortega, E.M.M. The Burr XII system of densities: Properties, regression model and applications. *J. Stat. Comput. Simul.* **2018**, *88*, 432–456. [CrossRef]
22. Nichols, M.D.; Padgett, W.J. A Bootstrap Control Chart for Weibull Percentiles. *Qual. Reliab. Eng. Int.* **2005**, *22*, 141–151. [CrossRef]
23. Smith, R.L.; Naylor, J.C. A Comparison of Maximum Likelihood and Bayesian Estimators for the Three- Parameter Weibull Distribution. *J. R. Stat. Soc. Ser. C Appl. Stat.* **1987**, *36*, 358. [CrossRef]
24. Silva, R.V.; de Andrade, T.A.; Maciel, D.B.; Campos, R.P.S.; Cordeiro, G.M. A New Lifetime Model: The Gamma Extended Frechet Distribution. *J. Stat. Theory Appl.* **2013**, *12*, 39–54. [CrossRef]
25. Aarset, M.V. How to Identify a Bathtub Hazard Rate. *IEEE Trans. Reliab.* **1987**, *R-36*, 106–108. [CrossRef]

mathematics

Article

On Designing of Bayesian Shewhart-Type Control Charts for Maxwell Distributed Processes with Application of Boring Machine

Fatimah Alshahrani [1], Ibrahim M. Almanjahie [2], Majid Khan [3,4,*], Syed M. Anwar [5], Zahid Rasheed [6] and Ammara N. Cheema [7]

1 Department of Mathematical Sciences, College of Science, Princess Nourah bint Abdulrahman University, P.O. Box 84428, Riyadh 11671, Saudi Arabia
2 Department of Mathematics, College of Science, King Khalid University, Abha 62223, Saudi Arabia
3 Department of Statistics, Government Postgraduate College Haripur, Haripur 22620, Pakistan
4 Department of Mathematics and Statistics, Riphah International University, Islamabad 46000, Pakistan
5 Department of Statistics, University of Azad Jammu and Kashmir, Muzaffarabad 13100, Pakistan
6 Department of Mathematics, Women University of Azad Jammu and Kashmir, Bagh 12500, Pakistan
7 Department of Mathematics, Air University Islamabad Pakistan, Islamabad 44000, Pakistan
* Correspondence: 285004majidkhan@gmail.com; Tel.: +92-33-3509-5095

Abstract: The quality characteristic(s) are assumed to follow the normal distribution in many control chart constructions, although this assumption may not hold in some instances. This study proposes the Bayesian-I and Bayesian-II Shewhart-type control charts for monitoring the Maxwell scale parameter in the phase II study. The posterior and predictive distributions are used to construct the control limits for the proposed Bayesian-I and Bayesian-II Shewhart-type control charts, respectively. Various performance indicators, including average run length, quadratic loss, relative average run length, and performance comparison index, are utilized to evaluate the performance of the proposed control charts. The Bayesian-I and Bayesian-II Shewhart-type control charts are compared to their competitive $CUSUM_V$, $EWMA_V$ and V control charts. Sensitivity analysis is also performed to study the effect of hyperparameter values on the performance behavior of the proposed control charts. Finally, real-life data is analyzed for the implementation of the proposed control charts.

Keywords: hyperparameters; Maxwell distribution; performance measures; posterior predictive distribution; sensitivity analysis

MSC: 62P30; 62C10; 62F15; 62F10

Citation: Alshahrani, F.; Almanjahie, I.M.; Khan, M.; Anwar, S.M.; Rasheed, Z.; Cheema, A.N. On Designing of Bayesian Shewhart-Type Control Charts for Maxwell Distributed Processes with Application of Boring Machine. *Mathematics* **2023**, *11*, 1126. https://doi.org/10.3390/math11051126

Academic Editor: Diana Mindrila

Received: 24 January 2023
Revised: 19 February 2023
Accepted: 21 February 2023
Published: 23 February 2023

1. Introduction

The statistical process control (SPC) kit is a collection of instruments used to monitor the variations in process parameters of the quality characteristic(s). Generally, these variations are categorized into the following two classes: natural (random) causes of variations and special (attributable) causes of variations. The natural causes of variations are harmless and acceptable to any stable process (IC state). On the contrary, the special causes of variations are harmful and create certain problems in the process (OOC state) that can deteriorate the quality of the product. The proper diagnosis and identification of the special causes of variations in the process parameters are essential for achieving high-quality products. Control charts are a statistical tool in the SPC that analyzes and monitor the special causes of variations in the process parameters to ensure the quality of products. The special cause of variations is termed a shift in the process parameters. The most familiar control chart is the classical Shewhart control chart developed by [1], also referred to as the memory-less control chart.

To monitor the shift of the process parameters efficiently, it is necessary to identify the nature of the distribution for the quality characteristic and then monitor the process parameter(s) shift via the control chart. The classical Shewhart-type control chart requires the assumptions of normality and independence of observations for the quality characteristic [2,3]. In literature, numerous authors designed the classical Shewhart-type control charts to detect the shifts in process parameters, assuming that the quality characteristic(s) follows the normal distribution. For instance, Al-Omari and Al-Nasser [4] proposed an efficient control chart based on robust extreme ranked set sampling to detect the shift in the process mean. Similarly, Al-Omari and Haq [5] suggested a double-rank set sampling-based Shewhart-type control chart for monitoring the shift in the process mean. Moreover, Haq and Al-Omari [6] designed an improved Shewhart-type control chart based on partially ordered judgment subset sampling to diagnose changes in the process mean. Equally, Shabbir and Awan [7] suggested the Shewhart-type control chart, which is based on the difference-in-difference estimator and detects a moderate shift in the process mean in the phase-II scenario.

The assumptions of normality for the quality characteristic may not be achieved in practice [8]. Therefore, using the normal distribution for non-normal data may result in a false alarm in process monitoring or may allow for the later detection of changes. In the literature, several researchers suggested control charts for quality characteristic which follows a skewed distribution. For example, Xie et al. [9] discussed the monitoring of shifts in the location parameter of the lognormal process. Likewise, Al-Oraini and Rahim [10] investigated the economic statistical design of $\overline{X}$ control chart when the (Gamma (λ, 2) distribution is considered its failure model. Similarly, Nichols and Padgett [11] offered the bootstrap control chart to monitor the percentiles of the Weibull distribution. Correspondingly, Guo and Wang [12] monitored the shape parameter of the Weibull distribution under type-II censored data. Moreover, Lio et al. [13] designed two parametric bootstrap control charts for monitoring the Burr Type-X percentiles. More comprehensive details about the control charts based on skewed distributions can be seen in the studies of [14–20].

The Maxwell (or Maxwell–Boltzmann) distribution is a familiar positively skewed distribution. The Maxwell distribution has a smoothly increasing hazard risk; therefore, it is commonly used in life-testing experiments and reliability analysis where the assumption of constant hazard risk, such as in exponential distribution, is not practical. The Maxwell distribution is widely used in statistical machines, physics, chemistry, and life testing experiment, but it is recently has been applied in the SPC techniques. For instance, Hossain et al. [21] suggested a Shewhart-type control chart, called a V control chart when the process variable follows Maxwell distribution. The design structure of the V control chart used the statistic V, which is known as the maximum likelihood estimate for the scale parameter of Maxwell distribution. Similarly, Hossain et al. [22] proposed a V statistic-based cumulative sum ($CUSUM$) control chart, also denoted as $CUSUM_V$, to detect the changes in the Maxwell process. Likewise, Hossain and Riaz [23] recommended the exponentially weighted moving average ($EWMA$) control chart based on V denoted by $EWMA_V$ for monitoring the Maxwell distribution, and results show that $EWMA_V$ control chart outperformed the existing $CUSUM_V$ and V control charts. Other control chart schemes based on Maxwell distribution can be seen in the studies, such as [24–26].

The Bayesian approach is commonly used in designing control charts to enhance process monitoring. For instance, Menzefricke [27] constructed the Bayesian control limits to monitor the mean of normal distribution. Similarly, Demirhan and Hamurkaroglu [3] suggested the Bayesian $\overline{X}$ control limits for exponentially distributed measurements. Likewise, Saghir [28] proposed the phase-I design scheme for the $\overline{X}$ control chart, which is based on the posterior distribution. Additionally, Raubenheimer and van der Merwe [29] offered the predictive distribution-based Bayesian c-control chart for monitoring nonconformities. Furthermore, Kumar and Chakraborti [30] recommended the Bayesian Shewhart t_r-control chart for monitoring the time between events. Moreover, Riaz et al. [31] designed the

Bayesian *EWMA* control chart with three loss functions to monitor the process mean shift. Further related works can be seen in [32–37].

As mentioned above, the Bayesian approach can enhance the performance behavior of conventional control charts to monitor the process parameters. Similarly, Hossain et al. [21] implemented a Shewhart-type control chart for monitoring changes in the Maxwell scale parameter. Inspired by the Bayesian approach with control charts, this study introduces the two Bayesian Shewhart-type control charts for monitoring the Maxwell scale parameter, called Bayesian-I and Bayesian-II Shewhart-type control charts in phase II case. The Bayesian approach along the Shewhart-type control charts is expected to further enhance the efficiency of the proposed control charts. The design structures of the Bayesian-I and Bayesian-II Shewhart-type control charts are based on the probability control limits. The Monte Carlo simulations are conducted, and the proposed Bayesian-I and Bayesian-II Shewhart-type control charts are compared to the $CUSUM_V$, $EWMA_V$, and $CUSUM$ control charts. The comparisons indicate that the proposed Bayesian-I and Bayesian-II Shewhart-type control charts outperform the $CUSUM_V$, $EWMA_V$, and V control charts.

The remainder of the paper is organized as follows: Section 2 presents the preliminaries. Besides, Section 3 contains the structures of the proposed control charts for monitoring the shifts in the Maxwell parameter. Furthermore, the simulation study is discussed in Section 4. Moreover, Section 5 illustrates the results and performance comparison of the proposed and existing control charts. Section 6 provides a real-life data analysis for the practical implementation of the proposed and existing control charts. Finally, a summary, conclusions, and recommendations are outlined in Section 7.

2. Preliminaries

This section is organized as follows. Section 2.1 describes the Maxwell distribution. Section 2.2 presents the distribution of the maximum likelihood estimate (V) of for the Maxwell scale parameter. The V control chart to monitor the changes in the Maxwell scale parameter is provided in Section 2.3. Section 2.4 deals with the methodology of the $CUSUM_V$ control chart for monitoring the Maxwell process scale shift. Finally, the $EWMA_V$ control chart is described in Section 2.5.

2.1. Maxwell Distribution

Suppose that X is a random variable having the Maxwell distribution with scale parameter σ^2, then its probability density function (PDF) and cumulative distribution function (CDF) are respectively given as follows:

$$f\left(X\middle|\sigma^2\right) = \sqrt{2/\pi}\sigma^{-3}X^2e^{-X^2/(2\sigma^2)}; \quad X, \sigma^2 > 0, \tag{1}$$

$$F\left(X\middle|\sigma^2\right) = \sqrt{2/\pi}\gamma\left(3/2, X^2/\left(2\sigma^2\right)\right), \tag{2}$$

The $\pi = 3.1429$ is constant, $\int_0^u u^{a-1}e^{-\lambda u}du = \lambda^{-1}\gamma(a, \lambda u)$, and $\gamma(a, \lambda u)$ is an incomplete gamma function, whereas $a > 0$ and $\lambda > 0$ are constants.

2.2. Distribution of Statistic V

Hossain et al. [21] used a sample of size n, which is randomly taken from Equation (1), to derive the maximum likelihood estimate for the scale parameter of the Maxwell distribution, which is given as follows:

$$\hat{\sigma}^2 = (3n)^{-1}\sum_{j=1}^{n} X_j^2, \tag{3}$$

and called it statistic V, that is $V = (3n)^{-1}\sum_{i=1}^{n} X_j^2$. Hossain et al. [22] showed that the transformations $T = X^2/\left(2\sigma^2\right)$ and $U = 3nV/\left(2\sigma^2\right)$ follow gamma distribution, that is, $T = X^2/\left(2\sigma^2\right) \sim G(3/2, 1)$ and $U = 3nV/\left(2\sigma^2\right) \sim G(3n/2, 1)$, respectively. Similarly,

Hossain et al. [23] defined the distribution of statistic V represented by the PDF, and it is given as follows:

$$g\left(V\middle|\sigma^2\right) = \frac{\left[3n/\left(2\sigma^2\right)\right]^{3n/2}}{\Gamma(3n/2)} V^{(3n/2)-1} e^{-3nV/(2\sigma^2)}; \quad V, \sigma^2 > 0. \tag{4}$$

The statistic V has the mean and variance, respectively given as follows:

$$E(V) = \sigma^2 \text{ and } Var(V) = 2\sigma^4/(3n),$$

Likewise, the αth quantile function of V is given as; $V_\alpha = \left[2\sigma^2/(3n)\right]F^{-1}(\alpha)$, where $F^{-1}(\cdot)$ the inverse CDF of $G(3n/2, 1)$. It is important to note that, here the basic objective is the monitoring of σ^2. If δ represents the shift in σ^2 then for IC and OOC situations, the following hypothesis can be formulated, respectively, as follows:

$$H_0 : \sigma^2 = \sigma_0^2; \text{or } \delta = 1 \text{ (the process is IC)},$$

$$H_1 : \sigma^2 = \sigma_1^2 = \delta\sigma_0^2;\ \delta > 0 \text{ but } \delta \neq 1 \text{ (the process is OOC)},$$

2.3. *V Control Chart*

Hossain et al. [21] suggested a V statistic-based Shewhart-type control chart, named as V control chart for monitoring changes in the Maxwell scale parameter. They derived LPL_V, CL_V, and UPL_V for the V control chart in the case of known and unknown σ^2. The basic design structure for the probability control limits of V control chart can be presented as follows:

$$\left.\begin{array}{l} LPL_V : V_{\alpha/2} = L_1\sigma^2 \\ CL_V : V_{0.5} = L_2\sigma^2 \\ UPL_V : V_{1-(\alpha/2)} = L_3\sigma^2 \end{array}\right\}, \tag{5}$$

where $L_1 = \left[\frac{2}{3n}\right]F^{-1}\left(\frac{\alpha}{2}\right)$, $L_2 = \left[\frac{2}{3n}\right]F^{-1}(0.5)$, and $L_3 = \left[\frac{2}{3n}\right]F^{-1}\left(1 - \frac{\alpha}{2}\right)$. The coefficients L_1, L_2, and L_3 are the quantiles of $G(3n/2, 1)$ multiplied by some constants.

Practically, the parameter σ^2 may be known then the probability control limits are defined as follows:

$$\left.\begin{array}{l} LPL_V = L_1\sigma_0^2 \\ CL_V = L_2\sigma_0^2 \\ UPL_V = L_3\sigma_0^2 \end{array}\right\}. \tag{6}$$

However, if σ^2 is unknown, then σ^2 can be estimated using the statistic V. So, in this case, the probability control limits for the V control chart are defined as follows:

$$\left.\begin{array}{l} LPL_V = L_1\overline{V} \\ CL_V = L_2\overline{V} \\ UPL_V = L_3\overline{V} \end{array}\right\}, \tag{7}$$

where $\overline{V}$ represents the average of the estimated V_i computed at each of the samples over time i.

The ARL is one of the measures that evaluate the performance of the control charts. Mathematically, the ARL can be defined as given below as follows:

$$ARL = \frac{1}{p_V}, \tag{8}$$

where p_V. is the power of the test, that is, the probability of rejecting the null hypothesis (H_0) when an alternative hypothesis (H_1) is true. For the V control chart, the power of the test is defined as follows:

$$p_V = P(V < LPL_V|H_1) + P(V > UPL_V|H_1).$$

It further can be expressed as follows:

$$p_V = P\left(V < \left(\frac{2\sigma_0^2}{3n}\right)F^{-1}(\alpha/2)\middle|\delta \neq 1\right) + P\left(V > \left(\frac{2\sigma_0^2}{3n}\right)F^{-1}(1-\alpha/2)\middle|\delta \neq 1\right).$$

Finally, the power of the test is given as follows:

$$p_V = 1 + \left[\Gamma\left(\frac{3n}{2}\right)\right]^{-1}\gamma\left(\frac{3n}{2}, \delta^{-1}F^{-1}(\alpha/2)\right) - \left[\Gamma\left(\frac{3n}{2}\right)\right]^{-1}\gamma\left(\frac{3n}{2}, \delta^{-1}F^{-1}(1-\alpha/2)\right). \tag{9}$$

In Equation (9) if there is no shift (i.e., $\delta = 1$) in the process, then, in this case, the power is equal to the false alarm rate α.

2.4. $CUSUM_V$ Control Chart

Hossain et al. [22] presented the design of the $CUSUM_V$ control chart for monitoring the process scale parameter shift. They designed the plotting statistic of the $CUSUM_V$ control chart, defined below as follows:

$$\left.\begin{aligned} C_i^+ &= \max\left(0, V_i - k + C_{i-1}^+\right) \\ C_i^- &= \min\left(0, V_i - k + C_{i-1}^-\right) \end{aligned}\right\}, \tag{10}$$

where $C_0^+ = C_0^- = 0$ are starting values and the k is referred to as the slack value that is defined as $k = -\frac{\ln(\delta)}{\sigma_1^{-2} - \sigma_0^{-2}}$. For a one-sided upper $CUSUM_V$ control chart, the process is considered to be OOC if the charting statistic exceeds then the threshold, h (i.e., control limit), that is, $C_i^+ > h$.

2.5. $EWMA_V$ Control Chart

Recently, Hossain and Riaz [23] proposed the $EWMA$ control chart that monitors the Maxwell scale parameter shift, which is denoted by the notation by $EWMA_V$. The charting statistic for the $EWMA_V$ control chart is represented by Z_i, which is based on V_i, and can be given as follows:

$$Z_i = \lambda V_i + (1-\lambda)Z_{i-1}, \tag{11}$$

where Z_0 is initial value set as; $Z_0 = \sigma_0^2$, and λ is the smoothing constant. The mean of the statistic Z_i is σ_0^2, while its variance is $\frac{\sigma_0^4}{3n}\left\{\frac{\lambda}{2-\lambda}\left(1-(1-\lambda)^{2i}\right)\right\}$. The lower control limit (LCL_E), center line (CL_E), and the upper control limit (UCL_E) for the $EWMA_V$ control chart are given as follows:

$$\left.\begin{aligned} LCL_E &= K_1\sigma_0^2 \\ CL_E &= \sigma_0^2 \\ UCL_E &= K_2\sigma_0^2 \end{aligned}\right\}. \tag{12}$$

where $K_1 = 1 - L_E\sqrt{\frac{2\lambda}{3n(2-\lambda)}\left\{1-(1-\lambda)^{2i}\right\}}$ and $K_2 = 1 + L_E\sqrt{\frac{2\lambda}{3n(2-\lambda)}\left\{1-(1-\lambda)^{2i}\right\}}$. For a very large value of i, K_1 and K_2 are reduced to $1 - L_E\sqrt{\frac{2\lambda}{3n(2-\lambda)}}$, and $1 + L_E\sqrt{\frac{2\lambda}{3n(2-\lambda)}}$, respectively. If σ_0^2 is unknown then it can be estimated by $\overline{V}$, hence Equation (12) can be rewritten, in this case, given as follows:

$$\left.\begin{aligned} LCL_E &= K_1\overline{V} \\ CL_E &= \overline{V} \\ UCL_E &= K_2\overline{V} \end{aligned}\right\}. \tag{13}$$

The $EWMA_V$ control chart provides the OOC signal if Z_i falls outside the upper or lower control limits.

3. Proposed Bayesian Shewhart-Type Control Charts

This section presents the schemes of the Bayesian-I and Bayesian-II Shewhart-type control charts for monitoring the Maxwell scale parameter shift. The proposed Bayesian-I Shewhart-type control chart is formulated using the posterior distribution, while the Bayesian-II Shewhart-type control chart is designed using the predictive distribution. The details are given in the following subsections.

3.1. Proposed Bayesian-I Shewhart-Type Control Chart

This subsection consists of constructing the posterior distribution for σ^2 given V using the conjugate prior. This posterior distribution is used to find the probability control limits for the Bayesian-I Shewhart-type control chart. The following theorems may be useful in this regard.

Theorem 1. *Given the prior distribution of σ^2 is inverted gamma (IG), that is, IG (a, b),*

$$g\left(\sigma^2 \middle| a, b\right) = \frac{b^a}{\Gamma(a)}\left(\sigma^2\right)^{-(a+1)} e^{-b/\sigma^2}; \quad \sigma^2,\ a, b > 0. \tag{14}$$

The posterior distribution of σ^2 given V is IG (3n/2 + a, b +3nV/2), that is,

$$g\left(\sigma^2 \middle| V\right) = \frac{(b+3nV/2)^{3n/2+a}}{\Gamma(3n/2+a)}\left(\sigma^2\right)^{-(3n/2+a+1)} e^{-(b+3nV/2)/\sigma^2}; \ \sigma^2 > 0.$$

Proof of Theorem 1. By definition, the posterior distribution of σ^2 given V is defined as follows:

$$g\left(\sigma^2 \middle| V\right) = \frac{g(\sigma^2|a,b)g(\sigma^2)}{\int_0^\infty g(\sigma^2|a,b)g(\sigma^2)d\sigma^2}. \tag{15}$$

From Equations (4), (14), and (15) the posterior distribution of σ^2 given V that can be specified as follows:

$$g\left(\sigma^2 \middle| V\right) = \frac{(b+3nV/2)^{3n/2+a}}{\Gamma(3n/2+a)}\left(\sigma^2\right)^{-(3n/2+a+1)} e^{-(b+3nV/2)/\sigma^2}; \quad \sigma^2 > 0, \tag{16}$$

which means that $\sigma^2|V \sim \mathrm{IG}(3n/2+a, b+3nV/2)$. □

Result 1. If a = 1 and a = 0 then the prior distribution of σ^2 becomes an improper prior known as the uniform prior, that is, as follows:

$$g\left(\sigma^2\right) \propto 1; \quad \sigma^2 > 0.$$

The posterior distribution in this case is as follows:

$$g\left(\sigma^2 \middle| V\right) = \frac{(3nV/2)^{3n/2+1}}{\Gamma(3n/2+1)}\left(\sigma^2\right)^{-(3n/2+2)} e^{-3nV/(2\sigma^2)}; \ \sigma^2 > 0,$$

which implies that $\sigma^2|V \sim \mathrm{IG}(3n/2+1, 3nV/2)$.

Result 2. If a = 0 and b = 0 then the prior distribution σ^2 becomes an improper prior referred to as Jeffreys prior, that is, as follows:

$$g\left(\sigma^2\right) \propto \frac{1}{\sigma^2}; \quad \sigma^2 > 0.$$

The posterior distribution in this case is the following:

$$g\left(\sigma^2\middle|V\right) = \frac{(3nV/2)^{3n/2}}{\Gamma(3n/2)}\left(\sigma^2\right)^{-(3n/2+1)}e^{-3nV/(2\sigma^2)}; \quad \sigma^2 > 0,$$

which suggests that $\sigma^2|V \sim \text{IG}(3n/2, 3nV/2)$.

Theorem 2. *Given the posterior distribution of* σ^2 *given V is IG (3n/2 + a, b + 3nV/2), that is*

$$g\left(\sigma^2|V\right) = \frac{(b+3\text{n}V/2)^{3n/2+a}}{\Gamma(3n/2+a)}\left(\sigma^2\right)^{-(3n/2+a+1)}e^{-(b+3\text{n}V/2)/\sigma^2}; \; \sigma^2 > 0.$$

the transformation $\theta = \frac{b+3nV/2}{\sigma^2}$ *given V has the PDF of a gamma distribution having shape parameter 3n/2 + a and scale parameter 1, that is,*

$$g(\theta|V) = \frac{1}{\Gamma(3n/2+a)}\theta^{3n/2+a-1}e^{-\theta}; \quad \theta > 0. \tag{17}$$

Proof of Theorem 2. For the PDF σ^2 given V in Equation (16), since $\theta = \frac{b+3nV/2}{\sigma^2} \Rightarrow \sigma^2 = \frac{b+3nV/2}{\theta}$ then the Jacobian of transformation, $J(\sigma^2 \to \theta) = \frac{b+3nV/2}{\theta^2}$. Hence using the relation, $g(\theta|V) = g(\sigma^2|V)|J|$ the PDF of θ given V is written by the following:

$$g(\theta|V) = \frac{1}{\Gamma(3n/2+a)}\theta^{3n/2+a-1}e^{-\theta}; \quad \theta > 0,$$

which indicates that $\theta|V \sim G(3n/2+a, 1)$. □

The probability control limits for the proposed Bayesian-I Shewhart-type control chart can be developed from Equation (16). The lower probability limit (LPL_1), central line (CL_1), and upper probability limit (UPL_1), at the desired false alarm rate α, based on the posterior distribution, can be given as follows:

$$\left.\begin{aligned} P(\sigma^2 \le LPL_1|V) &= \alpha/2 \\ P(\sigma^2 \le CL_1|V) &= 0.5 \\ P(\sigma^2 \le UPL_1|V) &= 1-\alpha/2 \end{aligned}\right\}. \tag{18}$$

where CL_1 is defined as the median of the posterior distribution. Under the transformation defined by Theorem 2, using the quantiles of the $G(3n/2+a, 1)$, the probability control limits LPL_1, CL_1, and UPL_1 can be designed as follows:

$$\left.\begin{aligned} LPL_1 &= (b+3nV/2)A_1 \\ CL_1 &= (b+3nV/2)A_2 \\ UPL_1 &= (b+3nV/2)A_3 \end{aligned}\right\}, \tag{19}$$

where $A_1 = 1/H^{-1}(1-\alpha/2)$, $A_2 = 1/H^{-1}(0.5)$, $A_3 = 1/H^{-1}(\alpha/2)$ and $H^{-1}(\cdot)$ the inverse CDF of $G(3n/2+a, 1)$. Table 1 contains the different values of A_1, A_2, and A_3 at various n, α with α = 8.5, 70. Let p_1 be the power of the test then in the case of the proposed Bayesian-I Shewhart-type control chart, it can be defined as follows:

$$p_1 = P\left(\sigma^2 < LPL_1\middle|V, \delta \ne 1\right) + P\left(\sigma^2 > UPL_1\middle|V, \delta \ne 1\right),$$

which can be solved to the following:

$$p_1 = 1+\left[\Gamma\left(\frac{3n}{2}+a\right)\right]^{-1}\gamma\left(\frac{1}{A_1},\frac{(b+3nV/2)}{\delta\sigma_0^2}\middle|\delta\neq 1\right)-\left[\Gamma\left(\frac{3n}{2}+a\right)\right]^{-1}\gamma\left(\frac{1}{A_3},\frac{(b+3nV/2)}{\delta\sigma_0^2}\middle|\delta\neq 1\right)$$

Table 1. A_1, A_2, and A_3 values for different n and a = 8.5, 70.

a	n	False Alarm Rate α								
		0.005			0.0027			0.002		
		A_1	A_2	A_3	A_1	A_2	A_3	A_1	A_2	A_3
8.5	1	0.04725	0.10345	0.29759	0.04510	0.10345	0.32434	0.04414	0.10345	0.33790
	2	0.04290	0.08955	0.23675	0.04105	0.08955	0.25601	0.04022	0.08955	0.26571
	3	0.03935	0.07895	0.19505	0.03773	0.07895	0.20963	0.03701	0.07895	0.21692
	4	0.03639	0.07059	0.16494	0.03495	0.07059	0.17639	0.03431	0.07059	0.18209
	5	0.03387	0.06383	0.14232	0.03259	0.06383	0.15156	0.03201	0.06383	0.15615
70	1	0.01025	0.01405	0.01998	0.01005	0.01405	0.02050	0.00995	0.01405	0.02075
	2	0.01007	0.01376	0.01949	0.00987	0.01376	0.02000	0.00978	0.01376	0.02024
	3	0.00990	0.01348	0.01903	0.00970	0.01348	0.01951	0.00961	0.01348	0.01974
	4	0.00973	0.01322	0.01858	0.00954	0.01322	0.01905	0.00945	0.01322	0.01928
	5	0.00957	0.01296	0.01816	0.00938	0.01296	0.01861	0.00930	0.01296	0.01883

Consequently, the OOC ARL, that is ARL_1. of the Bayesian-I Shewhart-type control chart is given as follows:

$$ARL_1 = \frac{1}{1+\left[\Gamma\left(\frac{3n}{2}+a\right)\right]^{-1}\gamma\left(\frac{1}{A_1},\frac{(b+3nV/2)}{\delta\sigma_0^2}\middle|\delta\neq 1\right)-\left[\Gamma\left(\frac{3n}{2}+a\right)\right]^{-1}\gamma\left(\frac{1}{A_3},\frac{(b+3nV/2)}{\delta\sigma_0^2}\middle|\delta\neq 1\right)} \quad (20)$$

3.2. Proposed Bayesian-II Shewhart-Type Control Chart

Let X_f be the future observation of random sample $X = (X_1, X_2, \ldots, X_n)$ taken from the Maxwell distribution then the predictive distribution of X_f given V is defined by Equation (24). Assuming that LPL_2, CL_2, and UPL_2 are the probability control limits for the Bayesian-II Shewhart-type control chart to monitor the process scale parameter shift, then the LPL_2, CL_2, and UPL_2 can be defined as follows:

$$\left.\begin{aligned} P\left[X_f \leq LPL_2 \middle| V\right] &= \alpha/2 \\ P\left[X_f \leq CL_2 \middle| V\right] &= 0.5 \\ P\left[X_f \leq UPL_2 \middle| V\right] &= 1-\alpha/2 \end{aligned}\right\}, \quad (21)$$

where CL_2 is regarded as a median of the posterior predictive distribution. To derive the probability control limits, that is, LPL_2, CL_2, and UPL_2 for the proposed Bayesian-II Shewhart-type control chart, the following theorems may help here.

Theorem 3. *Given the posterior distribution of σ^2 given V is IG ($3n/2 + a$, $b + 3nV/2$), that is,*

$$g\left(\sigma^2|V\right) = \frac{(b+3nV/2)^{3n/2+a}}{\Gamma(3n/2+a)}\left(\sigma^2\right)^{-(3n/2+a+1)}e^{-(b+3nV/2)/\sigma^2};\sigma^2>0$$

the predictive distribution of a future random variable X_f given V is expressed as follows:

$$g\left(X_f|V\right) = \frac{(b+3nV/2)^{3n/2+a}}{B(3n/2+a,3/2)}\frac{X_f^2}{\left(b+3nV/2+X_f^2/2\right)^{3n/2+a+3/2}};X_f>0$$

Proof of Theorem 3. The predictive distribution of X_f given V is defined as follows:

$$g\left(X_f\middle|V\right) = \int_0^\infty f\left(X_f\middle|\sigma^2\right)g\left(\sigma^2\middle|V\right)d\sigma^2. \tag{22}$$

From Equation (1), $f\left(X_f\middle|\sigma^2\right)$ is the PDF of Maxwell distribution, which can be written as follows:

$$f\left(X_f\middle|\sigma^2\right) = \sqrt{2/\pi}\sigma^{-3}X_f^2 e^{-X_f^2/2\sigma^2}; \quad X_f, \sigma^2 > 0. \tag{23}$$

Equations (16) and (22) provide the predictive PDF of X_f given V, which can be written as follows:

$$g\left(X_f\middle|V\right) = \frac{(b+3nV/2)^{3n/2+a}}{B(3n/2+a,3/2)} \frac{X_f^2}{\left(b+3nV/2+X_f^2/2\right)^{3n/2+a+3/2}}; \quad X_f > 0, \tag{24}$$

where $B(p,q) = \frac{\Gamma(p)\Gamma(q)}{\Gamma(p+q)}$ is a Beta function. □

Theorem 4. *Given the posterior predictive distribution* X_f given V, that is

$$g\left(X_f|V\right) = \frac{(b+3nV/2)^{3n/2+a}}{\sqrt{2}B(3n/2+a,3/2)} \frac{X_f^2}{\left(b+3nV/2+X_f^2/2\right)^{3n/2+a+3/2}}; X_f > 0,$$

the transformation $W_f = \frac{(b+3nV/2)}{\left(b+3nV/2+X_f^2/2\right)}$ *given V has the PDF of a Beta (3n/2 + a, 3/2) distribution.*

Proof of Theorem 4. In the PDF of X_f given V in Equation (24), since $W_f = \frac{(b+3nV/2)}{\left(b+3nV/2+X_f^2/2\right)} \Rightarrow$ $X_f = \sqrt{2(b+3nV/2)\left(W_f^{-1}-1\right)}$ then the Jacobian of transformation, $J\left(X_f \to W_f\right) = \sqrt{(b+3nV/2)/2}W_f^{-3/2}\left(1-W_f\right)^{-1/2}$. Hence, using the expression, $g\left(W_f\middle|V\right) = f\left(X_f\middle|V\right)|J|$ the PDF of W_f given V is expressed by the following:

$$g\left(W_f\middle|V\right) = \frac{1}{B(3n/2+a,3/2)}W_f^{3n/2+a-1}\left(1-W_f\right)^{1/2}; \quad 0 < W_f < 1. \tag{25}$$

□

Theorem 4 shows that W_f follows the beta distribution with parameters $3n/2+a$ and $3/2$. and the probability control limits LPL_2, CL_2 and UPL_2 may be expressed in terms of the quantiles of the beta distribution having parameters $3n/2+a$ and $3/2$. Thus Equations (22) and (25) provide the solution for Equation (21), which can be presented as given as follows:

$$\left.\begin{aligned} LPL_2 &= (b+3nV/2)B_1 \\ CL_2 &= (b+3nV/2)B_2 \\ UPL_2 &= (b+3nV/2)B_3 \end{aligned}\right\}, \tag{26}$$

where $B_1 = \left[\frac{1}{B_{1-\alpha}(3n/2+a,3/2)} - 1\right]$, $B_2 = \left[\frac{1}{B_{0.5}(3n/2+a,3/2)} - 1\right]$ and $B_3 = \left[\frac{1}{B_{\alpha/2}(3n/2+a,3/2)} - 1\right]$ and $B_p(\beta_1, \beta_2)$ represents the $100p$-quantiles of the beta distribution having parameters β_1 and β_2. Table 2 contains the different values of B_1, B_2, and B_3 for a specified n, α and a. If p_2 denotes the power of the test for the Bayesian-II Shewhart-type control chart, then it can be given as follows:

$$p_2 = P\left(X_f < LPL_2 \middle| \delta \neq 1\right) + P\left(X_f > UPL_2 \middle| \delta \neq 1\right),$$

that can be solved to

$$p_2 = 1 + G_{Beta}\left(\frac{2}{2 + (b + 3nV/2)B_1^2}; \beta_1, \beta_2 \middle| \delta \neq 1\right) - G_{Beta}\left(\frac{2}{2 + (b + 3nV/2)B_3^2}; \beta_1, \beta_2 \middle| \delta \neq 1\right)$$

where $G_{Beta}(\cdot; \beta_1, \beta_2)$ is the CDF of the beta distribution having parameters $\beta_1 = 3n/2 + a$ and $\beta_2 = 3/2$. Consequently, the ARL_1 for the Bayesian-II Shewhart-type control chart is given by the following:

$$ARL_1 = \frac{1}{1 + G_{Beta}\left(\frac{2}{2+(b+3nV/2)B_1^2}; \beta_1, \beta_2 \middle| \delta \neq 1\right) - G_{Beta}\left(\frac{2}{2+(b+3nV/2)B_3^2}; \beta_1, \beta_2 \middle| \delta \neq 1\right)} \quad (27)$$

Table 2. B_1, B_2, and B_3 values for different n and a = 25.5, 36.9.

		False Alarm Rate α								
a	n	0.005			0.0027			0.0027		
		B_1	B_2	B_3	B_1	B_2	B_3	B_1	B_2	B_3
25.5	1	0.00082	0.04423	0.29951	0.00054	0.04423	0.33103	0.00045	0.04423	0.34661
	2	0.00078	0.04188	0.28191	0.00052	0.04188	0.31137	0.00042	0.04188	0.32592
	3	0.00074	0.03977	0.26626	0.00049	0.03977	0.29391	0.00040	0.03977	0.30754
	4	0.00071	0.03786	0.25225	0.00047	0.03786	0.27829	0.00038	0.03786	0.29112
	5	0.00067	0.03612	0.23963	0.00045	0.03612	0.26424	0.00036	0.03612	0.27636
36.9	1	0.00058	0.03102	0.20310	0.00038	0.03102	0.22362	0.00031	0.03102	0.23371
	2	0.00056	0.02984	0.19483	0.00037	0.02984	0.21445	0.00030	0.02984	0.22408
	3	0.00054	0.02876	0.18721	0.00036	0.02876	0.20599	0.00029	0.02876	0.21522
	4	0.00052	0.02774	0.18016	0.00034	0.02774	0.19818	0.00028	0.02774	0.20702
	5	0.00050	0.02680	0.17362	0.00033	0.02680	0.19093	0.00027	0.02680	0.19943

4. Proposed Bayesian Shewhart-Type Control Charts

This section discusses the performance evaluation performance measures utilized to investigate the performance behavior of the proposed Bayesian-I and Bayesian-II Shewhart charts. Sections 4.1 and 4.2 defines the simulation study and average run length. In the same line, Section 4.3 defines the overall performance indicators. Section 4.3 discusses the Monte Carlo simulations. The sensitivity analysis is discussed in Section 4.4.

4.1. Simulation Study

The numerical results are obtained through the Monte Carlo simulation method by using R software. The sample is generated from the specified distribution, and then construct control limits and plotting statistics. The average run length properties are obtained under the assumptions of various parameters and different values of shifts such as 1, 1.25, 1.50, 1.72, 2, 2.25, 2.50, 3, 6. Moreover, different sets of hyperparameters are considered for the sensitivity analysis of the proposed control charts.

4.2. Average Run-Length

The average run length (ARL) is the familiar run-length characteristic that evaluates the performance behavior of the control charts. The ARL can be considered as the average number of sample points plotted on the control chart until the control chart indicates an OOC signal, where sample points are referred to as run-length (RL). The IC ARL (ARL_0)

and the OOC *ARL* (ARL_1) are the two types of *ARL* It is emphasized that if the operates in the IC state, then the ARL_0 should be larger to prevent frequent false alarms; but, for OOC states, the ARL_1 should be smaller to identify the process change as soon as possible. To enhance the efficiency of the control chart, it is important to attain a smaller ARL_1 for the control chart with predetermined ARL_0 at a desired level.

4.3. *Overall Performance Measure*

Although the *ARL* is the best-known measure to assess the performance behavior of the control charts at a single defined shift; however, there are alternative measures that can be used to assess the overall detection ability of a control chart. Extra quadratic loss (*EQL*), relative average run length (*RARL*), and performance comparison index (*PCI*) are among these measures. A control chart with smaller *EQL*, *RARL*, and *PCI* values is regarded to be superior. More detail about these performance measures is provided as follows

4.3.1. Extra Quadratic Loss

The *EQL* is the weighted *ARL* over the whole shift domain $(\delta_{min}, \delta_{max})$, where the square of the shift δ is used as a weight. The *EQL* mathematically can be defined as follows:

$$EQL = \frac{1}{\delta_{max} - \delta_{min}} \int_{\delta_{min}}^{\delta_{max}} \delta^2 ARL(\delta) d\delta$$

where δ_{min} is the minimum shift, δ_{max} is the maximum shift, and $ARL(\delta)$ is the *ARL* of a specific control chart at a shift δ.

4.3.2. Relative Average Run Length

The *RARL* is the ratio of the *ARL* of a certain control chart (i.e., $ARL(\delta)$ to the benchmark control chart (i.e., $(ARL_{bmk}(\delta))$). Mathematically the *RARL* can be defined as follows:

$$RARL = \frac{1}{\delta_{max} - \delta_{min}} \int_{\delta_{min}}^{\delta_{max}} \frac{ARL(\delta)}{ARL_{bmk}(\delta)} d\delta$$

A benchmark control chart is one, which has minimal *EQL*, or it can be regarded as some of the existing standard control charts.

4.3.3. Performance Comparison Index

The *PCI* is defined as the ratio between the *EQL* of the control chart to the *EQL* of the benchmark control chart. The *PCI*, mathematically, can be specified as follows:

$$PCI = \frac{EQL}{EQL_{bmk}}$$

The $PCI = 1$ for the benchmark control chart and the remaining control charts have $PCI > 1$.

4.4. *Sensitivity Analysis of Hyperparameters*

The Bayesian process monitoring largely depends on the form of the prior distribution. The informative and non-informative prior have a different impact on the control chart performance. In this study, the informative (conjugate) prior is considered for the Maxwell parameter σ^2, which is assumed to be inverted gamma with hyperparameters a and b (i.e., $\mathrm{IG}(a, b)$). The sensitivity analysis of hyperparameter values is performed to study the impact of an increase or decrease in hyperparameter values on the performance behavior of the Bayesian-I and Bayesian-II Shewhart-type control charts. Different pairs of hyperparameter values are chosen for this purpose, such as (a, b) = (0, 0), (1, 0), (8.5, 0.005), (25.5, 0.005), (25.5, 1.5), (40.9, 0.005), (70, 0.2), (2.5, 0.005), and (110, 0.005). The impact of various hyperparameter choices on the performance behavior of the proposed control chart is discussed as follows:

1. The Bayesian-I and Bayesian-II Shewhart-type control charts are very sensitive to hyperparameter values. A slight change in hyperparameters significantly affects the *ARL* performance. For example, for $n = 2$ and $\delta = 1.5$, the *ARL* for the proposed Bayesian-I Shewhart-type control chart is 211.94, if $(a, b) = (0, 0)$ and when $(a, b) = (1, 0)$ then *ARL* is 183.01 (see Table 3). Similarly, for the same $n = 2$, $\delta = 1.5$ the *ARL* for the proposed Bayesian-II Shewhart-type control chart is 221.54, if $(a, b) = (0, 0)$ and when $(a, b) = (1, 0)$ then *ARL* is 187.48 (see Table 4);
2. The detection ability of the proposed Bayesian-I and Bayesian-II Shewhart-type control chart improves when a gets larger and b becomes smaller at the same time. For example, for $n = 2$ and $\delta = 1.5$, the *ARL* for the proposed Bayesian-I Shewhart-type control chart is 34.00, if $(a, b) = (25.5, 1.5)$, whereas when $(a, b) = (40.9, 0.005)$ then *ARL* is reduced to 18.35 (see Table 3). Similarly, the *ARL* for the proposed Bayesian-II Shewhart-type control chart with $n = 2$, $\delta = 1.5$, is 51.064 when $(a, b) = (25.5, 1.5)$, while for $(a, b) = (40.9, 0.005)$ the *ARL* for the proposed Bayesian-II Shewhart-type control chart is 1.01 (see Table 4);
3. The constants A_1, A_2 and A_3 reduce as a increases. For instance, when $n = 2$ and $\alpha = 0.0027$, then $A_1 = 0.04105$, $A_2 = 0.08955$, and $A_3 = 0.25601$ if $a = 8.5$ and $a = 25.2$ then $A_1 = 0.02114$, $A_2 = 0.03539$, and $A_3 = 0.06592$ (see Table 1);
4. Similarly, the values of B_1, B_2, and B_3 decrease as a gets larger. For example, if $n = 2$, $\alpha = 0.0027$, then $B_1 = 0.00127$, $B_2 = 0.10600$, and $B_3 = 0.94696$. Likewise, when $a = 8.5$ and 25.5 then $B_1 = 0.00052$, $B_2 = 0.04188$, and $B_3 = 0.31137$ (see Table 2).

Table 3. Run length profile of the Bayesian-I Shewhart-type control chart with different sample sizes and hyperparameter values at $\alpha = 0.0027$ with $ARL \approx 370$.

	δ								
	1	**1.25**	**1.5**	**1.75**	**2**	**2.25**	**2.5**	**3**	**6**
n				$a = 0, b = 0$					
2	369.75	265.15	211.94	159.73	134.96	116.21	94.10	82.09	33.00
5	369.83	203.85	116.15	79.44	57.77	47.03	34.10	23.62	6.02
9	371.53	129.99	86.30	42.92	27.24	16.46	12.50	8.19	1.85
				$a = 1, b = 0$					
2	369.66	242.51	183.01	128.38	104.10	86.52	66.71	56.43	18.95
5	371.08	193.86	105.37	69.96	49.72	39.90	28.35	19.23	4.71
9	369.83	123.53	80.64	39.11	24.48	14.61	11.04	7.19	1.68
				$a = 8.5, b = 0.005$					
2	371.52	148.09	83.00	40.85	27.46	19.67	12.51	9.49	2.23
5	371.50	138.00	56.49	31.72	19.98	14.92	9.69	6.08	1.59
9	370.10	89.93	52.45	21.74	12.57	7.06	5.24	3.39	1.12
				$a = 25.5, b = 0.005$					
2	371.67	81.95	27.57	10.80	6.40	4.30	2.65	2.06	1.02
5	370.91	73.92	23.10	10.77	6.20	4.15	2.86	1.89	1.02
9	371.99	53.46	22.33	9.00	4.89	2.75	2.11	1.51	1.00
				$a = 25.5, b = 1.5$					
2	370.02	95.26	40.00	15.24	9.22	6.21	3.77	2.85	1.09
5	371.57	85.36	29.71	14.53	8.45	6.08	3.86	2.47	1.06
9	371.41	62.31	28.18	11.53	6.33	3.51	2.64	1.82	1.01
				$a = 40.9, b = 0.005$					
2	370.74	60.70	18.35	5.86	3.62	2.39	1.60	1.34	1.00
5	371.11	52.85	14.37	6.36	3.46	2.64	1.80	1.32	1.00
9	369.03	38.98	13.61	5.62	3.08	1.83	1.48	1.18	0.99

Table 3. *Cont.*

					δ				
					$a = 70, b = 0.2$				
2	369.21	29.36	8.58	2.65	1.70	1.32	1.09	1.03	1.00
5	370.48	37.24	7.01	3.05	1.86	1.47	1.17	1.04	1.00
9	370.33	23.42	6.90	2.91	1.73	1.22	1.10	1.02	1.00
					$a = 110, b = 0.005$				
2	369.44	23.78	3.78	1.86	1.25	1.11	1.05	1.00	1.00
5	373.26	16.07	3.47	1.79	1.18	1.06	1.02	1.00	1.00
9	369.37	14.12	3.25	1.76	1.22	1.04	1.01	1.00	1.00

Table 4. Run length profile of the Bayesian-II Shewhart-type control chart with different sample sizes and hyperparameter values at $\alpha = 0.0027$ with $ARL \approx 370$.

					δ				
	1	**1.25**	**1.5**	**1.75**	**2**	**2.25**	**2.5**	**3**	**6**
n					$\boldsymbol{a = 0, b = 0}$				
2	371.04	274.85	221.54	181.87	144.91	136.23	95.18	68.31	28.53
5	371.26	146.33	98.83	47.75	36.48	24.90	12.97	8.69	1.13
9	369.99	50.36	29.14	9.99	5.95	2.11	1.50	1.00	1.00
					$a = 1, b = 0$				
2	370.76	249.31	187.48	144.49	107.06	98.71	61.61	39.84	12.78
5	370.90	129.66	83.37	36.70	27.13	17.70	8.53	15.48	1.00
9	369.45	43.67	24.32	7.78	4.49	1.56	1.15	1.00	1.00
					$a = 8.5, b = 0.005$				
2	371.35	123.94	56.46	27.82	12.38	10.02	3.03	1.16	1.00
5	371.14	54.01	24.50	5.58	3.35	1.73	1.00	1.00	1.00
9	369.50	15.65	6.65	1.45	1.00	1.00	1.00	1.00	1.00
					$a = 25.5, b = 0.005$				
2	369.37	25.72	4.04	1.12	1.00	1.00	1.00	1.00	1.00
5	370.41	7.94	1.97	1.00	1.00	1.00	1.00	1.00	1.00
9	370.91	1.90	1.01	1.00	1.00	1.00	1.00	1.00	1.00
					$a = 25.5, b = 1.5$				
2	369.39	119.39	51.06	23.12	9.05	7.06	1.72	1.00	1.00
5	369.09	45.86	18.92	3.44	1.93	1.11	1.00	1.00	1.00
9	369.81	11.45	4.37	1.06	1.00	1.00	1.00	1.00	1.00
					$a = 40.9, b = 0.005$				
2	369.58	6.77	1.01	1.00	1.00	1.00	1.00	1.00	1.00
5	371.59	1.82	1.00	1.00	1.00	1.00	1.00	1.00	1.00
9	371.50	1.01	1.00	1.00	1.00	1.00	1.00	1.00	1.00
					$a = 70, b = 0.2$				
2	371.36	1.21	1.00	1.00	1.00	1.00	1.00	1.00	1.00
5	370.55	1.00	1.00	1.00	1.00	1.00	1.00	1.00	1.00
9	370.31	1.00	1.00	1.00	1.00	1.00	1.00	1.00	1.00
					$a = 2.5, b = 0.005$				
2	371.87	222.32	153.44	109.32	73.91	66.50	35.97	20.36	4.70
5	370.23	109.67	65.89	25.48	18.00	11.05	4.81	2.93	1.00
9	369.50	35.76	18.87	5.47	3.04	1.13	1.00	1.00	1.00

5. Performance Comparison and Illustration of Results

This section reports the finding and performance comparison of the Bayesian-I and Bayesian-II Shewhart-type control charts against the existing counterparts. Section 5.1 represents the comparison of the Bayesian-I and Bayesian-II Shewhart-type control charts with the $CUSUM_V$ control chart. Similarly, the Bayesian-I and Bayesian-II Shewhart-type control charts are compared with the $EWMA_V$ control chart in Section 5.2. Likewise, Section 5.1 offers the performance comparison of the Bayesian-I and Bayesian-II Shewhart-type control charts with the V control chart. Finally, Section 5.4 addresses the main outcomes of the study.

5.1. *Proposed versus* $\mathrm{CUSUM_V}$ *Control Chart*

The proposed Bayesian-I and Bayesian-II Shewhart-type control charts are compared against the $CUSUM_V$ control chart at $ARL_0 = 370$, and the results show that the proposed control charts outperformed the $CUSUM_V$ control chart. For example, at $n = 2$ and $\delta = 1.25, 1.5$, the ARL values for the proposed Bayesian-II Shewhart-type control chart ($a = 25.5$, $b = 0.005$) are 25.72, 9.74, while for the $CUSUM_V$ control chart, the ARL values are; 33.93, 18.71 (see Table 3 vs. Table 5). Along with the ARL values for the proposed Bayesian-I Shewhart-type control chart $a = 70$, $b = 0.2$ are 29.36, 23.59 (see Table 4). This indicates that the proposed Bayesian-I and Bayesian-II Shewhart-type control charts perform better against the $CUSUM_V$ control chart in monitoring the shift in the Maxwell scale parameter. Correspondingly, in overall performance comparison, the proposed Bayesian-I and Bayesian-II Shewhart-type control charts also demonstrate superior performance than the $CUSUM_V$ control chart, as their EQL, PCI, and $RARL$ values are smaller than that of $CUSUM_V$ control chart. For instance, for $n = 2$, the EQL, PCI, and $RARL$ values for the proposed Bayesian-I ($a = 110$, $b = 0.005$) and Bayesian-II ($a = 25.5$, $b = 0.005$) Shewhart-type control charts are, respectively, given as; 26.79, 1.00, 1.05, and 26.75, 1.00, 1.00, whereas, for the $CUSUM_V$ control the EQL, PCI, and $RARL$ values chart are provided as; 57.97, 2.17, 4.10 (see, Table 6).

Table 5. ARL values of $CUSUM_V$, $EWMA_V$ and V control charts.

		$CUSUM_V$	$EWMA_V$	V
n	δ	ARL	ARL	ARL
2	1	369.8	370.19	372.02
	1.25	33.93	63.02	116.26
	1.5	16.1	20.63	39.64
	1.75	10.88	9.52	18.67
	2	8.19	5.71	10.84
	2.25	6.39	3.95	7.08
	3	4.36	2.13	3.33
	6	2.13	1.14	1.37
5	1	373.98	371.90	374.69
	1.25	18.25	32.03	69.04
	1.5	9.15	8.22	17.55
	1.75	6.06	3.65	7.36
	2	4.58	2.17	4.19
	2.25	3.82	1.62	2.75
	3	2.51	1.21	1.51
	6	1.32	1.01	1.02
9	1	371.61	371.40	369.29
	1.25	12.65	18.28	42.4
	1.5	5.98	4.13	8.97
	1.75	4.11	1.94	3.76
	2	3.09	1.35	2.21
	2.25	2.64	1.13	1.60
	3	1.82	1.01	1.11
	6	1.04	1.00	1.00

Table 6. Overall performance comparison of the Bayesian-I and Bayesian-II Shewhart-type control charts versus the $CUSUM_V$ and V control charts.

n		$CUSUM_V$	$EWMA_V$ (λ = 0.75)	V	Bayesian-I (a = 110, b = 0.005)	Bayesian-II (a = 25.5, b = 0.005)
	EQL	57.97	40.60	57.49	26.79	26.75
2	*PCI*	2.17	1.52	2.15	1.00	1.00
	RRL	4.10	2.65	4.49	1.05	1.00
	EQL	38.31	29.53	36.20	26.20	25.14
5	*PCI*	1.52	1.17	1.44	1.04	1.00
	RARl	2.62	1.65	2.63	1.14	1.00
	EQL	30.91	26.53	29.93	25.92	24.57
9	*PCI*	1.26	1.08	1.22	1.05	1.00
	RARL	2.27	1.67	2.76	1.49	1.00

5.2. *Proposed versus* EWMA$_V$ *Control Chart*

The proposed Bayesian-I and Bayesian-II Shewhart-type control charts reveal an edge in performance over the $EWMA_V$ control chart. In detail, at $n = 2$ and $\delta = 1.25$, the Bayesian-I ($a = 110, b = 0.005$) Shewhart-type control chart provides *ARL* value of 23.78, while the Bayesian-II ($a = 25.5, b = 0.005$) Shewhart-type control chart bears *ARL* value of 25.72 and the $EWMA_V$ control chart delivers the *ARL* value of 32.03 (see Tables 2 and 3 vs. Table 5). Similarly, the proposed Bayesian-I and Bayesian-II Shewhart-type control charts also yield superior overall performance than the $EWMA_V$ control chart as the proposed Bayesian-I and Bayesian-II Shewhart-type control charts have minimum *EQL*, *PCI*, and *RARL* values than $EWMA_V$ control chart (see Table 6). For example, with $n = 5$, the Bayesian-I (a = 110, b = 0.005) Shewhart-type control chart has *EQL*, *PCI*, and *RARL* values of 26.20, 1.04, and 1.14, the Bayesian-II (a = 25.5, b = 0.005) Shewhart-type control charts address the *EQL*, *PCI*, and *RARL* values of 25.14, 1.00, and 1.00, whereas, for the $EWMA_V$ control chart generates the *EQL*, *PCI*, and *RARL* values of 29.53, 1.17, and 1.65 (see Table 6).

5.3. *Proposed versus V Control Chart*

The proposed Bayesian-I and Bayesian-II Shewhart-type control charts achieve better shift detection ability than the V control chart. For instance, at $ARL_0 = 370$, $n = 5$, and $\delta = 1.25$, 1.5, the ARL_1 values for the proposed Bayesian-I Shewhart-type control chart ($a = 70, b = 0.2$) are 37.24, 7.01, whereas the *ARL* values for the V control chart are 69.04 and 17.55 (see Table 4 vs. Table 5). This indicates the superiority of the Bayesian-I Shewhart-type control chart over V control chart. Similarly, when $n = 5$, and $\delta = 1.25$, 1.5, the proposed Bayesian-II Shewhart-type control chart ($a = 25.5, b = 0.005$) *ARL* values are 7.94, 1.97, which are smaller than V control chart. Likewise, for $n = 9$, the proposed Bayesian-I (a = 110, b = 0.005) and Bayesian-II (a = 25.5, b = 0.005) Shewhart-type control charts have smaller *EQL*, *PCI*, and *RARL* values (i.e., 25.92, 1.05, 1.49, and 25.57, 1.00, and 1.00) than the *EQL*, *PCI*, and *RARL* values of the V control chart (i.e., 26.53, 1.08, and 1.67); therefore, the proposed Bayesian-I and Bayesian-II Shewhart-type control charts have better overall performance relative to the V control chart (see Table 6).

5.4. *Main Finding of the Study*

Some important findings about the proposed Bayesian-I and Bayesian-II Shewhart-type control charts are given as follows:

1. The Bayesian-I and Bayesian-II Shewhart-type control charts are very sensitive to hyperparameter values. A slight change in hyperparameter values significantly affects the performance of the proposed Bayesian-I and Bayesian-II Shewhart-type control charts in terms of the *ARL* measure (see, Tables 3 and 4);

2. The detection ability of the proposed Bayesian-I and Bayesian-II Shewhart-type control charts improves when the hyperparameter a grows larger and b becomes smaller (see, Tables 3 and 4);
3. The proposed Bayesian-I and Bayesian-II Shewhart-type control charts have improved ARL performance than the $CUSUM_V$, $EWMA_V$ and V control charts, particularly when hyperparameters a and b increase (see, Table 5);
4. The Bayesian-II Shewhart-type control chart has enhanced detection ability than the Bayesian-I Shewhart-type control chart (see, Table 3 vs. Table 4).

6. Real Data Analysis

A boring machine is a tool used for making a wide hole in a fixed workpiece. These machines make use of a single steel cutting edge, carbide or diamond, or a small grinding wheel to make the hole cleaner, more accurate, and more specific. Boring machines with multiple spindles are typically used in a manufacturing plant where production is on a large scale. This study uses a real dataset by Hossain et al. [21] that addressed the failure rate of the vertical boring machine. This data set was also considered by Majumdar [38] to review the optimum maintenance approach for the vertical boring machine. Subsequently, Krishna and Malik [39] conducted a detailed statistical investigation to evaluate the distributions, which best fit this data set. They examined models such as exponential, gamma, Maxwell, lognormal, Weibull, and estimated the parameters of these models with the maximum likelihood method. In addition, various information criteria such as Akaike, second-order, Bayesian, and the Kolmogorov–Smirnov test have shown that the Maxwell distribution is the best fitted to this data set. Additionally, Hossain et al. [21] used the Kolmogorov–Smirnov test and showed that the data set followed the Maxwell distribution (p-value 0.4775) with σ = 1777.86. Hossain et al. [21] also verified that no larger change occurs for this data set. The failure time data for a vertical boring machine are specified as follows: 2802, 2937, 2136, 4359, 4020, 1781, 2816, 2655, 3886, 2296, 3158, 3695, 4155, 3811, 2380, 376, 2172, 3705, 2848, 4339, 2076, 2672, 3632, 1976, 1700, 1596, 1701, 3575, 3802, 4351, 4291, and 808.

Hossain et al. [22] first divided this data set into eight groups each of size four, and estimated σ^2 by using Equation (3), that is, $\overline{V}$ = 3.160782. They specified the various features of the $CUSUM_V$ control chart as; n = 4, σ_0 = 1777.86, h 13,400,000, and k 3,313,881. Using these constants, they simulated 22 OOC samples with an upward change of δ = 1.2. So, there are a total of 30 samples, out of which 8 are IC, and the rest of 22 samples are OOC (see Table 7). Following the rationale of Hossain et al. [22], in this study the $EWMA_V$, V, Bayesian-I, and Bayesian-II Shewhart-type control charts are constructed. At α = 0.002, n = 4, L_2 = 0.94503, and L_3 = 2.67246, the probability control limits LPL_V = 619,197 , CL_V = 3.160782, and UPL_V = 8.446242 of the V control chart are obtained. Similarly, with λ = 0.25 and L_E = 3.26, the $EWMA_V$ fixed limits are constructed, i.e., LCL_E = 1,570,819, CL_E = 3,160,786 and UCL_E = 4,750,753. Likewise, the probability control limits LPL_1 = 747,595, CL_1 = 1,597,923 and UPL_1 = 4,404,396 for the proposed Bayesian-I Shewhart-type control chart using α = 0.0027, n = 4 , a = 8.5, b = 0.005, A_1 = 0.01956, A_2 = 0.03199, and A_3 = 0.05763 are determined. Moreover, the LPL_2 = 6518, CL_2 = 526,155 and UPL_2 = 3,758,399 present the probability control limits of the proposed Bayesian-II Shewhart-type control chart, which is based on α = 0.0027, n = 4, a = 36.9, b = 0.005, B_1 = 0.00034, B_2 = 0.02774, and B_3 = 0.19818. The $CUSUM_V$, $EWMA_V$, V, Bayesian-I, and Bayesian-II Shewhart-type control charts are constructed using the aforementioned information. Figure 1 demonstrates the graphical properties of these control charts.

The proposed Bayesian-I and Bayesian-II Shewhart-type control charts are more sensitive than the $CUSUM_V$, $EWMA_V$, and V control charts, as the comparison reveals that the proposed Bayesian-I and Bayesian-II Shewhart-type control charts detect 6 and 11 OOC signals, respectively, and the $CUSUM_V$ control chart identifies 2 OOC points, while the $EWMA_V$ and V control charts fail to diagnose the OOC signal. Similarly, the proposed Bayesian-I and Bayesian-II Shewhart-type control charts detect the OOC points at sample numbers 9 and 8, respectively, while the $CUSUM_V$ control chart diagnoses the first OOC signal at sample number 29. This indicates that the proposed Bayesian-I and Bayesian-II Shewhart-type control charts are more efficient than the $CUSUM_V$, $EWMA_V$ and V control charts.

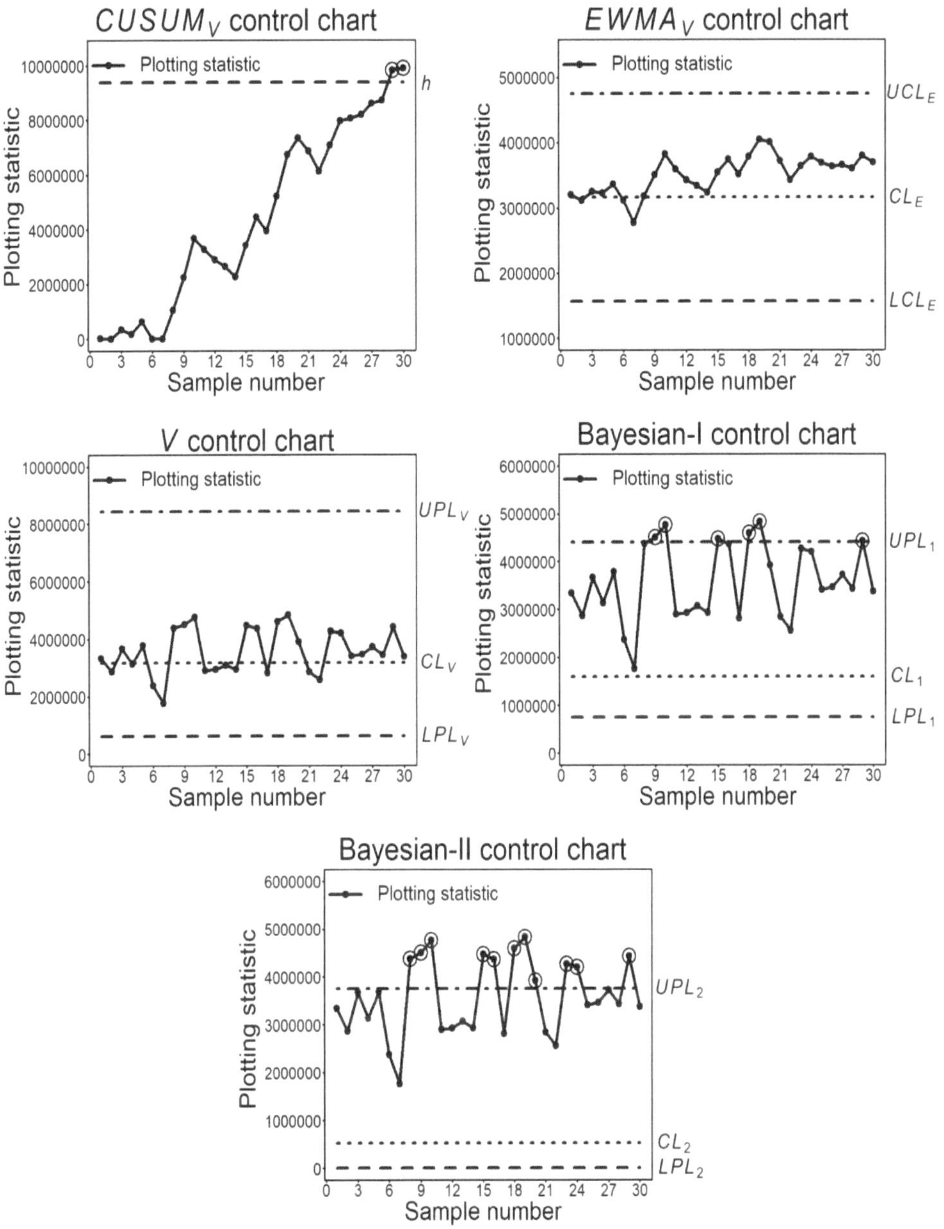

Figure 1. $CUSUM_V$, $EWMA_V$, V Bayesian-I, and Bayesian-II Shewhart-type control charts with a vertical boring machine failure data.

Table 7. Charting statistics for various control charts.

Sample Number	V_i	C_i^+	Z_i	Sample Number	V_i	C_i^+	Z_i
1	3,336,713	22,832	3,204,768	16	4,354,789	8,273,484	3,746,652
2	2,859,270	0	3,118,393	17	2,805,902	7,765,506	3,511,464
3	3,666,550	352,669	3,255,433	18	4,588,520	9,040,146	3,780,728
4	3,132,794	171,582	3,224,773	19	4,829,929	10,556,194	4,043,029
5	3,781,886	639,588	3,364,051	20	3,910,538	11,152,851	4,009,906
6	2,378,780	0	3,117,733	21	2,841,708	10,680,678	3,717,856
7	1,759,270	0	2,778,118	22	2,566,451	9,933,248	3,430,005
8	4,370,996	1,057,115	3,176,337	23	4,267,153	10,886,521	3,639,292
9	4,503,843	2,247,078	3,508,214	24	4,202,461	11,775,101	3,780,084
10	4,761,577	6,504,774	3,821,554	25	3,406,360	11,867,580	3,686,653
11	2,893,367	6,084,261	3,589,508	26	3,459,810	12,013,509	3,629,942
12	2,931,706	5,702,087	3,425,057	27	3,723,778	12,423,407	3,653,401
13	3,068,791	5,456,997	3,335,991	28	3,430,499	12,540,025	3,597,676
14	2,934,246	5,077,362	3,235,554	29	4,419,754	13,645,899	3,803,195
15	4,469,095	6,232,576	3,543,940	30	3,378,578	13,710,596	3,697,041

7. Summary, Conclusions, and Recommendation

This study is performed to propose two Bayesian Shewhart-type control charts, which are based on the probability control limits, to monitor the familiar Maxwell distribution. These control charts are called the Bayesian-I and Bayesian-II Shewhart-type control charts. The design structures for the proposed Bayesian-I and Bayesian-II Shewhart-type control charts are obtained using the posterior and posterior predictive distributions, respectively. These distributions are constructed under the assumption of conjugate prior for the scale parameter of Maxwell distribution, which is assumed to be an inverted gamma distribution with hyperparameters a and b. The performance of the suggested Bayesian-I and Bayesian-II Shewhart-type control charts is evaluated by computing the important performance evaluation measures such as *ARL*, *EQL*, *RARL*, and *PCI*. A comparative study is carried out among the proposed Bayesian-I and Bayesian-II Shewhart-type control charts, and some existing competitors are more sensitive than the $CUSUM_V$, $EWMA_V$ and V control charts. The sensitivity analysis is performed to study the effect of increasing and decreasing hyperparameter values on the performance behavior of the proposed Bayesian-I and Bayesian-II Shewhart-type control charts. The results derived from this study have shown that the proposed Bayesian-I and Bayesian-II Shewhart-type control charts perform well in the monitoring of the Maxwell scale parameter. A real-life data application is also provided for the practical implementation of the proposed Bayesian-I and Bayesian-II Shewhart-type control charts. Finally, it is recommended that the concept of the Bayesian-I and Bayesian-II Shewhart-type control charts can be extended to other distributions, where the manufacturing processes need to diagnose the small and large changes in the process parameters.

Author Contributions: Conceptualization, Z.R.; Methodology, S.M.A.; Software, Z.R.; Formal analysis, S.M.A. and A.N.C.; Investigation, F.A., M.K. and A.N.C.; Resources, M.K.; Data curation, M.K. and Z.R.; Writing—original draft, M.K. and S.M.A.; Writing—review & editing, F.A. and S.M.A.; Supervision, I.M.A.; Project administration, A.N.C.; Funding acquisition, F.A. and I.M.A. All authors have read and agreed to the published version of the manuscript.

Funding: This research was funded by Princess Nourah bint Abdulrahman University Researchers Supporting Project number (PNURSP2023R358), Princess Nourah bint Abdulrahman University, Riyadh, Saudi Arabia.

Data Availability Statement: All the relevant data is already included in the main manuscript.

Acknowledgments: The authors thank the reviewers for their valuable comments which increase the quality of this paper. They also thank and extend their appreciation to Princess Nourah bint Abdulrahman University for funding this work. through Princess Nourah bint Abdulrahman University Researchers Supporting Project number (PNURSP2023R358), Princess Nourah bint Abdulrahman University, Riyadh, Saudi Arabia.

Conflicts of Interest: The authors declare no conflict of interest.

References

1. Shewhart, W.A. *Economic Control of Quality of Manufactured Product*; D. Van Nostrand Company: New York, NY, USA, 1931.
2. Montgomery, D.C. *Introduction to Statistical Quality Control*, 7th ed.; John Wiley & Sons: Hoboken, NJ, USA, 2012.
3. Demirhan, H.; Hamurkaroglu, C. Bayesian X-bar control limits for exponentially distributed measurements. *J. Stat. Comput. Simul.* **2014**, *84*, 628–643. [CrossRef]
4. Al-Omari, A.I.; Al-Nasser, A.D. Statistical quality control limits for the sample mean chart using robust extreme ranked set sampling. *Econ. Qual. Control* **2011**, *26*, 73–89. [CrossRef]
5. Al-Omari, A.I.; Haq, A. Improved quality control charts for monitoring the process mean using double-ranked set sampling methods. *J. Appl. Stat.* **2012**, *39*, 745–763. [CrossRef]
6. Haq, A.; Al-Omari, A.I. A new Shewhart control chart for monitoring process mean based on partially ordered judgment subset sampling. *Qual. Quant.* **2015**, *49*, 1185–1202. [CrossRef]
7. Shabbir, J.; Awan, W.H. An Efficient Shewhart-Type Control Chart to Monitor Moderate Size Shifts in the Process Mean in Phase II. *Qual Reliab Engng Int* **2016**, *32*, 1597–1619. [CrossRef]
8. Khan, Z.; Gulistan, M.; Kadry, S.; Chu, Y.; Lane-Krebs, K. On Scale Parameter Monitoring of the Rayleigh Distributed Data Using a New Design. *IEEE Access* **2020**, *8*, 188390–188400. [CrossRef]
9. Xie, M.; Goh, T.N.; Kuralmani, V. *Statistical Models and Control Charts for High-Quality Processes*; Springer Science & Business Media: New York, NY, USA, 2002.
10. Al-Oraini, H.A.; Rahim, M. Economic statistical design of X control charts for systems with Gamma (λ, 2) in-control times. *Comput. Ind. Eng.* **2002**, *43*, 645–654. [CrossRef]
11. Nichols, M.D.; Padgett, W.J. A Bootstrap Control Chart for Weibull Percentiles. *Qual. Reliab. Eng. Int.* **2006**, *22*, 141–151. [CrossRef]
12. Guo, B.; Wang, B.X. Control Charts For Monitoring The Weibull Shape Parameter Based On Type-II Censored Sample. *Qual. Reliab. Eng. Int.* **2014**, *30*, 13–24. [CrossRef]
13. Lio, Y.L.; Tsai, T.-R.; Aslam, M.; Jiang, N. Control Charts for Monitoring Burr Type-X Percentiles. *Commun. Stat.-Simul. Comput.* **2014**, *43*, 761–776. [CrossRef]
14. Yang, Z.; Xie, M. Process monitoring of exponentially distributed characteristics through an optimal normalizing transformation. *J. Appl. Stat.* **2000**, *27*, 1051–1063. [CrossRef]
15. Chan, L.K.; Cui, H.J. Skewness correction X and R charts for skewed distributions. *Nav. Res. Logist.* **2003**, *50*, 555–573. [CrossRef]
16. Chen, Y.K. Economic design of X-bar control chart for Non-normal data using variable sampling policy. *Int. J. Prod. Econ.* **2004**, *92*, 61–74. [CrossRef]
17. Santiago, E.; Smith, J. Control charts based on the exponential distribution: Adapting runs rules for the t Chart. *Qual. Eng.* **2013**, *25*, 85–96. [CrossRef]
18. Ali, S.; Riaz, M. Cumulative quantity control chart for the mixture of inverse Rayleigh process. *Comput. Ind. Eng.* **2014**, *73*, 11–20. [CrossRef]
19. Sighir, A.; Lin, Z. Designing of Gini-Chart for Exponential, t, Logistic and Laplace Distributions. *Commun. Stat.-Simul. Comput.* **2015**, *44*, 2387–2409. [CrossRef]
20. Haghighi, F.; Castagliola, P. Control chart for monitoring the Weibull shape parameter under two competing risks. *Commun. Stat.-Simul. Comput.* **2018**, *48*, 2125–2137. [CrossRef]
21. Hossain, M.P.; Omar, M.H.; Riaz, M. New V control chart for the Maxwell distribution. *J. Stat. Comput. Simul.* **2016**, *87*, 594–606. [CrossRef]
22. Hossain, M.P.; Sanusi, R.A.; Omar, M.H.; Riaz, M. On designing Maxwell CUSUM control chart: An efficient way to monitor failure rates in boring processes. *Int. J. Adv. Manuf. Technol.* **2019**, *100*, 1923–1930. [CrossRef]
23. Hossain, M.P.; Riaz, M. On designing a new VEWMA control chart for efficient process monitoring. *Comput. Ind. Eng.* **2021**, *162*, 107751. [CrossRef]
24. Hossain, M.P.; Omar, M.H.; Riaz, M. Estimation of mixture Maxwell parameters and its possible industrial application. *Comput. Ind. Eng.* **2017**, *107*, 264–275. [CrossRef]
25. Saghir, A.; Ahmad, L.; Aslam, M. Modified EWMA control chart for transformed gamma data. *Commun. Stat.-Simul. Comput.* **2021**, *50*, 3046–3059. [CrossRef]
26. Godase, D.G.; Mahadik, S.B.; Rakitzis, A.C. The SPRT control charts for the Maxwell distribution. *Qual. Reliab. Eng. Int.* **2022**, *38*, 1713–1728. [CrossRef]
27. Menzefricke, U. On the evaluation of control chart limits based on predictive distributions. *Commun. Stat.-Theory Methods* **2002**, *31*, 1423–1440. [CrossRef]

28. Saghir, A. Phase-I Design Scheme for X-bar chart Based on Posterior Distribution. *Commun. Stat.-Theory Methods* **2014**, *44*, 644–655. [CrossRef]
29. Raubenheimer, L.; van der Merwe, A.J. Bayesian Control Chart for Nonconformities. *Qual. Reliab. Eng. Int.* **2015**, *31*, 1359–1366. [CrossRef]
30. Kumar, N.; Chakraborti, S. Bayesian Monitoring of Times Between Events: The Shewhart tr-Chart. *J. Qual. Technol.* **2017**, *49*, 136–154. [CrossRef]
31. Riaz, S.; Riaz, M.; Hussain, Z.; Abbas, T. Monitoring the performance of Bayesian EWMA control chart using loss functions. *Comput. Ind. Eng.* **2017**, *112*, 426–436. [CrossRef]
32. Menzefricke, U. Combined Exponentially Weighted Moving Average Charts for the Mean and Variance Based on the Predictive Distribution. *Commun. Stat.-Theory Methods* **2013**, *42*, 4003–4016. [CrossRef]
33. Riaz, M.; Ali, S. On process monitoring using location control charts under different loss functions. *Trans. Inst. Meas. Control* **2016**, *38*, 1107–1119. [CrossRef]
34. Raubenheimer, L.; van der Merwe, A.J. Bayesian process control for the Phase II Shewart-type p-chart. *Qual. Technol. Quant. Manag.* **2016**, *13*, 453–472. [CrossRef]
35. Rigdon, S.E.; Woodall, W.H. Using the predictive distribution to determine control limits for the Bayesian MEWMA chart. *Commun. Stat.-Simul. Comput.* **2017**, *46*, 7818–7826. [CrossRef]
36. Aslam, M.; Anwar, S.M. An improved Bayesian Modified-EWMA location chart and its applications in mechanical and sport industry. *PLoS ONE* **2020**, *15*, e0229422. [CrossRef]
37. Noor, S.; Noor-ul-Amin, M.; Mohsin, M.; Ahmed, A. Hybrid exponentially weighted moving average control chart using Bayesian approach. *Commun. Stat.-Theory Methods* **2022**, *51*, 3960–3984. [CrossRef]
38. Majumdar, S.K. An optimum maintenance strategy for a vertical boring machine system. *Oper. Res. Soc. India* **1993**, *30*, 344–365.
39. Krishna, H.; Malik, M. Reliability estimation in Maxwell distribution with progressively Type-II censored data. *J. Stat. Comput. Simul.* **2012**, *82*, 623–641. [CrossRef]

mathematics

MDPI

Article

Autocorrelation and Parameter Estimation in a Bayesian Change Point Model

Rui Qiang [1] and Eric Ruggieri [2,*]

[1] Department of Statistics, The Ohio State University, Columbus, OH 43210, USA
[2] Department of Mathematics and Computer Science, College of the Holy Cross, Worcester, MA 01610, USA
* Correspondence: eruggier@holycross.edu

Abstract: A piecewise function can sometimes provide the best fit to a time series. The breaks in this function are called change points, which represent the point at which the statistical properties of the model change. Often, the exact placement of the change points is unknown, so an efficient algorithm is required to combat the combinatorial explosion in the number of potential solutions to the multiple change point problem. Bayesian solutions to the multiple change point problem can provide uncertainty estimates on both the number and location of change points in a dataset, but there has not yet been a systematic study to determine how the choice of hyperparameters or the presence of autocorrelation affects the inference made by the model. Here, we propose Bayesian model averaging as a way to address the uncertainty in the choice of hyperparameters and show how this approach highlights the most probable solution to the problem. Autocorrelation is addressed through a pre-whitening technique, which is shown to eliminate spurious change points that emerge due to a red noise process. However, pre-whitening a dataset tends to make true change points harder to detect. After an extensive simulation study, the model is applied to two climate applications: the Pacific Decadal Oscillation and a global surface temperature anomalies dataset.

Keywords: change point analysis; prior distribution; model averaging; autocorrelation; PDO; temperature anomalies

MSC: 62F15; 62J05; 62P12

Citation: Qiang, R.; Ruggieri, E. Autocorrelation and Parameter Estimation in a Bayesian Change Point Model. *Mathematics* **2023**, *11*, 1082. https://doi.org/10.3390/math11051082

Academic Editor: Diana Mindrila

Received: 28 January 2023
Revised: 14 February 2023
Accepted: 17 February 2023
Published: 21 February 2023

1. Introduction

1.1. What Is a Change Point?

A change point is defined as the point at which the statistical properties of a model change. For example, suppose that a constant model, $Y = \mu + \epsilon$, is used to model the mean signal in a system. Here, a change to either the mean or the variance at any point in the time series indicates the existence of a change point. If a linear (e.g., trend) model is more appropriate, i.e., $Y = \beta_0 + \beta_1 t + \epsilon$, then a change in the slope (β_1), intercept (β_0), or variance of the error terms (ϵ) would indicate a change point in the data.

The problem is simple if the locations of the change points are known. In this case, a separate model can be fit to each section of the data. However, the problem quickly becomes intractable if the locations of the change points are unknown. For example, there are $\binom{250}{5}$ = 7,817,031,000 possible ways to place 5 change points among these 250 observations, and this is by no means a large dataset. Thus, our goal is to create an efficient change point model that can accurately determine the unknown location of change points in a dataset.

Change point analysis has been used in a variety of different settings. In finance, locating change points in a portfolio can help companies understand how their decisions affect their revenue and profit [1]. Change point models have also been used to study Bitcoin returns [2], stock market returns [3,4], and average annual wage growth [5]. In health care, change point models have been used to look at fMRI data [6], EEG signals [7], and visits to the emergency department [8]. Climate applications include the study of glacial records [9], precipitation data [10,11], and global temperature data [12]. Additional applications include social network analysis [13], speech processing [14], and bio-informatics [15,16], to name a few. Of further note are the summative works of [17,18], which provide a number of examples across a variety of fields.

1.2. Overview of Existing Approaches

Page published the first article concerning change points in 1954 [19]. This paper was motivated by a quality control problem in manufacturing and outlined a test for a single change point from a common parametric distribution. Now, the literature on change point models is vast. Broadly, change point detection algorithms can be classified as either batch (retrospectively analyze the data) or sequential (analyze the data as it comes in), and each category can be further categorized as either frequentist or Bayesian. In what follows, we give a brief overview of a few algorithms in each category to help place our model in an appropriate context.

Cumulative sum (CUSUM) statistics and likelihood ratio tests are two frequentist approaches to detecting change points. The CUSUM approach, introduced by [20], monitors either the mean or variance of the residuals and signals a change point if this cumulative sum begins to "drift" (see also [18,21]). A random set of residual errors would have a cumulative sum centered at zero, whereas a string of positive or negative residuals might indicate a break in the underlying model. For a likelihood ratio test (e.g., [22–24]), the null hypothesis of no change point is tested against the alternative hypothesis of a change point at each data point. Note that if the pre- and post-change parameters are assumed to be known, the CUSUM statistic becomes a sequential likelihood ratio test.

A popular approach to the multiple change point problem is binary segmentation, first introduced by [25]. Binary segmentation begins by searching the entire dataset (using any available method) for a single change point. If one is found, the data are split in two at the change point location and the process is repeated on each of the two smaller segments until no further change points are detected. This greedy algorithm is fast, but is not guaranteed to find the globally optimal solution, working best when the change points are well separated and the segment means are distinct [26]. Modern adaptations include circular binary segmentation [27], which pins the two ends of the dataset together to form a circle and introduces change points two at a time (i.e., two cuts in the circle), and wild binary segmentation [28], which considers a localized (rather than global) CUSUM statistic on random subsegments of the time series to identify change points.

Unlike binary segmentation, segment neighborhood algorithms (e.g., [9,29,30]) and the pruned exact linear time (PELT) algorithm [31] are guaranteed to find the global optimum solution to the multiple change point problem. Segment neighborhood algorithms use dynamic programming to recursively add change points to the time series, with the goal of minimizing a cost function such as squared error. PELT seeks to minimize an arbitrary cost function, plus a penalty function that helps to guard against overfitting, by recursively calculating the minimum cost at time s in terms of the minimal cost at time t, with $t < s$. If the number of change points increases linearly with the size of the dataset, the algorithm achieves linear complexity by removing calculations that are not relevant to finding the global minimum.

Bayesian approaches to the multiple change point problem have the advantage of being able to quantify the uncertainty in both the number and location of change points. MCMC approaches (e.g., [32–34]) are dominant on the Bayesian side, where the idea is often to make a proposal that changes the location of one change point (either adding, deleting, or moving its location) and then "accept" that proposal based on whether or not it produces a better fit to the data. As with all MCMC algorithms, convergence issues can exist due to strong correlations in the target distributions [35,36]. Dynamic programming (e.g., [35,37]) can instead be used to sample directly from the posterior distribution, avoiding issues with convergence.

Sequential Bayesian change point algorithms such as Bayesian Online Change Point Detection (BOCPD) [38] and particle filters (e.g., [39,40]) work by specifying a probability distribution over the length of each segment. BOCPD uses a recursive message passing algorithm to determine the probability distribution of the current "run length" given the observed data, a predictive model (e.g., i.i.d. Gaussian), and a hazard function (the probability of a change point at a given run length). For particle filters, each weighted "particle" represents one possible state of the system (in terms of the number and location of change points), so the number of particles grows exponentially with the length of the dataset. Resampling the particles at each step keeps those which are most probable and can be used to limit the computational burden [3], but this process introduces small errors which compound over time because particles that are removed cannot be brought back [36]. Alternately, Bayesian dynamic linear models (BDLM), also known as state-space models, are probabilistic models with time varying coefficients, which can include terms to model trends, seasonality, covariates, and autoregressive components to capture various features of a time series [41,42]. Broadly, BDLM use a learning process to sequentially revise the state of a priori knowledge as new data become available. In particular, a one-step-ahead prior distribution for the next state is updated after observing the data to create a posterior distribution at time t, and then the process is propagated forward in time. Changes to the (hidden) state of the system represent a change point in the system.

Readers looking for more information on change point analysis are directed to the summative works [17,18], which discuss a number of change point models using examples across a variety of fields. The changepoint.info website also maintains an extensive list of publications and software related to change point models.

In Section 2, we describe the Bayesian change point model of [37] which provides the methodological foundation for this study. While previous studies have considered variations of this original algorithm, compared the error and detection rates to other change point models, and offered suggestions on how to set the hyperparameters of the model (e.g., [43,44]), there has not been a systematic study to determine how the choice of parameters for the prior distribution can affect the inference made by the model. In addition, it is not known for certain how autocorrelation will affect the output of the model. Thus, Section 2 ends with an in-depth discussion of these two shortcomings of the Bayesian change point model. In Section 3, we discuss a pre-whitening technique to address autocorrelation and a model averaging technique to address parameter uncertainty. In both cases, an extensive simulation study is presented, first to show how the algorithm performs both in the presence of autocorrelation and after pre-whitening, and then to show how model averaging highlights the most probable solutions to the multiple change point problem. Section 4 presents a novel analysis of two climate datasets using these techniques. Discussion and conclusions are given in Section 5.

2. Materials and Methods

2.1. Description of the Bayesian Model

The Bayesian change point model described in this section assumes that the parameters of the model for any two segments of the data are independent (i.e., a product partition model [32]) and incorporates dynamic programming recursions to piece together the different subsets of the dataset in a computationally efficient way. Once complete, the model returns both the posterior distribution on the number and location of change points in the time series (which gives us probabilistic bounds on their location) and estimates of the parameters of the model between any two change points.

For each subset of the data, we assume a linear relationship between the response variable, Y, and a set of m known explanatory variables, $X_1, X_2, \ldots, X_m$. Thus, our model takes the form:

$$Y = \beta_0 + \beta_1 X_1 + \cdots + \beta_m X_m + \epsilon,$$

where β_i represents the regression coefficient corresponding to the ith explanatory variable, X_i. The explanatory variables are functions of time for a time series and can include terms that are constant, linear, periodic, etc. In addition, the random error terms, ϵ, are assumed to be independent normally distributed random variables with mean 0 and variance σ^2. For change point analysis, this model will be separately fit to each substring of the data separated by the change points, which implies that each substring has the same set of explanatory variables but its own set of regression coefficients. Here, we focus on the simplest versions of this model, i.e., the constant ($Y = \beta_0 + \epsilon$) and linear models ($Y = \beta_0 + \beta_1 X_1 + \epsilon$), but note that the ideas presented below can easily be applied to the more general case.

Suppose that a time series contains k change points, $c_1, c_2, \ldots, c_k$, defined as the location where the parameters of the model change. Generally, the value of k is unknown, and must be inferred from the data, along with the locations of the change points. In this setting, the parameters of our model are the regression parameters, so a change point can represent a change in the mean (the constant term, β_0), trend ($\beta_1, \ldots, \beta_m$), or even the variance of the data (the magnitude of the random error, ϵ). Since the goal is to fit a piecewise regression model to the dataset, each segment of the data will have a unique set of regression parameters.

Bayes' rule tells us:

$$P(A|B) = \frac{P(B|A)P(A)}{P(B)}$$

Define:

- $P(\beta, \sigma^2|Y, M)$ to be the *posterior distribution* of the regression parameters, β, and the error variance, σ^2, given the data, Y, and the model, M (e.g., constant, linear, etc.);
- $P(Y|\beta, \sigma^2, M)$ as the *likelihood* of the data given the regression parameters and the model;
- $P(\beta, \sigma^2|M)$ as the *prior distribution* of the regression parameters, given the model;
- $P(Y|M)$ as the *normalization constant*, or the probability of the data given the model,

so that Bayes' rule can be rewritten as:

$$posterior = \frac{likelihood * prior}{normalization\ constant} \rightarrow P\left(\beta, \sigma^2 \middle| Y, M\right) = \frac{P(Y|\beta, \sigma^2, M)P(\beta, \sigma^2|M)}{P(Y|M)}$$

In many applications, the quantity of interest is the posterior distribution and the normalization constant represents a nuisance quantity that is computationally difficult to evaluate. However, in our case, the normalization constant is exactly the quantity that we need for the first step of the Bayesian change point algorithm. Specifically, we aim to calculate the probability of the data for each possible substring given the model, after marginalizing out the parameters of the model. These calculations represent the building blocks that the algorithm pieces together in order to identify the "best" possible locations of change points. Assuming the error terms, ϵ, are i.i.d. ~N(0, σ^2) and a conjugate prior distribution is used for both β (multi-variate normal) and σ^2 (scaled-inverse χ^2), the normalization constant is relatively easy to evaluate as:

$$P(Y|M) = \frac{P(Y|\beta, \sigma^2, M)P(\beta, \sigma^2|M)}{P(\beta, \sigma^2|Y, M)}$$

Dynamic programming works by taking a complex problem (i.e., the multiple change point problem) and breaking it down into a series of simpler problems, the smallest of which (i.e., the placement of a single change point) can easily be solved. Consider a jigsaw puzzle. After dumping out the pieces, you begin by turning all of the pieces over so that the picture is facing upwards. Next, find two pieces that fit together, then add a third, and a fourth, etc., until you manage to complete the entire puzzle. The idea is the same here. After defining a general model for the data (i.e., linear, sinusoidal, etc.), the first step to solving the multiple change point problem (i.e., the completed jigsaw puzzle) is to determine the parameters of the model which best fit each section of the data (i.e., turn the pieces over). Change points are then identified one at a time (i.e., place two segments of the data together, then add a third, etc.) until we have a complete model for our dataset.

The Bayesian change point algorithm has three steps.

1. **Calculating the Probability Density of the Data $P(Y_{i,j} \mid M)$:**

The quantity $P(Y_{i:j}) = P(Y_{i:j} \mid M)$ is calculated for all possible substrings of the data, $Y_{i:j}$, with $1 \le i < j \le N$, where N is the number of observations in the dataset. Each calculated probability is then stored in an $N \times N$ matrix where the row index represents the starting point and the column index represents the ending point of the substring. Note that the exact form of this calculation depends on the nature of the underlying predictive model. The dependence on the model, M, is hereafter suppressed.

2. **Forward Recursion (Dynamic Programming):**

Using the probabilistic calculations from Step 1 as building blocks, we recursively piece together these segments, adding one change point at a time until the complete dataset has been modeled. Define $P_k(Y_{1:j})$ to be the probability that the first j data points contain k change points. Then, for $k \in \{1, 2, \ldots, k_{max}\}$:

$$P_k(Y_{1:j}) = \sum_{v<j} P_{k-1}(Y_{1:v}) P(Y_{v+1:j})$$

for $j = (k + 1){:}N$, where $P_0(Y_{1:v}) = P(Y_{1:v})$ is calculated in Step 1 of the algorithm. Here, our values are stored in a $k_{max} \times N$ matrix, where the row index represents the number of change points.

3. **Stochastic Backtrace via Bayes' Rule:**

Two additional prior distributions need to be specified in order to have a fully defined model. Specifically, we assume a uniform prior on the number of change points (i.e., $P(K = k) = 1/k_{max}$) and that all solutions with exactly k change points are equally likely, i.e., $P(c_1, \ldots, c_k | K = k) = 1/N_k$, where N_k is the number of possible solutions containing k change points. Note that if there are no restrictions on the distance between two change points, then $N_k = \binom{N}{k}$. This combinatorial prior accounts for the growing number of potential solutions as the number of change points increases. Taken together, our normalization constant becomes:

$$P(Y_{1:N}) = \sum_{k=0}^{k_{max}} \sum_{c_1 \ldots c_k} P_k(Y_{1:N}) * P(K = k, c_1 \ldots c_k),$$

with $P_k(Y_{1:N})$ calculated in Step 2. The parameters of interest can now be sampled directly from their respective posterior distributions. In particular, we can use Bayes' rule to:

3.1. **Sample a number of change points, k:**

$$P(k | Y_{1:N}) = \frac{P_k(Y_{1:N}) P(K = k, c_1 \ldots c_k)}{P(Y_{1:N})}$$

3.2. **Iteratively sample the locations of these k change points, $c_1, \ldots, c_k$:**

$$P(c_{k-1} | c_k) = \frac{P_{k-1}(Y_{1:c_{k-1}}) P(Y_{c_{k-1}:c_k})}{\sum_{v<c_k} P_{k-1}(Y_{1:v}) P(Y_{v+1:c_k})}$$

3.3 **Sample the regression parameters for the interval between adjacent change points c_k and c_{k+1}:**

Note that Step 3 must be repeated a large number of times to obtain an accurate representation of the joint posterior distribution of the number and location of change points, as well as the parameters of the regression model. See [37] for full implementation details.

2.2. *Shortcomings of the Existing Model*

2.2.1. Correlated Errors

Consistent with the majority of the literature on change point analysis, the Bayesian change point model described above assumes the error terms to be a white noise process. However, time series often exhibit "memory" at time scales longer than the measurement frequency [45]. A model runs the risk of flagging spurious change points if this internal variability is neglected, as positive autocorrelation can create a similar pattern to that of a shift in the mean or long-term trend [46–48]. Specifically, autocorrelated time series can exhibit intervals where the time series remains above or below its mean value for an extended period of time, which can be interpreted by a change point model that assumes independent data points as the time series having different "regimes" [47]. In summary, the algorithm can misinterpret internal variability as a change in the forced signal if autocorrelation is ignored [12].

One way to model the memory of a system is through a first-order autoregressive (AR(1)) process (e.g., [49]), where the memory of a system geometrically decays to zero over time. From here, model selection can be used to determine the most appropriate structure (e.g., [50]) or an information criterion can be used to distinguish between autocorrelation and true change points (e.g., [51]) for a regression model containing both a trend and an AR(1) component. An alternate approach is to pre-whiten the time series (e.g., [47,48,52,53]) before performing change point analysis. In Section 3.1, we look at how pre-whitening the data affects the Bayesian change point algorithm's ability to detect change points in simulated datasets.

2.2.2. Choosing Values for the Hyperparameters of the Model

Each of the calculations described in Section 2.1 is conditional on the model. The algorithm itself is general enough to handle nearly any type of model, but several modeling decisions must be made before data analysis can begin. In particular, a researcher needs to decide on:

- *Structure of the Model:* Examples include constant, linear, periodic, autoregressive, etc. Here, we use a linear function to model the data and assume that the error terms are i.i.d. ~$N(0, \sigma^2)$, so the likelihood function given this model follows a multivariate normal distribution.
- *Prior Distribution for Model Parameters:* The prior distribution encodes any prior information available about the parameters of interest. Here, we choose conjugate prior distributions for both the regression parameters, β and σ^2, mainly to obtain a closed form expression for $P(Y \mid M)$, the probability of the data given the model (calculated for every possible substring of the data in Step 1 of the algorithm). Here, $P(\sigma^2|M) \sim \textit{Scaled Inverse}\ \chi^2(v_0, \sigma_0^2)$ and $P(\beta|\sigma^2, M) \sim N(0, \sigma^2/k_0)$, where k_0 is a vector of the same length as β.
- *Prior Distribution on the Location of Change Points*: Here, we assume a non-informative prior on the number of change points, k, and their distribution in time (i.e., all change point solutions with exactly k change points are equally likely). Note that algorithms which base their inference on the "run length" (e.g., BOCPD [38] and particle filters (e.g., [39,40]) often encode their beliefs about the expected distance between change points with a geometric prior.

Five hyperparameters need to be set before starting the analysis:

- k_0 is a scale parameter that relates the variance of the regression parameters to the error variance, σ^2. In general, the value of k_0 can differ for each regression parameter, β_i, or be constant across all parameters. The practical effect is to act as a "penalty" against adding change points, where a smaller value of k_0 allows for larger values of the regression parameters (relative to the error variance), but also gives a larger penalty on introducing a change point. Allowing for large values of the regression parameters is especially important for the constant term in a long time series, as its value can differ significantly from zero. In Section 3.2, we consider different values of k_0 for the constant and trend terms in our model.

- v_0 and σ_0^2 act as pseudo-data for estimating the value of the residual variance, v_0, and pseudo-data points of variance σ_0^2. For example, setting v_0 equal to 1 and σ_0^2 equal to the variance of the data implies that we have one prior observation of the residual error whose magnitude is equal to the variance of the data.
- d_{min} represents the minimum distance between two consecutive change points. This hyperparameter can be set to any reasonable value for the problem of interest and normally does not affect the inference other than to prevent two change points from appearing in close proximity to one another. We recommend that d_{min} be at least twice as large as the number of regression parameters that need to be estimated.
- k_{max} represents the maximum number of allowed change points in the time series. The value of k_{max} should be at least as large as the expected maximum number of change points, but need not be any larger than n/d_{min}, where n is the number of observations in the dataset.
- One additional quantity that needs to be set by the researcher is the number of solutions sampled from the joint posterior distribution on the number and location of change points, as well as the parameters of the regression model fit between any two change points. Larger values of this parameter allow for a more accurate representation of the joint posterior distribution, and therefore a more accurate estimate of each quantity.

The choice of parameters for the prior distributions can have a significant impact on the overall inference. In this case, changing the values of k_0, v_0, and σ_0^2 can impact the number of change points that are detected, but not on their distribution within the dataset. In other words, changing the values of the prior parameters does not create a bias in the inferred location of a change point. Exploring how the values of these parameters affect the inference is the focus of Section 3.2.

3. Simulation Studies

3.1. Correcting for Autocorrelation

Autocorrelation in a times series can easily be misinterpreted as a change point by models which assume that the data are independent, including the Bayesian change point algorithm described in Section 2. Here, we use the pre-whitening technique described by [47] to try and mitigate the effect of autocorrelation. The idea is to remove the first-order autocorrelation using a bias-corrected estimate of the first-order autocorrelation:

$$y_t' = y_t - \hat{\rho}^c y_{t-1}$$

$$x_t' = x_t - \hat{\rho}^c x_{t-1}$$

for $t = 2, 3, \ldots, n$, where n represents the length of the time series, x_t and y_t represent the raw variables, x_t' and y_t' represent the pre-whitened variables at time t, and $\hat{\rho}^c$ is the bias-corrected estimate of the first-order correlation. Rodionov [47] notes that the situation becomes "complicated" if the time series contains both regime shifts and autocorrelation, as using all available data can lead to a misleading estimate of the value of ρ (since the first-order correlation used in pre-whitening is unknown and may also change over time).

A potential solution to this problem is to estimate the value of ρ using randomly selected subsegments of the dataset. If we set the size of these randomly selected subsegments appropriately, then the majority of them will not contain any change points. Rodionov [47] suggests that if change points occur at regular intervals of l years, then subsamples of size m should be selected so that m is less than or equal to $(l + 1)/3$. From here, $\hat{\rho}$ is chosen as the median of the first-order autocorrelation calculated from each subsegment of size m

(denoted $\hat{\rho}$). However, conventional estimators of ρ (e.g., OLS, maximum likelihood) are known to yield biased estimates of ρ for short subsamples of size m [54], so we can use a bias-corrected estimate of the first-order autocorrelation developed by [55]:

$$\hat{\rho}^c = \frac{(m-1)\,\hat{\rho}+1}{(m-4)}$$

We first aim to show that autocorrelation causes the Bayesian change point algorithm to detect change points when none actually exist and that the Bayesian change point algorithm can recover its predictive ability by pre-whitening the time series. Here, we consider a constant model with no change points ($Y = 1 + \epsilon$). Simulation of a linear model with no change points ($Y = 4 + 0.05X + \epsilon$) is included in the Appendix A. A total of 1000 datasets of length n = 200 were generated with an auto-regressive signal of level ρ = 0.1, 0.2, 0.3, . . . , 0.9 using the R function arima.sim(), for a total of 10,000 simulations. For this simulation, m is chosen to be 20, k_0 = 0.01, v_0 = 1, σ_0^2= var(Y), d_{min} = 5, and k_{max} = 20. Since the goal of this simulation is to see how autocorrelation affects the inference, optimizing these parameters is not critical. The Bayesian change point model calculates the posterior distribution of the number of change points for each dataset, so we use this distribution to determine the expected number of change points in the dataset and then average this quantity across all 1000 simulations. Table 1 gives the average number of detected change points for each value of ρ before and after pre-whitening, along with the number of datasets where the algorithm correctly identified zero change points. For smaller values of ρ, there appears to be little loss in the algorithm's predictive ability, but the quality of the inference quickly deteriorates as ρ increases (Table 1). It is also clear from these data that pre-whitening can help to eliminate spurious change points that arise from autocorrelation.

Table 1. Autocorrelation in a Constant Model. A total of 1000 datasets were generated for each value of the autocorrelation parameter, ρ. The average number of change points detected by the Bayesian change point model before and after pre-whitening is indicated for each value of ρ in addition to the number of datasets (out of 1000) where the algorithm correctly identified zero change points (i.e., the number of datasets where the expected number of change points < 0.5). Note that a value of ρ = 0 corresponds to white noise.

	ρ	0	0.1	0.2	0.3	0.4	0.5	0.6	0.7	0.8	0.9
Data with Autocorrelation	Estimated $\hat{\rho}^c$	N/A	0.101	0.212	0.320	0.425	0.521	0.632	0.729	0.823	0.920
	Change Points Detected	0.002	0.002	0.014	0.027	0.147	0.522	2.052	5.291	8.106	9.150
	# Correct	1000	1000	992	986	921	769	411	86	3	0
Pre-Whitened	Change Points Detected	N/A	0.002	0.004	0.002	0.006	0.004	0.007	0.005	0.022	0.080
	# Correct	N/A	999	998	1000	997	999	997	998	990	955

While pre-whitening can be used to help eliminate false positive change points, it also reduces the magnitude of the change between consecutive regimes, making it harder to detect true change points [47]. Our next simulation generates datasets of length n = 200 with a trend whose value changes by a random amount at randomly generated change points, according to the following process:

- The locations of the change points are selected as uniform random variables, $c_1 \sim Unif(60, 70)$, $c_2 \sim Unif(95,\ 100)$, and $c_3 \sim Unif(150, 160)$, creating four segments of varying length.
- The intercept for the model is selected $\beta_0 \sim Unif(-3,\ 3)$ and the trend for the first segment is selected $\beta_1 \sim N(0.2,\ 0.05^2)$, negated with probability 0.5.

- To avoid overly obvious change point locations, the function is made piecewise continuous. The change in trend from the first to the second line segment is selected N(0.75, 0.1^2), from the second to the third line segment N(0.6, 0.05^2), and from the third to the fourth segment N(0.5, 0.025^2). Each change in trend is negated with probability 0.5. Notice that by decreasing the potential magnitude of the change, successive change points become more difficult to detect.
- An auto-regressive signal of level $\rho = 0.1, 0.2, 0.3, \ldots, 0.9$ is generated using the R function arima.sim() and added to each dataset.

Figure 1 shows three versions of a representative dataset generated by this process: one with white noise, one with an autoregressive component using $\rho = 0.8$, and one after pre-whitening. This process ensures that each simulated dataset has a different set of change points and a different set of regression coefficients, making some of the change points more or less difficult to detect. Note that the data generation process is similar to a more extensive simulation study conducted by [44], which gives examples of the types of data generated and compares the speed and accuracy of detecting change points for several different change point models.

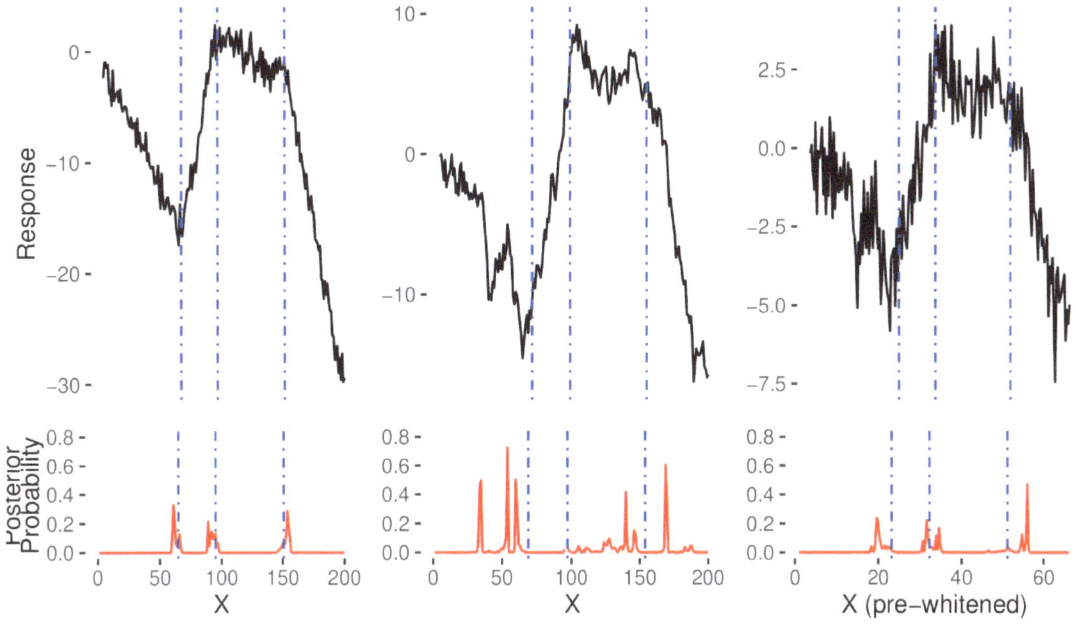

Figure 1. Detecting Change Points in the Presence of Autocorrelation. (**a**) A simulated dataset with 3 change points that contains only white noise. (**b**) Autocorrelation is added to (a) using $\rho = 0.8$. (**c**) The data in (**b**) after pre-whitening. The inferred location of change points is indicated below each figure, while their exact location is indicated by dotted vertical lines. Pre-whitening helps to eliminate spurious change points, but the location of the true change points becomes more difficult to correctly infer.

For each simulated dataset, we sample 500 sets of change points from the joint posterior distribution. To determine whether or not the Bayesian change point algorithm is successful in detecting change points after pre-whitening, define:

- **Position Uncertainty**: Amount of uncertainty allowed in the location of a detected change point while still considering it "accurate." For example, if the position uncertainty is 1, then we count the number of solutions sampled from the posterior distribution that detected a change point within 1 point of its true location.

- **Barrier Rate**: A barrier rate of B% means that if B% of the 500 simulated sets of change points contain a change point within the "position uncertainty" range, then we are considered to have successfully detected this change point.
- **Noise Level**: Refers to the residual variance, σ^2.

Two metrics will be used to measure the success of the algorithm:

- **True Positive Rate**: Proportion of the true change point locations that are detected.
- **Perfection Rate**: The proportion of datasets where the algorithm has successfully detected all three change points.

It is important to note that when the noise is large relative to the signal, the algorithm can be quite uncertain about the exact placement of a change point. As a result, if the algorithm knows that a change point should exist, but is uncertain about its location, it may appear to miss that change point when using relatively stringent detection criteria. In other words, changing either the position uncertainty or the barrier rate can impact the number of change points detected in a given simulation. However, since the goal of this simulation is to observe how autocorrelation can impact inference, the relative change in the true positive rate and the perfection rate is much more important than their absolute values.

For this simulation, m is chosen to be 20, $k_0 = (0.01, 0.01)$, $v_0 = 1$, $\sigma_0^2 = 1$, $d_{min} = 5$, and $k_{max} = 20$. Our position uncertainty is set to 7 and the barrier rate is set to 75%, with a noise level of 1. As before, we use the posterior distribution of the number of change points for each dataset to determine the expected number of change points in each dataset and then average this quantity across all 1000 simulations. Table 2 gives the average number of detected change points for each value of ρ before and after pre-whitening, along with values for the metrics described above, which help indicate the accuracy of detection.

Table 2. Autocorrelation in a Change Point Model with Linear Trend. A total of 1000 datasets containing a linear trend with 3 change points were generated for each value of the autocorrelation parameter, ρ. The average number of change points detected by the Bayesian change point model before and after pre-whitening is indicated for each value of ρ in addition to the true positive and perfection rate. Note that a value of $\rho = 0$ corresponds to white noise.

	ρ	0	0.1	0.2	0.3	0.4	0.5	0.6	0.7	0.8	0.9
	Estimated $\hat{\rho}^c$	N/A	0.110	0.201	0.289	0.381	0.463	0.547	0.624	0.701	0.754
With Autocorrelation	Number Detected	2.997	2.999	3.002	3.013	3.053	3.141	3.291	3.701	4.273	5.090
	True Positive Rate	0.972	0.965	0.956	0.945	0.937	0.894	0.838	0.779	0.681	0.596
	Perfection Rate	0.922	0.901	0.876	0.843	0.829	0.718	0.590	0.463	0.336	0.235
After Pre-Whitening	Number Detected	N/A	2.996	2.995	2.993	2.991	2.976	2.929	2.855	2.709	2.556
	True Positive Rate	N/A	0.962	0.949	0.927	0.892	0.795	0.667	0.536	0.354	0.284
	Perfection Rate	N/A	0.897	0.856	0.799	0.725	0.520	0.341	0.187	0.076	0.044

Figure 1 helps to illustrate several patterns that emerged from the simulation. First, change points are fairly easy to detect in the presence of white noise (Figure 1a). Since the magnitude of the change in trend is less than the amount of noise in the system, the Bayesian change point algorithm may have some uncertainty in the exact location of the change point (visualized by the mound-shaped density function centered at the location of each change point), but clearly identifies three regions where a change point exists. In this case, missing a change point is generally the result of stringent detection criteria, which requires the posterior distribution to be highly concentrated around the true location of the change point.

Second, values of $\rho < 0.5$ generally do not have a large impact on the inference made by the Bayesian change point algorithm, as we maintain a relatively high true positive rate and perfection rate. For values of $\rho \geq 0.5$, the number of change points detected by the algorithm increases (similar to the previous simulation study), but the true positive rate and perfection rate decrease. This indicates the emergence of spurious change points and/or greater uncertainty in the location of true change points, to the point where the posterior probability falls below the barrier rate (Figure 1b). The region from data points 1 to 69 (the location of the first change point) is a great example of how autocorrelation can create a pattern that looks like a change in the long-term trend [26,46,47]. Here, the data should appear as a downward sloping function, but we instead see an upward trend bracketed by regions of steep decline (Figure 1b). The upward feature in the middle of an otherwise downward signal introduces a pair of spurious change points into the model. Moreover, the pattern appears to be repeated just before the true location of the first change point so that, visually, the upward component of the signal now appears to begin before it actually should. Thus, while the Bayesian change point model will not receive credit for detecting a "true positive", it does appear to correctly identify the start of the upward trend in this dataset. This is exactly what [47] meant when he said that inferring change points is "complicated" in the presence of autocorrelation.

Finally, pre-whitening the data helps to eliminate the spurious change points, evidenced by the number of detected change points returning back down to the true value of three (Figure 1c). However, it also degrades our ability to detect changes in the trend due to the interdependence between the values of the autoregressive and slope parameters. Specifically, we now have much greater uncertainty in the location of change points and the potential for posterior distributions to be centered at the wrong spot (Figure 1c). As a result, there is not enough posterior mass to cross the barrier rate, so our true positive rate and perfection rate remain relatively low. A more lenient definition of "detection" (e.g., larger position uncertainty or lower barrier rate) can help account for the greater uncertainty, but would not be able to address the posterior distribution of a change point being centered at the wrong spot due to the phenomenon noted at the beginning of the dataset in Figure 1b.

3.2. Hyperparameters for the Bayesian Change Point Model

Bayesian methods are, in general, subjective, and the model described in Section 2 is no exception. Subjectivity arises through the researcher's choice of a prior distribution for the model, and their ability to use these distributions to code in prior information or beliefs about the parameters of interest. Our model assumes a conjugate prior distribution for both the error variance, σ^2, and the set of regression parameters, β, primarily so that the calculation required for Step 1 of the algorithm has a closed form solution. Specifically, we assume $\sigma^2 \sim \textit{Scaled} - \textit{Inverse}\ \chi^2(v_0, \sigma_0^2)$ and $\beta | \sigma^2 \sim N\left(0, \frac{\sigma^2}{k_0}\right)$.

One benefit of using conjugate prior distributions is that the parameters of these distributions are easily interpretable as prior observations. Generally, we set the parameters of our prior distribution to be as non-informative as possible. This allows the information contained in the data to dominate the inference. In this vein, a sensible choice for the $\textit{Scaled} - \textit{Inverse}\ \chi^2$ distribution might be $v_0 = 1$ and $\sigma_0^2 = \text{var}(Y)$, which implies that we have one prior observation of the residual variance whose value equals the variance of the dataset. The hyperparameter, k_0, relates the variance of the regression parameters to the residual variance. Different values of k_0 can be used for the slope and intercept parameters of the model, so k_0 can be thought of as a vector: $k_0 = (k_1, k_2)$. We want to allow the regression parameters to be larger than the error variance, so k_0 is generally chosen as a decimal value between 0 and 1. It is especially important that k_1 is a small value, as the intercept for the model may differ significantly from zero. A reasonable choice is $k_0 = (0.01, 0.01)$. This gives us four hyperparameters to choose. Unfortunately, the inference made by the Bayesian change point model is sensitive to the choice of these hyperparameters, which is a common feature of Bayesian analysis. The "best" choice for the values of k_0, v_0, and σ_0^2 remains an open question.

Figure 2 gives examples of "good" and "bad" choices for the prior parameters using a simulated dataset generated according to the process outlined in Section 3.1. Notice that when a poor choice for the parameter values is made, the model can infer either too few or too many change points, while a "good" set of parameters allows for a proper inference. It is easy to see what makes a "good" and "bad" choice of parameters on a simulated dataset where the locations of the change points are known, but much harder when the goal is to infer the unknown location of change points on a real dataset. Fortunately, our study shows that the "good" parameters also tend to produce the most probable solutions, so we can use Bayesian Model Averaging (BMA) to marginalize out the choice of the hyperparameters and arrive at the best overall solution [56].

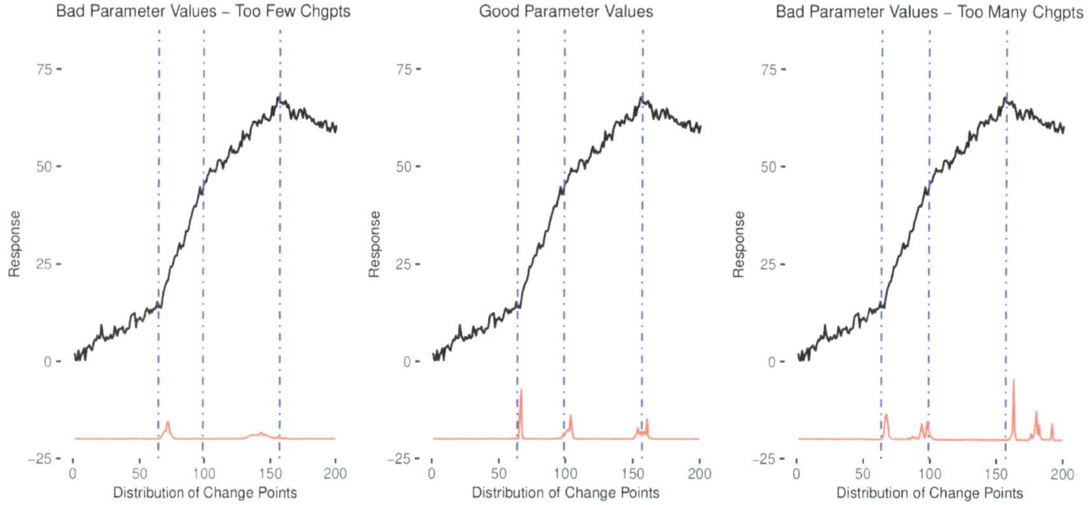

Figure 2. "Good" and "Bad" Parameter Choices. A representative dataset using the data generation process described in Section 3.1 is analyzed using different sets of values for the hyperparameters. Dotted vertical lines indicate the actual location of the change points. The **left** panel uses a set of parameters that produces a posterior distribution which infers too few change points while the **right** panel uses a set of parameters that produces a posterior distribution that infers too many change points. The **middle** panel correctly identifies the correct number of change points.

Define θ to be the parameters of interest and suppose that we have a set of possible models under consideration, $M_1, \ldots, M_m$. BMA is defined as:

$$P(\theta \mid Y) = \sum_{i=1}^{m} P(\theta|Y,\ M_i)P(M_i|Y)$$

In words, the posterior distribution of the parameters of interest is a weighted average of the posterior distribution of the parameters for each model, weighted by the likelihood of each model. This means that more probable models will have a stronger impact on the posterior distribution of the parameters of interest.

In this scenario, each model is defined by the chosen values of the hyperparameters $k_0 = (k_1, k_2)$, v_0, and σ_0^2. Therefore, we can equate the term "model" with a set of hyperparameters. The parameters of interest, θ, are the number and locations of the change points, along with the parameters of the regression model in each region of the data. Thus, our sampling procedure (Step 3 of the Bayesian change point algorithm), along with the use of conjugate prior distributions, makes the quantity $P(\theta|Y,\ M_i)$ easy to evaluate. Since each model is determined by its set of parameters, the term $P(M_i|Y)$ tells us how good a specific set of hyperparameters is and can be obtained through one final application of Bayes' rule:

$$P(M_i|Y) = \frac{P(Y|M_i)P(M_i)}{\sum_{j=1}^{m} P(Y|M_j)P(M_j)}$$

Note that $P(Y|M_i)$, the probability of the data given the model after marginalizing out the parameters of the model and the location of the change points, is calculated as part of Step 3 of the Bayesian change point algorithm. A priori, if we assume that all models (i.e., all sets of values for the hyperparameters) are equally likely, this expression reduces to:

$$P(M_i|Y) = \frac{P(Y|M_i)}{\sum_{j=1}^{m} P(Y|M_j)}$$

In the simulations that follow, we vary the values of $k_0 = (k_1, k_2)$, v_0, and σ_0^2 to determine the posterior distribution of each model and then combine this information with that model's distribution of change point locations to obtain the "model averaged" solution. The process is conceptually simple, but computationally intensive.

3.2.1. Changing the Values of k_1 and k_2

For this analysis, we generated a dataset according to the process outlined in Section 3.1 (Figure 3), assuming $\rho = 0.25$ and a noise level of 1. Values of v_0 and σ_0^2 are both fixed at 1 (i.e., one prior observation for the variance equal to 1). Here, we are interested in studying values of k_1 and k_2 between 10^{-4} and 10^{-1}, so we choose 16 equally spaced values for each parameter on the $\log_{10}$ scale. Figure 3 displays how the log probability of the data changes as we vary the values of k_1 and k_2 and the posterior distribution of change point locations for three sets of values of k_1 and k_2 (labeled A, B, and C). As the value of k_1 increases from 0.0001 to 0.1 (moving from point A to B to C), the log probability of the data decreases, while changing the value of k_2 from 0.001 to 0.1 (compare points A and B) does not have a significant impact on the log probability of the data. For this particular dataset, a small value of k_1 is necessary because it allows the intercept of the model to be significantly larger than the residual variance. Forcing the intercept to take on a small value also introduces a number of spurious change points, as we limit the set of potential regression parameters in each interval. Notice that points A and B have a relatively similar log probability, and show only subtle differences in their distribution of change point locations, whereas point C has a much lower log probability and a significantly different distribution of change points.

3.2.2. Changing the Values of v_0 and σ_0^2

We again generated a dataset according to the process outlined in Section 3.1 (Figure 4), assuming $\rho = 0.25$ and a noise level of 1. For this analysis, values of k_1 and k_2 are fixed at 0.001. Here, we study how values of v_0 and σ_0^2 affect the inference, so we choose v_0 = 1, 2, 4, 8, 16, and 32, and values of σ_0^2 = 0.1, 0.5, 1, 2, 5, 10, 20, and 50. The calculation required for Step 1 of the Bayesian change point algorithm calculates a quantity analogous to a posterior sum or squares, which is the sum of the prior variability, $v_0\sigma_0^2$, the variability of the regression parameters, and the residual sum of squares. When a change point is introduced into the model, this term appears twice (once for each region of the data), so a larger value of the product $v_0\sigma_0^2$ creates a barrier against additional change points. As we move from points A to B to C in Figure 4, this product increases, resulting in a posterior distribution with fewer detected change points. The locations of the change points are not altered, only our confidence in their existence. In addition, the value of v_0 does not affect the log probability of a model if we choose values of σ_0^2 similar to the actual residual variance (e.g., 0.5, 1, and 2), which is consistent with their interpretation as prior observations of the residual variance. On the other hand, choosing larger values of v_0 will quickly decrease the log probability of the model if the chosen value of σ_0^2 is inconsistent with the data. As a result, we always recommend choosing $v_0 = 1$, so that our prior distribution on the error variance has a minimal impact on the posterior inference.

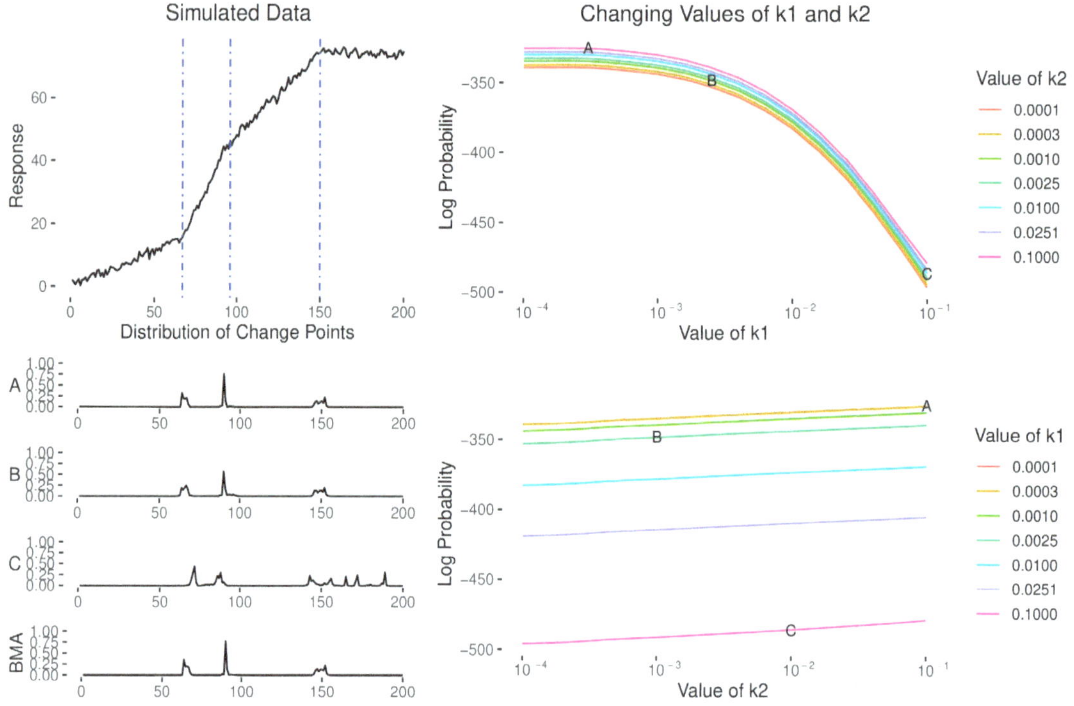

Figure 3. How k_1 and k_2 Affect the Posterior Distribution. The top left displays a dataset generated according to the process described in Section 3.1, along with the location of each change point, indicated as a dotted vertical line. The log probability of the data, i.e., $P(M_i|Y)$, is shown on the right for various combinations of v_0 and σ_0^2. Three sets of hyperparameters, labeled A, B, and C, are selected and their posterior distribution of the location of change points is shown in the bottom left, along with the posterior distribution for the BMA solution, which weights each solution according to its probability.

3.2.3. Applying BMA

Figures 3 and 4 show that the models of lower probability (i.e., those with a "bad" choice of values for the hyperparameters) often do not have the correct number of change points, inferring either too few or too many change points. Fortunately, BMA lets us keep all the benefits of a Bayesian solution to the multiple change point problem, in particular the uncertainty bounds on the number and locations of change points, while also helping to prevent a "bad" choice of hyperparameters. Here, the models weighted most heavily are those with the highest probability, which also tend to infer the correct number of change points. Figures 3 and 4 show the BMA solution to each simulation, which looks most similar to point A in each figure, the most probable of the three models shown for each simulation. This nicely illustrates how BMA can help to eliminate the effects of a "bad" choice of values for the hyperparameters.

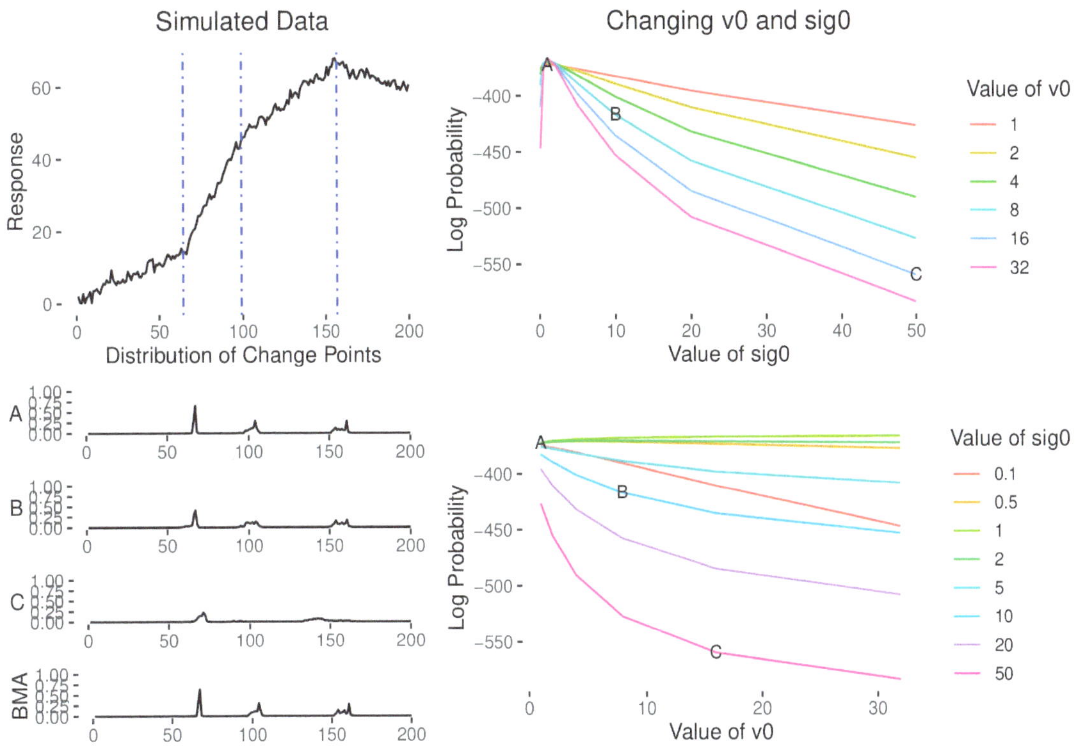

Figure 4. How v_0 and σ_0^2 Affect the Posterior Distribution. The **top** left displays a dataset generated according to the process described in Section 3.1, along with the location of each change point, indicated as a dotted vertical line. The log probability of the data, i.e., $P(M_i|Y)$, is shown on the right for various combinations of v_0 and σ_0^2. Three sets of hyperparameters, labeled A, B, and C, are selected and their posterior distribution of the location of change points is shown in the **bottom** left, along with the posterior distribution for the BMA solution, which weights each solution according to its probability.

4. Applications to Climate Data

4.1. Pacific Decadal Oscillation a Change in Mean

Pacific Decadal Oscillation (PDO) was first identified in the late 1990s [57] and describes sea surface temperature anomalies over the northeastern Pacific Ocean. Similar to El Niño/Southern Oscillation (ENSO), PDO oscillates between two states (positive and negative) that are correlated with by widespread variations in the Pacific Basin and North American climate [58]. The positive phase is characterized by cooler sea surface temperatures north of Hawaii and warmer than normal sea surface temperatures along the western coast of North America. The reverse is true in a negative phase. However, unlike ENSO, PDO is an aggregation of several independent processes rather than just a single climate phenomenon and the positive/negative phases can last for 20–30 years [59]. Researchers also believe that PDO can intensify or diminish the impacts of ENSO depending on whether or not they are in the same phase. If both ENSO and the PDO are in the same phase, then the impacts of El Niño/La Nina may be magnified. Conversely, if they are out of phase, then the effects may offset each other resulting in a milder ENSO event [60]. More information about PDO can be found in [61].

The PDO dataset can be downloaded from the National Centers for Environmental Information website (https://www.ncei.noaa.gov/access/monitoring/pdo/, accessed 15

August 2022). Annual means from 1854 to 2021 were calculated from monthly values for each year (Figure 5). This dataset has been previously analyzed for change points by other researchers (e.g., [12,47,57,62]), so it represents an interesting application of the approach described in this paper. Here, our goal is to fit a piecewise constant model to the PDO where the change points represent transitions between the positive and negative phases of PDO.

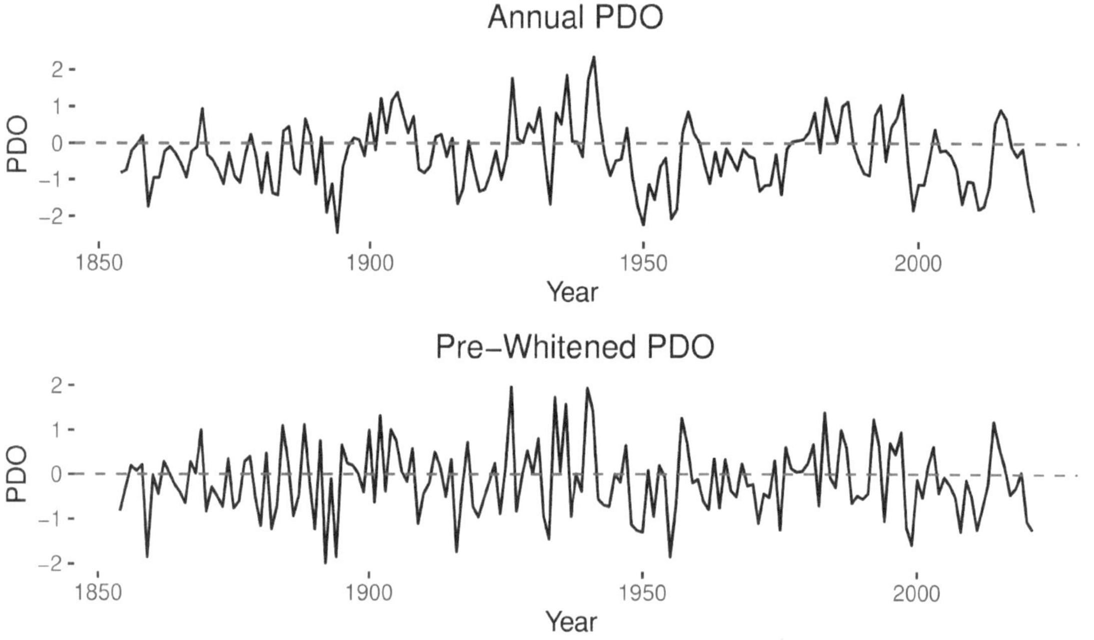

Figure 5. Change Points in PDO. The **top** panel shows the annual PDO values, while the **bottom** panel shows a pre-whitened version of this dataset. The horizontal dotted line at 0 is for reference to help identify positive and negative phases of the PDO. The Bayesian change point algorithm did not detect any change points in this dataset.

The autocorrelation function (R function acf()) shows that the residuals in the PDO are correlated, so we begin our analysis with the pre-whitening technique described in Section 3.1 to help eliminate change points due to autocorrelation rather than a change in the phase of the PDO. For our analysis, we set the value of m to be 8, since the positive and negative PDO phases are expected to last 20–30 years (l was chosen as 25). Following the procedure outlined in Section 3.1, we calculate $\hat{\rho} = 0.155$, and the bias-corrected estimate of the first-order autocorrelation as $\hat{\rho}^c = 0.53$ (consistent with 0.46 in [47], who studied a shorter time series), which is near the point at which false positive change points become a regular part of the inference (Table 1). At this point, the autocorrelation function shows no significant correlation in the residuals, so we can continue with change point analysis.

After pre-whitening, we set $v_0 = 1$, and then allowed potential values of k_1 and σ_0^2 to be the same as in Section 3.2. BMA was then used to accumulate these models to produce a single inference on the number and location of change points in the PDO data. The Bayesian change point algorithm does not detect any change points in the PDO. This result is surprising considering all the discussion of positive (e.g., 1925 to 1947 and 1976 to 1999) and negative phases (e.g., 1946 to 1976) of the PDO (see for example [57]), but consistent with the results of [12], who also identified a constant mean plus AR(1) model as the best fitting model for the PDO. This is not to say that positive and negative phases of the PDO do not exist—just that their magnitude and/or duration are not substantial enough to warrant the placement of a change point by either the Bayesian change point model or PELT [12].

Note that if we had instead analyzed the monthly rather than mean annual PDO values, the autocorrelation is much higher (ρ = 0.855). An autocorrelation this high will induce a large number of spurious change points if the model does not have an autoregressive component (Table 1). After pre-whitening the monthly PDO data, the Bayesian change point algorithm did not detect any change points (results not shown).

4.2. Global Surface Temperature Anomalies—A Change in Trend

The Earth's temperature has risen by an average of 0.14° Fahrenheit (0.08° Celsius) per decade since 1880, but the rate of warming has not been consistent over time. In fact, the rate of warming since 1981 is 0.32 °F (0.18 °C), more than twice the long-term average. This past year was the sixth-warmest on record (0.84 °C above the 20th century average), and the years 2013–2021 are nine of the ten warmest years on record [63,64]. Since the rate of warming fluctuates over time, a change point model with a linear trend seems most appropriate to model global surface temperature data.

Although surface temperature data are collected at stations across the globe, absolute temperature measurements can be difficult to take in certain geographic locations. Thus, temperature anomalies, or the departure from a reference value, are used instead and allow for a more effective and reliable comparison between different geographic locations. A global surface temperature anomalies dataset attempts to combine this temperature information into a measure of global surface temperatures. Several groups have created global surface temperature anomalies datasets, all with slightly different assumptions. One such time series is the HadCRUT5 dataset produced by the Met Office Hadley Centre [65], which begins in 1850 and has a reference period of 1961–1990. The data can be downloaded from https://www.metoffice.gov.uk/hadobs/hadcrut5/. Two others are MLOST, NOAA's Merged Land Ocean Global Surface Temperature Analysis Dataset (NOAA) [66], available at (https://www.ncei.noaa.gov/access/monitoring/global-temperature-anomalies/anomalies, accessed 15 August 2022), which starts in 1880 and has a reference period of 1901–2000, and GISTEMP, NASA's Goddard Institute for Space Studies Surface Temperature Analysis ([67], available at https://data.giss.nasa.gov/gistemp/, accessed 15 August 2022), which starts in 1880 and has a reference period of 1951–1980. The University Corporation for Atmospheric Research website (https://climatedataguide.ucar.edu/, accessed 15 August 2022) contains additional information on these and several related datasets.

As with the PDO data, the autocorrelation function was used to initially check for correlated residuals and then to verify that the pre-whitening technique was effective. For this analysis, we expect up to 4 change points across the 140-year record, so we set the value of m to be 12. Following the procedure outlined in Section 3.1, we calculate $\hat{\rho} = 0.154$, and the bias-corrected estimate of the first-order autocorrelation as $\hat{\rho}^c$ = 0.336, which is small enough so as to not have a major impact on the inference (Table 2). Nevertheless, we pre-whitened the data to help eliminate any change points that may arise due to autocorrelation. As with the PDO data, we set v_0 = 1, and then allowed the values of k_1 and k_2 to vary as in Section 3.2. Since the variance of the temperature anomaly datasets is so small (<0.1 after pre-whitening), we chose potential values for σ_0^2 = 0.05, 0.1, 0.5, 1, 2, 5, 10, and 20. BMA was then used to accumulate these models to produce a single inference on the number and location of change points in the three temperature anomaly datasets.

The Bayesian change point model with BMA detected only a single change point in the MLOST and GISTEMP datasets, and two change points in the HadCRUT5 data (Figure 6). This result is somewhat surprising and includes fewer change points than previous analyses (e.g., [68]). However, the authors note that these datasets are continually revised and updated. Repeating the analysis of [68] on the revised datasets using the same parameter values now produces an inference with fewer change points, so in that sense, the results are consistent with previous studies on the same dataset.

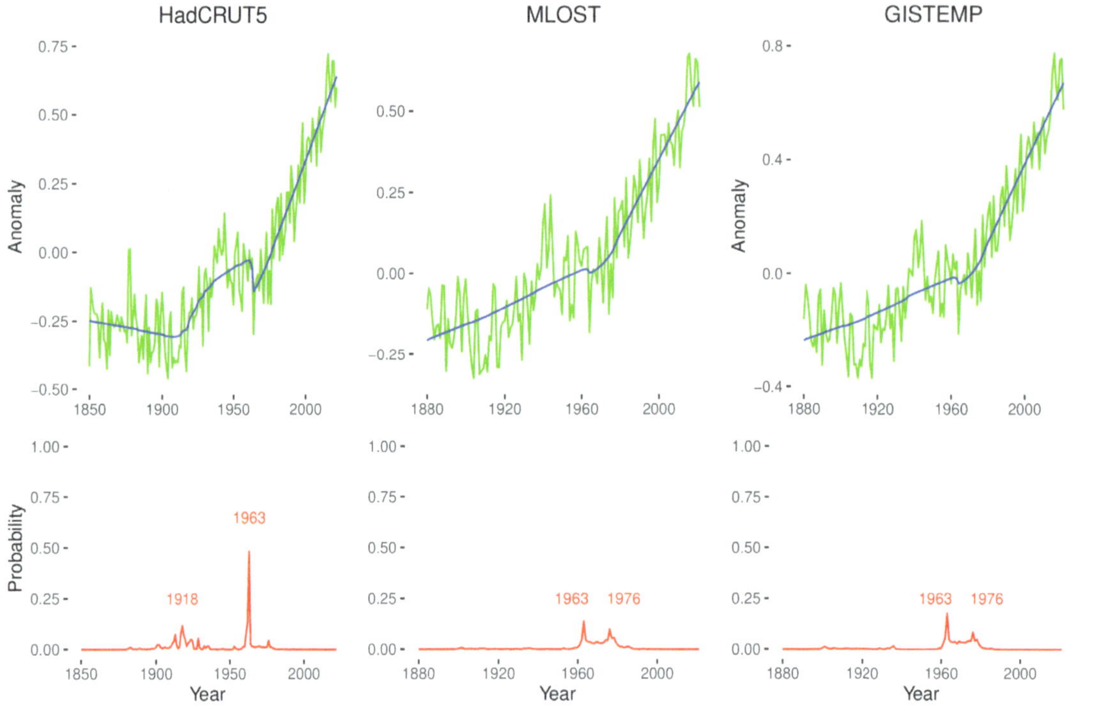

Figure 6. Change Points in the Temperature Anomaly Data. The **top** row displays the HadCRUT5, MLOST, and GISTEMP datasets, along with the model fitted by using BMA together with the Bayesian change point model. The **bottom** row displays the BMA posterior distribution for the locations of change points in each dataset. HadCRUT5 has two change points detected by the model, while MLOST and GISTEMP have a bimodal distribution for a single change point.

5. Discussion

This paper addresses two open questions related to the Bayesian change point model of [37], namely, how autocorrelation and the choice of values for the hyperparameters can affect the inference. When a change point model is used to analyze real data, the "true" number of change points is generally unknown. As a result, it is hard to know whether a model is giving accurate and precise results. To see how our model performs in difference scenarios, simulated data were generated which varied the number and location of change points, the regression coefficients in each section of the data, the variance of the residual error, and the magnitude of autocorrelation. Any change that reduces the signal-to-noise ratio of the dataset (e.g., larger values of ρ or σ^2, subtle changes in the regression parameters, etc.) makes change points harder to detect, and thus has an impact on the accuracy of the model. Specifically, a smaller signal-to-noise ratio manifests itself in the posterior distribution as greater uncertainty in the location of a change point or a complete lack of detection by the algorithm.

Autocorrelation is often present in real data, yet [37] assumes that the error terms are independent, mean 0, normally distributed random variables. Simulations show that the inference made by the Bayesian change point model is not strongly affected by low levels of first-order autocorrelation ($\rho < 0.5$, see Tables 1 and A1)—the algorithm is still able to detect the correct number of change points in the data. However, when the first-order autocorrelation is larger, it can create a similar pattern to that of a change in the mean or long-term trend (46–48), which can shift the inferred location of true change points (if the autocorrelation makes the pattern appear to start earlier or later than it actually does) and introduce spurious change points. To counter the effect of serial correlation, we pre-whiten the data using the Cochrane–Orcutt method with a bias-corrected estimate of the first-order autocorrelation. Results show that pre-whitening the data eliminates the spurious change points introduced by the autocorrelation. However, pre-whitening the data reduces the magnitude of the shift between adjacent segments, making true change points harder to detect. This is partially offset by a reduction in the variance, but not completely [47]. Both of our applications (PDO and global surface temperature anomalies) exhibit only first-order autocorrelation, so we did not study how the pre-whitening approach would fare on data with higher-order autocorrelation.

As with any Bayesian analysis, the inference can be sensitive to the choice of the prior distribution. In this case, we use conjugate priors for both the regression parameters and the error variance, so the subjectivity comes in through the values of the hyperparameters for these two distributions. As seen in Figures 3 and 4, changing the values of the hyperparameters generally affects the inferred number of change points, but not their location. In other words, we can think of the changing values of the hyperparameters as creating more or less stringent criteria for the algorithm to "detect" a change point. A "bad" choice of values for the hyperparameters can produce an inference with too few or too many inferred change points (Figure 2). To avoid this problem, we propose a BMA technique to weight each model's inference by the posterior probability of that model, so that "good" parameter choices (as defined by the posterior probability of the model) carry more weight than "bad" parameter choices. The result is an inference that takes into account multiple different potential sets of hyperparameters.

BMA partially eliminates the problem of the model being sensitive to the values of the hyperparameters. The problem is only partially eliminated because BMA is being conducted using only a finite set of potential values for the hyperparameters rather than considering all possible values. A Monte Carlo approach would fix this issue but at an increased computational cost. Since models (defined by the set of values of the hyperparameters) of similar probability tend to produce a similar inference (see the similarity of change point solutions for points A and B in Figure 3), and the model itself is not especially sensitive to small changes in parameters values, we did not feel that this increased computational burden would significantly improve our BMA solutions. The interested researcher could also try placing a non-uniform prior over the set of values for the hyperparameters if they believe certain values to be more likely than others.

A major limitation of the Bayesian change point model discussed in this paper is its run time, which can make BMA over a large parameter space prohibitive. Computational complexity is a common challenge for Bayesian methods, so this limitation is not unique to our model. However, we find inference produced by the Bayesian change point algorithm to be reliable and believe that the reduced subjectivity afforded by BMA to be an important step towards letting the data dictate which model is "best". Future research should focus on analyzing the impact that the range of parameter values has on the inference and on further reducing the compute time so that this approach can be applied to longer and more complex datasets.

Author Contributions: Both authors contributed equally to all aspects of this project. All authors have read and agreed to the published version of the manuscript.

Funding: This research received no external funding.

Data Availability Statement: R code to run the Bayesian Change Point algorithm and associated datasets can be obtained by contacting the corresponding author, eruggier@holycross.edu.

Acknowledgments: The authors would like to thank the Weiss Summer Research Program at the College of the Holy Cross for providing funding to support this project.

Conflicts of Interest: The authors declare no conflict of interest.

Appendix A. Correcting for Autocorrelation in the Presence of a Linear Trend Model

A second study was conducted to analyze how autocorrelation affects the Bayesian change point algorithm's ability to detect change points when none actually exist. Here, we consider a linear model with no change points ($Y = 4 + 0.05X + \epsilon$). An auto-regressive signal of level ρ = 0.1, 0.2, 0.3, ... , 0.9 is then generated using the R function arima.sim() and added to each dataset (a total of 10,000 simulations for each of the two models). The value of m is chosen to be 20, k_0 = (0.01, 0.01), v_0 = 1, σ_0^2 = 1, d_{min} = 5, and k_{max} = 20. Since the goal of this simulation is to see how autocorrelation affects the inference, optimizing these parameters is not critical. The Bayesian change point model calculates the posterior distribution of the number of change points for each dataset, which can be used to determine the expected number of change points in the dataset. Table A1 gives the average number of detected change points across the 1000 simulated datasets for each value of ρ before and after pre-whitening, along with the number of datasets where the algorithm correctly identified zero change points. As with Table 1, it is clear from these data that pre-whitening can help to eliminate spurious change points that arise from autocorrelation.

Table A1. Autocorrelation in a Linear Trend Model. A total of 1000 datasets were generated for each value of the autocorrelation parameter, ρ, for a model that includes a linear trend. The average number of change points detected by the Bayesian change point model before and after pre-whitening is indicated for each value of ρ, along with the number of datasets (out of 1000) where the algorithm correctly identified zero change points (i.e., the number of datasets where the expected number of change points <0.5). Note that a value of ρ = 0 corresponds to white noise.

	ρ	0	0.1	0.2	0.3	0.4	0.5	0.6	0.7	0.8	0.9
Data with Autocorrelation	Estimated $\hat{\rho}^c$	N/A	0.039	0.139	0.229	0.321	0.419	0.501	0.591	0.666	0.732
	Change Points Detected	$<10^{-3}$	$<10^{-3}$	$<10^{-3}$	$<10^{-3}$	0.003	0.013	0.087	0.472	1.861	3.577
	# Correct	1000	1000	1000	1000	998	990	928	700	203	18
Pre-Whitened	Change Points Detected	N/A	$<10^{-3}$	$<10^{-3}$	$<10^{-3}$	$<10^{-3}$	$<10^{-3}$	0.002	0.015	0.080	0.450
	# Correct	N/A	1000	1000	1000	1000	1000	998	989	940	693

For the linear trend model (Table A1), we estimate the value of the autoregressive parameter on the residuals of the model after accounting for the linear trend. The fact that we underestimate the value of the autocorrelation should not be surprising. Both [69] and [70] discuss the difficulty of jointly estimating the trend and autoregressive parameters of a model as they are highly interdependent. Note that if we estimated the value of ρ based on the entire dataset rather than short subsegments of the data, then the estimated value of ρ is much closer to the true value of ρ (results not shown). However, this approach is problematic in the presence of the change points.

References

1. Wurtz, D.; Chalabi, Y.; Setz, T. *New Directions in Active Portfolio Management: Stability Analytics, Risk Parity, Rating and Ranking, and Geometric Shape Factors*; ETH Econophysics Working and White Papers Series; Rmetrics: Zurich, Switzerland, 2013; p. 3.
2. Thies, S.; Molnár, P. Bayesian change point analysis of Bitcoin returns. *Financ. Res. Lett.* **2018**, *27*, 223–227. [CrossRef]
3. Chopin, N. Dynamic detection of change points in line time series. *Ann. Inst. Stat. Math.* **2007**, *59*, 349–366. [CrossRef]
4. Barnett, I.; Onnela, J.P. Change Point Detection in Correlation Networks. *Sci. Rep.* **2016**, *6*, 18893. [CrossRef]
5. Western, B.; Kleykamp, M. A Bayesian Change Point Model for Historical Time Series Analysis. *Political Anal.* **2004**, *12*, 354–374. [CrossRef]
6. Robinson, L.F.; Wager, T.D.; Lindquist, M.A. Change point estimation in multi-subject fMRI studies. *Neuroimage* **2010**, *49*, 1581–1592. [CrossRef]

7. Chen, G.; Lu, G.; Shang, W.; Xie, Z. Automated Change-Point Detection of EEG Signals Based on Structural Time-Series Analysis. *IEEE Access* **2019**, *7*, 180168–180180. [CrossRef]
8. Kass-Hout, T.A.; Xu, Z.; McMurray, P.; Park, S.; Buckeridge, D.L.; Brownstein, J.S.; Finelli, L.; Groseclose, S.L. Application of change point analysis to daily influenza-like illness emergency department visits. *J. Am. Med. Inform. Assoc.* **2012**, *19*, 1075–1081. [CrossRef] [PubMed]
9. Ruggieri, E.; Herbert, T.; Lawrence, K.T.; Lawrence, C.E. Change point method for detecting regime shifts in paleoclimatic time series: Application to d18O time series of the Plio-Pleistocene. *Paleoceanography* **2009**, *24*, PA1204. [CrossRef]
10. Gallagher, C.; Lund, R.; Robbins, M. Changepoint detection in daily precipitation data. *Environmetrics* **2012**, *23*, 407–419. [CrossRef]
11. Kim, C.; Suh, M.; Hong, K. Bayesian Changepoint Analysis of the Annual Maximum of Daily and Subdaily Precipitation over South Korea. *J. Clim.* **2009**, *22*, 6741–6757. [CrossRef]
12. Beaulieu, C.; Killick, R. Distinguishing trends and shifts from memory in climate data. *J. Clim.* **2018**, *31*, 9519–9543. [CrossRef]
13. Kendrick, L.; Musial, K.; Gabrys, B. Change point detection in social networks—Critical review with experiments. *Comput. Sci. Rev.* **2018**, *29*, 1–13. [CrossRef]
14. Desobry, F.; Davy, M.; Doncarli, C. An online kernel change detection algorithm. *IEEE Trans. Signal Process.* **2005**, *53*, 2961–2974. [CrossRef]
15. Liu, J.S.; Lawrence, C.E. Bayesian Inference on Biopolymer Models. *Bioinformatics* **1999**, *15*, 38–52. [CrossRef] [PubMed]
16. Maidstone, R.; Hocking, T.; Rigaill, G.; Fearnhead, P. On optimal multiple change-point algorithms for large data. *Stat. Comput.* **2017**, *27*, 519–533. [CrossRef]
17. Aminikhanghahi, S.; Cook, D.J. A survey of methods for time series change point detection. *Knowl. Inf. Syst.* **2017**, *51*, 339–367. [CrossRef]
18. Chen, J.; Gupta, A.K. *Parametric Statistical Change Point Analysis*; Birkhauser: New York, NY, USA, 2012. [CrossRef]
19. Page, E.S. Continuous Inspection Schemes. *Biometrika* **1954**, *41*, 100–115. [CrossRef]
20. Page, E. A test for a change in a parameter occurring at an unknown point. *Biometrika* **1955**, *42*, 523–527. [CrossRef]
21. Zeileis, A.; Leisch, F.; Hornik, K.; Kleiber, C. strucchange: An R Package for Testing for Structural Change in Linear Regression Models. *J. Stat. Softw.* **2002**, *7*, 1–38. [CrossRef]
22. Hawkins, D.M.; Qiu, P.; Kang, C.W. The Changepoint Model for Statistical Process Control. *J. Qual. Technol.* **2003**, *35*, 355–366. [CrossRef]
23. Kawahara, Y.; Sugiyama, M. Change-point detection in time-series data by direct density-ratio estimation. In *Proceedings of the 2009 SIAM International Conference on Data Mining, Sparks, NV, USA, 30 April–2 May 2009*; Society for Industrial and Applied Mathematics: Philadelphia, PA, USA, 2009; pp. 389–400. [CrossRef]
24. Ross, G.J. Parametric and Nonparametric Sequential Change Detection in R: The cpm Package. *J. Stat. Softw.* **2015**, *66*, 1–20. [CrossRef]
25. Scott, A.J.; Knott, M. A Cluster Analysis Method for Grouping Means in the Analysis of Variance. *Biometrics* **1974**, *30*, 507–512. [CrossRef]
26. Shi, X.; Gallagher, C.; Lund, R.; Killick, R. A comparison of single and multiple changepoint techniques for time series data. *Comput. Stat. Data Anal.* **2022**, *170*, 107433. [CrossRef]
27. Olshen, A.B.; Venkatraman, E.; Lucito, R.; Wigler, M. Circular binary segmentation for the analysis of array-based DNA copy number data. *Biostatistics* **2004**, *5*, 557–572. [CrossRef] [PubMed]
28. Fryzlewicz, P. Wild binary segmentation for multiple change-point detection. *Ann. Stat.* **2014**, *42*, 2243–2281. [CrossRef]
29. Auger, I.E.; Lawrence, C.E. Algorithms for the Optimal Identification of Segment Neighborhoods. *Bull. Math. Biol.* **1989**, *51*, 39–54. [CrossRef]
30. Bai, J.; Perron, P. Computation and Analysis of Multiple Structural Change Models. *J. Appl. Econom.* **2003**, *18*, 1–22. [CrossRef]
31. Killick, R.; Fearnhead, P.; Eckley, I.A. Optimal Detection of Changepoints With a Linear Computational Cost. *J. Am. Stat. Assoc.* **2012**, *107*, 1590–1598. [CrossRef]
32. Barry, D.; Hartigan, J.A. A Bayesian Analysis for Change Point Problems. *J. Am. Stat. Assoc.* **1993**, *88*, 309–319. [CrossRef]
33. Carlin, B.P.; Gelfand, A.E.; Smith, A.F.M. Hierarchical Bayesian analysis of changepoint problems. *Appl. Stat.* **1992**, *41*, 389–405. [CrossRef]
34. Green, P.J. Reversible jump Markov chain Monte Carlo computation and Bayesian model determination. *Biometrika* **1995**, *82*, 711–732. [CrossRef]
35. Fearnhead, P. Exact and Efficient Bayesian Inference for Multiple Changepoint problems. *Stat. Comput.* **2006**, *16*, 203–213. [CrossRef]
36. Whiteley, N.; Andrieu, C.; Doucet, A. Bayesian Computational Methods for Inference in Multiple Change-Point Models. 2011. Available online: http://www.maths.bris.ac.uk/~manpw/change_points_2011.pdf (accessed on 20 June 2022).
37. Ruggieri, E. A Bayesian Approach to Detecting Change Points in Climatic Records. *Int. J. Climatol.* **2013**, *33*, 520–528. [CrossRef]
38. Adams, R.P.; MacKay, D.J.C. Bayesian Online Changepoint Detection. 2007. Available online: http://arxiv.org/pdf/0710.3742.pdf (accessed on 20 June 2022).
39. Fearnhead, P.; Clifford, P. On-line inference for hidden Markov models via particle filters. *J. R. Stat. Soc. Ser. B* **2003**, *65*, 887–899. [CrossRef]

40. Fearnhead, P.; Liu, Z. On-line inference for multiple changepoint problems. *J. R. Stat. Soc. Ser. B* **2007**, *69*, 589–605. [CrossRef]
41. West, M.; Harrison, J. *Bayesian Forecasting and Dynamic Models*, 2nd ed.; Springer: New York, NY, USA, 1997.
42. Zhang, Y.M.; Wang, H.; Bai, Y.; Mao, J.X.; Chang, X.Y.; Wang, L.B. Switching Bayesian dynamic linear model for condition assessment of bridge expansion joints using structural health monitoring data. *Mech. Syst. Signal Process.* **2021**, *160*, 107879. [CrossRef]
43. Ruggieri, E. A Pruned, Recursive Solution to the Multiple Change Point Problem. *Comput. Stat.* **2018**, *33*, 1017–1045. [CrossRef]
44. Ruggieri, E.; Antonellis, M. An exact approach to Bayesian sequential change point detection. *Comput. Stat. Data Anal.* **2016**, *97*, 71–86. [CrossRef]
45. Hasselmann, K. Stochastic climate models Part I. Theory. *Tellus* **1976**, *28*, 473–485. [CrossRef]
46. von Storch, H. Misuses of statistical analysis in climate research. In *Analysis of Climate Variability*; Springer: Berlin/Heidelberg, Germany, 1999; pp. 11–26.
47. Rodionov, S.N. Use of prewhitening in climate regime shift detection. *Geophys. Res. Lett.* **2006**, *33*, L12707. [CrossRef]
48. Shi, X.; Beaulieu, C.; Killick, R.; Lund, R. Changepoint Detection: An Analysis of the Central England Temperature Series, J. *Clim.* **2022**, *35*, 2729–2742. [CrossRef]
49. Lund, R.; Wang, X.L.; Lu, Q.Q.; Reeves, J.; Gallagher, C.M.; Feng, Y. Changepoint detection in periodic and autocorrelated time series. *J. Clim.* **2007**, *20*, 5178–5190. [CrossRef]
50. Chatfield, C. *The Analysis of Time Series: An Introduction*, 7th ed.; Chapman & Hall/CRC Press: Boca Raton, FL, USA, 2003.
51. Beaulieu, C.; Chen, J.; Sarmiento, J.L. Change-point analysis as a tool to detect abrupt climate variations. *Philos. Trans. R. Soc. A Math. Phys. Eng. Sci.* **2012**, *370*, 1228–1249. [CrossRef]
52. Wang, X.L. Accounting for autocorrelation in detecting mean shifts in climate data series using the penalized maximal *t* or *F* test. *J. Appl. Meteor. Climatol.* **2008**, *47*, 2423–2444. [CrossRef]
53. Serinaldi, F.; Kilsby, C.G. The importance of prewhitening in change point analysis under persistence. *Stoch. Environ. Res. Risk Assess.* **2016**, *30*, 763–777. [CrossRef]
54. Shaman, P.; Stine, R. The bias of autoregressive coefficient estimators. *J. Am. Stat. Assoc.* **1988**, *83*, 842–848. [CrossRef]
55. Marriott, F.H.C.; Pope, J.A. Bias in the estimation of autocorrelations. *Biometrika* **1954**, *41*, 390–402. [CrossRef]
56. Hoeting, J.A.; Madigan, D.; Raftery, A.E.; Volinsky, C.T. Bayesian model averaging: A tutorial (with comments by M. Clyde, David Draper and E. I. George, and a rejoinder by the authors). *Statist. Sci.* **1999**, *14*, 382–417. [CrossRef]
57. Mantua, N.J.; Hare, S.R.; Zhang, Y.; Wallace, J.M.; Francis, R.C. A Pacific Interdecadal Climate Oscillation with Impacts on Salmon Production. *Bull. Am. Meteorol. Soc.* **1997**, *78*, 1069–1079. [CrossRef]
58. Dutton, J. "What is the Pacific Decadal Oscillation?" World Climate Service. 2021. Available online: https://www.worldclimateservice.com/2021/09/01/pacific-decadal-oscillation/ (accessed on 25 July 2022).
59. Zhang, Y.; Wallace, J.M.; Battisti, D.S. ENSO-like interdecadal variability: 1900–93. *J. Clim.* **1997**, *10*, 1004–1020. [CrossRef]
60. Wang, S.; Huang, J.; He, Y.; Guan, Y. Combined effects of the Pacific Decadal Oscillation and El Niño-Southern Oscillation on Global Land Dry–Wet Changes. *Sci. Rep.* **2014**, *4*, 6651. [CrossRef]
61. Mantua, N.J.; Hare, S.R. The Pacific decadal oscillation. *J. Oceanogr.* **2002**, *58*, 35–44. [CrossRef]
62. Schwing, F.B.; Jiang, J.; Mendelssohn, R. Coherency of multi-scale abrupt changes between the NAO, NPI, and PDO. *Geophys. Res. Lett.* **2003**, *30*, 1406. [CrossRef]
63. Lindsey, R.; Dahlman, L. Climate Change: Global Temperature. 2022. Available online: https://www.climate.gov/news-features/understanding-climate/climate-change-global-temperature (accessed on 22 July 2022).
64. NOAA National Centers for Environmental Information. State of the Climate: Global Climate Report for 2021. 2022. Available online: https://www.ncdc.noaa.gov/sotc/global/202113 (accessed on 28 July 2022).
65. Morice, C.P.; Kennedy, J.J.; Rayner, N.A.; Winn, J.P.; Hogan, E.; Killick, R.E.; Dunn, R.J.H.; Osborn, T.J.; Jones, P.D.; Simpson, I.R. An updated assessment of near-surface temperature change from 1850: The HadCRUT5 dataset. *J. Geophys. Res. Atmos.* **2021**, *126*, e2019JD032361. [CrossRef]
66. Smith, T.M.; Reynolds, R.W.; Peterson, T.C.; Lawrimore, J. Improvements to NOAA's historical merged land–ocean surface temperature analysis (1880–2006). *J. Clim.* **2008**, *21*, 2283–2296. [CrossRef]
67. GISTEMP Team. GISS Surface Temperature Analysis (GISTEMP), Version 4. NASA Goddard Institute for Space Studies; 2022. Available online: https://data.giss.nasa.gov/gistemp/ (accessed on 28 July 2022).
68. Yu, M.; Ruggieri, E. Change point analysis of global temperature records. *Int. J. Climatol.* **2019**, *39*, 3679–3688. [CrossRef]
69. Canjels, E.; Watson, M.W. Estimating deterministic trends in the presence of serially correlated errors. *Rev. Econ. Stat.* **1997**, *79*, 184–200. [CrossRef]
70. Roy, A.; Falk, B.; Fuller, W.A. Testing for trend in the presence of autoregressive error. *J. Am. Stat. Assoc.* **2004**, *99*, 1082–1091. [CrossRef]

MDPI AG
Grosspeteranlage 5
4052 Basel
Switzerland
Tel.: +41 61 683 77 34

Mathematics Editorial Office
E-mail: mathematics@mdpi.com
www.mdpi.com/journal/mathematics

www.ingramcontent.com/pod-product-compliance
Ingram Content Group UK Ltd.
Pitfield, Milton Keynes, MK11 3LW, UK
UKHW061544051026
15625UKWH00012B/604
* 9 7 8 3 7 2 5 8 3 4 9 9 0 *